Tailoring the Tomato Genome for NextGen Agriculture

Tailoring the Tomato Genome for NextGen Agriculture gives a comprehensive view of the modern revolution in tomato breeding—bridging classical genetics with modern molecular and data-driven innovations. Drawing on global germplasm resources from institutions such as the Tomato Genetics Resource Center (UC Davis) and USDA-GRIN, the book highlights how next-generation technologies are reshaping the improvement of one of the world's most important crops.

Eleven expert-authored chapters explore topics including genome editing (CRISPR/Cas), high-throughput phenotyping, genomic and marker-assisted selection, integrated data platforms like Tomato BreedBase, and breeding for fruit quality, abiotic and biotic stress tolerance, and controlled-environment agriculture. Each chapter combines theory, methodology, and real-world application—offering readers both depth and practical insight.

Intended for researchers, academics, graduate students, and professionals in plant breeding and biotechnology, this book serves as both a technical reference and a visionary roadmap for sustainable tomato improvement. It illustrates how future tomato breeding will advance global food security, climate resilience, and consumer satisfaction through data-driven, sustainable innovation.

PREFACE TO THE SERIES

Agriculture is going through a phase of transformation now. It has both challenges and opportunities. One of the major challenges is obviously food security. A population of about 9.3 billion projected by 2050 necessitates an increase of crop production by 60–70%. Another 10–15% increase would be needed to counter the reduction of crop production due to global warming and climate change. Nutrition security is another challenge, with more than 800 million people suffering from malnutrition. Depletion of fossil fuel and increasing dependence on biofuel from crop residues is another challenge. Loss of biodiversity and land degradation also add to the list of challenges. The cultivable area is also decreasing fast due to housing, transport, and industrialization. Above all, climate change and global warming have a huge impact on agriculture, health, and environment.

Several novel concepts and strategies are emerging to address the above challenges. For example, practice of precision farming will lead to precise and timely use of inputs resulting in higher yield and better quality of produce on the one hand and reduction of loss of inputs on the other. Similarly, urban agriculture and vertical agriculture will add a lot of cultivable areas in future. Protected farming will facilitate growing crops under controlled conditions leading to reduction in biotic and abiotic stresses and precise utilization of inputs.

Agriculture requires maintenance of the philosophy farmers have been following since inception over thousands of years which are being sold today in new packages with sophisticatedly coined labels such as traditional agriculture, natural agriculture, sustainable agriculture, conservation agriculture, integrated agriculture, regenerative agriculture, organic agriculture, etc.

In the changing scenarios, next generation agriculture will need designed crop varieties and breeds produced through designed practices. This book series aims to deliberate several novel concepts and innovative strategies developed for next generation agriculture and consequently benefit agricultural education, research, and outreach.

Chittaranjan Kole
Kolkata
January 10, 2024

Special Acknowledgement

To Phullare Kole for her outstanding assistance in editing this book series.

NextGen Agriculture: Novel Concepts and Innovative Strategies

Chittaranjan Kole

Series Editor

Allele Mining for Genomic Designing of Fruit Crops
Chittaranjan Kole, Kenta Shirasawa, and Anil Kumar Singh

Allele Mining for Genomic Designing of Cereal Crops
Chittaranjan Kole, Sharat Kumar Pradhan, and Vijay K Tiwari

Allele Mining for Genomic Designing of Vegetable Crops
Chittaranjan Kole, Tusar Kanti Behera, and Prashant Kaushik

Allele Mining for Genomic Designing of Oilseed Crops
Chittaranjan Kole, Manish Kumar Pandey, and Naveen Puppala

Sustainable Urban Agriculture: New Frontiers
Kheir Al-Kodmany, Madhav Govind, Sharmin Khan, and Chittaranjan Kole

Allele Mining for Genomic Designing of Grain Legume Crops
Chittaranjan Kole, C. Bharadwaj, and Abhimanyu Sarkar

The Vertical Farm: Scientific Advances and Technological Developments
Kheir Al-Kodmany, Andrew Keong Ng, Abel Tablada, and Chittaranjan Kole

Next Generation Food Crops for Human Health
Edited by Sangam Dwivedi, Rodomiro Ortiz, and Chittaranjan Kole

Smart Agriculture: Concepts, Strategies, and Case Studies
Edited by Verónica Saiz-Rubio, Francisco Rovira-Más, and Chittaranjan Kole

Tailoring the Tomato Genome for NextGen Agriculture
Edited by Dilip R. Panthee and Chittaranjan Kole

For more information, please visit our website: https://www.routledge.com/Nextgen-Agriculture/book-series/NGA

Tailoring the Tomato Genome for NextGen Agriculture

Edited by

Dilip R. Panthee and Chittaranjan Kole

CRC Press
Taylor & Francis Group
Boca Raton London

CRC Press is an imprint of the
Taylor & Francis Group, an **informa** business

Designed cover image: Shutterstock

First edition published 2026
by CRC Press
2385 NW Executive Center Drive, Suite 320, Boca Raton FL 33431

and by CRC Press
4 Park Square, Milton Park, Abingdon, Oxon, OX14 4RN

CRC Press is an imprint of Taylor & Francis Group, LLC

ISBN: 978-1-032-82040-8 (hbk)
ISBN: 978-1-032-82830-5 (pbk)
ISBN: 978-1-003-50653-9 (ebk)

DOI: 10.1201/9781003506539

Typeset in Times
by Apex CoVantage, LLC

Dedicated to

Dr. Norman Ernest Borlaug
The Father of Green Revolution, Nobel Peace Prize winner
and the Founder of the World Food Prize

Who inspired our research and academic works with his generous appreciations and advice.

And to
tomato growers, students, and
researchers worldwide.

Contents

Preface

Tomato (*Solanum lycopersicum*) is the second-most widely consumed vegetable crop in the world, after the potato, and is cultivated across approximately 5.2 million hectares, yielding over 189 million metric tons annually. This production supports a global industry valued at approximately $1.4 billion annually (FAOSTAT). Rich in antioxidants such as lycopene, as well as essential vitamins and minerals, tomatoes are a major contributor to human nutrition (Fraser and Bramley 2004; Sharma et al. 2023). As a result, global demand—particularly in the United States—continues to grow (Kim 2016; Mondal et al. 2016). Tomato also plays a dual role as both an economically vital crop and a model organism for genetic and genomic research, due in part to its relatively small genome size of approximately 950 Mbp. This has facilitated the development and application of numerous molecular tools aimed at improving yield, stress resilience, and fruit quality. These tools have proven instrumental in addressing modern challenges, including climate change, emerging pests and pathogens, and shifting consumer demands, all of which require more precise, rapid, and data-driven breeding approaches.

Comprehensive genetic resources for tomato are maintained at several major institutions. The Tomato Genetics Resource Center (TGRC) at the University of California, Davis, houses a broad range of wild and cultivated accessions. The USDA-GRIN curates additional germplasm collections in Geneva, New York, and the National Plant Germplasm System in Fort Collins, Colorado. Collectively, these resources support the global tomato research and breeding community. This book explores the modern tools and strategies being used to unlock the genetic potential of the tomato, including high-throughput phenotyping, genome editing, and integrated data management platforms. Each chapter provides in-depth insights into these technologies and their applications. The volume is intended to serve researchers, academics, graduate students, and professionals in private seed companies working to advance tomato breeding in the face of 21st-century challenges.

The advent of high-throughput genomic and phenomic technologies has revolutionized the paradigms of crop improvement. Today's tomato breeders are equipped with an unprecedented array of tools, including next-generation sequencing (NGS), genome-wide association studies (GWAS), RNA sequencing, marker-assisted and genomic selection, genome editing technologies such as CRISPR/Cas, and integrated data platforms. These innovations enable the dissection of complex traits at a molecular level and facilitate the introgression of desirable alleles with speed and accuracy. This edited volume—comprising 11 chapters written by experts in molecular breeding, genomics, phenotyping, data management, and applied biotechnology—offers a comprehensive and timely synthesis of current advancements in tomato improvement. Collectively, the chapters cover fundamental genomic tools, applied breeding strategies, and future-oriented topics such as vertical farming. This introduction provides an overview of the book's thematic structure and highlights the significance and interconnectedness of its components.

Chapter 1 explores germplasm management in the context of digitization and global exchange. Access to diverse, well-characterized genetic resources is fundamental to trait discovery and pre-breeding efforts. The chapter underscores the pivotal role of genebanks, passport data, and international collaboration in maintaining and utilizing germplasm collections effectively. Chapter 2 introduces *Tomato BreedBase*, a robust data management platform designed to integrate genotypic, phenotypic, environmental, and metadata information. With tools for data visualization, statistical analysis, and genomic selection, the platform supports informed decision-making in multi-environment trials. Moreover, it facilitates collaborative breeding efforts and ensures data traceability—critical features for modern breeding programs that adhere to the FAIR (Findable, Accessible, Interoperable, and Reusable) principles. Building on this, Chapter 3 addresses high-throughput phenotyping (HTP), a rapidly advancing field that remains a major bottleneck in breeding pipelines. Utilizing imaging technologies, drones, and sensor-based systems, HTP enables non-destructive, real-time trait assessment in both controlled and field environments. The integration of HTP with genotypic data significantly strengthens the predictive power of selection algorithms. Centralized, interoperable data systems are essential for maximizing the impact of genomic tools in modern breeding.

Chapter 4 delves into the foundation of molecular breeding—namely, the discovery and application of reliable molecular markers linked to key traits. It provides an in-depth overview of QTL mapping and marker development, highlighting progress in SSRs, SNPs, indels, and their application in marker-assisted selection (MAS). These tools serve as proxies for trait selection, streamlining breeding cycles and accelerating variety development. Expanding on trait mapping strategies, Chapter 5 presents Bulked Segregant Analysis sequencing (BSA-seq) as a powerful and efficient method for identifying major QTLs. By combining pooled phenotyping with deep sequencing, BSA-seq enables rapid mapping of traits, even in early segregating populations. This approach is particularly effective for traits governed by a few loci with large phenotypic effects. Chapter 6 emphasizes allele mining strategies, highlighting the value of uncovering natural genetic variation and regulatory mechanisms to enhance key traits. These efforts often reveal rare or underutilized alleles that can be incorporated through conventional breeding or molecular methods, offering new avenues for trait improvement. With the advent of precise genome-editing tools, Chapter 7 discusses CRISPR/Cas-based technologies and their transformative role in tomato breeding. Genome editing allows for targeted, heritable modifications without the introduction of foreign DNA. Applications include knocking out susceptibility genes and fine-tuning regulatory elements to create disease-resistant, high-yielding, and shelf-stable cultivars.

Moving beyond stress resilience and yield, Chapter 8 shifts the focus to fruit quality—encompassing flavor, firmness, shelf life, and nutritional content. As consumer preferences evolve toward healthier and more flavorful produce, breeding for quality traits has become a central priority. Chapters 9 and 10 cover advances in breeding for biotic and abiotic stress tolerance, respectively. Genomic tools have enabled precise resistance gene identification, pyramiding strategies, and enhanced trait stability. These chapters showcase how classical breeding and genomic innovations converge to deliver stress-resilient varieties, while also identifying current limitations in translating lab-based discoveries into field performance. The success of these efforts depends heavily on the availability of high-quality genotypic and phenotypic data. Finally, Chapter 11 addresses breeding for controlled environment agriculture (CEA), where traits such as light-use efficiency, compact growth, and optimized resource utilization are prioritized. These efforts are crucial for adapting tomato breeding to emerging production systems that emphasize sustainability and year-round cultivation.

This book encapsulates the transformative journey of tomato improvement—from phenotypic selection in open fields to data-intensive, trait-targeted, and environment-responsive breeding. By integrating genomics, phenomics, bioinformatics, and precision technologies, modern tomato breeding is poised to address global food security, climate resilience, and market demands simultaneously. The future of tomato improvement will likely rest on three pillars: (1) *precision breeding* enabled by multi-omics and AI-assisted selection; (2) *sustainability* through stress-tolerant and input-efficient varieties; and (3) *consumer-centric breeding*, focusing on nutrition, taste, and health benefits. Interdisciplinary collaborations, open data sharing, and continued innovation in tools and strategies will be essential in achieving these goals. By compiling cutting-edge knowledge and methodologies, this volume aims to serve not only as a technical reference for breeders, researchers, and students but also as a strategic guide for shaping the next era of tomato improvement.

The 11 chapters of this book have been authored by 20 scientists from three countries. We express our thanks to them for their elegant contributions. We also extend our gratitude to Randy Brehm,

Senior Editor, Life Sciences and Medicine, and Tom Connelly, Editorial Assistant, Life Sciences of the CRC Press/Taylor and Francis Group LLC, for their outstanding cooperation from inception to publication of this book. Finally, we express our thanks to our family members for their support throughout the process of editing this book.

Dilip R. Panthee
Mills River, North Carolina, USA

Chittaranjan Kole
Kolkata, West Bengal, India

Editors' Biographies

Dr. Dilip R. Panthee is Associate Professor and Tomato Breeder at the Department of Horticultural Science at North Carolina State University (NCSU). His expertise is in plant breeding, genetics, genomics, QTL mapping, molecular marker development, and marker-assisted selection. His tomato breeding program uses conventional and molecular breeding approaches to improve tomato disease resistance, yield, and fruit quality. He has released 13 hybrid tomato cultivars and 19 tomato breeding lines. He has published over 95 journal articles, 10 book chapters, and more than 85 abstracts and research reports. He has mentored more than a dozen graduate students, post-doctoral research scientists, research scholars, and visiting scientists. He has received 10 professional awards for contributing to the scientific community, including the Fulbright Specialist Award, Julian C. Miller Sr. Distinguished Research Award, and Vegetable Breeding Excellence Award.

Prof. Chittaranjan Kole is an internationally renowned academician with an impressive professional career spanning over 40 years. His lifetime original scientific contributions, specifically in the fields of plant genomics and biotechnology, have been globally recognized. His scholarly publications include 170 research articles, mostly published in peer-reviewed international journals, 35 book chapters, and over 200 books with globally reputed publishers. His academic contributions have been acknowledged by seven Nobel Laureates, and his research contributions have been recognized with several international awards and fellowships. Prof. Kole has worked as a faculty member, Head of Department, Dean, and Vice Chancellor in three universities in India and as Visiting Professor in three universities in the USA. Currently, Prof. Kole is acting as the Founding President of Genome India International and of two other international organizations, including the International Climate Resilient Crop Genomics Consortium and the International Consortium on Phytomedomics and Nutriomics.

Contributors

Krishna Bhattarai
Texas A&M AgriLife Research and Extension Center
Department of Horticultural Sciences
Texas A&M University
Dallas, Texas USA

Kedong Da
Department of Horticultural Sciences
North Carolina State University
Raleigh, North Carolina USA

Weilong He
Department of Biological and Agricultural Engineering
Cornell University
Ithaca, New York USA

Gograj Singh Jat
ICAR-Indian Agricultural Research Institute
Pusa Campus
New Delhi, India

Md Jahid Hasan Jone
Department of Horticultural Science
North Carolina State University
Mountain Horticultural Crops Research and Extension Center
Mills River, North Carolina USA

Suman Lata
ICAR-Indian Agricultural Research Institute
Pusa Campus
New Delhi, India

Lukas A. Mueller
Boyce Thompson Institute
Ithaca, New York USA

Reshav Naik
ICAR-Indian Agricultural Research Institute
Pusa Campus
New Delhi, India

Anna Nelson
Department of Horticultural Sciences
North Carolina State University
Raleigh, North Carolina USA

Chet Ram
ICAR-Central Institute for Arid Horticulture
Bikaner, Rajasthan, India

Hanuman Ram
ICAR-Central Institute for Arid Horticulture
Bikaner, Rajasthan, India

Manoj Sapkota
Department of Horticulture
University of Kentucky
Lexington, Kentucky USA

Sadikshya Sharma
Department of Viticulture and Enology
University of California at Davis
Davis, California USA

Muhammad Irfan Siddique
Department of Horticultural Science
North Carolina State University
Mountain Horticultural Crops Research and Extension Center
Mills River, North Carolina USA

Avinash Tomer
ICAR-Indian Agricultural Research Institute
Pusa Campus
New Delhi, India

Yasin Topcu
Batı Akdeniz Agricultural Research Institute
Antalya, Turkey

Lirong Xiang
Department of Biological and Agricultural Engineering
Cornell University
Ithaca, New York USA

Ramesh Kumar Yadav
ICAR-Indian Agricultural Research Institute
Pusa Campus
New Delhi, India

Jianbo Zhang
Department of Horticultural Science
North Carolina State University
Mountain Horticultural Crops Research and Extension Center
Mills River, North Carolina USA

Zhenghua Zhang
Department of Biological and Agricultural Engineering
Cornell University
Ithaca, New York USA

Abbreviations

Acronym	Abbreviation
2D	two-dimensional
3D	three-dimensional
ABA	abscisic acid
APX	ascorbate peroxidase
CAT	catalase
CCD	charged-coupled device
CMOS	complementary metal–oxide–semiconductor
CNN	convolutional neural network
CRISPR	clustered regularly interspaced short palindromic repeats
DC	direct current
DeepLab	deep convolutional neural network for semantic image segmentation
DL	deep learning
DREB	dehydration-responsive element binding
EST	expressed sequence tag
F1	first filial generation
FAPAR	fraction of absorbed photosynthetically active radiation
Faster R-CNN	faster region-based convolutional neural network
GAI	green area index
GPS	Global Positioning System
GWAS	genome-wide association study
H–B	Horsfall–Barrett
HSP	heat shock protein
HTPP	high-throughput plant phenotyping
LiDAR	light detection and ranging
LSTM	long short-term memory
MAPK	mitogen-activated protein kinase
MAS	marker-assisted selection
Mask R-CNN	mask region-based convolutional neural network
MAV	manned aerial vehicle
ML	machine learning
MVS	multi-view stereo
NDVI	normalized difference vegetation index
QTL	quantitative trait loci
RGB	red–green–blue (color model)
RNN	recurrent neural network
ROS	reactive oxygen species
ROS	Robot Operating System
SfM	structure-from-motion
SNP	Single Nucleotide Polymorphism
SOS	salt overly sensitive
SSR	simple sequence repeat
T-DNA	transfer DNA
TERRA-REF	Transportation Energy Resources from Renewable Agriculture Phenotyping Reference Platform
ToF	Time-of-Flight
UAV	unmanned aerial vehicle

UDASeg	unsupervised domain-adaptive semantic segmentation
UE5	Unreal Engine 5
U-Net	U-shaped convolutional neural network for biomedical image segmentation
VTOL	vertical takeoff and landing
YOLO	You Only Look Once
ZFN	Zinc-Finger Nucleases

1 Tomato Germplasm Management at the National and International Levels in the Current Era

Dilip R. Panthee

1.1 INTRODUCTION

Tomato (*Solanum lycopersicum*) ranks among the most widely cultivated and economically important vegetable crops globally, boasting a rich evolutionary history rooted in the Andean regions of South America and early domestication in Mesoamerica (Peralta and Spooner 2007). Its global spread has been shaped by centuries of human selection, trade, and breeding, leading to thousands of cultivated varieties and wild relatives adapted to various environments. As global agriculture confronts escalating challenges from climate change, emerging pests and diseases, and shifting consumer demands, the strategic management of tomato germplasm has become more vital than ever (Figas et al. 2018). Germplasm—comprising landraces, wild species, heirloom cultivars, and improved lines—serves as the essential genetic reservoir for developing resilient, productive, and nutritious tomato varieties (Ramírez-Ojeda et al. 2022). This chapter offers a comprehensive overview of tomato germplasm management, tracing its historical origins, domestication pathways, and global dissemination, while exploring the status and strategies of germplasm collection, characterization, conservation, and utilization at both national and international levels. It emphasizes the importance of leveraging wild genetic resources, modern genomic tools, and collaborative networks to protect and enhance tomato genetic diversity for present and future generations.

1.2 THE ORIGIN

The cultivated tomato (*Solanum lycopersicum*) originates in the western regions of South America, primarily in the Andean areas of present-day Peru, Ecuador, and northern Chile. This region is considered the center of diversity for wild tomato relatives, including species such as *Solanum pimpinellifolium*, genetically closest to the modern cultivated tomato (Peralta and Spooner 2007). These wild species thrive in various ecological zones and exhibit a rich genetic pool that has played a critical role in tomato domestication and subsequent breeding efforts. Initially, wild tomato species bore small, cherry-sized fruits and grew in the highland and coastal environments of the Andes (Peralta and Spooner 2007; Ramírez-Ojeda et al. 2022). Over time, early inhabitants of these regions began selecting for favorable traits such as fruit size, palatability, and plant vigor, setting the stage for domestication. The domestication process is believed to have advanced significantly in Mesoamerica, particularly in southern Mexico. Mesoamerican farmers continued to select for larger (Figure 1.1), tastier fruits, and by the time of European contact, tomatoes were already a well-established crop in the region (Blanca et al. 2012). The Spanish conquistadors encountered tomatoes during their exploration and conquest of the Americas in the 16th century and introduced them to Europe (Peralta and Spooner 2007).

DOI: 10.1201/9781003506539-1

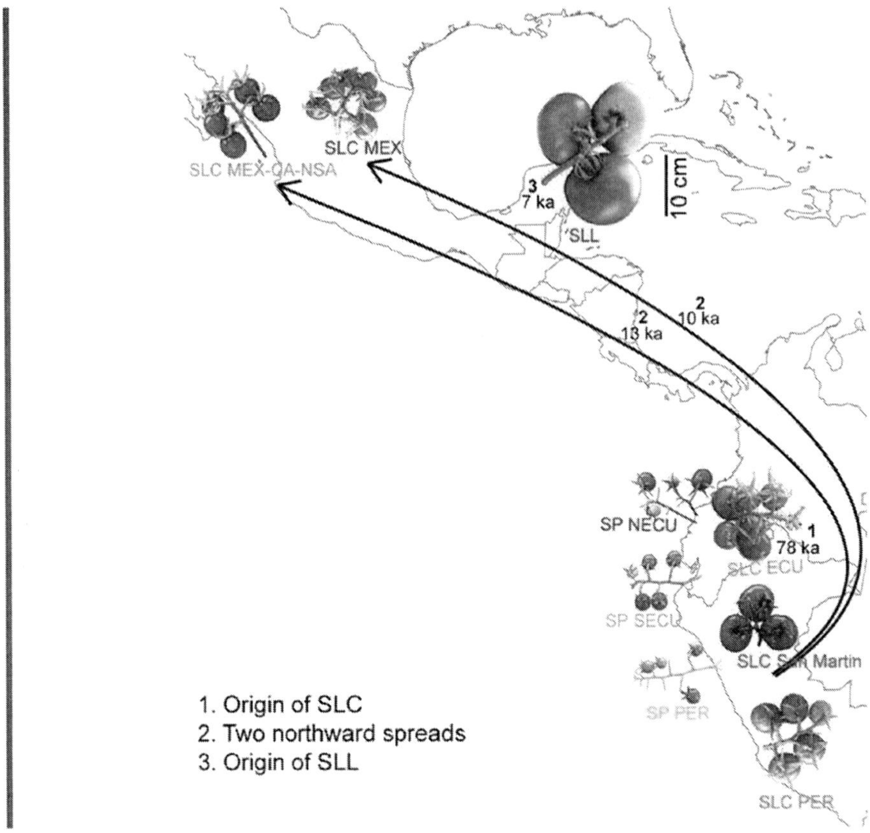

FIGURE 1.1 Distribution of various species of tomato (*Solanum spp*), including *S. lycopersicum* var *lycopersicum* (SLL), *S. lycopersicum* var *cerasiforme* (SLC).

Adopted from (Razifard et al. 2020).

Initially, tomatoes were met with suspicion in Europe due to their association with the nightshade family (*Solanaceae*), which includes several toxic species. As a result, tomatoes were first grown primarily as ornamental plants. However, their culinary potential was realized over time, particularly in Mediterranean countries such as Italy and Spain, where tomatoes became integral to regional cuisines. From Europe, tomatoes spread to Asia and other parts of the world, where they were adapted to various climates and cultural practices (Peralta and Spooner 2007; Ramírez-Ojeda et al. 2022). Today, tomatoes are among the most widely cultivated and consumed vegetable crops globally, grown in diverse environments. Their domestication history reflects a complex interplay of natural selection, human ingenuity, and cultural exchange that has shaped the fruit into a staple of global agriculture and cuisine.

1.3 EVOLUTION

The evolution of the cultivated tomato (*Solanum lycopersicum*) is a story shaped by natural selection, geographic dispersion, human intervention, and complex genetic divergence, with its roots deeply embedded in the rich biodiversity of the Andean region of South America. The genus *Solanum*, one of the largest and most diverse genera in the plant kingdom, includes over a thousand species, and within this genus, the tomato belongs to the section *Lycopersicon*, which encompasses both domesticated tomatoes and their wild relatives (Nakazato et al. 2010; Pons et al. 2022). The closest wild relatives are native to western South America, particularly the coastal and montane regions of Peru, Ecuador, and northern Chile. These regions, characterized by diverse microclimates and topographical variation, provided an ideal environment for evolutionary radiation and natural hybridization events that gave rise to a wide spectrum of wild tomato species, including *Solanum pimpinellifolium*, *S. cheesmaniae*, *S. galapagense*, *S. chmielewskii*, *S. habrochaites*, *S. peruvianum*, and *S. pennellii*. These species display remarkable variation in fruit size, shape, color, taste, trichome density, salt tolerance, drought tolerance, and disease resistance—traits that have continued to interest plant breeders and evolutionary biologists alike (Caicedo and Schaal 2004).

The evolutionary path of *S. lycopersicum* is most closely linked to *S. pimpinellifolium*, a small-fruited, red-berried species native to the coastal lowlands of northern Peru and Ecuador. Genetic and genomic analyses have consistently shown that *S. pimpinellifolium* shares the greatest similarity with the cultivated tomato and likely represents its wild progenitor (Caicedo and Schaal 2004; Peralta and Spooner 2007). This species produces small, sweet fruits and demonstrates relatively short generation cycles and ease of crossability with *S. lycopersicum*, making it a key player in understanding the early steps of tomato domestication. From the Andean highlands and lowlands, ancestral tomato forms, perhaps semi-domesticated populations of *S. pimpinellifolium* or early *S. lycopersicum* types, migrated northward into Central America and ultimately into Mexico (Avina-Padilla et al. 2022). This geographic movement, likely facilitated by human dispersal and trade, exposed tomato populations to new ecological pressures and cultivation practices, thereby further driving evolutionary changes.

The wild relatives of *Solanum lycopersicum* (cultivated tomato) belong primarily to the *Solanum* section *Lycopersicon* and are native to South America, particularly the Andean regions of Peru, Ecuador, and Chile (Figure 1.2). These species represent a vital reservoir of genetic diversity, offering valuable traits for tomato breeding, including disease resistance, abiotic stress tolerance, and improved fruit quality.

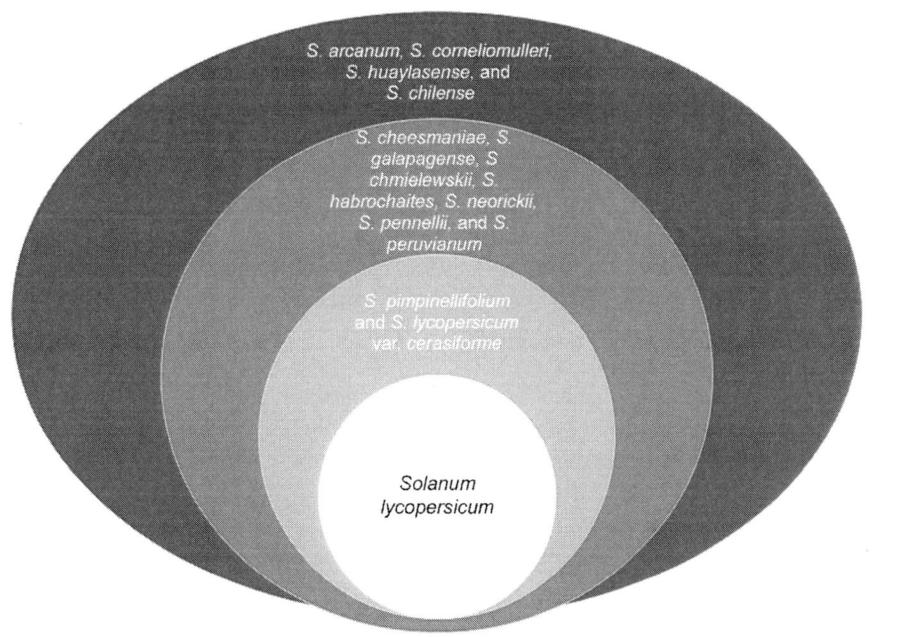

FIGURE 1.2 Wild relatives of tomato (*Solanum lycopersicum*). The farther the distance from *S. lycopersicum*, the more difficult to make crosses.

The primary gene pool includes two closest wild relatives that are easily crossable with *S. lycopersicum*:

- *Solanum pimpinellifolium*—the closest wild ancestor of the cultivated tomato, characterized by small red fruits and notable resistance to diseases and environmental stress.
- *S. lycopersicum* var. *cerasiforme*, a semi-domesticated cherry tomato type, often regarded as a transitional form between wild and cultivated types.

The secondary gene pool, which requires moderate effort for crossing, includes seven species:

- *S. cheesmaniae*—endemic to the Galápagos Islands; known for salt and pest resistance.
- *S. galapagense*—also from the Galápagos; exhibits high trichome density and insect resistance.
- *S. chmielewskii*—native to the Andean highlands; offers high sugar content and frost tolerance.
- *S. habrochaites*—rich in glandular trichomes; provides resistance to insects and fungal pathogens.
- *S. neorickii*—adapted to arid conditions; resistant to drought and nematodes.
- *S. pennellii*—highly tolerant to salinity and drought; widely used in introgression studies.
- *S. peruvianum*—self-incompatible with broad stress resistance; a key source of disease and pest resistance.

The tertiary gene pool comprises four species with significant crossing barriers, often requiring advanced biotechnological approaches:

- *S. arcanum*—native to Peru and Ecuador; known for resistance to multiple tomato viruses and pathogens.
- *S. corneliomulleri*—a rare species with promising resistance traits.
- *S. huaylasense*—a high-altitude species with notable stress tolerance.
- *S. chilense*—extremely drought-tolerant and valuable for breeding in arid environments.

By the time the tomato reached Mesoamerica, it had already diverged significantly from its Andean ancestors. In the Mexican region, where it encountered well-established agricultural systems and human selection pressures, further domestication took place (Städler et al. 2005). Mesoamerican farmers, likely the Maya and Aztecs, among others, began selecting for larger fruit size, improved flavor, and more favorable growth habits, such as determinate growth and synchronous ripening. The result was *S. lycopersicum* var. *cerasiforme*, an intermediate domesticated form that exhibits traits of both wild and cultivated tomatoes. This variety, sometimes seen as a bridge between *S. pimpinellifolium* and fully domesticated *S. lycopersicum* var. *lycopersicum*, played a critical role in shaping the modern tomato. Morphological and genetic studies suggest that *S. lycopersicum* var. *cerasiforme* evolved under both natural and artificial selection and was possibly the tomato type encountered by Spanish explorers in the 16th century during their conquests in the Americas (Blanca et al. 2012).

When Spanish conquistadors brought tomato seeds back to Europe around the early 1500s, they were met with curiosity but also skepticism. As a member of the nightshade family (*Solanaceae*), which includes toxic plants such as belladonna and henbane, the tomato was initially suspected of being poisonous. As a result, it was first cultivated as an ornamental plant in European gardens. Over time, however, particularly in Mediterranean climates such as those of Italy and Spain, the tomato was gradually adopted for culinary use (Pons et al. 2022). European growers began selecting for new traits, such as thicker flesh, lower acidity, and increased size, which gave rise to many of the heirloom varieties still prized today. This post-domestication diversification in Europe further contributed to the morphological and genetic breadth of the tomato gene pool, although it also introduced a genetic bottleneck, as relatively few genotypes were initially introduced from the Americas (Pons et al. 2022). This reduction in genetic diversity has had long-term implications, particularly in terms of susceptibility to pests and diseases, which breeders have tried to address by reintroducing genes from wild relatives.

The wild relatives of the tomato remain critically important in modern breeding programs. Species such as *S. habrochaites* and *S. pennellii*, which are native to dry and high-altitude environments, exhibit tolerance to abiotic stresses like drought and salinity. *S. peruvianum* and *S. chmielewskii* provide resistance to multiple pathogens, including *Phytophthora infestans*, *Fusarium* spp., and Tomato yellow leaf curl virus (TYLCV). Breeders often rely on these wild species for resistance genes, particularly in the context of climate change and the increasing demand for sustainable agriculture (Caicedo and Schaal 2004; Seong et al. 2020). However, introgressing traits from wild species into cultivated tomatoes presents challenges, as it often involves overcoming genetic incompatibilities, linkage drag, and reduced fertility. Advanced genomic tools, such as marker-assisted selection, genome-wide association studies (GWAS), and CRISPR/Cas9 genome editing, have increasingly enabled breeders to precisely identify and incorporate desirable alleles from wild relatives into elite germplasm, without introducing undesirable traits (Jahan et al. 2025; Tyagi et al. 2021).

Recent advances in molecular phylogenetics and genomics have further refined our understanding of tomato evolution. Whole-genome sequencing of multiple wild and cultivated accessions has revealed the complex patterns of introgression, gene duplication, and selection that underlie key domestication and improvement traits. For instance, studies have identified key genes such as *fw2.2* and *lc* that control fruit size, *rin* for ripening, and *sun* and *ovate* for fruit shape—all of which underwent strong selection during domestication (Rodriguez et al. 2011). Moreover, the availability of high-quality reference genomes for both *S. lycopersicum* and its wild relatives has facilitated comparative studies that uncover how structural variants, such as inversions and transposon insertions, have influenced trait evolution.

In recent years, efforts have also been made to conserve and utilize tomato wild relatives more effectively. Gene banks and international initiatives, such as the Crop Wild Relatives Project, aim to safeguard these genetic resources in the face of habitat destruction and climate change (del Rio et al. 1997; Steiner et al. 1997). Field studies and expeditions continue to uncover new accessions and populations of wild tomatoes, enriching the pool of traits available for future breeding (Chetelat 2001; Gordillo et al. 2008). At the same time, there is growing recognition of the importance of participatory breeding, where farmers and indigenous communities collaborate in selecting and developing varieties suited to their local conditions, often drawing upon traditional knowledge and historical relationships with these crops.

1.4 DOMESTICATION

The domestication of the tomato (*Solanum lycopersicum*) is a complex evolutionary and cultural process that traces its origins to the wild species native to western South America, particularly the Andean regions of present-day Peru, Ecuador, and northern Chile (Dixon et al. 2024; Gasparini et al. 2024). Among the wild relatives, *Solanum pimpinellifolium*, a small-fruited species, is considered the closest ancestor of the cultivated tomato due to its high genetic similarity. These wild tomatoes likely produced small, red, and acidic fruits that were not initially cultivated for human consumption (Vicente et al. 2023). However, early human populations in the Andes are thought to have used these fruits, perhaps initially as a minor food source, and eventually began selecting for desirable traits such as reduced bitterness, increased fruit size, and improved palatability. This early selection marks the beginning of a gradual domestication pathway that would lead to the development of the tomato as we know it today (Salazar-Mendoza et al. 2023). The initial stages of tomato domestication likely involved the movement of semi-domesticated forms northward into Central America and ultimately into Mesoamerica, where further domestication occurred. In Mesoamerica, early cultivators, particularly the indigenous peoples of southern Mexico, played a critical role in transforming the tomato into a viable crop. These communities selected for traits such as larger fruit size, more uniform ripening, reduced fruit cracking, and determinate growth habits—characteristics that distinguish modern domesticated tomatoes from their wild ancestors. Archaeological evidence and linguistic records suggest that the tomato was cultivated by civilizations, such as the Aztecs, as early as 500 BCE. The Nahuatl word "xitomatl" (meaning "plump thing with a navel") not only underscores its cultural importance but also hints at its morphological characteristics. This Mesoamerican phase of domestication represents a critical turning point, as it consolidated key domestication traits that enabled the tomato to be reliably grown and consumed as a staple food (Sun et al. 2017).

The tomato was brought back to Europe with the arrival of Spanish explorers in the Americas during the early 16th century. Although regarded with suspicion, partly due to its botanical relation to toxic nightshades, interest in the plant grew, especially in Mediterranean regions where it adapted well to the climate. European growers further selected tomato varieties for culinary use, developing numerous landraces with unique flavors, colors, and shapes (Sun et al. 2017). The tomato was introduced from Europe to other parts of the world, including Asia and Africa, where it became rapidly integrated into local cuisines and agricultural systems. Modern domesticated tomatoes exhibit a significant divergence from their wild ancestors in terms of size, shape, and traits such as sweetness, acidity, shelf life, and disease resistance. Domestication of tomatoes is a prime example of how human selection and plant adaptation intersect. Advances in genomics have revealed that domestication involved key genetic changes, particularly in loci controlling fruit morphology, ripening, and stress tolerance (Alseekh et al. 2021; Ferrero et al. 2020; Welter and Steggall 1993). Today, ongoing breeding efforts continue to draw from wild species to improve traits such as disease resistance, climate resilience, and nutritional content, highlighting the continued importance of understanding and leveraging tomato domestication in modern agriculture.

1.5 GLOBAL DEVELOPMENT OF TOMATO PRODUCTION AND PROCESSING

The global development of tomato production and processing has been a remarkable transformation in both agriculture and industry, spanning centuries and influenced by regional climates, technological innovations, market demands, and evolving culinary traditions. From its origins in South America, where the wild ancestors of the tomato (*Solanum lycopersicum*) were first domesticated, tomato cultivation spread rapidly across continents following its introduction to Europe in the 16th century by Spanish explorers (Pereira et al. 2021; Razifard et al. 2020). Initially grown as a curiosity, the tomato eventually gained prominence in Mediterranean agriculture, particularly in Italy and Spain, where it became a culinary staple and was integrated into a wide range of traditional dishes (van Andel et al. 2022). As European settlers introduced the tomato to colonies in Asia, Africa, and the Americas, its cultivation adapted to diverse agroecological conditions, resulting in the emergence of numerous landraces and varieties with traits suited to local climates, soils, and consumer preferences. By the 19th and early 20th centuries, tomatoes were widely cultivated across the globe, particularly in countries with warm temperatures and long growing seasons (Pereira et al. 2021; Razifard et al. 2020). As demand grew, production systems evolved from small-scale, seasonal gardening into intensive, large-scale farming operations supported by scientific breeding, mechanization, and irrigation.

The industrialization of tomato production paralleled broader trends in global agriculture during the 20th century, driven by increasing urbanization, improved transportation infrastructure, and the rise of food processing industries. Countries like the United States, Italy, China, and later Brazil, Turkey, and India emerged as major producers due to their favorable climates, established agricultural sectors, and investment in research and development (Rogo et al. 2024; Swarup et al. 2021). In the United States, California became the epicenter of commercial tomato production, thanks to its Mediterranean climate, fertile Central Valley, and robust agricultural infrastructure. The development of mechanized harvesting in the 1960s revolutionized the industry, especially for processing tomatoes, which are cultivated for products like ketchup, paste, juice, sauces, and canned tomatoes.

Breeders responded by developing varieties with firm fruit, uniform ripening, and concentrated setting, which were suitable for mechanical harvesting, thereby drastically reducing labor costs and increasing efficiency. Parallel advancements in irrigation technology, pest management, and fertilizer use further optimized yields, making large-scale tomato farming both profitable and scalable. Similar advancements occurred in Italy, where tomato production became tightly integrated with the pasta and canning industries, reinforcing the country's reputation for high-quality processed tomato products such as passata and peeled tomatoes (Arnoux et al. 2021; Lin et al. 2014).

Processing has become a vital segment of the tomato value chain as consumer demand has shifted toward convenience foods and shelf-stable products. Advancements in canning technology, aseptic processing, vacuum concentration, and cold-chain logistics enabled the establishment of tomato processing facilities worldwide. These innovations allowed fresh tomatoes to be transformed into a wide range of processed goods with extended shelf lives and consistent quality, which could be distributed globally. Countries like China aggressively scaled up tomato production and processing capacity in the late 20th and early 21st centuries, becoming one of the top exporters of tomato paste, particularly to African and Middle Eastern markets. Simultaneously, Turkey, Iran, and Egypt emerged as regional powerhouses in the tomato processing industry, supporting both domestic consumption and exports (Pereira et al. 2021; Razifard et al. 2020; van Andel et al. 2022). In many developing countries, the growth of tomato processing also created employment opportunities, stimulated rural economies, and reduced post-harvest losses by converting perishable fruits into value-added products. However, this growth was not without challenges; the sector has faced issues related to price volatility, trade imbalances, food safety standards, and environmental concerns, such as water use and waste management.

Global tomato production and processing have also been shaped by market segmentation and consumer preferences. Fresh market tomatoes—those sold whole and unprocessed—represent a significant portion of global production and are typically grown in greenhouses or open fields, depending on regional conditions (Dumas et al. 1998). Greenhouse tomato production, particularly in high-latitude countries such as the Netherlands, Canada, and Russia, has flourished due to the ability to control temperature, humidity, and light. The adoption of hydroponic and soilless growing systems in high-tech greenhouses has allowed year-round production of high-quality tomatoes with minimal pesticide use and precise nutrient delivery. In contrast, field-grown tomatoes dominate production in warmer regions, with field operations designed for either fresh consumption or processing (Dumas et al. 1998). Consumer preferences have also driven diversification in tomato types, including cherry, grape, plum, beefsteak, and heirloom varieties, each catering to different culinary uses and aesthetic desires. In the processing sector, different tomato cultivars are utilized based on product requirements—for example, thick-walled plum tomatoes are ideal for paste and sauce production due to their low water content and high solids content.

The role of technology and innovation in shaping the modern tomato industry is undeniable. Plant breeding, both traditional and molecular, has played a critical role in improving yield, disease resistance, shelf life, and stress tolerance. The release of hybrid cultivars dramatically increased productivity and uniformity in both fresh and processing markets. In recent decades, marker-assisted selection and genome editing techniques such as CRISPR have allowed for the precise development of tomatoes with improved nutritional profiles (e.g., higher lycopene content), longer shelf lives (e.g., delayed ripening genes), and resilience to pests, diseases, and climate change (Dutta et al. 2023; Saini et al. 2025). Precision agriculture technologies, including remote sensing, drones, and artificial intelligence, are being increasingly integrated into tomato production to optimize planting, irrigation, and harvesting decisions. These innovations are especially critical as climate variability threatens traditional growing regions with increased temperatures, irregular rainfall, and emerging pathogens. In response, global breeding programs are working to develop heat-tolerant, drought-resilient, and virus-resistant tomato lines to future-proof the industry.

International trade and globalization have further influenced the structure of the tomato processing industry. Trade flows of tomato paste, canned tomatoes, and sauces have created a competitive global market, with a few major producers—namely China, the European Union, and the United

States—dominating exports (Dimyati 2004). However, concerns over market saturation, trade disputes, and tariffs have led to calls for more equitable trade practices and market diversification. At the same time, there is increasing emphasis on sustainability and environmental responsibility in tomato production and processing. Water conservation strategies, such as drip irrigation and deficit irrigation, are being adopted to address water scarcity in arid growing regions. Processing facilities are investing in waste management systems to recycle tomato pomace and wastewater, with an ever-increasing interest in utilizing renewable energy to power their operations (Burke et al. 2009; Łącka et al. 2024). Certifications such as Global G.A.P., Fair Trade, and organic labeling are gaining traction as consumers demand transparency, traceability, and ethical sourcing practices.

Today, tomatoes are the most widely grown vegetable crop globally, with an estimated annual production exceeding 180 million metric tons. China leads in total production, followed by India, the United States, Turkey, and Egypt (FAOSTAT 2022). Meanwhile, processed tomato products form a multibillion-dollar global industry, with tomato paste being the most traded processed tomato commodity. The continued growth and evolution of tomato production and processing reflect broader trends in agriculture, including the integration of science and technology, the globalization of food systems, and the increasing importance of sustainable practices. As urban populations grow and dietary habits shift toward plant-based and processed foods, the tomato industry stands at a crossroads, balancing the need for efficiency and profitability with the imperatives of environmental stewardship, nutritional quality, and equitable development. With ongoing research, investment, and policy support, the global tomato sector is poised to meet these challenges while continuing to play a central role in the global food economy.

1.6 GERMPLASM COLLECTION

The status of tomato (*Solanum lycopersicum*) germplasm collection at both the national and international levels represents a vital component of global agricultural biodiversity preservation and the future of tomato breeding and research. Tomatoes, one of the world's most widely grown and consumed crops, are subject to constant demand for varieties that cater to diverse climatic conditions, production systems, and market preferences. As such, the collection, conservation, and distribution of tomato germplasm, which encompasses both cultivated varieties and wild relatives, is of paramount importance for ensuring food security, improving crop resilience, and meeting the challenges posed by climate change, pests, diseases, and evolving market demands (Chen et al. 2025; Pascual et al. 2013; Xu et al. 2024). National and international institutions have played a crucial role in maintaining diverse collections, providing essential resources for breeders and researchers working to improve tomato productivity and quality.

At the international level, several organizations have dedicated efforts to collecting, preserving, and distributing tomato germplasm, thereby creating a global network that facilitates access to genetic resources. The most prominent of these organizations is the International Board for Plant Genetic Resources (IBPGR), now known as Bioversity International, which has long recognized the importance of tomato as a food and horticultural crop. Through collaboration with national governments and research institutions, Bioversity International has supported the establishment of regional and global germplasm collections, providing a repository for diverse tomato species. A significant contribution to the international tomato germplasm collection is made by the Global Crop Diversity Trust (Crop Trust), which collaborates with the Food and Agriculture Organization (FAO) to ensure the security and accessibility of genetic resources worldwide. Crop Trust's seed banks, including the Svalbard Global Seed Vault in Norway, provide a safeguard for the long-term storage of tomato germplasm, which can be utilized in the event of natural disasters, conflicts, or other crises that threaten crop diversity (Alatawi et al. 2024).

One of the central international repositories for tomato germplasm is the International Tomato Genebank (ITG), located at the World Vegetable Center (WorldVeg) in Taiwan (Engle et al. 2003). The ITG holds a vast collection of tomato varieties and wild relatives, totaling more than 13,000 accessions from over 70 countries. This collection includes not only cultivated tomatoes but also species such as *Solanum pimpinellifolium*, *S. habrochaites*, *S. pennellii*, and *S. chilense*, which are critical sources of valuable traits, including disease resistance, drought tolerance, and enhanced nutritional content (Ayenan et al. 2021). The WorldVeg gene bank is part of the larger CGIAR (Consultative Group on International Agricultural Research) network, which plays a crucial role in the collection, conservation, and use of plant genetic resources (Alatawi et al. 2022a; Alatawi et al. 2022b; Alatawi et al. 2024). The CGIAR's gene banks serve as central hubs for tomato germplasm, making it accessible to researchers, breeders, and farmers worldwide. Additionally, the International Seed Testing Association (ISTA) and other international organizations provide frameworks for certifying, exchanging, and utilizing tomato germplasm, further facilitating the global movement of genetic resources.

At the national level, individual countries have established their germplasm collections, often in collaboration with international organizations, to safeguard local genetic diversity and support national breeding programs. In the United States, the National Plant Germplasm System (NPGS), part of the US Department of Agriculture (USDA), maintains an extensive collection of tomato germplasm, including over 10,000 accessions (Chen et al. 2025; Xu et al. 2024). The USDA's Agricultural Research Service (ARS) oversees the Tomato Genetic Resources Unit (TGRU), which is responsible for conserving and distributing tomato germplasm. The TGRU holds a diverse array of wild relatives, landraces, and cultivars, making it a critical resource for US tomato breeders working on disease resistance, yield improvement, and climate adaptation (Wang et al. 2016). The USDA's NPGS collection is a crucial resource for both public and private sector breeding programs, providing a foundation for developing improved varieties for fresh and processing markets.

In Europe, tomato germplasm collection is also well-established, with multiple countries contributing to the broader European repository network. The European Cooperative Programme for Plant Genetic Resources (ECPGR) coordinates the conservation and distribution of plant genetic resources within Europe. Countries such as Italy, Spain, and the Netherlands have large national collections that focus on preserving traditional and heirloom tomato varieties, as well as introducing wild species for breeding. For example, the Centro di Ricerca per la Patologia Vegetale (CRPV) in Italy maintains a collection of Italian tomato landraces that are renowned for their flavor, texture, and adaptability to Mediterranean climates (Mavlyanova 2011; Villand et al. 1998). Similarly, Spain's National Germplasm Bank (BNG) at the Spanish National Institute for Agricultural and Food Research and Technology (INIA) has played a pivotal role in collecting tomato germplasm, with a focus on local varieties adapted to Spanish agro-climatic conditions. In Latin America, the region's historical significance as the center of origin and diversity of the tomato crop has resulted in rich collections maintained by both national and international institutions. Mexico, Peru, and Ecuador have extensive national tomato germplasm collections, as these countries represent the primary centers of origin for many of the wild relatives of the cultivated tomato (Canul-Ku et al. 2025; Ramírez-Ojeda et al. 2022). The Mexican National Institute of Forestry, Agriculture, and Livestock Research (INIFAP) maintains a collection that includes native species and landraces, which are critical for breeding programs aimed at enhancing disease resistance and improving the quality of tomatoes for both domestic and international markets (Mavlyanova 2011; Villand et al. 1998). Similarly, the International Potato Center (CIP), based in Peru, is a significant partner in the conservation and utilization of Andean tomato germplasm, working to enhance traits such as pest resistance, fruit quality, and drought tolerance.

In Asia, the status of tomato germplasm collection is marked by active efforts in countries such as China and India, where tomato production is a major agricultural activity. In China, the National Germplasm Bank of Vegetables, based at the Institute of Vegetables and Flowers in Beijing, houses a comprehensive collection of tomato varieties, including wild relatives and landraces, which are utilized in breeding programs to improve yield, disease resistance, and stress tolerance (Chen et al. 2025; Wang et al. 2019; Xu et al. 2024). India's National Bureau of Plant Genetic Resources (NBPGR) has similarly invested in tomato germplasm conservation, focusing on collecting accessions from regional agro-ecosystems and ensuring the availability of genetic resources for research and breeding purposes (Singh and Goswami 2014; Singh et al. 2018). The Indian Institute of Horticultural

Research (IIHR) in Bangalore plays a crucial role in collecting and characterizing tomato germplasm, with a focus on enhancing the resilience of Indian varieties to diseases and environmental stresses. In sub-Saharan Africa, efforts to collect and conserve tomato germplasm are more recent but growing in importance (Table 1.1). The World Vegetable Center (WorldVeg), through

TABLE 1.1
Number of Tomato Accessions Collected by Various Institutions in Various Parts of the World

Agency	Country/Region	Approximate Number of Tomato Accessions	Remarks
International Tomato Genebank (ITG)	WorldVeg Center, Taiwan (Global)	13,000	Holds a diverse collection of cultivated and wild tomato species from over 70 countries.
USDA National Plant Germplasm System (NPGS)	United States	10,000	Includes a comprehensive collection of tomato varieties, wild relatives, and landraces.
Centro di Ricerca per la Patologia Vegetale (CRPV)	Italy	1,500	Focused on Italian tomato landraces, particularly for quality and flavor.
National Germplasm Bank (BNG)	Spain	2,000	Includes local Spanish varieties and wild relatives for breeding and research.
National Institute of Forestry, Agriculture, and Livestock Research (INIFAP)	Mexico	2,500	Collects significant varieties of Mexican tomato landraces and wild species.
International Potato Center (CIP)	Peru	2,000	Contains Andean tomato germplasm, focusing on disease resistance and climate adaptability.
National Germplasm Bank of Vegetables	China	5,000	Houses a large collection of Chinese tomato varieties, wild relatives, and landraces.
Indian Institute of Horticultural Research (IIHR)	India	3,000	Maintains tomato germplasm focused on disease resistance and environmental stress tolerance.
National Bureau of Plant Genetic Resources (NBPGR)	India	1,500	Focuses on the collection and conservation of diverse tomato varieties across India.
World Vegetable Center (WorldVeg)	Africa (Regional Centers)	1,500	Collects germplasm from across Africa, with a focus on pest and disease resistance.
Agricultural Research Service (ARS)	United States (Global)	10,000	Includes wild tomato species and cultivated varieties, with a focus on disease resistance.
The European Cooperative Programme for Plant Genetic Resources (ECPGR)	Europe (Regional Network)	4,000	Maintains accessions from European countries, with a focus on traditional and heirloom varieties.
National Germplasm Bank of Vegetables (IVT)	Vietnam	500	Houses a smaller collection but focuses on local and regional varieties, including wild species.
National Seed Bank (Svalbard Global Seed Vault)	Norway (Global)	Variable (more than 1,000)	Stores backup collections of tomato germplasm from multiple countries, with a focus on security.

Notes:

1. Approximate Numbers: The numbers provided are estimates and are subject to change as germplasm collections continually expand.
2. Wild Relatives & Landraces: The collections often include both wild relatives (such as *Solanum pennellii*, *S. chilense*, and *S. pimpinellifolium*) and cultivated varieties, with a special focus on species that contribute traits like disease resistance, drought tolerance, and improved nutritional quality.

Regional and Global Contributions: Many agencies contribute to national and international efforts, ensuring that collections are accessible to global breeders and researchers.

its regional offices in East and West Africa, has been instrumental in collecting and distributing tomato germplasm, with a particular focus on varieties resistant to pests such as Tuta absoluta and diseases like Tomato yellow leaf curl virus (TYLCV). National research programs in countries such as Kenya, Ghana, and Nigeria are increasingly engaging in conserving tomato genetic resources to ensure the availability of high-yielding, pest-resistant, and nutritious tomato varieties suited to the region's farming systems (Corrado et al. 2014; Délices et al. 2021; Teresa and Elzbieta 2000).

These agencies play a crucial role in preserving the genetic diversity of the tomato crop, ensuring that breeders and researchers worldwide have access to a rich pool of resources for developing improved tomato varieties.

1.7 GERMPLASM CHARACTERIZATION

Characterization of tomato (*Solanum lycopersicum*) germplasm is a foundational process in crop improvement, conservation, and utilization, enabling researchers and breeders to understand the diversity within and between accessions, identify useful traits, and associate them with genetic markers or genomic regions (Mellidou et al. 2020). Germplasm characterization using both phenotypic and molecular approaches has emerged as a complementary and synergistic strategy for effectively evaluating the immense diversity present in cultivated tomatoes, landraces, and wild relatives. Phenotypic characterization involves the detailed assessment of observable traits—morphological, agronomic, physiological, and biochemical—that can vary across genotypes (Athinodorou et al. 2021). These traits include plant habit, leaf morphology, flowering time, fruit shape, color, size, yield components, resistance to pests and diseases, stress tolerance, and nutritional quality. Morphological descriptors established by the International Plant Genetic Resources Institute (IPGRI), now Bioversity International, and national germplasm systems such as the USDA's National Plant Germplasm System (NPGS), have provided standardized protocols for phenotyping. In practice, phenotypic evaluations are conducted under controlled environments, field trials, or multi-location testing to capture the expression of traits under different environmental conditions and to estimate heritability and genotype-by-environment interactions (Délices et al. 2021; Fozia et al. 2024). However, phenotypic traits are often influenced by environmental conditions, making it challenging to distinguish between genetic and environmental effects. Thus, while phenotyping is indispensable for evaluating horticulturally and agriculturally important traits, its precision is greatly enhanced when complemented by molecular tools that offer insights at the genetic and genomic levels.

Molecular characterization of tomato germplasm involves using DNA-based markers and genomic tools to assess genetic diversity, population structure, phylogenetic relationships, and alleles associated with key traits (Babu et al. 2015). This approach provides high-resolution data not influenced by environmental variability, making it a powerful tool for germplasm management and improvement. Over the years, various molecular markers have been employed in tomato research, including Random Amplified Polymorphic DNA (RAPD), Amplified Fragment Length Polymorphism (AFLP), Restriction Fragment Length Polymorphism (RFLP), Simple Sequence Repeats (SSR or microsatellites), and, more recently, Single Nucleotide Polymorphisms (SNPs) (Öztürk 2022; Singh et al. 2018). Each marker system has its strengths and limitations, but collectively, they have enabled researchers to assess genetic distances among accessions, construct phylogenetic trees, perform genetic fingerprinting, and identify redundant or unique accessions within extensive germplasm collections. SSRs, for example, have been widely used in tomato for their co-dominant nature, reproducibility, and high polymorphism information content (Goncalves et al. 2008; Ray 2010; Tranchida-Lombardo et al. 2019). They have been instrumental in constructing linkage maps and identifying quantitative trait loci (QTLs) associated with yield, fruit quality, and disease resistance. The advent of SNP markers and high-throughput genotyping platforms such as genotyping-by-sequencing (GBS), SNP arrays (e.g., SolCAP 7.5K and 10K SNP arrays), and whole-genome resequencing has revolutionized molecular characterization by enabling genome-wide association studies (GWAS), population structure analysis, and genome selection in tomato breeding (Herraiz et al. 2015; Pratta et al. 2011).

Phenotypic and molecular characterization efforts have been significant in studying wild relatives and traditional landraces of tomato, which often harbor unique traits not found in commercial cultivars. For instance, *S. pimpinellifolium*, the closest wild relative of cultivated tomato, has been extensively characterized for fruit and plant traits, including small fruit size, high soluble solids content, and disease resistance. Molecular markers have confirmed its basal position in the tomato phylogenetic tree and its contribution to the domesticated gene pool. Similarly, accessions of *S. habrochaites* and *S. peruvianum* have been profiled for glandular trichomes and secondary metabolites that confer resistance to whiteflies, thrips, and fungal pathogens. Combining phenotypic data with marker information enables the identification of marker-trait associations, facilitating marker-assisted selection (MAS) and accelerating breeding pipelines. For example, resistance to tomato yellow leaf curl virus (TYLCV), Fusarium wilt, Verticillium wilt, nematodes, and late blight has been mapped using molecular markers derived from resistant wild species (Mei et al. 2023). In cases involving quantitative traits, QTL mapping has enabled the dissection of complex traits, including fruit size, shape, sugar and acid content, firmness, shelf life, and responses to abiotic stress (Gosal et al. 2010). The development of introgression lines (ILs) from wild species, where individual genomic segments are introduced into a common cultivated background, has become a powerful strategy for characterizing germplasm at high resolution and identifying beneficial alleles.

Another significant advancement in tomato germplasm characterization is the integration of high-throughput phenotyping (HTP) platforms and bioinformatics tools (Araus and Cairns 2014). HTP systems, which include imaging-based technologies, drones, multispectral sensors, and automated phenotyping chambers, allow the non-destructive, rapid, and precise measurement of phenotypic traits over time and across environments. Traits such as canopy temperature, chlorophyll fluorescence, plant architecture, fruit development, and stress responses can now be quantified with unprecedented accuracy. When paired with molecular data, these phenotypes enable the implementation of phenomics and functional genomics approaches to explore gene function and regulatory networks (Montes et al. 2007). In turn, this supports the identification of key regulatory genes, candidate loci, and allelic variants that contribute to phenotypic diversity in tomato germplasm. Moreover, genome-wide tools such as RNA-sequencing, epigenomics, proteomics, and metabolomics provide deeper layers of characterization that link genotype to phenotype and environment, especially in the context of developmental processes and stress responses. These omics technologies are increasingly being integrated into germplasm characterization efforts to provide a holistic view of trait expression and regulation.

Digital databases and bioinformatics platforms have greatly enhanced the accessibility and utility of phenotypic and molecular characterization data. Key resources such as the Sol Genomics Network (SGN), Tomato Genetics Resource Center (TGRC), Genesys, GRIN-Global, and the European Search Catalogue for Plant Genetic Resources (EURISCO) allow researchers to access, analyze, and share germplasm characterization data globally. These databases often contain passport information, phenotypic descriptors, marker data, genome sequences, and breeding history, facilitating informed decisions in germplasm selection and utilization. For instance, a breeder seeking drought-tolerant tomato lines can query accessions from arid environments with associated traits and SNP markers, accelerating the identification of donor lines. Likewise, researchers can identify unique or underutilized germplasm for trait discovery or conservation priority.

Despite this progress, challenges persist in characterizing tomato germplasm. Phenotypic characterization is time-consuming, labor-intensive, and prone to environmental variation, whereas molecular data generation can be expensive and requires specialized computational expertise. Moreover, integrating multi-omics data and translating it into actionable breeding decisions requires advanced analytical pipelines and interdisciplinary collaboration. Nevertheless, the growing investment in phenomics, genotyping platforms, machine learning, and open data initiatives addresses many challenges. Collaborative projects, such as the Tomato 150 Genome Project, G2P-SOL (Genotype to Phenotype), and Tomato Compass, pave the way for a deeper characterization and utilization of tomato genetic resources. These initiatives aim to document diversity and mobilize it for crop improvement by linking phenotype to genotype in a predictive framework.

1.8 GERMPLASM UTILIZATION

Utilizing tomato (*Solanum lycopersicum*) germplasm across different parts of the world is a dynamic and multifaceted process that underpins the global tomato industry, which ranks among the most important horticultural sectors worldwide. Tomato germplasm, encompassing traditional landraces, modern cultivars, and wild relatives, is the foundation for breeding programs, research initiatives, genetic improvement, and conservation strategies to enhance productivity, resilience, nutritional quality, and adaptability to climate and biotic stresses (Engle et al. 2003). In North America, particularly in the United States, the use of germplasm has been intensively focused on improving fruit quality, shelf life, disease resistance, and mechanical harvestability. The United States Department of Agriculture (USDA), through the National Plant Germplasm System (NPGS), maintains a vast collection of tomato germplasm, including heirlooms and wild accessions, which are widely distributed to breeders and researchers (Panthee et al. 2013). American breeding programs have leveraged wild species such as *Solanum pimpinellifolium, S.* habrochaites, and *S. pennellii* to introduce resistance genes against Fusarium wilt, Verticillium wilt, root-knot nematodes, Tomato yellow leaf curl virus (TYLCV), and abiotic stresses like salinity and drought (Borja-Bravo et al. 2013). Marker-assisted selection (MAS) and genomic selection are routinely employed to accelerate breeding pipelines. Introgression lines developed at institutions such as Cornell University and the University of California, Davis, have been utilized worldwide for mapping and gene discovery.

In Latin America, particularly in Peru, Ecuador, and Mexico, the utilization of tomato germplasm assumes both traditional and modern roles (Okogbenin et al. 2007). As the region represents the center of origin and diversification of tomatoes and their wild relatives, indigenous and local communities have long cultivated diverse landraces selected for specific traits, such as flavor, pest resistance, or adaptation to local microclimates. In Mexico, traditional *jitomates* are still grown for local markets, and national research institutes such as INIFAP (Instituto Nacional de Investigaciones Forestales, Agrícolas y Pecuarias) work with native germplasm to enhance yield, nutritional content, and resilience (Razifard et al. 2020). There is growing interest in participatory breeding, which involves smallholder farmers in preserving landraces while incorporating beneficial traits from wild species. Furthermore, international collaborations through centers such as the International Potato Center (CIP) and the International Center for Tropical Agriculture (CIAT) have supported the collection and characterization of tomato germplasm in Andean countries, ensuring its availability for global breeding efforts. Efforts have also been made to document and protect indigenous knowledge of local varieties through community seed banks and digital repositories.

Europe represents another central hub for the utilization of tomato germplasm, where historical and contemporary breeding efforts have shaped the diversity of tomato types found in Mediterranean countries. In Italy, Spain, and Greece, landraces like *San Marzano*, *Piennolo del Vesuvio*, and *Costoluto Fiorentino* are not only cultural icons but also genetic resources rich in traits for flavor, fruit shape, and stress tolerance (Corrado et al. 2014). European breeders often utilize local landraces to develop niche market varieties specifically tailored for regional cuisines and gourmet markets. Simultaneously, European public and private breeding programs—facilitated by institutions such as the Centre for Genetic Resources (CGN) in the Netherlands, INRAE in France, and the James Hutton Institute in the UK—have utilized germplasm collections for pre-breeding and trait introgression (Andolfo et al. 2021; D'Agostino et al. 2013). Disease resistance to late blight (*Phytophthora infestans*), tomato spotted wilt virus (TSWV), and Cladosporium leaf mold are among the key breeding targets. The European Union's Horizon research programs have fostered collaborative projects like G2P-SOL (Genotype to Phenotype), which integrates phenotypic and genomic data across European germplasm banks to optimize their use in breeding. Additionally, the increasing demand for sustainable, organic, and heirloom varieties among consumers has encouraged breeders to revive old landraces and cross them with elite material to combine flavor and robustness (Gonias et al. 2019; Mellidou et al. 2020).

In Asia, particularly in China and India, two of the world's largest tomato producers, germplasm utilization is focused on both enhancing commercial hybrid performance and addressing challenges related to climate change, pests, and diseases. In China, government-supported institutes such as the Institute of Vegetables and Flowers under the Chinese Academy of Agricultural Sciences (CAAS) maintain and characterize thousands of tomato accessions, including local landraces and introduced wild relatives. Chinese breeders have focused on traits such as heat and cold tolerance, bacterial wilt resistance, and enhanced shelf life, particularly in varieties intended for export or long-distance transportation. In India, the Indian Institute of Horticultural Research (IIHR) and other institutions under the Indian Council of Agricultural Research (ICAR) utilize diverse germplasm from both local sources and international gene banks to develop high-yielding hybrids and open-pollinated varieties suited for diverse agro-ecological zones (Jha et al. 2022). The adoption of hybrids has expanded rapidly across India, and traits such as bacterial wilt and tomato leaf curl virus resistance are in high demand. Moreover, community-level programs in India have emphasized the value of farmer-selected landraces in preserving diversity while improving productivity, especially under organic and low-input farming systems. Countries like Japan and South Korea have also invested in flavor and nutritional quality, incorporating germplasm with higher levels of lycopene, beta-carotene, and volatile compounds to develop premium tomato cultivars for both domestic and export markets (Das et al. 2023).

In the Middle East and North Africa (MENA) region, the utilization of tomato germplasm has been crucial for breeding heat- and drought-tolerant varieties suitable for arid and semi-arid environments. Countries like Egypt, Morocco, and Jordan have collaborated with international centers, such as the International Center for Agricultural Research in the Dry Areas (ICARDA) and the World Vegetable Center (WorldVeg), to access and utilize germplasm collections for enhancing abiotic stress tolerance and shelf life. Cultivation in protected environments, such as greenhouses and plastic tunnels, is widespread; therefore, developing cultivars with resistance to greenhouse pests and diseases is also a priority (Elbaz et al. 2016). In Israel, a global leader in agricultural innovation, extensive research on tomato physiology and genetics has leveraged germplasm to develop high-tech breeding solutions, including drought-resistant rootstocks and flavor-improved hybrids, utilizing advanced marker-assisted breeding techniques. The Israeli gene bank and private seed companies are deeply involved in screening wild relatives for novel traits and transferring these into elite backgrounds using modern genomics (Salmeron et al. 1994; Salmeron et al. 1996).

In sub-Saharan Africa, the utilization of tomato germplasm is driven by the need to improve yield stability, pest and disease resistance, and postharvest quality in smallholder farming systems. Tomatoes are a major income-generating crop for millions of farmers, but they face constraints such as bacterial wilt, TYLCV, Tuta absoluta infestation, and limited access to quality seed. The World Vegetable Center (WorldVeg) and national agricultural research systems (NARS) have spearheaded efforts to introduce improved germplasm and train breeders in utilizing diverse genetic materials (Ziervogel and Zermoglio 2009). WorldVeg's open-access breeding lines, developed from a broad base of Asian, African, and South American germplasm, have been adopted and adapted in countries like Tanzania, Nigeria, and Ghana. Participatory varietal selection (PVS) approaches are increasingly used to ensure that farmer preferences, market traits, and resilience are prioritized in breeding programs. Additionally, collaborations with international organizations such as FAO and CGIAR have enabled African breeders to access and utilize germplasm stored in global gene banks for local improvement, thereby bridging the genetic resource gap (Tripathi et al. 2022).

Globally, the successful utilization of tomato germplasm depends on the availability of diverse accessions and infrastructure, capacity building, and policy support. Advances in molecular biology, phenotyping, and bioinformatics have opened new avenues for the systematic evaluation of germplasm. International projects, such as the Tomato Genome Consortium, G2P-SOL, and Tomato

Compass, have sought to map the genetic architecture of global tomato diversity. Moreover, frameworks like the International Treaty on Plant Genetic Resources for Food and Agriculture (ITPGRFA) and the Nagoya Protocol regulate access and benefit-sharing, encouraging equitable and responsible use of germplasm worldwide (Morioka 2018; Xia 2023). However, technical capacity and funding disparities mean that germplasm remains underutilized in many regions. Strengthening regional networks, enhancing local breeding programs, and investing in gene bank management and phenotyping platforms are crucial for realizing the full potential of global tomato germplasm.

1.9 CONCLUSION

The global status of tomato germplasm management reflects a remarkable convergence of historical insight, scientific innovation, and international collaboration. From its Andean origins to its global spread, the tomato has undergone profound evolutionary and cultural transformations, supported by deliberate efforts to conserve and harness its genetic diversity. Today, an extensive network of gene banks, research institutions, and community-driven initiatives ensures the preservation, characterization, and utilization of tomato germplasm. Advances in molecular biology, high-throughput phenotyping, and bioinformatics have revolutionized our capacity to explore and exploit the genetic potential of both wild and cultivated accessions. Yet, the continued erosion of traditional landraces, habitat loss, climate uncertainties, and the complexities of access and benefit-sharing pose ongoing challenges. Meeting these challenges will require sustained investment, inclusive policies, and greater integration between scientific, institutional, and farmer-led efforts. Ultimately, robust germplasm management is not only a safeguard for the tomato's genetic legacy but a strategic necessity for ensuring global food security, environmental resilience, and agricultural innovation in a rapidly changing world.

REFERENCES

Alatawi I, Bhattarai G, Xiong HZ, Phiri TM, Chiwina K, Chen YL, Bluhm B, Joshi N, Panthee D, Shi AN (2022a) Characterization of genetic diversity of tomato germplasm accessions. *HortScience* 57:S162–S163.

Alatawi I, Bhattarai G, Xiong HZ, Ravelombola W, Phiri TM, Chiwina K, Chen YL, Bluhm B, Joshi N, Panthee D, Shi AN (2022b) Evaluation of tomato germplasm for salt tolerance at seedling stage. *HortScience* 57:S212–S212.

Alatawi I, Xiong H, Alkabkabi H, Chiwina K, Luo Q, Ling K-S, Qu Y, Du R, Shi A (2024) Genetic diversity and population structure of tomato (*Solanum lycopersicum*) from the USDA-GRIN germplasm collection. *Agronomy* 15. https://doi.org/10.3390/agronomy15010022.

Alseekh S, Scossa F, Wen WW, Luo J, Yan JB, Beleggia R, Klee HJ, Huang SW, Papa R, Fernie AR (2021) Domestication of crop metabolomes: Desired and unintended consequences. *Trends in Plant Science* 26:650–661. https://doi.org/10.1016/j.tplants.2021.02.005.

Andolfo G, D'Agostino N, Frusciante L, Ercolano MR (2021) The tomato interspecific NB-LRR gene arsenal and its impact on breeding strategies. *Genes (Basel)* 12. https://doi.org/10.3390/genes12020184.

Araus JL, Cairns JE (2014) Field high-throughput phenotyping: The new crop breeding frontier. *Trends in Plant Science* 19:52–61. https://doi.org/10.1016/j.tplants.2013.09.008.

Arnoux S, Fraisse C, Sauvage C (2021) Genomic inference of complex domestication histories in three Solanaceae species. *Journal of Evolutionary Biology* 34:270–283. https://doi.org/10.1111/jeb.13723.

Athinodorou F, Foukas P, Tsaniklidis G, Kotsiras A, Chrysargyris A, Delis C, Kyratzis AC, Tzortzakis N, Nikoloudakis N (2021) Morphological diversity, genetic characterization, and phytochemical assessment of the Cypriot tomato germplasm. *Plants (Basel)* 10. https://doi.org/10.3390/plants10081698.

Avina-Padilla K, Zamora-Macorra EJ, Ochoa-Martinez DL, Alcantar-Aguirre FC, Hernandez-Rosales M, Calderon-Zamora L, Hammond RW (2022) Mexico: A landscape of viroid origin and epidemiological relevance of endemic species. *Cells* 11. https://doi.org/10.3390/cells11213487.

Ayenan MaT, Danquah A, Agre PA, Hanson P, Asante IK, Danquah EY (2021) Genomic and phenotypic diversity of cultivated and wild tomatoes with varying levels of heat tolerance. *Genes (Basel)* 12. https://doi.org/10.3390/genes12040503.

Babu AN, Jogaiah S, Ito S, Nagaraj AK, Tran LSP (2015) Improvement of growth, fruit weight and early blight disease protection of tomato plants by rhizosphere bacteria is correlated with their beneficial traits and induced biosynthesis of antioxidant peroxidase and polyphenol oxidase. *Plant Science* 231:62–73. https://doi.org/10.1016/j.plantsci.2014.11.006.

Blanca J, Canizares J, Cordero L, Pascual L, Diez MJ, Nuez F (2012) Variation revealed by SNP genotyping and morphology provides insight into the origin of the tomato. *PLoS ONE* 7:e48198. https://doi.org/10.1371/journal.pone.0048198.

Borja-Bravo M, García-Salazar JA, Skaggs RK (2013) Mexican fresh tomato exports in the North American market: A case study of the effects of productivity on competitiveness. *Canadian Journal of Plant Science* 93:839–850.

Burke MB, Lobell DB, Guarino L (2009) Shifts in African crop climates by 2050, and the implications for crop improvement and genetic resources conservation. *Global Environmental Change* 19:317–325. https://doi.org/10.1016/j.gloenvcha.2009.04.003.

Caicedo AL, Schaal BA (2004) Population structure and phylogeography of Solanum pimpinellifolium inferred from a nuclear gene. *Molecular Ecology* 13:1871–1882. https://doi.org/10.1111/j.1365-294X.2004.02191.x.

Canul-Ku J, González-Pérez E, Barrios-Gómez EJ, Anaya-López JL, Hernández-Meneses E, Rangel-Estrada SE (2025) Evaluation of tomato progenies and their native progenitors from central and southern Mexico. *Interciencia* 50:16–22.

Chen X, Liu Y, Zheng F, Cheng G, Guo M, Li J, Wang X (2025) Construction of a core collection of tomato (*Solanum lycopersicum*) germplasm based on phenotypic traits and SNP markers. *Scientia Horticulturae* 339. https://doi.org/10.1016/j.scienta.2024.113855.

Chetelat R (2001) *Plant Exploration in Northern Chile to Collect Wild Tomato Species, with Emphasis on Solanum Lycopersicoides and s. Sitiens*. Final report 6/28/2001. TGRC. http://tgrc.ucdavis.edu/chile.html.

Corrado G, Caramante M, Piffanelli P, Rao R (2014) Genetic diversity in Italian tomato landraces: Implications for the development of a core collection. *Scientia Horticulturae* 168:138–144. https://doi.org/10.1016/j.scienta.2014.01.027.

D'Agostino N, Golas T, Van De Geest H, Bombarely A, Dawood T, Zethof J, Driedonks N, Wijnker E, Bargsten J, Nap JP, Mariani C, Rieu I (2013) Genomic analysis of the native European Solanum species, s. Dulcamara. *BMC Genomics* 14:356. https://doi.org/10.1186/1471-2164-14-356.

Das T, Anand U, Pal T, Mandal S, Kumar M, Radha, Gopalakrishnan AV, Lastra JMP, Dey A (2023) Exploring the potential of CRISPR/Cas genome editing for vegetable crop improvement: An overview of challenges and approaches. *Biotechnology & Bioengineering* 120:1215–1228. https://doi.org/10.1002/bit.28344.

Délices G, Raúl Leyva-Ovalle O, Mota-Vargas C, Núñez-Pastrana R, Andrés-Meza P, Herrera-Corredor JA (2021) Morphological characterization of wild populations of Solanum lycopersicum var. Cerasiforme in the tomato domestication area. *Emirates Journal of Food and Agriculture* 33:303–313.

Del Rio AH, Bamberg JB, Huaman Z (1997) Assessing changes in genetic diversity of potato gene banks. 1. Effects of seed increase. *Theoretical and Applied Genetics* 95:191–198.

Dimyati A (2004) Current development and future prospects in horticulture: Asian perspective. In *International Symposium on Harnessing the Potential of Horticulture in the Asian-Pacific Region*. Coolum, Australia, pp. 25–31.

Dixon MM, Afkairin A, Davis JG, Chitwood-Brown J, Buchanan CM, Ippolito JA, Manter DK, Vivanco JM (2024) Tomato domestication rather than subsequent breeding events reduces microbial associations related to phosphorus recovery. *Scientific Reports* 14:9934. https://doi.org/10.1038/s41598-024-60775-3.

Dumas Y, Bussières P, Battilani A, Cornillon P, Losada MHP, Branthôme X, Di Lucca G, Bues R, Dadomo M, Machado R, Christou M, San Martín C, Lyannaz JP, Koutsos T, Ho LC (1998) Qualitom a European research programme to build and test a technical itinerary model in processing tomato cropping. In *6th International ISHS Symposium on the Processing Tomato/Workshop on Irrigation and Fertigation of Processing Tomato*. Pamplona, Spain, pp. 151–158.

Dutta TK, Ray S, Phani V (2023) The status of the CRISPR/Cas9 research in plant-nematode interactions. *Planta* 258:103. https://doi.org/10.1007/s00425-023-04259-0.

Elbaz M, Hanson P, Fgaier S, Laarif A, Havey M (2016) Evaluation of tomato entries with different combinations of resistance genes to tomato yellow leaf curl disease in Tunisia. *Plant Breeding* 135:525–530. https://doi.org/10.1111/pbr.12375.

Engle LM, Shanmugasundaram S, Hanson P (2003) Evaluation and utilization of vegetable genetic resources for the tropics. In PL Forsline, et al. (Eds.), *Plant Genetic Resources: The Fabric of Horticulture's Future. Acta Horticulturae, 623*, pp. 263–274.

FAOSTAT (2022) *World Production (Tonnes) of Tomatoes in 2020*. Food and Agriculture Organization of the United Nations (FAO), Rome, Italy. http://www.fao.org/faostat/en/#data.

Ferrero V, Baeten L, Blanco-Sanchez L, Planello R, Diaz-Pendon JA, Rodriguez-Echeverria S, Haegeman A, De La Pena E (2020) Complex patterns in tolerance and resistance to pests and diseases underpin the domestication of tomato. *New Phytologist* 226:254–266. https://doi.org/10.1111/nph.16353.

Figas MR, Prohens J, Raigon MD, Pereira-Dias L, Casanova C, Garcia-Martinez MD, Rosa E, Soler E, Plazas M, Soler S (2018) Insights into the adaptation to greenhouse cultivation of the traditional Mediterranean long shelf-life tomato carrying the ALC mutation: A multi-trait comparison of landraces, selections, and hybrids in open field and greenhouse. *Frontiers in Plant Science* 9:1774. https://doi.org/10.3389/fpls.2018.01774.

Fozia, Zahid A, Gao Y, Ramzan M, Ahmed MaA, Stępień P, Yousef AF (2024) Morphological, phenotypic and nutritional characterization of exotic and local tomato germplasm under tunnel. *Food Bioscience* 59. https://doi.org/10.1016/j.fbio.2024.104241.

Gasparini K, Figueiredo YG, Araujo WL, Peres LE, Zsogon A (2024) De novo domestication in the solanaceae: Advances and challenges. *Current Opinion in Biotechnology* 89:103177. https://doi.org/10.1016/j.copbio.2024.103177.

Goncalves LSA, Rodrigues R, Sudre CP, Bento CD, Moulin MM, De Araujo ML, Daher RF, Pereira TNS, Pereira MG (2008) Genetic divergence among tomato accessions using RAPD markers and its comparison with multicategoric descriptors. *Horticultura Brasileira* 27:364–370.

Gonias ED, Ganopoulos I, Mellidou I, Bibi AC, Kalivas A, Mylona PV, Osanthanunkul M, Tsaftaris A, Madesis P, Doulis AG (2019) Exploring genetic diversity of tomato (*Solanum lycopersicum* L.) germplasm of genebank collection employing SSR and SCAR markers. *Genetic Resources and Crop Evolution* 66:1295–1309. https://doi.org/10.1007/s10722-019-00786-6.

Gordillo LF, Stevens MR, Millard MA, Geary B (2008) Screening two Lycopersicon peruvianum collections for resistance to tomato spotted wilt virus. *Plant Disease* 92:694–704. https://doi.org/10.1094/PDIS-92-5-0694.

Gosal SS, Wani SH, Kang MS (2010) Biotechnology and crop improvement. *Journal of Crop Improvement* 24:153–217. https://doi.org/10.1080/15427520903584555.

Herraiz FJ, Vilanova S, Andújar I, Torrent D, Plazas M, Gramazio P, Prohens J (2015) Morphological and molecular characterization of local varieties, modern cultivars and wild relatives of an emerging vegetable crop, the pepino (*Solanum muricatum*), provides insight into its diversity, relationships and breeding history. *Euphytica* 206:301–318. https://doi.org/10.1007/s10681-015-1454-8.

Jahan I, Angon PB, Mou SJ, Zannat M, Antu UB, Alam MM, Sweety Akivha A, Islam MS, Sakil MA (2025) The potential of CRISPR-Cas genome editing technologies to mitigate biotic stress in plants. *CABI Agriculture and Bioscience* 6. https://doi.org/10.1079/ab.2025.0005.

Jha UC, Nayyar H, Parida SK, Deshmukh R, Von Wettberg EJB, Siddique KHM (2022) Ensuring global food security by improving protein content in major grain legumes using breeding and 'omics' tools. *International Journal of Molecular Sciences* 23. https://doi.org/10.3390/ijms23147710.

Łącka I, Suproń B, Szczepaniak I (2024) Does climate change and energy consumption affect the food security of European Union countries? Empirical evidence from a panel study. *Energies* 17. https://doi.org/10.3390/en17133237.

Lin T, Zhu G, Zhang J, Xu X, Yu Q, Zheng Z, Zhang Z, Lun Y, Li S, Wang X, Huang Z, Li J, Zhang C, Wang T, Zhang Y, Wang A, Zhang Y, Lin K, Li C, Xiong G, Xue Y, Mazzucato A, Causse M, Fei Z, Giovannoni JJ, Chetelat RT, Zamir D, Stadler T, Li J, Ye Z, Du Y, Huang S (2014) Genomic analyses provide insights into the history of tomato breeding. *Nature Genetics* 46:1220–1226. https://doi.org/10.1038/ng.3117.

Mavlyanova R (2011) Strategic approaches for research and promotion of underutilized vegetable crops for food security in Central Asia and the Caucasus. In *2nd International Symposium on Underutilized Plant Species—Crops for the Future—Beyond Food Security*, Kuala Lumpur, Malaysia, pp. 541–547.

Mei Y, Cai L, Wang Y, Li F, Yang X, Yang J, Zhou X (2023) Molecular characterization and pathogenicity of an infectious clone of tomato leaf curl New Delhi virus isolate infecting Cucumis melo. *Stress Biology* 3:51. https://doi.org/10.1007/s44154-023-00128-8.

Mellidou I, Krommydas K, Nianiou-Obeidat I, Ouzounidou G, Kalivas A, Ganopoulos I (2020) Exploring morpho-physiological profiles of a collection of tomato (*Solanum lycopersicum*) germplasm using multivariate statistics. *Plant Genetic Resources: Characterization and Utilization* 18:88–97. https://doi.org/10.1017/s1479262120000088.

Montes JM, Melchinger AE, Reif JC (2007) Novel throughput phenotyping platforms in plant genetic studies. *Trends in Plant Science* 12:433–436. https://doi.org/10.1016/j.tplants.2007.08.006.

Morioka H (2018) Research compliance with the Nagoya protocol. *Japanese Journal of Applied Entomology and Zoology* 62:21–28. https://doi.org/10.1303/jjaez.2018.21.

Nakazato T, Warren DL, Moyle LC (2010) Ecological and geographic modes of species divergence in wild tomatoes. *American Journal of Botany* 97:680–693. https://doi.org/10.3732/ajb.0900216.

Okogbenin E, Porto MCM, Egesi C, Mba C, Espinosa E, Santos LG, Ospina C, Marin J, Barrera E, Gutierrez J, Ekanayake I, Iglesias C, Fregene MA (2007) Marker-assisted introgression of resistance to cassava mosaic disease into Latin American germplasm for the genetic improvement of cassava in Africa. *Crop Science* 47:1895–1904.

Öztürk HI (2022) Morphological and molecular characterization of some tomato (*Solanum lycopersicum* L.) genotypes collected from Erzincan Province of Turkey. *Molecular Biology Reports* 49:7111–7121. https://doi.org/10.1007/s11033-022-07310-7.

Panthee DR, Labate JA, Robertson LD (2013) Evaluation of tomato accessions for flavour and flavour-contributing components. *Plant Genetic Resources-Characterization and Utilization* 11:106–113. https://doi.org/10.1017/s1479262112000421.

Pascual L, Xu J, Biais B, Maucourt M, Ballias P, Bernillon S, Deborde C, Jacob D, Desgroux A, Faurobert M, Bouchet JP, Gibon Y, Moing A, Causse M (2013) Deciphering genetic diversity and inheritance of tomato fruit weight and composition through a systems biology approach. *Journal of Experimental Botany* 64:5737–5752. https://doi.org/10.1093/jxb/ert349.

Peralta IE, Spooner DM (2007) History, origin and early cultivation of tomato (Solanaceae). In *Genetic Improvement of Solanaceous Crops*, vol. 2, 1–27. Science Publishers Inc., Enfield New Hampshire, USA.

Pereira L, Zhang L, Sapkota M, Ramos A, Razifard H, Caicedo AL, Van Der Knaap E (2021) Unraveling the genetics of tomato fruit weight during crop domestication and diversification. *Theoretical and Applied Genetics* 134:3363–3378. https://doi.org/10.1007/s00122-021-03902-2.

Pons C, Casals J, Palombieri S, Fontanet L, Riccini A, Rambla JL, Ruggiero A, Figás MD, Plazas M, Koukounaras A, Picarella ME, Sulli M, Fisher J, Ziarsolo P, Blanca J, Cañizares J, Cammareri M, Vitiello A, Batelli G, Kanellis A, Brouwer M, Finkers R, Nikoloudis K, Soler S, Giuliano G, Grillo S, Grandillo S, Zamir D, Mazzucato A, Causse M, Diez MJ, Prohens J, Monforte AJ, Granell A (2022) Atlas of phenotypic, genotypic and geographical diversity present in the European traditional tomato. *Horticulture Research* 9. https://doi.org/10.1093/hr/uhac112.

Pratta GR, Rodriguez GR, Zorzoli R, Valle EM, Picardi LA (2011) Phenotypic and molecular characterization of selected tomato recombinant inbred lines derived from the cross Solanum lycopersicum x s. Pimpinellifolium. *Journal of Genetics* 90:229–237. https://doi.org/10.1007/s12041-011-0063-0.

Ramírez-Ojeda G, Rodríguez-Pérez JE, Rodríguez-Guzmán E, Sahagún-Castellanos J, Chávez-Servia JL, Peralta IE, Barrera-Guzmán LA (2022) Distribution and climatic adaptation of wild tomato (*Solanum lycopersicum* L.) populations in Mexico. *Plants-Basel* 11. https://doi.org/10.3390/plants11152007.

Ray DP (2010) Molecular characterization of six released tomato varieties using inter simple sequence repeat markers. *Indian Journal of Horticulture* 67:328–332.

Razifard H, Ramos A, Della Valle AL, Bodary C, Goetz E, Manser EJ, Li X, Zhang L, Visa S, Tieman D, Van Der Knaap E, Caicedo AL (2020) Genomic evidence for complex domestication history of the cultivated tomato in Latin America. *Molecular Biology and Evolution* 37:1118–1132. https://doi.org/10.1093/molbev/msz297.

Rodriguez GR, Munos S, Anderson C, Sim SC, Michel A, Causse M, Gardener BB, Francis D, Van Der Knaap E (2011) Distribution of SUN, OVATE, LC, and FAS in the tomato germplasm and the relationship to fruit shape diversity. *Plant Physiology* 156:275–285. https://doi.org/10.1104/pp.110.167577.

Rogo U, Simoni S, Fambrini M, Giordani T, Pugliesi C, Mascagni F (2024) Future-proofing agriculture: De novo domestication for sustainable and resilient crops. *International Journal of Molecular Sciences* 25. https://doi.org/10.3390/ijms25042374.

Saini H, Devrani A, Synrem G, Priyanka, Salem KFM (2025) Application of CRISPR technology in plant improvement: An update review. *Advances in Agriculture* 2025. https://doi.org/10.1155/aia/4578877.

Salazar-Mendoza P, Magalhaes DM, Lourençao AL, Bento JMS (2023) Differential defensive and nutritional traits among cultivated tomato and its wild relatives shape their interactions with a specialist herbivore. *Planta* 257. https://doi.org/10.1007/s00425-023-04108-0.

Salmeron JM, Barker SJ, Carland FM, Mehta AY, Staskawicz BJ (1994) Tomato mutants altered in bacterial disease resistance provide evidence for a new locus controlling pathogen recognition. *Plant Cell* 6:511–520. https://doi.org/10.2307/3869931.

Salmeron JM, Oldroyd GE, Rommens CM, Scofield SR, Kim HS, Lavelle DT, Dahlbeck D, Staskawicz BJ (1996) Tomato Prf is a member of the leucine-rich repeat class of plant disease resistance genes and lies embedded within the Pto kinase gene cluster. *Cell* 86:123–133. https://doi.org/10.1016/s0092-8674(00)80083-5.

Seong K, Seo E, Witek K, Li M, Staskawicz B (2020) Evolution of NLR resistance genes with noncanonical N-terminal domains in wild tomato species. *New Phytologist* 227:1530–1543. https://doi.org/10.1111/nph.16628.

Singh B, Goswami A (2014) Analysis of molecular diversity in tomato (*Solanum lycopersicum*) genotypes using SSR primers. *Indian Journal of Agricultural Sciences* 84:1004–1010.

Singh M, Vaishali A, Singh AK, Kumar A, Pandey KD (2018) Molecular diversity of tomato germplasm (Lycopersicum esculentum L.) using lycopene specific markers. *Biocatalysis and Agricultural Biotechnology* 16:340–346. https://doi.org/10.1016/j.bcab.2018.08.017.

Städler T, Roselius K, Stephan W (2005) Genealogical footprints of speciation processes in wild tomatoes: Demography and evidence for historical gene flow. *Evolution* 59:1268–1279.

Steiner AM, Ruckenbauer P, Goecke E (1997) Maintenance in genebanks, a case study: Contaminations observed in the nürnberg oats of 1831. *Genetic Resources and Crop Evolution* 44:533–538.

Sun L, Chen J, Xiao K, Yang WC (2017) Origin of the domesticated horticultural species and molecular bases of fruit shape and size changes during the domestication, taking tomato as an example. *Horticultural Plant Journal* 3:125–132. https://doi.org/10.1016/j.hpj.2017.07.007.

Swarup S, Cargill EJ, Crosby K, Flagel L, Kniskern J, Glenn K (2021) Genetic diversity is indispensable for plant breeding to improve crops. *Crop Science* 61:839–852. https://doi.org.prox.lib.ncsu.edu/10.1002/csc2.20377.

Teresa K, Elzbieta H (2000) Tomato genetic resources and activities in Poland. *Acta Physiologiae Plantarum* 22:238–242. https://doi.org/10.1007/s11738-000-0020-7.

Tranchida-Lombardo V, Mercati F, Avino M, Punzo P, Fiore MC, Poma I, Patanè C, Guarracino MR, Sunseri F, Tucci M, Grillo S (2019) Genetic diversity in a collection of Italian long storage tomato landraces as revealed by SNP markers array. *Plant Biosystems* 153:288–297. https://doi.org/10.1080/11263504.2018.1478900.

Tripathi L, Dhugga KS, Ntui VO, Runo S, Syombua ED, Muiruri S, Wen Z, Tripathi JN (2022) Genome editing for sustainable agriculture in Africa. *Front Genome Ed* 4:876697. https://doi.org/10.3389/fgeed.2022.876697.

Tyagi S, Kumar R, Kumar V, Won SY, Shukla P (2021) Engineering disease resistant plants through CRISPR-Cas9 technology. *Gm Crops & Food-Biotechnology in Agriculture and the Food Chain* 12:125–144. https://doi.org/10.1080/21645698.2020.1831729.

Van Andel T, Vos RA, Michels E, Stefanaki A (2022) Sixteenth-century tomatoes in Europe: Who saw them, what they looked like, and where they came from. *PeerJ* 9. https://doi.org/10.7717/peerj.12790.

Vicente MH, Macleod K, Zhu F, Rafael DD, Figueira A, Fernie AR, Mohareb F, Kevei Z, Thompson AJ, Zsögön A, Peres LEP (2023) The ORGAN SIZE (ORG) locus modulates both vegetative and reproductive gigantism in domesticated tomato. *Annals of Botany* 132:1233–1248. https://doi.org/10.1093/aob/mcad150.

Villand J, Skroch PW, Lai T, Hanson P, Kuo CG, Nienhuis J (1998) Genetic variation among tomato accessions from primary and secondary centers of diversity. *Crop Science* 38:1339–1347.

Wang T, Li HT, Zhu H, Qi SY, Zhang YM, Zhang ZJ, Zou QD (2019) Comparative analyses of genetic variation in a tomato (*Solanum lycopersicum* L.) germplasm collection with single nucleotide polymorphism and insertion-deletion markers. *Russian Journal of Genetics* 55:204–211. https://doi.org/10.1134/s1022795419020182.

Wang T, Zou QD, Qi SY, Wang XF, Wu YY, Liu N, Zhang YM, Zhang ZJ, Li HT (2016) Analysis of genetic diversity and population structure in a tomato (*Solanum lycopersicum* L.) germplasm collection based on single nucleotide polymorphism markers. *Genetics and Molecular Research* 15. https://doi.org/10.4238/gmr.15038209.

Welter SC, Steggall JW (1993) Contrasting the tolerance of wild and domesticated tomatoes to herbivory—agroecological implications. *Ecological Applications* 3:271–278. https://doi.org/10.2307/1941830.

Xia N (2023) Access and benefit-sharing in China: Exploring the extent to which China fulfils the obligations of the Nagoya Protocol. *Queen Mary Journal of Intellectual Property* 13:97–120.

Xu W, Gong C, Mai P, Li Z, Sun B, Li T (2024) Genetic diversity and population structure analysis of 418 tomato cultivars based on single nucleotide polymorphism markers. *Frontiers in Plant Science* 15. https://doi.org/10.3389/fpls.2024.1445734.

Ziervogel G, Zermoglio F (2009) Climate change scenarios and the development of adaptation strategies in Africa: Challenges and opportunities. *Climate Research* 40:133–146. https://doi.org/10.3354/cr00804.

2 SGN and BreedBase *Platforms for Genomics and Breeding*

Lukas A. Mueller

2.1 INTRODUCTION

With the sequencing of the first plant genomes in the early 2000s, many plant genome databases were established to disseminate the resulting data in the research community. Early databases included TAIR (the Arabidopsis Information Resource, for the major plant model and first plant to be fully sequenced, *Arabidopsis thaliana* (The Arabidopsis Genome Initiative, 2000), https://arabidopsis.org/), maizeGDB (for the model system and important agricultural crop *Zea mays* (Schnable et al., 2009), https://maizegdb.org), Gramene (initially a database for grasses centered around the rice genome (International Rice Genome Sequencing Project & Sasaki, 2005), later expanded to all plants, https://gramene.org/), and SGN (the Solanaceae Genomics Network, a database for the Solanaceae family, centered around the tomato genome (The Tomato Genome Consortium, 2012), https://solgenomics.net/).

While TAIR and MaizeGDB focused on a single model plant species, Gramene and SGN covered a whole clade of plants, with comparative genomics approaches anchoring to a high-quality genome sequence. Initially, these databases catered mainly to researchers with an interest in basic plant biology, for the identification of genes and pathways involved in their field of interest. Subsequently, genome-based approaches also became a game changer in plant breeding. However, for plant breeding applications, a single genome reference turned out to be insufficient; the genotypes of many different lines have to be known in some detail to make meaningful selections. In addition, phenotypic information has to be available in great detail for each line for selection purposes. This type of data is usually collected in randomized field experiments.

At SGN, an early effort was made to establish a database that could integrate both genotypic and phenotypic information. For phenotypic data, so-called trait ontologies were established that defined and structured all the traits for a given crop. Images are also an important data type for documenting phenotypes; therefore, support to incorporate images was added early on.

Marker-assisted breeding became an important technology, especially for hard-to-score single-gene or oligogenic traits, such as many disease resistance traits. As genotyping technologies evolved, higher-density marker systems became available that enabled new approaches, such as genomic selection.

In the early 2010s, the SGN code base was used as a starting point to create Cassavabase (https://cassavabase.org/), a global database supporting cassava breeding. In the process, several new functionalities were added to the system, including field layout randomization capabilities, interfacing with data collection tools such as Field Book (Rife & Poland, 2014) and genomic selection workflows (Jannink et al., 2010). The resulting system was made available for any breeding program as an open-source, crop-agnostic system, named Breedbase (https://breedbase.org/) (Morales et al., 2022). The vision for Breedbase was to create a "digital ecosystem" for plant breeding, covering the entire breeding cycle. In this self-contained system, data would be collected electronically,

DOI: 10.1201/9781003506539-2

transferred to the database in almost real time, and analyzed with built-in tools to select the best progeny for the next cycle. Through the work of the last decade, this vision has largely been realized.

2.2 DATA EXCHANGE IN THE DIGITAL ECOSYSTEM—THE BRAPI STANDARD

Early on in the process of implementing Breedbase, it became clear that a new level of communication protocol between various digital breeding databases or tools would be necessary to enable seamless integration of workflows. Breeding data is complex and often located in distinct databases and, further, needs to be aggregated on the fly; analysis tools would like to access information on different types of servers; and phenotyping apps need to exchange data with a central database. For all these use cases, a standardized information exchange protocol or Application Programming Interface would be a huge advantage, as interface logic would have to be implemented only once and could be re-used by many different tools or databases, allowing users to mix and match different tools while preventing the "re-invention of the wheel" problem in software development (very similar solutions are tailored to a specific environment and re-written from the ground up for other environments). Several breeding database and breeding tool providers came together to define such a standard, now known as the Breeding API (or BrAPI, https://brapi.org/) (Selby et al., 2019), with early contributors from Breedbase (Morales et al., 2022), BMS (Ribaut et al., 2012), Germinate (Lee et al., 2005), and PhenoApps (Röckel et al., 2022) projects. The so-called "BrAPPs" are applications that can obtain all their data using the breeding API and are therefore run with any BrAPI-enabled data source (Selby et al., 2019). For many programming languages, modules are available to facilitate BrAPI queries, one of the most important of which is QBMS (https://github.com/icarda-git/QBMS) for the R language, due to the popularity of R in breeding informatics. The BrAPI standard has evolved over several years and is now professionally managed, versioned, and comes with development tools, such as SwaggerHub and testing frameworks. The BrAPI standard has benefited greatly from the twice-yearly hackathons—which are legendary not only for producing new and useful code but for driving the entire breeding informatics field forward while creating an atmosphere of collaboration and creativity.

2.3 DATA ACQUISITION AND LOADING

Data acquisition and subsequent storage in a database can only work efficiently if the database has already defined objects within it for the data to be collected. For data uploads, the data can easily be associated with the proper objects (Ellerbrock et al., 2025; Morales et al., 2022). For example, for field data collection, field randomization would be created in the database first, then the field planted, barcoded, and field collection tools, such as tablets running the Field Book app (Figure 2.1) (Rife & Poland, 2014), preloaded with the field layout and traits to collect. Apps such as Field Book (Rife & Poland, 2014) can then scan the barcode in the field associated with the desired observation units, which can be plots, plants, or tissue samples, to be associated with predefined database identifiers. Data collected with this approach, sometimes referred to as "round-tripping," can trivially be loaded into the database (Ellerbrock et al., 2025). We have found this process to be vastly superior to "ad-hoc" collection of data, which suffers from ad-hoc and ill-defined observation unit names and leads to extremely difficult, expensive, and curation-intensive matching of names in order to load data back into a database.

Breedbase, and many other breeding databases, support the Field Book app for phenotyping data collection; the Coordinate app for tissue sample collection in 96-well format, particularly tuned to genotyping samples; and the Intercross app, which supports data collection and guidance in crossing workflows, all of which are collectively known as PhenoApps (Röckel et al., 2022). Other BrAPI-based apps are available that should work with most databases, such as the GridScore app (Raubach et al., 2022) from the James Hutton Institute, giving users more choices and flexibility in data collection.

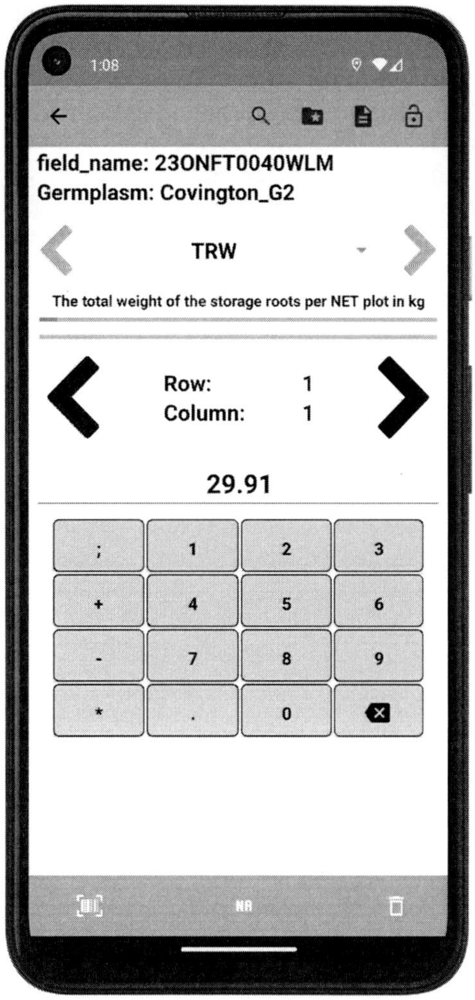

FIGURE 2.1 The Field Book (Rife & Poland, 2014) data acquisition tool that can be used with Breedbase.

2.4 TRAIT ONTOLOGIES

Trait ontologies contribute to standardization and harmonization of data, which leads to increased data re-use and integration, resulting in better prediction accuracies and breeding outcomes. Like other aspects of data collection mentioned earlier, the traits that will be measured must be defined in the database well before data collection. Traits are organized into ontologies—structured lists of hierarchical terms where the top level defines the ontology and child terms define trait categories or individual traits. Multiple parentage in this data structure is possible but follows important constraints, such as the true path rule. The relationship between terms has a type: the *is_a* relationship designates relationships between traits or trait categories, while the leaf nodes of the ontology are called *variables*, which have a *variable_of* relationship with their parent trait term but also define a method and a scale, such that all the parameters for measuring the data are fully defined. The *trait* type entries in the ontologies only represent concepts, for example, "plant height," but an entry of the *variable* type would be "plant height, measured with yardstick, in cm." As a result, no phenotypic data can be associated with the traits; all data is associated with variables. Many Breedbase instances use ontologies created by the CropOntology project (https://cropontology.org/) (Shrestha et al., 2010); for example, Cassavabase uses the CO_334 cassava ontology. More than 41 crop-specific ontologies are available from that project alone.

Clear procedures need to be in place for adding new terms to an ontology. In collaboration with the CropOntology project, Breedbase offers request forms (https://submit.rtbbase.org/) for users to suggest new terms (Morales et al., 2022). The proposed terms are then checked by curators for duplication or overlap with existing terms and general appropriateness, and an identifier is associated with the new term. Trait requests need to be addressed expeditiously by the curators to prevent delays in phenotyping activities. Still, breeders will want to record new, unexpected observations in the field. For this case, a "notes" term is available that can be used to record ad-hoc observations that may be developed into a formal trait later.

In Breedbase, trait variables can be combined with other ontologies to create so-called post-composed terms (Menda et al., 2025). For example, if the data to be measured has a time component, terms could be post-composed with a time ontology. For the example of plant height, post-composed terms could be created for "plant height at 1 week," "plant height at 2 weeks," etc. This is especially useful if the user would like to consistently compare the values within the weekly intervals. If the goal is to look at specific time series, the traits can be defined as *time_series* and no post-composing is necessary. Traits of type *time_series* require a timestamp to be associated with the measurement. Other uses for post-composing are the definition of more complex sample preparation schemes. For example, traits related to the quality of a prepared dish might depend on whether the food was cooked, steamed, grilled, etc., potentially with different cooking times. Such attributes can also be post-composed if the necessary orthogonal ontologies are created.

2.5 DATA STORAGE BACKEND

Data storage has to be reliable and scalable, and allow data to be actively managed, meaning it should be straightforward to add, remove, or update data and allow for regular back-ups. While there are many new types of NoSQL databases in use today, relational databases are still a great way to achieve all these goals. In Breedbase, most data are stored in a PostgreSQL database (https://postgresql.org/), an advanced open-source relational database. Breedbase uses a database schema called Chado ND, developed in conjunction with other databases (Jung et al., 2011), with some extensions. Certain data types, such as images, are stored in the file system, with only the metadata being present in the database. Other data types, such as genotypic data (Morales et al., 2020) or NIRS spectra (Hershberger et al., 2021), are stored as special data structures within a field of type jsonb, which allows for storage of key/value pairs, and are compressed and indexed for fast access.

2.6 DATA QUERYING

Most users will query Breedbase through the web interface, where searches are available for most data types, and which are linked from the "Search" menu (Morales et al., 2022). For example, to search for accessions, a user would choose the "Accessions and plots" search from that menu. The input form has options for name searches, along with more advanced search options, such as searching by accession attributes or the submitter of the records. The search will return a list of matches, which can be selected to get to the detail page of the accession. Similar searches are available for trials, traits, genotyping protocols, users, etc. A special type of search is the "Wizard," which can search trial data by selecting elements from different data dimensions, including trial year, location, breeding program, traits, accession names, plot names, etc. (Figure 2.2). Selecting the first dimension type, elements of that dimension are displayed in a column and can be selected; a second dimension can then be chosen, which will display all the elements that intersect with the first. The Wizard has up to four dimensions to choose from; the intersect can be stored as a so-called dataset and used to specify input data for many tools on Breedbase, such as the solGS tool (Tecle et al., 2014), the GWAS tool, or the heritability tool.

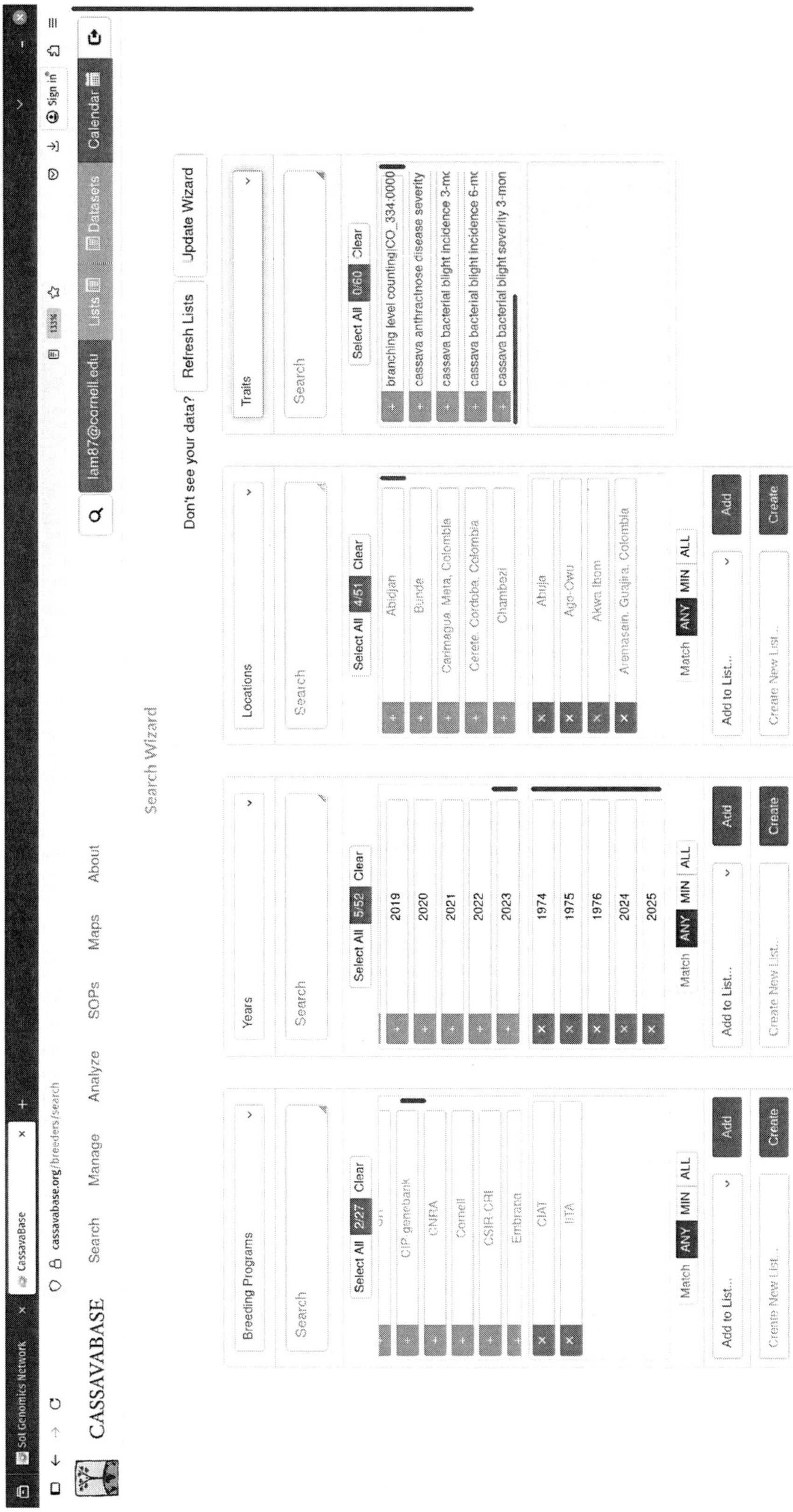

FIGURE 2.2 Screenshot of the wizard tool.

The aforementioned breeding API (BrAPI) can be used to access information in the database, particularly, if programmatic access is needed. If a login requirement is set up for the BrAPI, the user rights will depend on the user account with which it is accessed. Certain users have read-only capability, while other users have both read and write capability.

Breedbase can also be queried directly from programming languages that have BrAPI-specific code libraries. For example, the R language has a library called QBMS (https://github.com/icarda-git/QBMS) with which the contents can be read directly from R, implemented through BrAPI calls to the database.

The SQL interface upon which all the querying is based is not intended for end users but can be used only by database administrators. Direct SQL access can also be used programmatically to write loading or data retrieval scripts. As there are no restrictions when using SQL, extreme care should be taken when using it, as potential data loss may result.

2.7 DATA ANALYSIS

Breedbase offers a wide range of analysis tools, in most instances available under the "Analysis" menu. Available breeding data related analyses for phenotypic data include PCA, clustering, correlation, stability, and heritability. For genotypic data, PCA and clustering are available. Because Breedbase contains both genotypic and phenotypic data that are interlinked with each other, it becomes relatively straightforward to implement tools that require both types of data, such as genome-wide association studies (GWAS) or genomic prediction analyses (Jannink et al., 2010). GWAS can identify genomic regions that contribute to a specific trait, while genomic predictions use correlations of phenotypic and genotypic data to predict phenotypic data from genotypic data, which in many cases can contribute to speeding up of the breeding cycle, potentially allowing for more gain per unit of time. For genomic selection purposes, Breedbase implements two different paradigms. The first is the prediction of genetically estimated breeding values (GEBVs), as implemented in the solGS tool (Tecle et al., 2014). The second is the Genomic Prediction of Cross Performance (GPCP) tool, which, in addition to predicting the GEBVs, also analyzes each potential combination of genotypes to estimate the potential of each combination and returns a table sorted by the combination score.

2.8 GENOME DATA

As many modern breeding paradigms are based on genome data, a solid database infrastructure for genome information that ties in with the breeding data can be key to advancing the understanding of the underlying biological processes and genes. Breedbase, having been developed from SGN (Bombarely et al., 2011; Fernandez-Pozo et al., 2015), a major genome database for the Solanaceae, comes with all the tools that such databases rely on, including genome viewers (JBrowse2) (Diesh et al., 2023), BLAST (Altschul et al., 1990), a comparative map viewer (Mueller et al., 2008), and a community-based gene curation system (Menda et al., 2008).

2.9 CONCLUSIONS

Advanced breeding methods in tomato breeding and other crops can be managed through Breedbase; SGN, as an instance of Breedbase, has all the functionality to manage breeding programs for any Solanaceae species. Individual Breedbase instances can also be generated for private breeding data. Breedbase can be used for any breeding program but is particularly suited to breeding using genomic selection methods and manages the complexity of trial creation, data acquisition, and analysis.

ACKNOWLEDGMENTS

The author would like to thank all the current and former lab members who made Breedbase possible; the breeders and staff of the NextGen Cassava, Africa Yam, SweetGAINs, and Breeding

Better Bananas project, who provided valuable direction, feedback, and use cases; Quana Bice for critically reading the manuscript; and the NSF, USDA, and BMGF for funding various projects related to SGN and Breedbase.

REFERENCES

Altschul, S. F.; Gish, W.; Miller, W.; Myers, E. W.; Lipman, D. J. Basic Local Alignment Search Tool. *Journal of Molecular Biology* 1990, *215* (3), 403–410. https://doi.org/10.1016/S0022-2836(05)80360-2.

The Arabidopsis Genome Initiative. Analysis of the Genome Sequence of the Flowering Plant *Arabidopsis thaliana*. *Nature* 2000, *408*, 796–815.

Bombarely, A.; Menda, N.; Tecle, I. Y.; Buels, R. M.; Strickler, S.; Fischer-York, T.; Pujar, A.; Leto, J.; Gosselin, J.; Mueller, L. A. The Sol Genomics Network (Solgenomics.Net): Growing Tomatoes Using Perl. *Nucleic Acids Research* 2011, *39* (Database), D1149–D1155. https://doi.org/10.1093/nar/gkq866.

Diesh, C.; Stevens, G. J.; Xie, P.; De Jesus Martinez, T.; Hershberg, E. A.; Leung, A.; Guo, E.; Dider, S.; Zhang, J.; Bridge, C.; Hogue, G.; Duncan, A.; Morgan, M.; Flores, T.; Bimber, B. N.; Haw, R.; Cain, S.; Buels, R. M.; Stein, L. D.; Holmes, I. H. JBrowse 2: A Modular Genome Browser with Views of Synteny and Structural Variation. *Genome Biology* 2023, *24* (1), 74. https://doi.org/10.1186/s13059-023-02914-z.

Ellerbrock, B. J.; Simoes, C. C.; Karaikal, S. K.; Nyaga, C. M.; Mueller, L. A. Sweetpotato Breeding in the Genomic Age: Harnessing Databases, Bioinformatics, Digital Tools, and Genomic Insights. In *The Sweetpotato Genome*; Yencho, G. C., Olukolu, B. A., Isobe, S., Eds.; Compendium of Plant Genomes; Springer International Publishing: Cham, 2025; pp. 135–145. https://doi.org/10.1007/978-3-031-65003-1_11.

Fernandez-Pozo, N.; Menda, N.; Edwards, J. D.; Saha, S.; Tecle, I. Y.; Strickler, S. R.; Bombarely, A.; Fisher-York, T.; Pujar, A.; Foerster, H.; Yan, A.; Mueller, L. A. The Sol Genomics Network (SGN)—from Genotype to Phenotype to Breeding. *Nucleic Acids Research* 2015, *43* (D1), D1036–D1041. https://doi.org/10.1093/nar/gku1195.

Hershberger, J.; Morales, N.; Simoes, C. C.; Ellerbrock, B.; Bauchet, G.; Mueller, L. A.; Gore, M. A. Making Waves in Breedbase: An Integrated Spectral Data Storage and Analysis Pipeline for Plant Breeding Programs. *The Plant Phenome Journal* 2021, *4* (1), e20012. https://doi.org/10.1002/ppj2.20012.

International Rice Genome Sequencing Project & Takuji Sasaki. The Map-based Sequence of the Rice Genome. *Nature* 2005, *436*, 793–800.

Jannink, J. L.; Lorenz, A. J.; Iwata, H. Genomic Selection in Plant Breeding: From Theory to Practice. *Brief Funct Genomics* 2010, *9* (2), 166–177. https://doi.org/10.1093/bfgp/elq001.

Jung, S.; Menda, N.; Redmond, S.; Buels, R. M.; Friesen, M.; Bendana, Y.; Sanderson, L.-A.; Lapp, H.; Lee, T.; MacCallum, B.; Bett, K. E.; Cain, S.; Clements, D.; Mueller, L. A.; Main, D. The Chado Natural Diversity Module: A New Generic Database Schema for Large-Scale Phenotyping and Genotyping Data. *Database* 2011, *2011*, bar051. https://doi.org/10.1093/database/bar051.

Lee, J. M.; Davenport, G. F.; Marshall, D.; Ellis, T. H. N.; Ambrose, M. J.; Dicks, J.; van Hintum, T. J. L.; Flavell, A. J. GERMINATE. A Generic Database for Integrating Genotypic and Phenotypic Information for Plant Genetic Resource Collections. *Plant Physiology* 2005, *139* (2), 619–631. https://doi.org/10.1104/pp.105.065201.

Menda, N.; Buels, R. M.; Tecle, I.; Mueller, L. A. A Community-Based Annotation Framework for Linking Solanaceae Genomes with Phenomes. *Plant Physiology* 2008, *147* (4), 1788–1799. https://doi.org/10.1104/pp.108.119560.

Menda, N.; Ellerbrock, B. J.; Simoes, C. C.; Karaikal, S. K.; Nyaga, C.; Flores-Gonzalez, M.; Tecle, I. Y.; Lyon, D.; Agbona, A.; Agre, P. A.; Peteti, P.; Akech, V.; Asiimwe, A.; Fauvelle, E.; Meghar, K.; Tran, T.; Dufour, D.; Cooper, L.; Laporte, M.-A.; Arnaud, E.; Mueller, L. A. Post-Composing Ontology Terms for Efficient Phenotyping in Plant Breeding. *Database* 2025, *2025*, baaf020. https://doi.org/10.1093/database/baaf020.

Morales, N.; Bauchet, G. J.; Tantikanjana, T.; Powell, A. F.; Ellerbrock, B. J.; Tecle, I. Y.; Mueller, L. A. High Density Genotype Storage for Plant Breeding in the Chado Schema of Breedbase. *PLoS ONE* 2020, *15* (11), e0240059. https://doi.org/10.1371/journal.pone.0240059.

Morales, N.; Ogbonna, A. C.; Ellerbrock, B. J.; Bauchet, G. J.; Tantikanjana, T.; Tecle, I. Y.; Powell, A. F.; Lyon, D.; Menda, N.; Simoes, C. C.; Saha, S.; Hosmani, P.; Flores, M.; Panitz, N.; Preble, R. S.; Agbona, A.; Rabbi, I.; Kulakow, P.; Peteti, P.; Kawuki, R.; Esuma, W.; Kanaabi, M.; Chelangat, D. M.; Uba, E.; Olojede, A.; Onyeka, J.; Shah, T.; Karanja, M.; Egesi, C.; Tufan, H.; Paterne, A.; Asfaw, A.; Jannink,

J.-L.; Wolfe, M.; Birkett, C. L.; Waring, D. J.; Hershberger, J. M.; Gore, M. A.; Robbins, K. R.; Rife, T.; Courtney, C.; Poland, J.; Arnaud, E.; Laporte, M.-A.; Kulembeka, H.; Salum, K.; Mrema, E.; Brown, A.; Bayo, S.; Uwimana, B.; Akech, V.; Yencho, C.; De Boeck, B.; Campos, H.; Swennen, R.; Edwards, J. D.; Mueller, L. A. Breedbase: A Digital Ecosystem for Modern Plant Breeding. *G3 GenesGenomesGenetics* 2022, *12* (7), jkac078. https://doi.org/10.1093/g3journal/jkac078.

Mueller, L. A.; Mills, A. A.; Skwarecki, B.; Buels, R. M.; Menda, N.; Tanksley, S. D. The SGN Comparative Map Viewer. *Bioinformatics* 2008, *24* (3), 422–423. https://doi.org/10.1093/bioinformatics/btm597.

QBMS. https://github.com/icarda-git/QBMS.

Raubach, S.; Schreiber, M.; Shaw, P. D. GridScore: A Tool for Accurate, Cross-Platform Phenotypic Data Collection and Visualization. *BMC Bioinformatics* 2022, *23* (1), 214. https://doi.org/10.1186/s12859-022-04755-2.

Ribaut, J.-M.; Delannay, X.; McLaren, G.; Okono, F. Molecular Breeding Molecular Breeding (MB) Platforms Molecular Breeding (MB) Platforms (MBP) in World Agriculture. In *Encyclopedia of Sustainability Science and Technology*; Meyers, R. A., Ed.; Springer New York: New York, 2012; pp. 6692–6720. https://doi.org/10.1007/978-1-4419-0851-3_237.

Rife, T. W.; Poland, J. A. Field Book: An Open-Source Application for Field Data Collection on Android. *Crop Science*. 2014, *54* (4), 1624–1627. https://doi.org/10.2135/cropsci2013.08.0579.

Röckel, F.; Schreiber, T.; Schüler, D.; Braun, U.; Krukenberg, I.; Schwander, F.; Peil, A.; Brandt, C.; Willner, E.; Gransow, D.; Scholz, U.; Kecke, S.; Maul, E.; Lange, M.; Töpfer, R. PhenoApp: A Mobile Tool for Plant Phenotyping to Record Field and Greenhouse Observations. *F1000Research* 2022, *11*, 12. https://doi.org/10.12688/f1000research.74239.2.

Schnable, P. S.; Ware, D.; Fulton, R. S.; Stein, J. C.; Wei, F.; et al. The B73 Maize Genome: Complexity, Diversity, and Dynamics. *Science* 2009, *326* (5956), 1112–1115. https://doi.org/10.1126/science.1178534.

Selby, P.; Abbeloos, R.; Backlund, J. E.; Basterrechea Salido, M.; Bauchet, G.; Benites-Alfaro, O. E.; Birkett, C.; Calaminos, V. C.; Carceller, P.; Cornut, G.; Vasques Costa, B.; Edwards, J. D.; Finkers, R.; Yanxin Gao, S.; Ghaffar, M.; Glaser, P.; Guignon, V.; Hok, P.; Kilian, A.; König, P.; Lagare, J. E. B.; Lange, M.; Laporte, M.-A.; Larmande, P.; LeBauer, D. S.; Lyon, D. A.; Marshall, D. S.; Matthews, D.; Milne, I.; Mistry, N.; Morales, N.; Mueller, L. A.; Neveu, P.; Papoutsoglou, E.; Pearce, B.; Perez-Masias, I.; Pommier, C.; Ramírez-González, R. H.; Rathore, A.; Raquel, A. M.; Raubach, S.; Rife, T.; Robbins, K.; Rouard, M.; Sarma, C.; Scholz, U.; Sempéré, G.; Shaw, P. D.; Simon, R.; Soldevilla, N.; Stephen, G.; Sun, Q.; Tovar, C.; Uszynski, G.; Verouden, M.; The BrAPI Consortium. BrAPI—an Application Programming Interface for Plant Breeding Applications. *Bioinformatics* 2019, *35* (20), 4147–4155. https://doi.org/10.1093/bioinformatics/btz190.

Shrestha, R.; Arnaud, E.; Mauleon, R.; Senger, M.; Guy, F.; Davenport, G. F.; Hancock, D.; Norman Morrison, N.; Bruskiewich, R.; McLaren, G. Multifunctional Crop Trait Ontology for Breeders' Data: Field Book, Annotation, Data Discovery and Semantic Enrichment of the Literature. *AoB PLANTS* 2010, plq008. https://doi.org/10.1093/aobpla/plq008.

Tecle, I. Y.; Edwards, J. D.; Menda, N.; Egesi, C.; Rabbi, I. Y.; Kulakow, P.; Kawuki, R.; Jannink, J.-L.; Mueller, L. A. solGS: A Web-Based Tool for Genomic Selection. *BMC Bioinformatics* 2014, *15* (1), 398. https://doi.org/10.1186/s12859-014-0398-7.

The Tomato Genome Consortium. The Tomato Genome Sequence Provides Insights into Fleshy Fruit Evolution. *Nature* 2012, *485*, 635–641.

3 High-Throughput Field-Based Phenotyping and Its Application in Tomato Improvement

Weilong He, Zhenghua Zhang, Dilip R. Panthee, and Lirong Xiang

3.1 INTRODUCTION

By 2050, the global population is expected to surge to nearly 10 billion, creating an urgent need for unprecedented increases in agricultural productivity to mitigate the impending threat of global food shortages (Hickey et al., 2019). This challenge is further compounded by escalating environmental and biological pressures, such as climate change, soil degradation, water scarcity, and pest outbreaks, all of which continue to threaten crop yields (Raza et al., 2019). Addressing these multifaceted challenges is not merely about increasing yields; it is essential for maintaining global food security and ensuring that agricultural systems are resilient in the face of growing environmental volatility. Plant breeding programs aim to select high-performance future cultivars, ensuring they possess the necessary traits for thriving in diverse and challenging conditions. Central to this endeavor is the advancement of plant phenotyping technologies, which enable the precise measurement and analysis of these crucial traits, driving the next generation of agricultural innovation.

For a long time, plant phenotyping has been a critical component of crop breeding, with efficient and accurate phenotyping being essential for selecting the desired genotypes (Hickey et al., 2019). Plant phenotyping involves precise measurement and analysis of plant traits such as growth, morphology, physiology, and biochemical properties, which are crucial for identifying stress-resistant genetic markers and optimizing crop performance under diverse environmental conditions (L. Li et al., 2014; Z. Li et al., 2020). However, traditional phenotyping methods are limited by their inherently labor-intensive, time-consuming, and costly nature. Additionally, these conventional methods often struggle with issues of accuracy and scalability, making them inadequate for large-scale agricultural applications where efficiency and precision are critical. Considering these challenges, advanced phenotyping technologies have emerged as a cornerstone of modern agricultural research, offering vital insights needed to develop more resilient crop varieties and enhance agricultural productivity.

The emergence of high-throughput plant phenotyping (HTPP) systems represents a significant paradigm shift in the field, offering a transformative approach to overcoming the limitations of traditional phenotyping. HTPP systems leverage cutting-edge technologies, including autonomous or semi-autonomous platforms equipped with advanced multi-modal sensor arrays, to rapidly and nondestructively measure vast arrays of morphological, physiological, biochemical, and performance traits (Jangra et al., 2021).

In recent years, the field of plant phenotyping has seen remarkable progress, leading to the development of various HTPP platforms, including stationary, vehicle-based, autonomous, and portable systems (Jiang et al., 2018; D. Li et al., 2021; Xiang et al., 2023; C. Zhang et al., 2020; Z. Zhang

DOI: 10.1201/9781003506539-3

et al., 2024). Stationary platforms are ideal for high-precision measurements in laboratories and greenhouses, although their range is limited. Vehicle-based platforms, on the other hand, are suited for large-scale field data collection but require skilled operators. Autonomous mobile platforms can navigate complex terrains independently, offering powerful capabilities, though they come with high costs and potential safety concerns. Portable platforms are known for their flexibility, making them suitable for smaller-scale studies, though they may be less efficient in large-scale data processing. Aerial phenotyping technologies, utilizing satellites, low-altitude aircraft, and drones (Feng et al., 2021; He, Gage et al., 2024; Herr et al., 2023; Sagan et al., 2019), have expanded the spatial scope of phenotyping, making them suitable for monitoring large agricultural fields, although they face challenges related to resolution, operation time, and payload capacity. Recent HTPP systems have integrated artificial intelligence (AI), machine learning (ML), and deep learning (DL) technologies, significantly enhancing data processing and analysis capabilities. This integration has made plant phenotyping more intelligent and efficient, supporting crop breeding and precision agriculture management, and providing crucial tools for addressing global food security challenges.

Machine learning (ML) and deep learning (DL) algorithms have played a pivotal role in improving phenotyping accuracy and predictive capabilities, due to their ability to identify complex patterns and relationships in large datasets (Pound et al., 2017; Singh et al., 2016). Convolutional neural networks (CNNs) and recurrent neural networks (RNNs) have demonstrated exceptional performance in tasks such as image classification and time-series data analysis, which are critical in agricultural contexts. These models can learn from the high-dimensional data generated by HTPP systems, enabling more precise identification of stress factors and more accurate yield predictions. The application of these technologies extends beyond simple data processing; they also foster a deeper understanding of plant-environment interactions, paving the way for the development of crops that are more resilient to environmental changes.

This chapter aims to provide an in-depth review of the latest advancements in HTPP methodologies, with a particular focus on imaging technologies, commonly used imaging sensors for high-throughput phenotyping, phenotyping platforms, and the integration of machine learning algorithms in the phenotyping process. Through a case study on tomato disease phenotyping, this chapter showcases the transformative potential of these technologies. The case study not only highlights the capabilities of HTPP systems in identifying disease-resistant traits but also underscores the comparative advantages and limitations of different approaches. Through this analysis, the chapter offers a comprehensive overview of the current state of high-throughput phenotyping while exploring its prospects in crop management. The goal is to demonstrate how these advanced phenotyping technologies can be leveraged to enhance agricultural productivity and contribute to global food security in the coming decades.

3.2 PROXIMAL SENSING AND ALGORITHMS

The process that transforms raw sensor data into meaningful phenotypic data is essentially machine perception. Plant traits that can be characterized using HTPP are largely limited by the available proximal sensing technologies. Before reviewing specific sensors and their associated algorithms, it is worth categorizing plant traits that can be quantified with proximal sensing at different levels of integration (i.e., plot, plant, or organ). Plot-level traits are often generic and can be applied to a wide range of crop species and crop growth stages. Such traits include plant morphological measurements (e.g., canopy height, canopy width, and leaf area) and plant physiological indicators (e.g., vegetation indices and canopy temperature), which require low-level machine perception. All plot-level traits can be adopted as plant-level traits if the sensor data of individual plants can be delineated, namely, plant segmentation. Plant segmentation itself also enables an important plant-level trait, stand count. Other plant-level traits are based on the measurement of a particular plant organ. In the case of cereal crops, for instance, plant height is measured from the ground to the flag leaf of a plant before the panicle becomes visible or to the panicle apex afterwards. Since each plant has only one panicle and one flag leaf, machine perception of these special organs enables the characterization of plant-level traits. Stem diameter is another example of plant-level trait. If the plant of interest can grow multiple organs of the same type (e.g., leaves, branches, internodes, flowers, tomato fruits, bean pods, corn ears, cotton bolls, grape berries), accurate machine perception of these organs allows for organ-level phenotyping and lays the foundation for higher levels of integration. Due to the small sizes and occlusion issues in the crop fields, machine perception at the organ level remains a challenging research area. Foliar disease assessment is another type of organ-level trait, which can be based on multi- or hyper-spectral analysis or visual perception. Some plant organs are components of larger plant organs, for instance, the spikelets on a wheat spike.

The majority of proximal sensors used in HTPP are imaging-based. A thorough review of imaging techniques for plant phenotyping was given by L. Li et al. (2014). Five key imaging techniques were identified: visible imaging, fluorescence imaging, thermal imaging, spectroscopic imaging, and LiDAR. Visible imaging has been used to characterize various plant morphological and architectural traits, including those of the leaf (H. Zhang et al., 2023), stem (Holmes et al., 2020), panicle (Holmes et al., 2020), root (Z. Xu et al., 2022), and seed (de Medeiros et al., 2020). Fluorescence imaging is typically used to measure photosynthesis and plant stress response (Bresson et al., 2015). Thermal imaging can measure surface temperature, which is related to stomatal conductance and water stress (Brewer et al., 2022; Z. Zhou et al., 2021). Spectroscopic imaging, such as hyperspectral imaging, can indicate water and health statuses of leaves and canopies (Beć et al., 2020; Liang et al., 2017; Pandey et al., 2017). LiDAR is mostly used to estimate canopy height, leaf surface area, volume, and biomass (S. Sun et al., 2018a; H. Wang et al., 2023; L. Zhou et al., 2020). In addition to imaging sensors, other proximal sensors often provide an average response in the field of view. Such sensors include ultrasonic and laser distance sensors, NDVI spectrometers (e.g., Crop Circle), and infrared radiometers.

3.2.1 RGB Vision

Digital Red–Green–Blue (RGB) cameras are the most accessible sensors for plant phenotyping. Silicon-based Charge-Coupled Device (CCD) and complementary metal–oxide–semiconductor (CMOS) sensors are inherently sensitive to visible and near-infrared wavelengths; however, most RGB cameras use an IR-cut filter, and the captured RGB images primarily represent the visible spectrum. The three-color channels are typically realized by placing a Bayer filter mosaic in alignment with the photosensor array. RGB cameras have been widely used on most HTPP platforms (Rong et al., 2023; Zhao et al., 2020). The RGB images with intensity values in red, green, and blue color channels provide the spatial information and color features of a scene. Applications based on RGB vision include disease detection and classification (Shruthi et al., 2022), canopy cover estimation (J. Liu & Pattey, 2010), and weed detection (Abdalla et al., 2019; Rai et al., 2023). Suitable for outdoor applications, RGB cameras have the advantages of high spatial resolutions, fast acquisition rates, and low costs. However, the use of RGB imaging in the field is limited by complex background, wind, occlusion, and lighting variation. The disparities in image quality have affected the performance of the imaging analysis and pose challenges for algorithm development. Recent advancements in deep convolutional neural networks (CNNs) have provided state-of-the-art solutions to overcome these challenges in RGB image analysis. For example, DeepSeedling was developed to count cotton plant seedlings in the field (Jiang et al., 2019a). The system-derived seedling counts were highly correlated ($R^2 = 0.98$) with the ground truth. Besides plant counting, the applications of CNNs were further extended to count plant organs such as wheat spikes (Zaji et al., 2023), corn tassels (Yu et al., 2024), rice panicles (Chen et al., 2023), and cotton bolls (Xiao et al., 2024). Several studies have developed CNN-based approaches for detection and analysis of wheat spikes (Sadeghi-Tehran et al., 2019; Xiong et al., 2019; Z. Yang et al., 2024). Despite the challenging field conditions, the proposed deep learning models achieved high detection accuracy and satisfactory performance. A recent review provides a comprehensive overview of the applications of CNNs in plant phenotyping applications (Jiang & Li, 2020).

3.2.2 Stereo Vision

Stereo vision, which reconstructs a 3D model from multi-view images, offers an inexpensive, efficient, and reasonably accurate solution for infield phenotyping of 3D plant morphology. Multi-view stereo systems combine multiple-view stereo (MVS) and structure-from-motion (SfM) to reconstruct 3D models of plants from images. Nguyen et al. (2016) developed an infield phenotyping system that utilized 16 color stereo vision cameras for capturing multi-view images of an eggplant. An arc-shaped superstructure was designed to deploy the cameras and texture-enhancing structured illumination modules. The system can estimate plant height, leaf number, leaf area, and plant biomass. The MVS–SfM method can generate dense point clouds of plants, but the major limitation is the intensive computation required for dense 3D reconstruction. Binocular stereo vision is being considered an alternative option to overcome the limitations of the MVS–SfM technique. The technique only needs two cameras to compute depth information. Previous work has demonstrated the potential of stereo vision in measuring plant architecture outdoors. For example, Sodhi et al. (2017) developed a vertically moving boom system to scan infield energy sorghum plants with stereo cameras. Phenotypic traits such as leaf length, leaf width, and stem diameter were computed. Xiang et al. (2023) developed a stereo-vision-based robotic system for tall dense canopy crops in the field. Stereo cameras were positioned at multiple heights on an extension rig to perform side-viewing imaging. Several morphological traits were quantified and found to be accurate and highly repeatable. Stereo vision has the advantages of high resolutions and low costs; however, the performance is affected by textureless regions, sunlight variation, and wind (Bao et al., 2018; Dandrifosse et al., 2020).

3.2.3 Time-of-Flight Cameras

A Time-of-Flight (ToF) camera is another widely used depth imaging sensor for infield plant phenotyping. Distances between an object and an image sensor are measured by calculating the time or phase shift of a light pulse that has travelled to and from the object. ToF cameras were successfully used for plant architecture characterization (Bao, Tang, Srinivasan et al., 2018; Jiang et al., 2016), biomass estimation (Mortensen et al., 2018), fruit detection (Gené-Mola et al., 2019), crop recognition, and crop monitoring (Rui et al., 2024). Low costs and high frame rates are the advantages of ToF cameras, whereas the major weakness is their poor performance under strong sunlight. To alleviate this issue, a sunshade was adopted to reduce the sunlight intensity in outdoor applications (Gai et al., 2020; Mortensen et al., 2018). Recent advancements in commercial-grade depth sensors provide an inexpensive solution for measuring plant architecture. The Microsoft Kinect v2 camera was one of the most commonly used depth sensors that, with upgraded color and depth resolutions, performed well in plant phenotyping. Bao et al. (2019) investigated the utility of side-viewing Kinect v2 cameras for characterization of corn plant architecture under field conditions at approximately sunset. A skeleton-based algorithm was developed to analyze the plant architecture and extract shoot architecture phenotypes. This work showed that the system was robust and accurate in acquiring several important traits, including plant height, plant orientation, and leaf angle. Andújar et al. (2016) described a method for 3D modeling of cauliflower plants with a Kinect-based automated system. This method relied on the creation of a 3D point cloud in real time and the subsequent generation of a solid surface model. Strong correlations between measurements from the models and actual structural parameters were found.

3.2.4 Range Sensing

Light detection and ranging (LiDAR) is one of the most representative remote-sensing techniques for field-based phenotyping. LiDAR measures distances by illuminating an object and analyzing the reflected light. In recent years, LiDAR has been widely integrated with field-based phenotyping systems and used for canopy reconstruction (Medic et al., 2023), plant area density estimation (Hosoi & Omasa, 2009), plant height measurements (T. Liu et al., 2024; X. Wang et al., 2018), biomass estimation (Du et al., 2023), nitrogen uptake assessment (Hütt et al., 2023), and plant growth analysis (S. Sun et al., 2018). For example, Qiu et al. (2019) reconstructed 3D models of corn plants by mounting a 3D LiDAR on a mobile ground robot. In their study, the robot with the 3D LiDAR moved in alleyways to collect data on corn plant rows. The point clouds were registered and merged by using the Iterative Closest Points algorithm with the assistance of landmarks in the field. Morphological traits, including row spacing and plant height, were derived from the 3D models. Besides canopy-level morphological traits, high-resolution 3D LiDAR sensors have also provided the potential for organ-level traits, such as cotton bolls (S. Sun et al., 2018), sorghum panicles (Malambo et al., 2019), and corn stems (S. Jin et al., 2020). A terrestrial LiDAR sensor mounted on a sprayer has allowed the automated detection and morphological measurement of sorghum panicles using density-based clustering of LiDAR data (Malambo & Heatwole, 2020). Cotton plant internode detection is feasible using terrestrial LiDAR data collected under field conditions (S. Sun et al., 2019). A Laplacian-based skeletonization algorithm and minimum-spanning-tree-based method were developed to analyze the structure of cotton plants. Thus, LiDAR sensors can be incorporated into different plant phenotyping systems to characterize a wide range of phenotypes.

LiDAR provides detailed 3D characterization of plant canopies by actively emitting laser pulses and measuring their returns (e.g., time-of-flight), generating dense point clouds for accurate estimation of structural parameters. Some shortcomings of LiDAR sensors for plant phenotyping include (1) low temporal resolution (Dandrifosse et al., 2020), meaning that it takes a long scanning time to obtain a dense point cloud of a plant canopy; (2) lack of color information in the measurement, but this can be solved by integrating other kinds of imaging sensors; and (3) for field-based plant phenotyping, LiDAR is sensitive to the noise caused by wind, insects, rain, and small panicles in the air (Perez-Sanz et al., 2017). Despite these shortcomings, some new technologies such as high-density LiDAR, Full-waveform LiDAR, and Hyperspectral LiDAR have shown the next-generation potential of LiDAR-based plant phenotyping (Lin, 2015).

An ultrasonic sensor is a range detector, which has long been utilized for the evaluation of plant height (Sharma et al., 2016; Sui et al., 2012). The working principle of ultrasonic sensors is similar to that of LiDARs, whereas the difference is that an ultrasonic sensor emits sound waves instead of a light source to measure distance. Ultrasonic sensors have the advantages of low costs, high sampling rates, and simple communication interface. However, the sensors can suffer from materials and surfaces that absorb sound waves, and they are only suitable for short-range sensing. Yuan et al. (2018) developed a multi-sensor phenotyping platform equipped with ultrasonic sensors and LiDARs. The study of measuring wheat plant height using the system showed that LiDAR provided better results than ultrasonic sensors. Although more costly, the measurement produced by LiDAR is more accurate due to its higher spatial resolution (Qiu et al., 2018).

3.2.5 Spectral and Thermal Sensing

Some other imaging techniques have been used in numerous plant phenotyping platforms, including spectral sensors and thermal imaging. Spectral sensors assess the spectral response from objects in a number of spectral bands, including visible (400–700 nm), near infrared (700–1200 nm), and short-wavelength infrared regions (1400–3000 nm). Some commercial spectral sensors, such as GreenSeeker (Trimble Inc., CA, USA) and Crop Circle (Holland Scientific Inc., NE, USA), were widely used by researchers to form vegetative indices such as normalized difference vegetation index (NDVI) (Barker et al., 2016; Bayati & Fotouhi, 2018; Crain et al., 2016). The high-resolution spectral measurements make it a promising tool for the detection of disease and abiotic stresses (Behmann et al., 2015; Corti et al., 2017; Mahlein et al., 2017; J. Sun et al., 2017; Thomas et al., 2018). Spectral image analysis is usually performed using appropriate statistical/machine learning algorithms to select significant features and build robust prediction models. These algorithms include least squares regression, support vector machines, principal components analysis, random forests, neural networks, etc. Furthermore, ensemble learning can be implemented to aggregate the strengths of multiple machine learning approaches (Moghimi et al., 2018). Thermal imaging, which can detect emitted infrared radiation in the mid- and long-wave infrared range from 3 to 12 um, allows for measuring temperature distribution on plant canopies. These different sensors have been integrated into phenotyping platforms to assess various agronomic traits. For example, Field Scanalyzer, an automated gantry-based field phenotyping platform, carries a dedicated sensor array consisting of hyperspectral, thermal infrared, RGB, and chlorophyll fluorescence cameras, as well as 3D laser scanners (Virlet et al., 2016). The sensor array enables detailed non-invasive measurements of wheat plants, including NDVI, canopy temperature, wheat head detection, plant height, and chlorophyll fluorescence. BreedVision is another multi-sensor platform developed for field-based plant phenotyping. Various optical sensors like hyperspectral imaging, 3D imagers, and color cameras were used to collect spectral and morphological information of plants, and phenotypic traits, including plant moisture content, tiller density, and lodging, were measured from the multi-sensor fusion models (Busemeyer et al., 2013).

3.3 PHENOTYPING PLATFORMS

Phenotyping platforms are crucial tools for measuring and analyzing the physical and biochemical characteristics, or phenotypes, of crops (X. Jin et al., 2021). These platforms range from simple imaging devices to complex systems integrating a variety of sensors, control systems, and data analysis tools. They are critical in plant breeding, genetics, and other agricultural applications because they allow precise measurement of characteristics such as plant height, leaf area, and biomass under different environmental conditions (Feng et al., 2021; X. Jin et al., 2021; Ma et al., 2022; R. Xu & Li, 2022). Phenotyping platforms are primarily divided into two categories: ground-based and aerial platforms, each offering unique advantages and being customized for specific applications (Figure 3.1).

3.3.1 Ground Platforms

Ground platforms are essential tools in plant phenotyping, offering a range of capabilities to measure and analyze the phenotypes of crops at the ground level (Atefi et al., 2021; Bao et al., 2021). Ground platforms can be further categorized into several types based on their structure and mobility, including phenotyping poles, phenotyping vehicles, and fixed platforms.

3.3.2 Phenotyping Poles

Phenotyping poles are among the simplest forms of ground phenotyping platforms. They are typically made from stainless steel, aluminum, and carbon fibers (X. Jin et al., 2021). These poles can be either fixed or mobile: fixed phenotyping poles are stationary poles that are usually installed in the field, similar to weather stations. They are equipped with sensors such as low-cost RGB cameras to capture images at regular intervals, and these images are then transmitted to a server for analysis. An example of this type is the PhenoCam (Richardson et al., 2018), which uses fixed cameras to monitor canopy cover, fraction of absorbed photosynthetically active radiation (FAPAR), and the green area index (GAI) over time. However, their limitation is the restricted coverage area, making them suitable for small experimental plots but costly to deploy across larger fields. Researchers manually operate mobile phenotyping poles that can be moved around the field freely. The distance of the sensors to the plants (typically within 1–3 meters) allows for high-resolution imaging, which is beneficial for detailed studies on plant structure and development. Despite their effectiveness, mobile phenotyping poles are labor-intensive, particularly when covering large fields, which can limit the feasibility of large-scale studies (X. Jin et al., 2021).

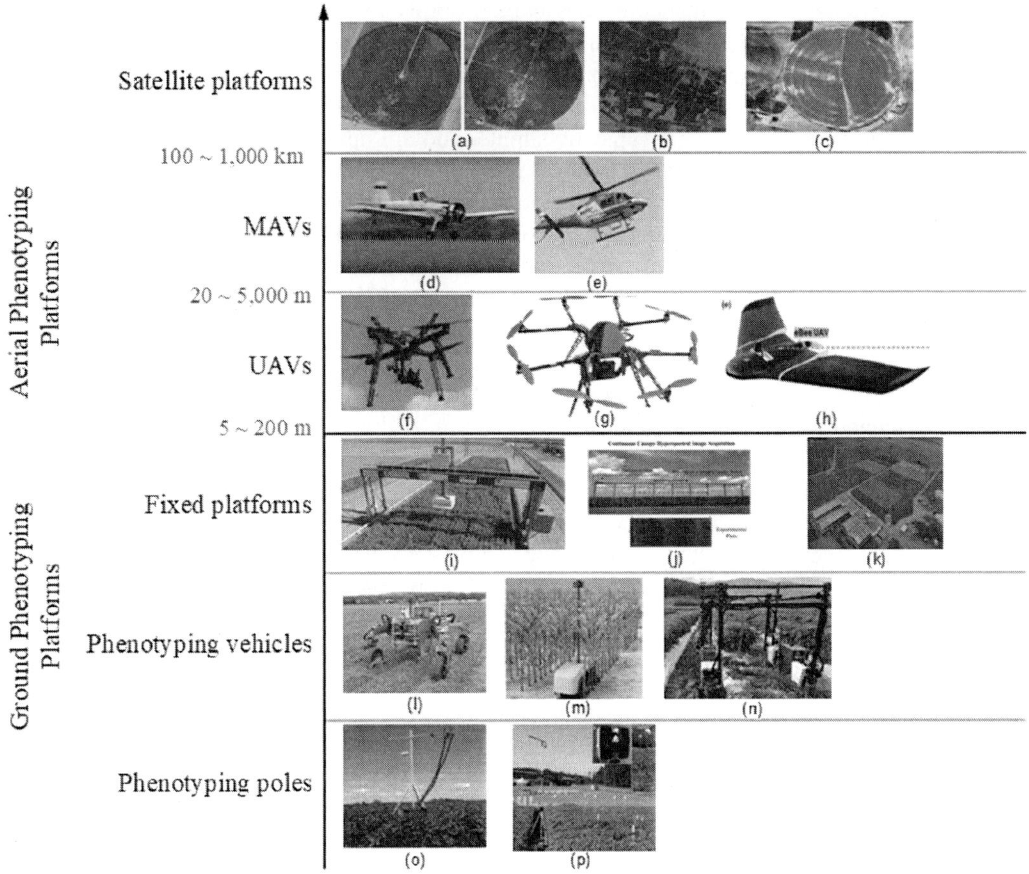

FIGURE 3.1 Aerial phenotyping platforms, including satellite platforms ((a) (C. Yang, 2020), (b) (Nguyen et al., 2020), (c) (X. Jin et al., 2021)); MAVs ((d) (X. Jin et al., 2021), (e) (X. Jin et al., 2021)) and UAVs ((f) (Xie & Yang, 2020), (g) (Bendig et al., 2015), (h) (Sagan et al., 2019)). Ground phenotyping platforms, including fixed platforms ((i) (Abebe et al., 2023), (j) (J. Zhang et al., 2023), (k) (Kirchgessner et al., 2016)); phenotyping vehicles ((l) (Biber et al., 2012), (m) (Bao et al., 2021), (n)); and phenotyping poles ((o) (X. Jin et al., 2021), (p) (Abebe et al., 2023)).

3.3.3 Phenotyping Vehicles

Phenotyping vehicles offer greater flexibility and efficiency compared to phenotyping poles. These mobile platforms can be equipped with various sensors to capture a wide range of phenotypic data. (R. Xu & Li, 2022). Depending on the mode of operation, phenotyping vehicles can be categorized into automatic, manual, and tractor-mounted systems.

Automatic phenotyping vehicles are advanced, autonomous platforms integrating multiple sensors, GPS systems, and control systems. They can independently traverse fields and collect comprehensive phenotypic data, including RGB, multispectral, hyperspectral images, thermal data, and 3D point clouds (Comar et al., 2012). The Vinobot robotic phenotyping platform, developed by the University of Missouri (Shafiekhani et al., 2017), is equipped with depth cameras for capturing depth information and environmental sensors, including temperature and light intensity sensors. This platform is designed to autonomously collect phenotypic data in agricultural fields, following predefined navigation routes, which significantly reduces the need for manual labor and enhances operational efficiency. The platform is particularly suited for field crops such as sorghum and maize.

The primary advantage of automatic vehicles is their ability to continuously monitor crops with minimal human intervention, making them ideal for large-scale studies that require consistent and repeatable data collection. However, the complexity and high cost of these systems can be prohibitive for some research budgets (Feng et al., 2021; Guzmán et al., 2016; Z. Zhang et al., 2020).

Manual phenotyping vehicles are typically simpler and are operated manually by researchers. They can also carry various sensors, but their data collection capabilities may be more limited than automatic vehicles. Examples of such systems include the Phenomobile Lite (Jimenez-Berni et al., 2018), which integrates data acquisition systems operated by a laptop. This platform can accommodate RGB, multispectral, hyperspectral, and thermal sensors, enabling the collection of various crop phenotyping data. Manual phenotyping vehicles are more affordable and effective in smaller or more controlled environments. However, they are less efficient and require more human resources, which can limit their use in phenotyping studies (Bai et al., 2016; Kumar et al., 2020; Meacham-Hensold et al., 2020; Thompson, Conrad et al., 2018; Thompson, Thorp et al., 2018).

Tractor-mounted phenotyping systems are standard agricultural tractors equipped with sensor arrays, GPS systems, data processing systems, etc. (Barmeier & Schmidhalter, 2017). They are more commonly used in large-scale agricultural settings where they can cover extensive field areas. Researchers at Iowa State University have developed a plant phenotyping platform mounted on a John Deere 1026R compact tractor (Bao & Tang, 2016). The system has several GRAS-20S4C-C depth cameras for plant phenotyping data acquisition. It also features a Topcon 350 automatic navigation system and an AGI-4 steering controller, enabling fully automated phenotyping data collection. With the integration of the antenna, receiver, and steering controller, the system achieves a positioning accuracy of 2 cm. Additionally, a rotating platform is mounted at the front of the tractor to support a stereo camera, allowing the adjustment of the distance between the camera and the crops. The system simultaneously captures stereo images of two rows and stitches multiple images together to collect comprehensive crop phenotype data. These systems are relatively cost-effective and can operate over large fields, but they may cause soil compaction and have limited use in fields with delicate or densely planted crops (Andrade-Sanchez et al., 2013; Barker et al., 2016; Redden & Colgan, 2017).

3.3.4 Fixed Platforms

Fixed platforms are highly specialized phenotyping systems that provide continuous monitoring of crops in a designated area. These platforms are usually installed in fixed locations within the field and are equipped with various sensors, including RGB cameras, multispectral and hyperspectral cameras, thermal imagers, and LiDARs (R. Xu & Li, 2022). Fixed platforms can be further divided into gantry-based and cable-suspended systems.

Gantry-based systems consist of a fixed track or frame structure spanning the field. Sensors are mounted on a motorized carriage that moves along the gantry, allowing for precise and repeated measurements of crops. For example, J. Zhang et al. (2023) developed a fixed gantry platform in a cornfield. This platform can capture near-infrared (VNIR) and visible hyperspectral images of corn at consecutive times. These systems are designed for high-resolution, high-frequency data collection and are particularly useful for detailed studies on plant development and stress responses. They provide stability and minimize the effects of environmental factors like wind, making them ideal for experiments that require consistent data over time. A representative gantry-based plant phenotyping platform, such as the Field Scanalyzer from LemnaTec (Atefi et al., 2021), integrates various sensors to monitor plant growth. The platform's 3D topside imaging capability enables the measurement of parameters such as leaf area, leaf inclination, and canopy structure, with the advantage of performing these measurements continuously. By employing a fixed track system in the field and moving the gantry crane via a drive motor, the platform achieves fully automated, high-throughput measurements across a range of 10 m × 110 m (Virlet et al., 2016). Another example is the TERRA-REF platform, which allows optical observations of crops to be acquired at ultra-high spatial resolution (Babaeian, Sidike et al., 2019). Located in Maricopa, Arizona, this state-of-the-art instrument consists of a 30-ton steel gantry and suspended instrument box containing various imaging systems and sensors. The robot moves autonomously along two 200-meter-long tracks, continuously imaging the soil and crops below (Babaeian, Sadeghi et al., 2019).

Cable-suspended systems are supported by towers positioned around the field's perimeter, with sensors suspended on cables that can move in three dimensions (Kirchgessner et al., 2016; Salafian, 2017; R. Xu & Li, 2022). This configuration allows for flexible positioning of sensors at different heights and locations within the field, enabling comprehensive monitoring of plant phenotypes. Cable-suspended systems are particularly useful for large-scale phenotyping studies that require extensive spatial coverage and the ability to monitor crops from multiple angles. A cable-suspended field phenotyping platform was established at the Plant Science Research Station in Zurich, Switzerland, which can be used for automated and precise plant phenotyping data collection (Kirchgessner et al., 2016). Four 24 m-high towers were built in a 2.5-acre rectangular field, with pulleys at the top of the towers to guide the cables. The cables were wrapped around a winch at the bottom of each tower. The sensor head is positioned anywhere above the field, with a maximum positioning height of 6 m and a displacement speed as fast as 2 m/s. The sensors on board include two DSLR cameras, a laser scanner, and a thermal imager.

3.3.5 Aerial Platforms

Aerial phenotyping platforms have become increasingly important in plant phenotyping research due to their ability to overcome some of the limitations associated with ground-based platforms (Chapman & Mesbahi, 2015; Tattaris et al., 2016). These platforms enable the rapid collection of data over large areas, making them particularly valuable for large-scale agricultural monitoring and research. Aerial platforms can be broadly categorized into three main types: unmanned aerial vehicles (UAVs), manned aerial vehicles (MAVs), and satellite-based platforms.

3.3.6 Unmanned Aerial Vehicles (UAVs)

UAVs, commonly known as drones, have gained significant popularity as plant phenotyping platforms due to their versatility, ease of deployment, and ability to collect high-resolution data over large areas relatively quickly (C. Zhang & Kovacs, 2012). UAVs are equipped with a variety of sensors to measure and analyze different plant phenotypes, such as plant height (Anthony et al., 2014; Holman et al., 2016), canopy temperature (Zarco-Tejada et al., 2012), and vegetation indices (Holman et al., 2019; H. Liu et al., 2017; Tattaris et al., 2016). UAVs used in plant phenotyping are primarily of two types: multirotor and fixed-wing drones.

Multirotor drones are driven by four to eight rotors, offering the ability to hover and fly at various altitudes. Multirotor drones are highly maneuverable and can be operated in confined spaces, making them suitable for detailed field surveys where precision is required. They are capable of vertical takeoff and landing (VTOL), requiring minimal space for deployment (Peña et al., 2013). An example of a multirotor drone used in phenotyping is the MK-Oktokopter, which is equipped with RGB sensors for capturing detailed images of crop canopies (Bendig et al., 2015). However, these drones typically have limited flight times and payload capacities, which restricts the number of sensors they can carry and the duration of their operations.

Fixed-wing drones are designed for longer flight durations and can cover larger areas at higher speeds than multi-rotor drones. They are well-suited for surveying extensive fields, providing a broader perspective on crop health and growth (Hernandez et al., 2016). However, they cannot hover and require more space for takeoff and landing (Herwitz et al., 2004; Link et al., 2013). Fixed-wing drones generally carry fewer sensors due to their payload restrictions, and high-speed flight may result in lower image resolution or motion blur, which can be mitigated by using high-shutter-speed imaging sensors (Sankaran et al., 2015; Shi et al., 2016).

3.3.7 Manned Aerial Vehicles (MAVs)

MAVs are another type of aerial platform used in plant phenotyping. These platforms, including airplanes and helicopters, can carry more sophisticated and heavier sensors than UAVs. MAVs can cover vast areas in a single flight, making them ideal for large-scale agricultural monitoring and research (X. Jin et al., 2021).

Despite their advantages, MAVs are generally more expensive to operate and require specialized personnel for piloting and maintenance. Due to their higher flight altitudes and speeds, the lower resolution of images captured by MAVs can be a limitation for detailed phenotypic analysis. Additionally, MAVs are less flexible than UAVs in terms of flight paths and require larger areas for takeoff and landing. For these reasons, very few researchers currently use MAVs as phenotyping platforms.

3.3.8 Satellite-Based Platforms

Satellite-based phenotyping platforms can monitor plant traits over vast geographic areas. These platforms are beneficial for large-scale, global agricultural monitoring and for applications requiring frequent, repeated measurements across large areas. They are less affected by weather conditions compared to UAVs, as many satellites can capture data through clouds using radar and other non-optical sensors. Satellites can provide data in multiple spectral bands, as different bands of light have various sensitivities to crop structure (Lopez-Sanchez et al., 2012), leaf pigmentation, crop yield (X. Jin et al., 2021; Schwalbert et al., 2020), disease detection (C. Yang, 2020), and moisture content (Er-Raki et al., 2010; H. Wang et al., 2019). They can provide images with meter-scale resolution, allowing crop phenotypes to be estimated over a very wide range.

3.4 DEEP LEARNING IN PLANT PHENOTYPING

In recent years, plant phenotyping has increasingly shifted from traditional image processing to deep learning methods. Traditional methods rely on manually crafted features, which struggle to capture the complexity and variability of plant traits, such as irregular leaf shapes or disease symptoms. In contrast, deep learning models automatically learn intricate, high-level features from large datasets, offering greater accuracy and robustness. With advances in computational power and data availability, deep learning has proven more effective in handling the scale and complexity of plant phenotyping, driving its adoption in the field. This section introduces the application of deep learning in HTPP, focusing on three key areas: 2D image analysis, time series analysis, and 3D data processing.

3.4.1 Image Classification

In the domain of high-throughput plant phenotyping, image classification using deep learning techniques has emerged as a pivotal method for automating the identification of plant traits. By leveraging convolutional neural networks (CNNs) (Kamilaris & Prenafeta-Boldú, 2018), researchers can efficiently classify vast amounts of image data into predefined categories, such as distinguishing between healthy and diseased plants (Abdullahi et al., 2017) or categorizing different disease types (Lu et al., 2021). This automation not only enhances the speed and accuracy of phenotypic assessments but also facilitates large-scale screening of plant populations, which is essential for modern breeding programs.

3.4.2 Object Detection

Object detection in high-throughput plant phenotyping involves the application of deep learning models to identify and localize specific traits within complex images. Techniques such as YOLO and Faster R-CNN have proven instrumental in accurately detecting phenotypic features like fruits, leaves, and flowers (Gao et al., 2020; Mathew & Mahesh, 2022). These models enable precise quantification of traits at the organ level, thereby providing detailed insights into plant development and enabling more informed selection in breeding programs. Moreover, object detection allows for real-time monitoring of plants in dynamic environments, significantly advancing the capabilities of phenotyping platforms.

3.4.3 Instance Segmentation

Instance segmentation, a sophisticated extension of object detection, plays a crucial role in high-throughput plant phenotyping by not only identifying but also delineating individual plant structures within an image. Utilizing state-of-the-art models such as Mask R-CNN, researchers can achieve pixel-level precision in segmenting leaves, stems, and fruits, enabling detailed morphological analysis (Jia et al., 2020; Shinoda et al., 2023). This granular level of segmentation supports the accurate measurement of trait variation within and across plant populations, thereby enhancing our understanding of phenotypic diversity and its genetic underpinnings.

3.4.4 Semantic Segmentation

Semantic segmentation is a powerful tool in high-throughput plant phenotyping, allowing for the classification of every pixel in an image into predefined categories, such as different plant organs or background (He, Li et al., 2024). This technique is particularly valuable for separating crops from weeds, assessing canopy structure, and evaluating leaf area index. The adoption of deep learning models like U-Net and DeepLab in semantic segmentation tasks has significantly improved the accuracy and efficiency of phenotypic data extraction, contributing to more precise and scalable phenotyping approaches (S. Zhang & Zhang, 2023; Zhu et al., 2023).

3.4.5 Time-Series Analysis

Time series prediction in high-throughput plant phenotyping leverages deep learning models to forecast the future states of phenotypic traits based on historical data. Techniques such as long short-term memory (LSTM) networks enable the prediction of growth patterns, stress responses, and yield outcomes, providing valuable insights for crop management and breeding (Patle et al., 2022; Tong et al., 2022). Accurate time series predictions are crucial for anticipating phenotypic changes under varying environmental conditions, thus supporting the development of resilient crop varieties.

3.4.6 3D Phenotyping Analysis

The integration of 3D phenotyping analysis into high-throughput plant phenotyping represents a significant advancement in capturing the complex spatial structures of plants. By applying deep learning techniques to 3D point clouds or volumetric data, researchers can reconstruct detailed models of plant architecture, including canopy structure, leaf morphology (Jiang et al., 2019b; Xiang & Wang, 2023), and root systems (Todo et al., 2021). This 3D approach allows for a more comprehensive assessment of phenotypic traits, facilitating the study of plant growth dynamics and the interaction between plant structure and function in response to environmental stimuli.

3.5 A CASE STUDY: DEVELOPMENT OF A HIGH-THROUGHPUT PHENOTYPING SYSTEM FOR TOMATO FOLIAR DISEASE PHENOTYPING

Background and Major Gaps in Tomato Disease Phenotyping

Tomato (*Solanum lycopersicum* L.) is one of the most economically significant vegetable crops worldwide, with a market value of $93.9 billion in 2018 and an estimated production of 180.8 million tons in 2019 (Bento-Da-Silva et al., 2023). However, tomato plants are highly susceptible to over 200 different types of diseases including fungi, viruses, and bacteria, which can substantially reduce both yield and fruit quality (Adhikari et al., 2017; Campos et al., 2021). Among these threats, foliar diseases pose one of the most significant challenges to tomato production systems, potentially yield losses can reach up to 100% in severe cases (Panno et al., 2021).

Understanding the impact of disease stress on tomatoes is crucial for safeguarding yields and meeting the growing demand for food production. Typically, the disease phenotypes of tomatoes are characterized by two important traits: disease type and disease severity. Disease severity can be assessed at both the individual leaf level and the whole-plant level. The present case study focuses on high-throughput quantitative assessment of disease severity at the plant level. The most common approach to obtaining ideal phenotypic data related to plant diseases and yield involves continuous monitoring in the field. Traditional phenotyping methods rely on visual inspection by breeders to detect and diagnose diseases. However, non-experts often interpret symptoms based on their own experience, leading to the potential for misidentification. This process is labor-intensive, prone to subjective bias and visual fatigue, and generally inefficient, severely limiting the scale of experiments and the frequency of data collection. Moreover, manual detection is effective for estimating vertical resistance (conferred by single major R genes) but less effective at identifying horizontal resistance (conferred by the additive effects of multiple minor genes) (Bhattarai et al., 2017; Panthee et al., 2017). Horizontal resistance, which is more durable and thus more desirable, is challenging for pathogens to overcome due to the involvement of multiple genes. However, the nature of additive gene effects results in a range of resistance levels in the phenotype, which may be difficult to distinguish through manual detection. Developing high-throughput field phenotyping systems can enhance the reliability and reproducibility of potential horizontal resistance phenotypes compared to manual phenotyping, enabling breeders to address this gap and improve the durability of resistance in tomatoes and other crops.

In recent years, traditional machine learning has been successfully applied to crop disease phenotyping (Domingues et al., 2022; Ngugi et al., 2024). However, most studies (Deng et al., 2023; Kanda et al., 2022; Shoaib et al., 2022) on tomato diseases have focused on single leaf images obtained under controlled laboratory conditions, with well-controlled lighting and backgrounds. While segmentation of single leaves and local leaf recognition in laboratory settings often achieve high accuracy, these methods are rarely transferable to complex field conditions. Advances in deep learning algorithms have spurred interest in exploring plant-level disease detection under field conditions. Recent developments have combined deep learning with traditional phenotyping techniques, utilizing handheld (Fuentes et al., 2018; Saeed et al., 2023; Tang et al., 2023) and drone imaging systems (Abdulridha et al., 2020; Abdulridha et al., 2020) to improve disease detection in the field. However, these approaches face operational limitations, such as low spatial resolution of captured images and limited visibility of (1) lateral phenotypic features (leaf spots, shapes, textures, fruit size, and shape) and (2) the lower parts of the plant canopy, where diseases often originate. Furthermore, previous deep learning studies on tomato disease phenotyping have primarily focused on classification tasks, limited to disease type identification. There has been almost no research exploring quantitative assessment of disease severity using deep learning models, which requires precise semantic segmentation models.

Apart from the complexity and inapplicability of algorithm models, there is a critical challenge common to any supervised deep learning task: the requirement for large, annotated datasets (Alzubaidi et al., 2021; Rasmussen et al., 2022). For segmentation algorithms, such as in our previous study, pixel-level annotation of a single image of a tomato plant in complex scenes can take up to five hours. Further, training a model often requires tens of thousands of annotated images, making manual image labeling nearly impossible. In precision agriculture, where data scarcity in complex agricultural scenarios is a common issue, improving network structures generally yields limited benefits. Some researchers have turned to data generation techniques that can automatically annotate datasets (Paulin & Ivasic-Kos, 2023). Specifically, the recent Unreal Engine 5 (UE5) not only allows for the creation of realistic 3D scenes (dynamic lighting, meteorological conditions, etc.) but also offers potential for agricultural robot simulation and control. Integrating robotic simulators into 3D simulations can enhance the clarity and realism of perspectives in synthetic images and provide execution times to reach these positions, resulting in more realistic and diverse synthetic datasets. Additionally, data collection workflows should also consider different perspectives from robotic data collection to identify effective workflows and optimal experimental designs.

3.5.1 Overview of the System

To address the challenges in tomato disease phenotyping, we developed a robotic phenotyping system (Figure 3.2) that comprises a physical ground robot and a digital twin system. These components are used to collect real-world and synthetic data, respectively. The real-world dataset, which serves as the target for our models, was collected using a customized Husky robot at the Mountain Horticultural Crops Research and Extension Center in Mills River, North Carolina, USA. This ground robot was deployed in tomato fields affected by foliar diseases such as early blight and bacterial spot. The HuskyBot, equipped with stereo cameras and GPS, was designed to capture georeferenced side-view images. With a width of 0.66 meters, the HuskyBot was specifically designed

(a) Physical system (b) Virtual system

FIGURE 3.2 Digital synthetic tomato dataset twin system. (a) Physical system under Field conditions, including Battery, HuskyBot, and tomato plants affected by disease; (b) Virtual system in Unreal Engine, including simulated HuskyBot and virtual tomato plants affected by foliar disease. GPS, Global Positioning System.

to navigate crop rows spaced 1.14 meters apart, adhering to standard agricultural practices. An integrated PhenoStereo system, featuring dual RGB cameras and customized lighting, enabled the capture of high-quality images under various field conditions, facilitating three-dimensional reconstruction.

The synthetic dataset, which serves as the source for our model, consists of photorealistic images generated within Unreal Engine 5 (UE5). This virtual system mirrors real-world settings, simulating the Husky robot's operations within a three-dimensional tomato field affected by diseases. The UE5 environment allows for the automated generation of images through customizable settings, including background, disease distribution, tomato plant density, time of day, and weather conditions, thereby mimicking real-world scenarios.

3.5.2 Virtual Plant Image Acquisition System

The accurate creation of realistic 3D models of tomato plants is essential for training robust AI models in plant science. Utilizing Unreal Engine 5 (UE5), we synthesized both healthy and disease-affected tomato plants by manipulating parameters like size, shape, color, and disease localization. Our pipeline (Figure 3.3) integrates a Robot Operating System (ROS) with a UE5 simulator, where a virtual HuskyBot equipped with a robotic arm and camera simulates data collection in a configurable environment (Figure 3.4). The system, built on ROS-UE5 communication via ROS Bridge, automates image capture and mask labeling, enhancing dataset diversity for semantic segmentation models in plant science.

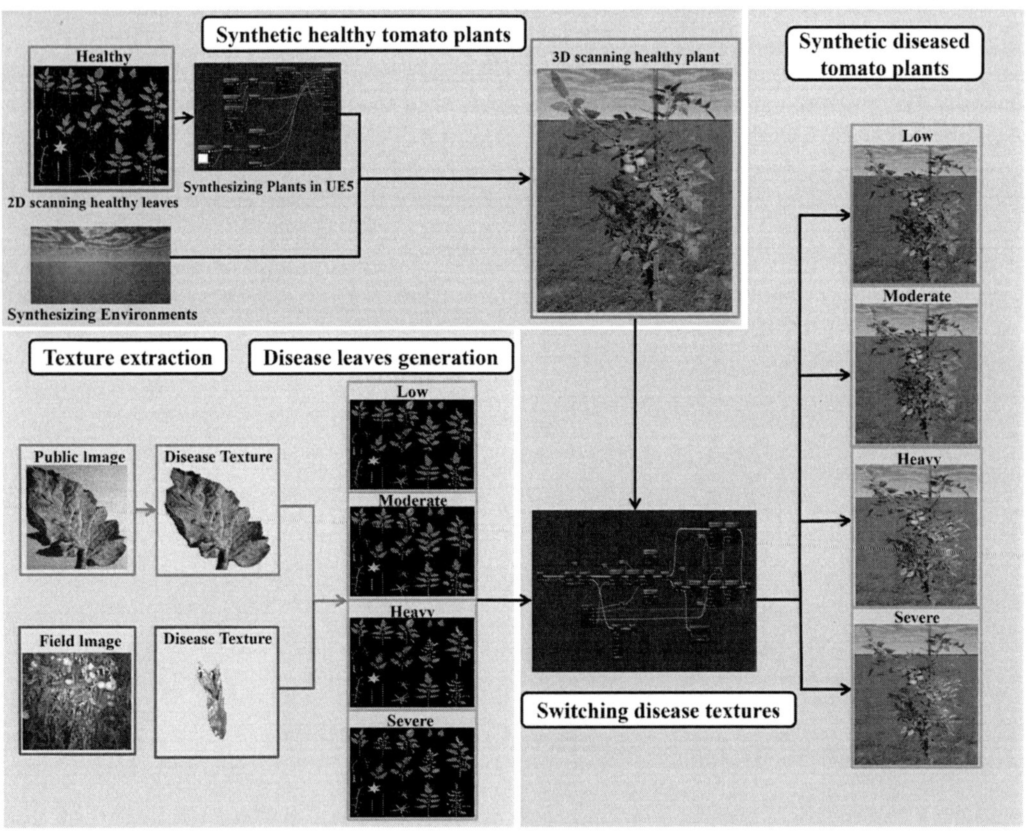

FIGURE 3.3 Schematic diagram of the process for synthesizing the 3D plant generation pipeline. 2D, two dimensional; 3D, three dimensional; UE5, Unreal Engine 5.

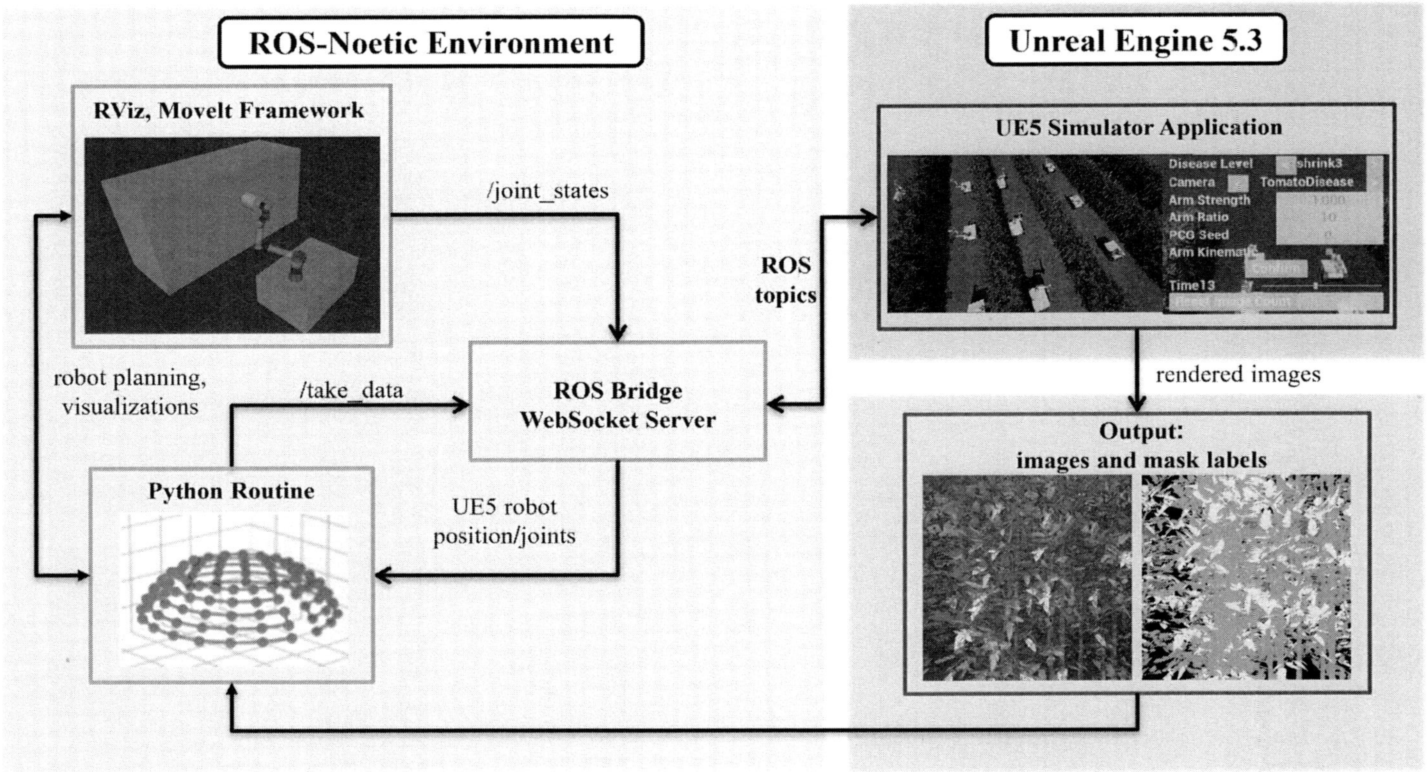

FIGURE 3.4 Synthetic robot dataset collection system in the Unreal Engine 5 (UE5) environment, simulating real-world data collection processes to create an extensive synthetic dataset. ROS, Robot Operating System; Rviz, Robot Operating System Visualization.

A summary of the tomato disease dataset is illustrated in Table 3.1. The synthetic dataset comprises 1,296 images with auto-generated segmentation masks (Figure 3.5), and the real field dataset contains 504 stereo images without segmentation masks and 56 stereo images with segmentation masks (Figure 3.6). The severity of tomato diseases was assessed based on the percentage of disease-affected areas using the Horsfall–Barratt (H–B) scale (1 to 12) (Table 3.2) (Couture, 1980).

3.5.3 Integrated Segmentation Model with 3D Depth Denoising and Calibration System

Most of the previous studies about tomato disease have focused on images of single tomato leaves under controlled laboratory conditions. These methods often fail in field conditions due to significant

TABLE 3.1
Information on Tomato Disease Dataset

Dataset	Ground Type	Disease Stages	# of Images	Mask Colormap
Real field dataset (Target dataset)	Brown soil and white mulch film	Healthy	104	56 manually labeled masks:
		Low	104	1. Green: health
		Medium	104	2. Yellow: disease
		Mild wilting	124	3. Red: fruits
		Severe wilting	124	4. Black: Background
3D synthetic dataset (Source dataset)	Simulated brown soil	Healthy	120	Auto-generated masks:
		Low	120	1. Green: health
		Medium	120	2. Yellow: disease
		Mild wilting	144	3. Red: fruits
		Severe wilting	144	4. Black: Background
	Simulated white mulch film	Healthy	120	
		Low	120	
		Medium	144	
		Mild wilting	144	
		Severe wilting	144	

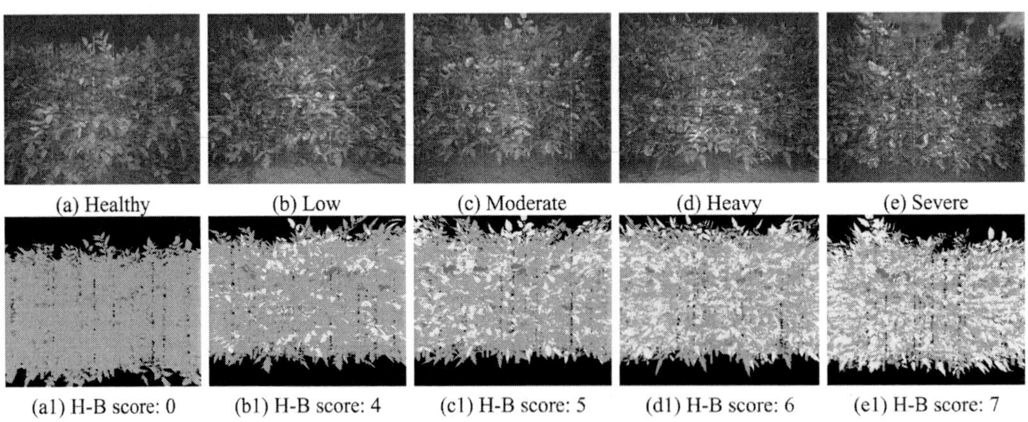

FIGURE 3.5 Examples of synthesized images at different disease stages in the dataset. Various disease stages were simulated in the Unreal Engine 5 (UE5) environment. The second row presents the corresponding segmentation masks for the images in the first row. H-B, Horsfall–Barratt.

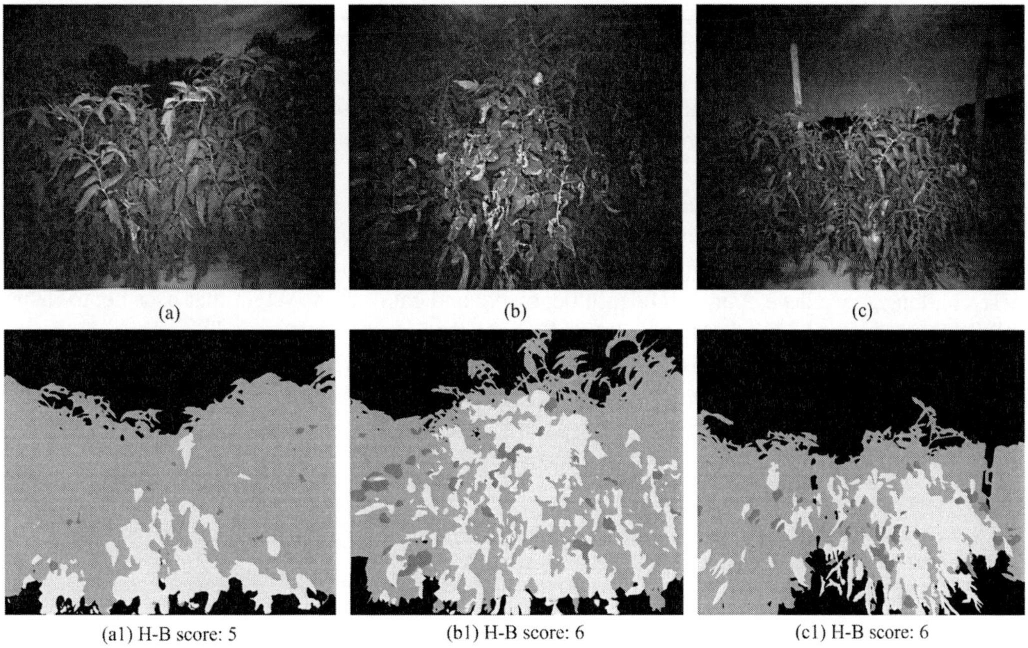

FIGURE 3.6 Manually labeled field images for system evaluation. (a)–(c) Manually annotated images, and (a1)–(c1) the corresponding annotated segmentation masks. H–B, Horsfall–Barratt.

TABLE 3.2
Horsfall–Barratt Scale Table: The percentage of leaf area affected by the disease is estimated, and a categorical number (H-B scale) is assigned for easy data collection. The size of the area interval is variable, and the midpoint can be used in data analyses

H-B Score	% Leaf Area Affected	Size of Interval	Midpoint
1	0	0	0
2	0^{+}–3	3	1.5
3	3^{+}–6	3	4.5
4	6^{+}–12	6	9.0
5	12^{+}–25	13	18.5
6	25^{+}–50	25	37.5
7	50^{+}–75	25	62.5
8	75^{+}–87	13	81.5
9	87^{+}–94	6	91.0
10	94^{+}–97	3	96.2
11	97^{+}–100	3	98.5
12	100	0	100

environmental differences. Our approach allows disease segmentation algorithms to function effectively under field conditions. The supervised disease segmentation algorithm, trained in the source domain, is adapted to the target domain without labeling information through an unsupervised domain-adaptive semantic segmentation algorithm.

Our innovative algorithm combines semantic segmentation, unsupervised learning, and domain adaptation, achieving precise segmentation of tomato leaf diseases even in complex and unstable environments without manual labeling. As shown in Figure 3.7, the unsupervised learning input for the target domain consists of real field images, while the supervised learning input for the source domain includes synthetic images and corresponding mask labels from the virtual plant image acquisition system.

Field conditions, with their complex environments and varying lighting, introduce significant noise, challenging the model's generalization and transferability. Although simulating various environmental scenarios can reduce noise impact, it cannot cover all potential variations. To overcome these challenges, we developed a 3D depth denoising and calibration system that uses depth maps derived from stereo vision to add a third dimension to the images.

In the first phase, the stereo vision system helps the UDASeg model reduce noise by using depth information to remove background elements from segmented disease images (Figure 3.8). The

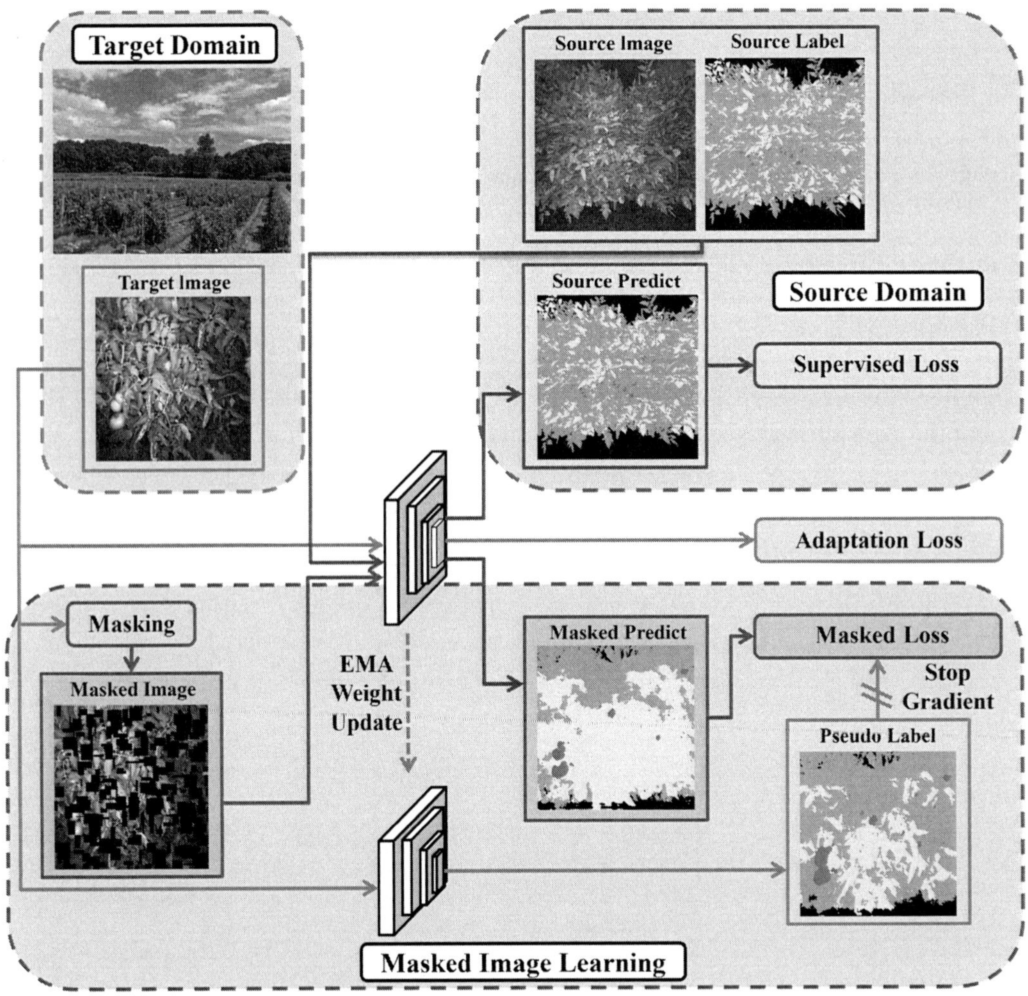

FIGURE 3.7 UDA semantic segmentation (UDASeg) algorithmic framework for tomato disease segmentation. The segmentation model was trained based on a synthetic dataset to accurately segment tomato diseases and ultimately provide an accurate assessment of tomato diseases. EMA, Exponential Moving Average.

second phase involves using depth information to correct geometric errors caused by camera perspective distortion. This distortion occurs because objects closer to the camera appear larger and occupy more pixels in 2D images (Figure 3.9). The pixel area of a tomato leaf image is approximately inversely proportional to the square of the distance (depth) between the leaf and the camera under a pinhole camera assumption. By calibrating the pixel area with the square of the pixel depth, we can reduce imaging errors introduced by perspective distortion to a certain extent.

3.5.4 Field Phenotyping Results and Conclusion

The tomato images captured by HuskyBot in the field present a challenging scenario for disease detection due to their complex backgrounds. As shown in Figure 3.10, the developed model

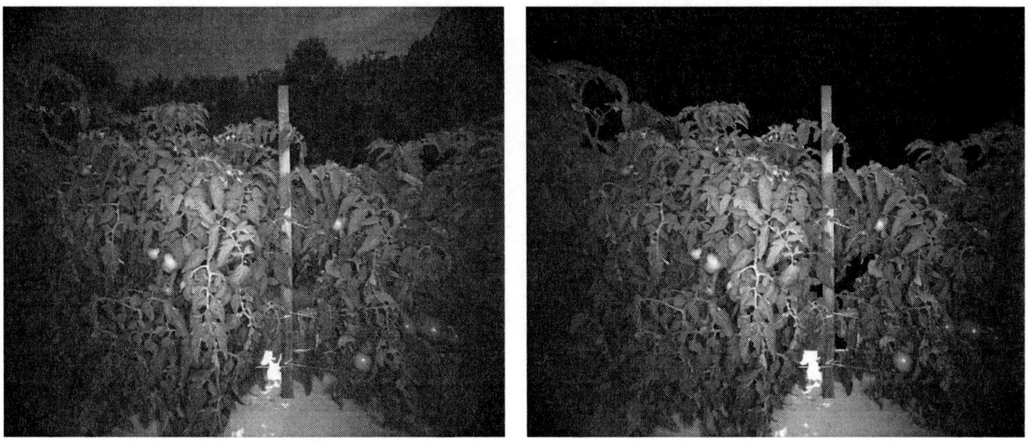

(a) Before background removal (b) After background removal

FIGURE 3.8 Comparison of images before (a) and after (b) background noise removal.

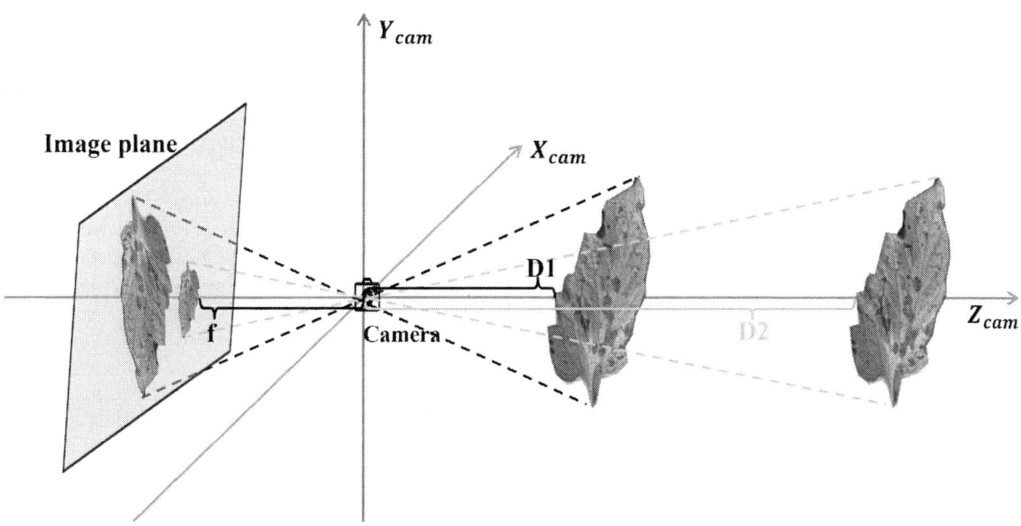

FIGURE 3.9 Schematic representation of the error introduced by the camera's perspective distortion.

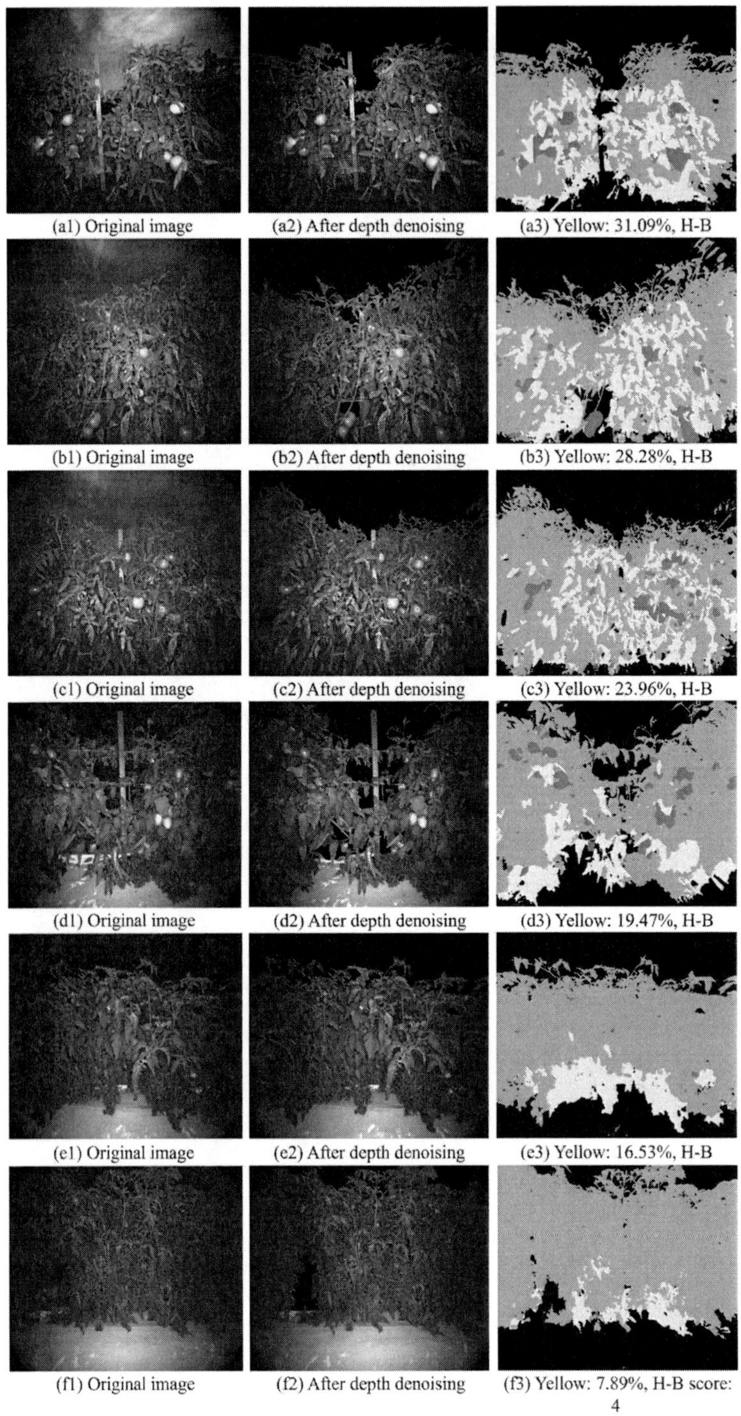

FIGURE 3.10 Semantic segmentation of disease phenotyping results. The first column is the original image; the second column is the image after depth-based denoising, note that the background was filtered to black pixels; and the third column is the image output from the UDASeg model, where the green pixels are the healthy leaf region, the yellow pixels are the disease-affected leaf region, the red pixels are the tomato fruit, and the black pixels are the background region.

performs robustly across various conditions, effectively segmenting tomato foliar diseases despite diverse symptoms, severity levels, background noise, and leaf density. The final column of segmentation results in Figure 3.10 demonstrates accurate detection of disease-affected leaves. The proportion of diseased leaf area increases progressively from the first to the last row, and the model's outputs align well with the actual disease conditions. These results underscore the model's effectiveness in real field settings, highlighting its potential for practical application in agricultural disease management.

This project emphasizes the critical role of high-throughput plant phenotyping systems that apply deep learning and synthetic data generation techniques to overcome the challenges of tomato disease phenotyping under field conditions. By leveraging UE5, we generated photorealistic field images that accurately simulate complex agricultural scenarios, reducing the labor and resources needed for data collection and labeling. Our customized UDASeg model, which combines advanced unsupervised domain adaptation with semantic segmentation, represents a significant breakthrough in automating the accurate assessment of tomato diseases without the need for manually annotated datasets. This approach not only lowers costs but also sets a new standard for using synthetic data in plant science, with the potential to revolutionize agricultural research and management. Our research contributes significantly to the field, offering scalable solutions that can be applied to disease detection in other crops, thereby supporting global efforts toward sustainable agriculture and food security.

3.6 SUMMARY

In this chapter, we provided an in-depth overview of HTPP technologies, focusing on their evolution and integration with advanced proximal sensors and data processing algorithms. We discussed key sensing technologies, including RGB vision, stereo vision, LiDAR, and spectral sensing, emphasizing their critical roles in enhancing the accuracy and efficiency of phenotypic data collection across different scales and environments. Furthermore, we explored various phenotyping platforms, including ground-based systems like phenotyping poles, vehicles, and fixed platforms, as well as aerial platforms such as UAVs and satellites. The chapter also included a case study on tomato disease phenotyping, demonstrating the application of a novel robotic system coupled with a domain-adaptive segmentation algorithm and synthetic data generation techniques to overcome challenges in field conditions. This case study exemplifies the potential of combining AI and synthetic data in modern phenotyping, illustrating how these advancements contribute to more accurate and scalable next-generation agriculture.

REFERENCES

Abdalla, A., Cen, H., El-manawy, A., & He, Y. (2019). Infield oilseed rape images segmentation via improved unsupervised learning models combined with supreme color features. *Computers and Electronics in Agriculture*, *162*(C), 1057–1068. https://doi.org/10.1016/j.compag.2019.05.051

Abdullahi, H., Sheriff, R., & Mahieddine, F. (2017). *Convolution Neural Network in Precision Agriculture for Plant Image Recognition and Classification* (p. 3). https://doi.org/10.1109/INTECH.2017.8102436

Abdulridha, J., Ampatzidis, Y., Kakarla, S. C., & Roberts, P. (2020). Detection of target spot and bacterial spot diseases in tomato using UAV-based and benchtop-based hyperspectral imaging techniques. *Precision Agriculture*, *21*(5), 955–978. https://doi.org/10.1007/s11119-019-09703-4

Abdulridha, J., Ampatzidis, Y., Qureshi, J., & Roberts, P. (2020). Laboratory and UAV-based identification and classification of tomato yellow leaf curl, bacterial spot, and target spot diseases in tomato utilizing hyperspectral imaging and machine learning. *Remote Sensing*, *12*(17), Article 17. https://doi.org/10.3390/rs12172732

Abebe, A. M., Kim, Y., Kim, J., Kim, S. L., & Baek, J. (2023). Image-based high-throughput phenotyping in horticultural crops. *Plants*, *12*(10), Article 10. https://doi.org/10.3390/plants12102061

Adhikari, P., Oh, Y., & Panthee, D. R. (2017). Current status of early blight resistance in tomato: An update. *International Journal of Molecular Sciences*, *18*(10), 2019. https://doi.org/10.3390/ijms18102019

Alzubaidi, L., Zhang, J., Humaidi, A. J., Al-Dujaili, A., Duan, Y., Al-Shamma, O., Santamaría, J., Fadhel, M. A., Al-Amidie, M., & Farhan, L. (2021). Review of deep learning: Concepts, CNN architectures, challenges, applications, future directions. *Journal of Big Data*, *8*(1), 53. https://doi.org/10.1186/s40537-021-00444-8

Andrade-Sanchez, P., Gore, M. A., Heun, J. T., Thorp, K. R., Carmo-Silva, A. E., French, A. N., Salvucci, M. E., & White, J. W. (2013). Development and evaluation of a field-based high-throughput phenotyping platform. *Functional Plant Biology: FPB*, *41*(1), 68–79. https://doi.org/10.1071/FP13126

Andújar, D., Ribeiro, A., Fernandez-Quintanilla, C., & Dorado, J. (2016). Using depth cameras to extract structural parameters to assess the growth state and yield of cauliflower crops. *Computers and Electronics in Agriculture*, *122*, 67–73. https://doi.org/10.1016/j.compag.2016.01.018

Anthony, D., Elbaum, S., Lorenz, A., & Detweiler, C. (2014). On crop height estimation with UAVs. *2014 IEEE/RSJ International Conference on Intelligent Robots and Systems*, 4805–4812. https://doi.org/10.1109/IROS.2014.6943245

Atefi, A., Ge, Y., Pitla, S., & Schnable, J. (2021). Robotic technologies for high-throughput plant phenotyping: Contemporary reviews and future perspectives. *Frontiers in Plant Science*, *12*, 611940. https://doi.org/10.3389/fpls.2021.611940

Babaeian, E., Sadeghi, M., Jones, S. B., Montzka, C., Vereecken, H., & Tuller, M. (2019). Ground, proximal, and satellite remote sensing of soil moisture. *Reviews of Geophysics*, *57*(2), 530–616. https://doi.org/10.1029/2018RG000618

Babaeian, E., Sidike, P., Newcomb, M. S., Maimaitijiang, M., White, S. A., Demieville, J., Ward, R. W., Sadeghi, M., LeBauer, D. S., Jones, S. B., Sagan, V., & Tuller, M. (2019). A new optical remote sensing technique for high-resolution mapping of soil moisture. *Frontiers in Big Data*, *2*, 37. https://doi.org/10.3389/fdata.2019.00037

Bai, G., Ge, Y., Hussain, W., Baenziger, P. S., & Graef, G. (2016). A multi-sensor system for high throughput field phenotyping in soybean and wheat breeding. *Computers and Electronics in Agriculture*, *128*, 181–192. https://doi.org/10.1016/j.compag.2016.08.021

Bao, Y., Gai, J., Xiang, L., & Tang, L. (2021). Field robotic systems for high-throughput plant phenotyping: A review and a case study. In J. Zhou & H. T. Nguyen (Eds.), *High-Throughput Crop Phenotyping* (pp. 13–38). Springer International Publishing. https://doi.org/10.1007/978-3-030-73734-4_2

Bao, Y., & Tang, L. (2016). Field-based robotic phenotyping for sorghum biomass yield component traits characterization using stereo vision. *IFAC-PapersOnLine*, *49*(16), 265–270. https://doi.org/10.1016/j.ifacol.2016.10.049

Bao, Y., Tang, L., Breitzman, M., Salas Fernandez, M., & Schnable, P. (2018). Field-based robotic phenotyping of sorghum plant architecture using stereo vision. *Journal of Field Robotics*, *36*. https://doi.org/10.1002/rob.21830

Bao, Y., Tang, L., Srinivasan, S., & Schnable, P. (2018). *Field-Based Architectural Traits Characterisation of Maize Plant Using Time-of-Flight 3D Imaging*. https://dr.lib.iastate.edu/handle/20.500.12876/1797

Bao, Y., Tang, L., Srinivasan, S., & Schnable, P. (2019). Field-based architectural traits characterisation of maize plant using time-of-flight 3D imaging. *Biosystems Engineering*, *178*, 86–101. https://doi.org/10.1016/j.biosystemseng.2018.11.005

Barker, J., Zhang, N., Sharon, J., Steeves, R., Wang, X., Wei, Y., & Poland, J. (2016). Development of a field-based high-throughput mobile phenotyping platform. *Computers and Electronics in Agriculture*, *122*(C), 74–85. https://doi.org/10.1016/j.compag.2016.01.017

Barmeier, G., & Schmidhalter, U. (2017). High-throughput field phenotyping of leaves, leaf sheaths, culms and ears of spring barley cultivars at anthesis and dough ripeness. *Frontiers in Plant Science*, *8*. https://doi.org/10.3389/fpls.2017.01920

Bayati, M., & Fotouhi, R. (2018). A mobile robotic platform for crop monitoring. *Advances in Robotics & Automation*, *7*. https://doi.org/10.4172/2168-9695.1000186

Beć, K. B., Grabska, J., Bonn, G. K., Popp, M., & Huck, C. W. (2020). Principles and applications of vibrational spectroscopic imaging in plant science: A review. *Frontiers in Plant Science*, *11*. https://doi.org/10.3389/fpls.2020.01226

Behmann, J., Mahlein, A.-K., Rumpf, T., Römer, C., & Plümer, L. (2015). A review of advanced machine learning methods for the detection of biotic stress in precision crop protection. *Precision Agriculture*, *16*, 239–260. https://doi.org/10.1007/s11119-014-9372-7

Bendig, J., Yu, K., Aasen, H., Bolten, A., Bennertz, S., Broscheit, J., Gnyp, M., & Bareth, G. (2015). Combining UAV-based plant height from crop surface models, visible, and near infrared vegetation indices for biomass monitoring in barley. *International Journal of Applied Earth Observation and Geoinformation, 39*, 79–87. https://doi.org/10.1016/j.jag.2015.02.012

Bento-Da-Silva, E. P., Mendonça, S. R., & de Moraes, M. G. (2023). Trends and gaps in tomato grafting literature: A systematic approach. *Spanish Journal of Agricultural Research, 21*(3), e0904.

Bhattarai, K., Louws, F. J., Williamson, J. D., & Panthee, D. R. (2017). Resistance to *Xanthomonas perforans* race T4 causing bacterial spot in tomato breeding lines. *Plant Pathology, 66*(7), 1103–1109. https://doi.org/10.1111/ppa.12656

Biber, P., Weiss, U., Dorna, M., & Albert, A. (2012). *Navigation System of the Autonomous Agricultural Robot "BoniRob"*. https://www.semanticscholar.org/paper/Navigation-System-of-the-Autonomous-Agricultural-%E2%80%9C-Biber-Weiss/e85b58686704f355f1d755bb4757a3e1909e4e78

Bresson, J., Vasseur, F., Dauzat, M., Koch, G., Granier, C., & Vile, D. (2015). Quantifying spatial heterogeneity of chlorophyll fluorescence during plant growth and in response to water stress. *Plant Methods, 11*(1), 23. https://doi.org/10.1186/s13007-015-0067-5

Brewer, K., Clulow, A., Sibanda, M., Gokool, S., Odindi, J., Mutanga, O., Naiken, V., Chimonyo, V. G. P., & Mabhaudhi, T. (2022). Estimation of maize foliar temperature and stomatal conductance as indicators of water stress based on optical and thermal imagery acquired using an unmanned aerial vehicle (UAV) platform. *Drones, 6*(7), Article 7. https://doi.org/10.3390/drones6070169

Busemeyer, L., Mentrup, D., Möller, K., Wunder, E., Alheit, K., Hahn, V., Maurer, H. P., Reif, J. C., Würschum, T., Müller, J., Rahe, F., & Ruckelshausen, A. (2013). BreedVision—A multi-sensor platform for non-destructive field-based phenotyping in plant breeding. *Sensors, 13*(3), Article 3. https://doi.org/10.3390/s130302830

Campos, M. D., Félix, M. do R., Patanita, M., Materatski, P., & Varanda, C. (2021). High throughput sequencing unravels tomato-pathogen interactions towards a sustainable plant breeding. *Horticulture Research, 8*(1), 1–12. https://doi.org/10.1038/s41438-021-00607-x

Chapman, A., & Mesbahi, M. (2015). UAV swarms: Models and effective interfaces. In K. P. Valavanis & G. J. Vachtsevanos (Eds.), *Handbook of Unmanned Aerial Vehicles* (pp. 1987–2019). Springer Netherlands. https://doi.org/10.1007/978-90-481-9707-1_34

Chen, Y., Xin, R., Jiang, H., Liu, Y., Zhang, X., & Yu, J. (2023). Refined feature fusion for in-field high-density and multi-scale rice panicle counting in UAV images. *Computers and Electronics in Agriculture, 211*, 108032. https://doi.org/10.1016/j.compag.2023.108032

Comar, A., Burger, P., de Solan, B., Baret, F. D. R., Daumard, F., & Hanocq, J.-F. O. (2012). A semi-automatic system for high throughput phenotyping wheat cultivars in-field conditions: Description and first results. *Functional Plant Biology: FPB, 39*(11), 914–924. https://doi.org/10.1071/FP12065

Corti, M., Marino Gallina, P., Cavalli, D., & Cabassi, G. (2017). Hyperspectral imaging of spinach canopy under combined water and nitrogen stress to estimate biomass, water, and nitrogen content. *Biosystems Engineering, 158*, 38–50. https://doi.org/10.1016/j.biosystemseng.2017.03.006

Couture, L. (1980). Assessment of severity of foliage diseases of cereals in co-operative evaluation tests. *Canadian Plant Disease Survey*. https://www.semanticscholar.org/paper/Assessment-of-severity-of-foliage-diseases-of-in-Couture/63734ce2a5f84f809e69c9968632a38aa4d49434

Crain, J., Wei, Y., Barker, J., Thompson, S., Alderman, P., Reynolds, M., Zhang, N., & Poland, J. (2016). Development and deployment of a portable field phenotyping platform. *Crop Science, 56*. https://doi.org/10.2135/cropsci2015.05.0290

Dandrifosse, S., Bouvry, A., Leemans, V., Dumont, B., & Mercatoris, B. (2020). Imaging wheat canopy through stereo vision: Overcoming the challenges of the laboratory to field transition for morphological features extraction. *Frontiers in Plant Science, 11*. https://doi.org/10.3389/fpls.2020.00096

de Medeiros, A. D., Capobiango, N. P., da Silva, J. M., da Silva, L. J., da Silva, C. B., & dos Santos Dias, D. C. F. (2020). Interactive machine learning for soybean seed and seedling quality classification. *Scientific Reports, 10*(1), 11267. https://doi.org/10.1038/s41598-020-68273-y

Deng, Y., Xi, H., Zhou, G., Chen, A., Wang, Y., Li, L., & Hu, Y. (2023). An effective image-based tomato leaf disease segmentation method using MC-UNet. *Plant Phenomics, 5*, 0049. https://doi.org/10.34133/plantphenomics.0049

Domingues, T., Brandão, T., & Ferreira, J. C. (2022). Machine learning for detection and prediction of crop diseases and pests: A comprehensive survey. *Agriculture, 12*(9), Article 9. https://doi.org/10.3390/agriculture12091350

Du, L., Pang, Y., Wang, Q., Huang, C., Bai, Y., Chen, D., Lu, W., & Kong, D. (2023). A LiDAR biomass index-based approach for tree- and plot-level biomass mapping over forest farms using 3D point clouds. *Remote Sensing of Environment, 290*, 113543. https://doi.org/10.1016/j.rse.2023.113543

Er-Raki, S., Chehbouni, A., & Duchemin, B. (2010). Combining satellite remote sensing data with the FAO-56 dual approach for water use mapping in irrigated wheat fields of a semi-arid region. *Remote Sensing, 2*(1), Article 1. https://doi.org/10.3390/rs2010375

Feng, L., Chen, S., Zhang, C., Zhang, Y., & He, Y. (2021). A comprehensive review on recent applications of unmanned aerial vehicle remote sensing with various sensors for high-throughput plant phenotyping. *Computers and Electronics in Agriculture, 182*, 106033. https://doi.org/10.1016/j.compag.2021.106033

Fuentes, A. F., Yoon, S., Lee, J., & Park, D. S. (2018). High-performance deep neural network-based tomato plant diseases and pests diagnosis system with refinement filter bank. *Frontiers in Plant Science, 9*. https://doi.org/10.3389/fpls.2018.01162

Gai, J., Tang, L., & Steward, B. L. (2020). Automated crop plant detection based on the fusion of color and depth images for robotic weed control. *Journal of Field Robotics, 37*(1), 35–52. https://doi.org/10.1002/rob.21897

Gao, F., Fu, L., Zhang, X., Majeed, Y., Li, R., Karkee, M., & Zhang, Q. (2020). Multi-class fruit-on-plant detection for apple in SNAP system using Faster R-CNN. *Computers and Electronics in Agriculture, 176*, 105634. https://doi.org/10.1016/j.compag.2020.105634

Gené-Mola, J., Vilaplana, V., Rosell-Polo, J., Morros Rubió, J. R., Ruiz-Hidalgo, J., & Gregorio Lopez, E. (2019). Multi-modal deep learning for Fuji apple detection using RGB-D cameras and their radiometric capabilities. *Computers and Electronics in Agriculture, 162*, 689–698. https://doi.org/10.1016/j.compag.2019.05.016

Guzmán, R., Ariño, J., Navarro, R., Lopes, C., Graça, J., Reyes, M., Barriguinha, A., & Braga, R. (2016). Autonomous hybrid GPS/reactive navigation of an unmanned ground vehicle for precision viticulture-VINBOT. In *Intervitis Interfructa Hortitechnica – Technology for Wine, Juice and Special Crops, Stuttgart, Germany November 27–30 November 2016* (pp. 1–11). http://hdl.handle.net/10400.5/23060

He, W., Gage, J. L., Rellán-Álvarez, R., & Xiang, L. (2024). Swin-roleaf: A new method for characterizing leaf azimuth angle in large-scale maize plants. *Computers and Electronics in Agriculture, 224*, 109120. https://doi.org/10.1016/j.compag.2024.109120

He, W., Li, X., Panthee, D., Zhang, Z., Chen, Y., & Xiang, L. (2024). High-throughput robotic phenotyping for quantifying tomato disease severity enabled by synthetic data and domain-adaptive semantic segmentation. In *2024 Anaheim, California July 28–31, 2024*. https://doi.org/10.13031/aim.202400747

Hernandez, J. G., Gonzalez-Ferreiro, E., Sarmento, A., Silva, J., Nunes, A., Correia, A. C., Fontes, L., Tomé, M., & Diaz-Varela, R. (2016). Short communication. Using high resolution UAV imagery to estimate tree variables in Pinus pinea plantation in Portugal. *Forest Systems, 25*(2), Article 2. https://doi.org/10.5424/fs/2016252-08895

Herr, A. W., Adak, A., Carroll, M. E., Elango, D., Kar, S., Li, C., Jones, S. E., Carter, A. H., Murray, S. C., Paterson, A., Sankaran, S., Singh, A., & Singh, A. K. (2023). Unoccupied aerial systems imagery for phenotyping in cotton, maize, soybean, and wheat breeding. *Crop Science, 63*(4), 1722–1749. https://doi.org/10.1002/csc2.21028

Herwitz, S. R., Johnson, L. F., Dunagan, S. E., Higgins, R. G., Sullivan, D. V., Zheng, J., Lobitz, B. M., Leung, J. G., Gallmeyer, B. A., Aoyagi, M., Slye, R. E., & Brass, J. A. (2004). Imaging from an unmanned aerial vehicle: Agricultural surveillance and decision support. *Computers and Electronics in Agriculture, 44*(1), 49–61. https://doi.org/10:1016/j.compag.2004.02.006

Hickey, L. T., N Hafeez, A., Robinson, H., Jackson, S. A., Leal-Bertioli, S. C. M., Tester, M., Gao, C., Godwin, I. D., Hayes, B. J., & Wulff, B. B. H. (2019). Breeding crops to feed 10 billion. *Nature Biotechnology, 37*(7), 744–754. https://doi.org/10.1038/s41587-019-0152-9

Holman, F. H., Riche, A. B., Castle, M., Wooster, M. J., & Hawkesford, M. J. (2019). Radiometric calibration of 'commercial off the shelf' cameras for UAV-based high-resolution temporal crop phenotyping of reflectance and NDVI. *Remote Sensing, 11*(14), Article 14. https://doi.org/10.3390/rs11141657

Holman, F. H., Riche, A. B., Michalski, A., Castle, M., Wooster, M. J., & Hawkesford, M. J. (2016). High throughput field phenotyping of wheat plant height and growth rate in field plot trials using UAV based remote sensing. *Remote Sensing, 8*(12), Article 12. https://doi.org/10.3390/rs8121031

Holmes, W. S., Po-Leen Ooi, M., Kuang, Y. C., Simpkin, R., Lopez-Ubiria, I., Vidiella, A., Blanchon, D., Gupta, G. S., & Demidenko, S. (2020). Classifying cannabis sativa flowers, stems and leaves using statistical machine learning with near-infrared hyperspectral reflectance imaging. In *2020 IEEE International Instrumentation and Measurement Technology Conference (I2MTC)*, 1–6. https://doi.org/10.1109/I2MTC43012.2020.9129531

Hosoi, F., & Omasa, K. (2009). Estimating vertical plant area density profile and growth parameters of a wheat canopy at different growth stages using three-dimensional portable lidar imaging. *ISPRS Journal of Photogrammetry and Remote Sensing*, *64*, 151–158. https://doi.org/10.1016/j.isprsjprs.2008.09.003

Hütt, C., Bolten, A., Hüging, H., & Bareth, G. (2023). UAV LiDAR metrics for monitoring crop height, biomass and nitrogen uptake: A case study on a winter wheat field trial. *PFG—Journal of Photogrammetry, Remote Sensing and Geoinformation Science*, *91*(2), 65–76. https://doi.org/10.1007/s41064-022-00228-6

Jangra, S., Chaudhary, V., Yadav, R. C., & Yadav, N. R. (2021). High-throughput phenotyping: A platform to accelerate crop improvement. *Phenomics*, *1*(2), 31–53. https://doi.org/10.1007/s43657-020-00007-6

Jia, W., Tian, Y., Luo, R., Zhang, Z., Lian, J., & Zheng, Y. (2020). Detection and segmentation of overlapped fruits based on optimized mask R-CNN application in apple harvesting robot. *Computers and Electronics in Agriculture*, *172*, 105380. https://doi.org/10.1016/j.compag.2020.105380

Jiang, Y., & Li, C. (2020). Convolutional neural networks for image-based high-throughput plant phenotyping: A review. *Plant Phenomics (Washington, D.C.)*, *2020*, 4152816. https://doi.org/10.34133/2020/4152816

Jiang, Y., Li, C., & Paterson, A. H. (2016). High throughput phenotyping of cotton plant height using depth images under field conditions. *Computers and Electronics in Agriculture*, *130*, 57–68. https://doi.org/10.1016/j.compag.2016.09.017

Jiang, Y., Li, C., Paterson, A. H., & Robertson, J. S. (2019a). DeepSeedling: Deep convolutional network and Kalman filter for plant seedling detection and counting in the field. *Plant Methods*, *15*(1), 141. https://doi.org/10.1186/s13007-019-0528-3

Jiang, Y., Li, C., Robertson, J. S., Sun, S., Xu, R., & Paterson, A. H. (2018). GPhenoVision: A ground mobile system with multi-modal imaging for field-based high throughput phenotyping of cotton. *Scientific Reports*, *8*(1), 1213. https://doi.org/10.1038/s41598-018-19142-2

Jiang, Y., Li, C., Takeda, F., Kramer, E. A., Ashrafi, H., & Hunter, J. (2019b). 3D point cloud data to quantitatively characterize size and shape of shrub crops. *Horticulture Research*, *6*(1), 1–17. https://doi.org/10.1038/s41438-019-0123-9

Jimenez-Berni, J. A., Deery, D. M., Rozas-Larraondo, P., Condon, A. (Tony) G., Rebetzke, G. J., James, R. A., Bovill, W. D., Furbank, R. T., & Sirault, X. R. R. (2018). High throughput determination of plant height, ground cover, and above-ground biomass in wheat with LiDAR. *Frontiers in Plant Science*, *9*. https://doi.org/10.3389/fpls.2018.00237

Jin, S., Su, Y., Gao, S., Wu, F., Ma, Q., Xu, K., Ma, Q., Hu, T., Liu, J., Pang, S., Guan, H., Zhang, J., & Guo, Q. (2020). Separating the structural components of maize for field phenotyping using terrestrial LiDAR data and deep convolutional neural networks. *IEEE Transactions on Geoscience and Remote Sensing*, *58*(4), 2644–2658. https://doi.org/10.1109/TGRS.2019.2953092

Jin, X., Zarco-Tejada, P. J., Schmidhalter, U., Reynolds, M. P., Hawkesford, M. J., Varshney, R. K., Yang, T., Nie, C., Li, Z., Ming, B., Xiao, Y., Xie, Y., & Li, S. (2021). High-throughput estimation of crop traits: A review of ground and aerial phenotyping platforms. *IEEE Geoscience and Remote Sensing Magazine*, *9*(1), 200–231. https://doi.org/10.1109/MGRS.2020.2998816

Kamilaris, A., & Prenafeta-Boldú, F. X. (2018). A review of the use of convolutional neural networks in agriculture. *The Journal of Agricultural Science*, *156*(3), 312–322. https://doi.org/10.1017/S0021859618000436

Kanda, P. S., Xia, K., Kyslytysna, A., & Owoola, E. O. (2022). Tomato leaf disease recognition on leaf images based on fine-tuned residual neural networks. *Plants*, *11*(21), 2935. https://doi.org/10.3390/plants11212935

Kirchgessner, N., Liebisch, F., Yu, K., Pfeifer, J., Friedli, M., Hund, A., & Walter, A. (2016). The ETH field phenotyping platform FIP: A cable-suspended multi-sensor system. *Functional Plant Biology: FPB*, *44*(1), 154–168. https://doi.org/10.1071/FP16165

Kumar, D., Kushwaha, S., Delvento, C., Liatukas, Ž., Vivekanand, V., Svensson, J. T., Henriksson, T., Brazauskas, G., & Chawade, A. (2020). Affordable phenotyping of winter wheat under field and controlled conditions for drought tolerance. *Agronomy*, *10*(6), Article 6. https://doi.org/10.3390/agronomy10060882

Li, D., Quan, C., Song, Z., Li, X., Yu, G., Li, C., & Muhammad, A. (2021). High-throughput plant phenotyping platform (HT3P) as a novel tool for estimating agronomic traits from the lab to the field. *Frontiers in Bioengineering and Biotechnology*, *8*. https://doi.org/10.3389/fbioe.2020.623705

Li, L., Zhang, Q., & Huang, D. (2014). A review of imaging techniques for plant phenotyping. *Sensors (Basel, Switzerland)*, *14*(11), 20078–20111. https://doi.org/10.3390/s141120078

Li, Z., Guo, R., Li, M., Chen, Y., & Li, G. (2020). A review of computer vision technologies for plant phenotyping. *Computers and Electronics in Agriculture*, *176*, 105672. https://doi.org/10.1016/j.compag.2020.105672

Liang, Z., Pandey, P., Stoerger, V., Xu, Y., Qiu, Y., Ge, Y., & Schnable, J. C. (2017). *Conventional and Hyperspectral Time-Series Imaging of Maize Lines Widely Used in Field Trials* (p. 169045). bioRxiv. https://doi.org/10.1101/169045

Lin, Y. (2015). LiDAR: An important tool for next-generation phenotyping technology of high potential for plant phenomics? *Computers and Electronics in Agriculture*, *119*, 61–73. https://doi.org/10.1016/j.compag.2015.10.011

Link, J., Senner, D., & Claupein, W. (2013). Developing and evaluating an aerial sensor platform (ASP) to collect multispectral data for deriving management decisions in precision farming. *Computers and Electronics in Agriculture*, *94*, 20–28. https://doi.org/10.1016/j.compag.2013.03.003

Liu, H., Zhu, H., & Wang, P. (2017). Quantitative modelling for leaf nitrogen content of winter wheat using UAV-based hyperspectral data. *International Journal of Remote Sensing*, *38*(8–10), 2117–2134. https://doi.org/10.1080/01431161.2016.1253899

Liu, J., & Pattey, E. (2010). Retrieval of leaf area index from top-of-canopy digital photography over agricultural crops. *Agricultural and Forest Meteorology*, *150*, 485–1490. https://doi.org/10.1016/j.agrformet.2010.08.002

Liu, T., Zhu, S., Yang, T., Zhang, W., Xu, Y., Zhou, K., Wu, W., Zhao, Y., Yao, Z., Yang, G., Wang, Y., Sun, C., & Sun, J. (2024). Maize height estimation using combined unmanned aerial vehicle oblique photography and LIDAR canopy dynamic characteristics. *Computers and Electronics in Agriculture*, *218*, 108685. https://doi.org/10.1016/j.compag.2024.108685

Lopez-Sanchez, J. M., Cloude, S. R., & Ballester-Berman, J. D. (2012). Rice phenology monitoring by means of SAR polarimetry at X-band. *IEEE Transactions on Geoscience and Remote Sensing*, *50*(7), 2695–2709. https://doi.org/10.1109/TGRS.2011.2176740

Lu, J., Tan, L., & Jiang, H. (2021). Review on convolutional neural network (CNN) applied to plant leaf disease classification. *Agriculture*, *11*(8), Article 8. https://doi.org/10.3390/agriculture11080707

Ma, Z., Rayhana, R., Feng, K., Liu, Z., Xiao, G., Ruan, Y., & Sangha, J. S. (2022). A review on sensing technologies for high-throughput plant phenotyping. *IEEE Open Journal of Instrumentation and Measurement*, *1*, 1–21. https://doi.org/10.1109/OJIM.2022.3178468

Mahlein, A. K., Kuska, M. T., Thomas, S., Bohnenkamp, D., Alisaac, E., Behmann, J., Wahabzada, M., & Kersting, K. (2017). Plant disease detection by hyperspectral imaging: From the lab to the field. *Advances in Animal Biosciences*, *8*(2), 238–243. https://doi.org/10.1017/S2040470017001248

Malambo, L., & Heatwole, C. D. (2020). Automated training sample definition for seasonal burned area mapping. *ISPRS Journal of Photogrammetry and Remote Sensing*, *160*, 107–123. https://doi.org/10.1016/j.isprsjprs.2019.11.026

Malambo, L., Popescu, S. C., Horne, D. W., Pugh, N. A., & Rooney, W. L. (2019). Automated detection and measurement of individual sorghum panicles using density-based clustering of terrestrial lidar data. *ISPRS Journal of Photogrammetry and Remote Sensing*, *149*, 1–13. https://doi.org/10.1016/j.isprsjprs.2018.12.015

Mathew, M. P., & Mahesh, T. Y. (2022). Leaf-based disease detection in bell pepper plant using YOLO v5. *Signal, Image and Video Processing*, *16*(3), 841–847. https://doi.org/10.1007/s11760-021-02024-y

Meacham-Hensold, K., Fu, P., Wu, J., Serbin, S., Montes, C. M., Ainsworth, E., Guan, K., Dracup, E., Pederson, T., Driever, S., & Bernacchi, C. (2020). Plot-level rapid screening for photosynthetic parameters using proximal hyperspectral imaging. *Journal of Experimental Botany*, *71*(7), 2312–2328. https://doi.org/10.1093/jxb/eraa068

Medic, T., Bömer, J., & Paulus, S. (2023). Challenges and recommendations for 3D plant phenotyping in agriculture using terrestrial lasers scanners. In *ISPRS Annals of the Photogrammetry, Remote Sensing and Spatial Information Sciences*, *X-1-W1-2023*, 1007–1014. ISPRS Geospatial Week 2023-2-7 September 2023, Cairo, Egypt. https://doi.org/10.5194/isprs-annals-X-1-W1-2023-1007-2023

Moghimi, A., Yang, C., & Marchetto, P. M. (2018). Ensemble feature selection for plant phenotyping: A journey from hyperspectral to multispectral imaging. *IEEE Access*, *6*, 56870–56884. https://doi.org/10.1109/ACCESS.2018.2872801

Mortensen, A. K., Bender, A., Whelan, B., Barbour, M. M., Sukkarieh, S., Karstoft, H., & Gislum, R. (2018). Segmentation of lettuce in coloured 3D point clouds for fresh weight estimation. *Computers and Electronics in Agriculture, 154*, 373–381. https://doi.org/10.1016/j.compag.2018.09.010

Naik, H. S., Zhang, J., Lofquist, A., Assefa, T., Sarkar, S., Ackerman, D., Singh, A., Singh, A. K., & Ganapathysubramanian, B. (2017). A real-time phenotyping framework using machine learning for plant stress severity rating in soybean. *Plant Methods, 13*(1), 23. https://doi.org/10.1186/s13007-017-0173-7

Ngugi, H. N., Ezugwu, A. E., Akinyelu, A. A., & Abualigah, L. (2024). Revolutionizing crop disease detection with computational deep learning: A comprehensive review. *Environmental Monitoring and Assessment, 196*(3), 302. https://doi.org/10.1007/s10661-024-12454-z

Nguyen, T. T., Hoang, T. D., Pham, M. T., Vu, T. T., Nguyen, T. H., Huynh, Q.-T., & Jo, J. (2020). Monitoring agriculture areas with satellite images and deep learning. *Applied Soft Computing, 95*, 106565. https://doi.org/10.1016/j.asoc.2020.106565

Nguyen, T. T., Slaughter, D. C., Maloof, J. N., & Sinha, N. (2016). *Plant Phenotyping Using Multi-View Stereo Vision with Structured Lights* (vol. 9866, p. 986608). Autonomous Air and Ground Sensing Systems for Agricultural Optimization and Phenotyping. https://doi.org/10.1117/12.2229513

Pandey, P., Ge, Y., Stoerger, V., & Schnable, J. C. (2017). High throughput in vivo analysis of plant leaf chemical properties using hyperspectral imaging. *Frontiers in Plant Science, 8*. https://doi.org/10.3389/fpls.2017.01348

Panno, S., Davino, S., Caruso, A. G., Bertacca, S., Crnogorac, A., Mandić, A., . . . & Matić, S. (2021). A review of the most common and economically important diseases that undermine the cultivation of tomato crop in the mediterranean basin. *Agronomy, 11*(11), 2188.

Panthee, D. R., Piotrowski, A., & Ibrahem, R. (2017). Mapping quantitative trait loci (QTL) for resistance to late blight in tomato. *International Journal of Molecular Sciences, 18*(7), 1589. https://doi.org/10.3390/ijms18071589

Patle, K. S., Saini, R., Kumar, A., & Palaparthy, V. S. (2022). Field evaluation of smart sensor system for plant disease prediction using LSTM network. *IEEE Sensors Journal, 22*(4), 3715–3725. https://doi.org/10.1109/JSEN.2021.3139988

Paulin, G., & Ivasic-Kos, M. (2023). Review and analysis of synthetic dataset generation methods and techniques for application in computer vision. *Artificial Intelligence Review, 56*(9), 9221–9265. https://doi.org/10.1007/s10462-022-10358-3

Peña, J. M., Torres-Sánchez, J., Castro, A. I. de, Kelly, M., & López-Granados, F. (2013). Weed mapping in early-season maize fields using object-based analysis of unmanned aerial vehicle (UAV) images. *PLoS ONE, 8*(10), e77151. https://doi.org/10.1371/journal.pone.0077151

Perez-Sanz, F., Navarro, P. J., & Egea-Cortines, M. (2017). Plant phenomics: An overview of image acquisition technologies and image data analysis algorithms. *GigaScience, 6*(11), gix092. https://doi.org/10.1093/gigascience/gix092

Pound, M. P., Atkinson, J. A., Townsend, A. J., Wilson, M. H., Griffiths, M., Jackson, A. S., Bulat, A., Tzimiropoulos, G., Wells, D. M., Murchie, E. H., Pridmore, T. P., & French, A. P. (2017). Deep machine learning provides state-of-the-art performance in image-based plant phenotyping. *GigaScience, 6*(10), gix083. https://doi.org/10.1093/gigascience/gix083

Qiu, Q., Sun, N., Bai, H., Wang, N., Fan, Z., Wang, Y., Meng, Z., Li, B., & Cong, Y. (2019). Field-based high-throughput phenotyping for maize plant using 3D LiDAR point cloud generated with a "Phenomobile". *Frontiers in Plant Science, 10*, 554. https://doi.org/10.3389/fpls.2019.00554

Qiu, R., Wei, S., Zhang, M., Li, H., Sun, H., Liu, G., & Li, M. (2018). Sensors for measuring plant phenotyping: A review. *International Journal of Agricultural and Biological Engineering, 11*(2), Article 2. https://doi.org/10.25165/ijabe.v11i2.2696

Rai, N., Zhang, Y., Ram, B. G., Schumacher, L., Yellavajjala, R. K., Bajwa, S., & Sun, X. (2023). Applications of deep learning in precision weed management: A review. *Computers and Electronics in Agriculture, 206*, 107698. https://doi.org/10.1016/j.compag.2023.107698

Rasmussen, C. B., Kirk, K., & Moeslund, T. B. (2022). The challenge of data annotation in deep learning—a case study on whole plant corn silage. *Sensors, 22*(4), Article 4. https://doi.org/10.3390/s22041596

Raza, A., Razzaq, A., Mehmood, S. S., Zou, X., Zhang, X., Lv, Y., & Xu, J. (2019). Impact of climate change on crops adaptation and strategies to tackle its outcome: A review. *Plants, 8*(2), 34. https://doi.org/10.3390/plants8020034

Redden, L. K., & Colgan, M. (2017). *Method for Automatic Phenotype Measurement and Selection* (United States Patent US9658201B2). https://patents.google.com/patent/US9658201B2/en

Richardson, A. D., Hufkens, K., Milliman, T., Aubrecht, D. M., Chen, M., Gray, J. M., Johnston, M. R., Keenan, T. F., Klosterman, S. T., Kosmala, M., Melaas, E. K., Friedl, M. A., & Frolking, S. (2018). Tracking vegetation phenology across diverse North American biomes using PhenoCam imagery. *Scientific Data*, *5*(1), 180028. https://doi.org/10.1038/sdata.2018.28

Rong, J., Zhou, H., Zhang, F., Yuan, T., & Wang, P. (2023). Tomato cluster detection and counting using improved YOLOv5 based on RGB-D fusion. *Computers and Electronics in Agriculture*, *207*, 107741. https://doi.org/10.1016/j.compag.2023.107741

Rui, Z., Zhang, Z., Zhang, M., Azizi, A., Igathinathane, C., Cen, H., Vougioukas, S., Li, H., Zhang, J., Jiang, Y., Jiao, X., Wang, M., Ampatzidis, Y., Oladele, O. I., Ghasemi-Varnamkhasti, M., & Radi, R. (2024). High-throughput proximal ground crop phenotyping systems—A comprehensive review. *Computers and Electronics in Agriculture*, *224*, 109108. https://doi.org/10.1016/j.compag.2024.109108

Sadeghi-Tehran, P., Virlet, N., Ampe, E. M., Reyns, P., & Hawkesford, M. J. (2019). DeepCount: In-field automatic quantification of wheat spikes using simple linear iterative clustering and deep convolutional neural networks. *Frontiers in Plant Science*, *10*. https://doi.org/10.3389/fpls.2019.01176

Saeed, A., Abdel-Aziz, A. A., Mossad, A., Abdelhamid, M. A., Alkhaled, A. Y., & Mayhoub, M. (2023). Smart detection of tomato leaf diseases using transfer learning-based convolutional neural networks. *Agriculture*, *13*(1), Article 1. https://doi.org/10.3390/agriculture13010139

Sagan, V., Maimaitijiang, M., Sidike, P., Eblimit, K., Peterson, K. T., Hartling, S., Esposito, F., Khanal, K., Newcomb, M., Pauli, D., Ward, R., Fritschi, F., Shakoor, N., & Mockler, T. (2019). UAV-based high resolution thermal imaging for vegetation monitoring, and plant phenotyping using ICI 8640 P, FLIR Vue Pro R 640, and thermoMap cameras. *Remote Sensing*, *11*(3), Article 3. https://doi.org/10.3390/rs11030330

Salafian, I. (2017). Development of the end-effector of a cable-driven parallel manipulator for automated crop sensing. *Department of Mechanical and Materials Engineering: Dissertations, Theses, and Student Research*. https://digitalcommons.unl.edu/mechengdiss/128

Sankaran, S., Khot, L. R., Espinoza, C. Z., Jarolmasjed, S., Sathuvalli, V. R., Vandemark, G. J., Miklas, P. N., Carter, A. H., Pumphrey, M. O., Knowles, N. R., & Pavek, M. J. (2015). Low-altitude, high-resolution aerial imaging systems for row and field crop phenotyping: A review. *European Journal of Agronomy*, *70*, 112–123. https://doi.org/10.1016/j.eja.2015.07.004

Schwalbert, R. A., Amado, T., Corassa, G., Pott, L. P., Prasad, P. V. V., & Ciampitti, I. A. (2020). Satellite-based soybean yield forecast: Integrating machine learning and weather data for improving crop yield prediction in southern Brazil. *Agricultural and Forest Meteorology*, *284*, 107886. https://doi.org/10.1016/j.agrformet.2019.107886

Shafiekhani, A., Kadam, S., Fritschi, F., & DeSouza, G. (2017). Vinobot and vinoculer: Two robotic platforms for high-throughput field phenotyping. *Sensors*, *17*(1), 214. https://doi.org/10.3390/s17010214

Sharma, L. K., Bu, H., Franzen, D. W., & Denton, A. (2016). Use of corn height measured with an acoustic sensor improves yield estimation with ground based active optical sensors. *Computers and Electronics in Agriculture*, *124*, 254–262. https://doi.org/10.1016/j.compag.2016.04.016

Shi, Y., Thomasson, J. A., Murray, S. C., Pugh, N. A., Rooney, W. L., Shafian, S., Rajan, N., Rouze, G., Morgan, C. L. S., Neely, H. L., Rana, A., Bagavathiannan, M. V., Henrickson, J., Bowden, E., Valasek, J., Olsenholler, J., Bishop, M. P., Sheridan, R., Putman, E. B., . . . Yang, C. (2016). Unmanned aerial vehicles for high-throughput phenotyping and agronomic research. *PLoS ONE*, *11*(7), e0159781. https://doi.org/10.1371/journal.pone.0159781

Shinoda, R., Kataoka, H., Hara, K., & Noguchi, R. (2023). Transformer-based ripeness segmentation for tomatoes. *Smart Agricultural Technology*, *4*, 100196. https://doi.org/10.1016/j.atech.2023.100196

Shoaib, M., Hussain, T., Shah, B., Ullah, I., Shah, S. M., Ali, F., & Park, S. H. (2022). Deep learning-based segmentation and classification of leaf images for detection of tomato plant disease. *Frontiers in Plant Science*, *13*. https://doi.org/10.3389/fpls.2022.1031748

Shruthi, U., Nagaveni, V., Arvind, C. S., & Sunil, G. L. (2022). Tomato plant disease classification using deep learning architectures: A review. In A. B. Reddy, B. V. Kiranmayee, R. R. Mukkamala, & K. Srujan Raju (Eds.), *Proceedings of Second International Conference on Advances in Computer Engineering and Communication Systems* (pp. 153–169). Springer Nature. https://doi.org/10.1007/978-981-16-7389-4_15

Singh, A., Ganapathysubramanian, B., Singh, A. K., & Sarkar, S. (2016). Machine learning for high-throughput stress phenotyping in plants. *Trends in Plant Science*, *21*(2), 110–124. https://doi.org/10.1016/j.tplants.2015.10.015

Sodhi, P., Vijayarangan, S., & Wettergreen, D. (2017). In-field segmentation and identification of plant structures using 3D imaging. In *2017 IEEE/RSJ International Conference on Intelligent Robots and Systems (IROS)*, 5180–5187. https://doi.org/10.1109/IROS.2017.8206407

Sui, R., Fisher, D. K., & Reddy, K. N. (2012). Cotton yield assessment using plant height mapping system. *Journal of Agricultural Science*, *5*(1), Article 1. https://doi.org/10.5539/jas.v5n1p23

Sun, J., Shi, S., Gong, W., Yang, J., Du, L., Song, S., Chen, B., & Zhang, Z. (2017). Evaluation of hyperspectral LiDAR for monitoring rice leaf nitrogen by comparison with multispectral LiDAR and passive spectrometer. *Scientific Reports*, *7*, 40362. https://doi.org/10.1038/srep40362

Sun, S., Li, C., Paterson, A. H., Adhikari, J., Robertson, J., & Meng, C. (2019, January 1). *Automated Plant Node Detection Using Terrestrial LiDAR Data Under Field Conditions*. https://doi.org/10.13031/aim.201900598

Sun, S., Li, C., Paterson, A. H., Jiang, Y., Xu, R., Robertson, J. S., Snider, J. L., & Chee, P. W. (2018). In-field high throughput phenotyping and cotton plant growth analysis using LiDAR. *Frontiers in Plant Science*, *9*. https://doi.org/10.3389/fpls.2018.00016

Tang, Z., He, X., Zhou, G., Chen, A., Wang, Y., Li, L., & Hu, Y. (2023). A precise image-based tomato leaf disease detection approach using PLPNet. *Plant Phenomics*, *5*, 0042. https://doi.org/10.34133/plantphenomics.0042

Tattaris, M., Reynolds, M. P., & Chapman, S. C. (2016). A direct comparison of remote sensing approaches for high-throughput phenotyping in plant breeding. *Frontiers in Plant Science*, *7*. https://doi.org/10.3389/fpls.2016.01131

Thomas, S., Behmann, J., Steier, A., Kraska, T., Muller, O., Rascher, U., & Mahlein, A.-K. (2018). Quantitative assessment of disease severity and rating of barley cultivars based on hyperspectral imaging in a non-invasive, automated phenotyping platform. *Plant Methods*, *14*(1), 45. https://doi.org/10.1186/s13007-018-0313-8

Thompson, A. L., Conrad, A., Conley, M. M., Shrock, H., Taft, B., Miksch, C., Mills, T., & Dyer, J. M. (2018). Professor: A motorized field-based phenotyping cart. *HardwareX*, *4*, e00025. https://doi.org/10.1016/j.ohx.2018.e00025

Thompson, A. L., Thorp, K. R., Conley, M., Andrade-Sanchez, P., Heun, J. T., Dyer, J. M., & White, J. W. (2018). Deploying a proximal sensing cart to identify drought-adaptive traits in upland cotton for high-throughput phenotyping. *Frontiers in Plant Science*, *9*. https://doi.org/10.3389/fpls.2018.00507

Todo, C., Ikeno, H., Yamase, K., Tanikawa, T., Ohashi, M., Dannoura, M., Kimura, T., & Hirano, Y. (2021). Reconstruction of conifer root systems mapped with point cloud data obtained by 3D laser scanning compared with manual measurement. *Forests*, *12*(8), Article 8. https://doi.org/10.3390/f12081117

Tong, Y.-S., Lee, T.-H., & Yen, K. S. (2022). Deep learning for image-based plant growth monitoring: A review. *International Journal of Engineering and Technology Innovation*, *12*, 225–246. https://doi.org/10.46604/ijeti.2022.8865

Virlet, N., Sabermanesh, K., Sadeghi-Tehran, P., & Hawkesford, M. J. (2016). Field Scanalyzer: An automated robotic field phenotyping platform for detailed crop monitoring. *Functional Plant Biology*, *44*(1), 143–153. https://doi.org/10.1071/FP16163

Wang, H., Magagi, R., Goïta, K., Trudel, M., McNairn, H., & Powers, J. (2019). Crop phenology retrieval via polarimetric SAR decomposition and Random Forest algorithm. *Remote Sensing of Environment*, *231*, 111234. https://doi.org/10.1016/j.rse.2019.111234

Wang, H., Zhang, W., Yang, G., Lei, L., Han, S., Xu, W., Chen, R., Zhang, C., & Yang, H. (2023). Maize ear height and ear–plant height ratio estimation with LiDAR data and vertical leaf area profile. *Remote Sensing*, *15*(4), Article 4. https://doi.org/10.3390/rs15040964

Wang, X., Singh, D., Marla, S., Morris, G., & Poland, J. (2018). Field-based high-throughput phenotyping of plant height in sorghum using different sensing technologies. *Plant Methods*, *14*(1), 53. https://doi.org/10.1186/s13007-018-0324-5

Xiang, L., Gai, J., Bao, Y., Yu, J., Schnable, P. S., & Tang, L. (2023). Field-based robotic leaf angle detection and characterization of maize plants using stereo vision and deep convolutional neural networks. *Journal of Field Robotics*, *40*(5), 1034–1053. https://doi.org/10.1002/rob.22166

Xiang, L., & Wang, D. (2023). A review of three-dimensional vision techniques in food and agriculture applications. *Smart Agricultural Technology*, *5*, 100259. https://doi.org/10.1016/j.atech.2023.100259

Xiao, S., Fei, S., Ye, Y., Xu, D., Xie, Z., Bi, K., Guo, Y., Li, B., Zhang, R., & Ma, Y. (2024). 3D reconstruction and characterization of cotton bolls in situ based on UAV technology. *ISPRS Journal of Photogrammetry and Remote Sensing*, *209*, 101–116. https://doi.org/10.1016/j.isprsjprs.2024.01.027

Xie, C., & Yang, C. (2020). A review on plant high-throughput phenotyping traits using UAV-based sensors. *Computers and Electronics in Agriculture*, *178*, 105731. https://doi.org/10.1016/j.compag.2020.105731

Xiong, H., Cao, Z., Lu, H., Madec, S., Liu, L., & Shen, C. (2019). TasselNetv2: In-field counting of wheat spikes with context-augmented local regression networks. *Plant Methods*, *15*(1), 150. https://doi.org/10.1186/s13007-019-0537-2

Xu, R., & Li, C. (2022). A review of high-throughput field phenotyping systems: Focusing on ground robots. *Plant Phenomics*, *2022*, 2022/9760269. https://doi.org/10.34133/2022/9760269

Xu, Z., York, L. M., Seethepalli, A., Bucciarelli, B., Cheng, H., & Samac, D. A. (2022). Objective phenotyping of root system architecture using image augmentation and machine learning in alfalfa (Medicago sativa L.). *Plant Phenomics*, *2022*. https://doi.org/10.34133/2022/9879610

Yang, C. (2020). Remote sensing and precision agriculture technologies for crop disease detection and management with a practical application example. *Engineering*, *6*(5), 528–532. https://doi.org/10.1016/j.eng.2019.10.015

Yang, Z., Yang, W., Yi, J., & Liu, R. (2024). WH-DETR: An efficient network architecture for wheat spike detection in complex backgrounds. *Agriculture*, *14*(6), Article 6. https://doi.org/10.3390/agriculture14060961

Yu, X., Yin, D., Xu, H., Pinto Espinosa, F., Schmidhalter, U., Nie, C., Bai, Y., Sankaran, S., Ming, B., Cui, N., Wu, W., & Jin, X. (2024). Maize tassel number and tasseling stage monitoring based on near-ground and UAV RGB images by improved YoloV8. *Precision Agriculture*, *25*(4), 1800–1838. https://doi.org/10.1007/s11119-024-10135-y

Yuan, W., Li, J., Bhatta, M., Shi, Y., Baenziger, P. S., & Ge, Y. (2018). Wheat height estimation using LiDAR in comparison to ultrasonic sensor and UAS. *Sensors*, *18*(11), Article 11. https://doi.org/10.3390/s18113731

Zaji, A., Liu, Z., Xiao, G., Bhowmik, P., Sangha, J. S., & Ruan, Y. (2023). AutoOLA: Automatic object level augmentation for wheat spikes counting. *Computers and Electronics in Agriculture*, *205*, 107623. https://doi.org/10.1016/j.compag.2023.107623

Zarco-Tejada, P. J., González-Dugo, V., & Berni, J. A. J. (2012). Fluorescence, temperature and narrow-band indices acquired from a UAV platform for water stress detection using a micro-hyperspectral imager and a thermal camera. *Remote Sensing of Environment*, *117*, 322–337. https://doi.org/10.1016/j.rse.2011.10.007

Zhang, C., & Kovacs, J. M. (2012). The application of small unmanned aerial systems for precision agriculture: A review. *Precision Agriculture*, *13*(6), 693–712. https://doi.org/10.1007/s11119-012-9274-5

Zhang, C., Marzougui, A., & Sankaran, S. (2020). High-resolution satellite imagery applications in crop phenotyping: An overview. *Computers and Electronics in Agriculture*, *175*, 105584. https://doi.org/10.1016/j.compag.2020.105584

Zhang, H., Wang, L., Jin, X., Bian, L., & Ge, Y. (2023). High-throughput phenotyping of plant leaf morphological, physiological, and biochemical traits on multiple scales using optical sensing. *The Crop Journal*, *11*(5), 1303–1318. https://doi.org/10.1016/j.cj.2023.04.014

Zhang, J., Ma, D., Wei, X., & Jin, J. (2023). Visible and near-infrared hyperspectral diurnal variation calibration for corn phenotyping using remote sensing. *Remote Sensing*, *15*(12), Article 12. https://doi.org/10.3390/rs15123057

Zhang, S., & Zhang, C. (2023). Modified U-Net for plant diseased leaf image segmentation. *Computers and Electronics in Agriculture*, *204*, 107511. https://doi.org/10.1016/j.compag.2022.107511

Zhang, Z., He, W., Wu, F., Quesada, L., & Xiang, L. (2024). Development of A bionic hexapod robot with adaptive gait and clearance for enhanced agricultural field scouting. In *2024 Anaheim, California July 28–31, 2024*. https://doi.org/10.13031/aim.202400650

Zhang, Z., Kayacan, E., Thompson, B., & Chowdhary, G. (2020). High precision control and deep learning-based corn stand counting algorithms for agricultural robot. *Autonomous Robots*, *44*(7), 1289–1302. https://doi.org/10.1007/s10514-020-09915-y

Zhao, J., Kechasov, D., Rewald, B., Bodner, G., Verheul, M., Clarke, N., & Clarke, J. L. (2020). Deep learning in hyperspectral image reconstruction from single RGB images—a case study on tomato quality parameters. *Remote Sensing*, *12*(19), Article 19. https://doi.org/10.3390/rs12193258

Zhou, L., Gu, X., Cheng, S., Yang, G., Shu, M., & Sun, Q. (2020). Analysis of plant height changes of lodged maize using UAV-LiDAR data. *Agriculture*, *10*(5), Article 5. https://doi.org/10.3390/agriculture10050146

Zhou, Z., Majeed, Y., Diverres Naranjo, G., & Gambacorta, E. M. T. (2021). Assessment for crop water stress with infrared thermal imagery in precision agriculture: A review and future prospects for deep learning applications. *Computers and Electronics in Agriculture*, *182*, 106019. https://doi.org/10.1016/j.compag.2021.106019

Zhu, S., Ma, W., Lu, J., Ren, B., Wang, C., & Wang, J. (2023). A novel approach for apple leaf disease image segmentation in complex scenes based on two-stage DeepLabv3+ with adaptive loss. *Computers and Electronics in Agriculture*, *204*, 107539. https://doi.org/10.1016/j.compag.2022.107539

4 QTL Mapping, Molecular Markers Development, and Their Application in Tomato Crop Improvement

Muhammad Irfan Siddique, Jianbo Zhang, and Dilip R. Panthee

4.1 INTRODUCTION

The tomato, *Solanum lycopersicum* (native to the Andes in South America), was domesticated in Mexico and introduced to Europe in the 16th century. Since then, it has become the world's second-most cultivated vegetable crop (Paran and Van Der Knaap 2007; Du et al. 2025). Tomatoes are generally classified into two types: fresh-market and processing. FAO statistics indicate that approximately 190 million tons are produced annually, with around 39 million tons dedicated to processing (FAO FAOSTAT 2022). The United States ranks as the second-largest tomato producer in the world, after China, with an approximate total output of 10.4 million tons (FAO FAOSTAT 2022). The primary objective of tomato breeders is to develop tomatoes with optimal fruit quality, uniform fruit color, extended shelf life, and resistance to diseases and pests. However, in recent years, consumers have grown more critical of the diminished flavor of tomatoes. Therefore, improving fruit quality has become an urgent and challenging task for tomato breeders, as well as maintaining disease resistance and ensuring high yields (Du et al. 2025). Furthermore, tomato production, especially for fresh-market tomatoes, is labor-intensive. Developing new tomato varieties that are suitable for simplified cultivation practices becomes increasingly crucial as labor costs rise. Achieving these breeding goals depends on understanding the genetic foundations and molecular mechanisms underlying these traits (Du et al. 2025).

The phenotypic variation observed in many traits of agricultural and evolutionary importance is quantitative, resulting from the combined effects of multiple segregating loci. These loci can interact with each other and environmental factors, making it especially difficult to understand the genetic architecture and molecular basis of these traits. (Falconer 1996; Grandillo and Cammareri 2016). Before molecular markers became available, the genetics of complex traits was analyzed broadly through "quantitative genetics" (Mather 1941), with no insight into the number or location of the underlying genes, which (Mather and Jinks 2013) referred to as polygenes. The foundational milestones for mapping polygenes date back to 1923, when Sax linked seed size in beans, a trait inherited quantitatively, to seed-coat pigmentation, a discrete monogenic trait (Sax 1923). Later, Thoday (1961) developed the core method for using marker genes in segregating populations to locate and analyze individual polygenes. A researcher then coined the term "quantitative trait locus" (QTL) to refer to a genetic locus where different alleles segregate and significantly influence a polygenic trait (Geldermann 1975; Grandillo and Cammareri 2016).

However, in the 1980s, when isozyme markers began to be widely used for QTL analysis in tomato (*Solanum lycopersicum*) (Tanksley et al. 1982; Vallejos and Tanksley 1983; Weller et al. 1988) and in maize (Edwards et al. 1987), this marked the start and became the foundation of

DOI: 10.1201/9781003506539-4

modern QTL mapping. Many factors affect the ability to detect QTL, such as trait heritability, gene action, the type of mapping population, marker coverage, the number and effects of QTL, and the distance between marker loci and QTL influencing the trait (Tanksley 1993; Mackay et al. 2009; Grandillo and Cammareri 2016). Early tomato QTL mapping primarily used morphological and isozyme markers in F_2 and backcross populations, such as BC_1. While various plant and fruit traits were studied, the limited number of informative isozyme markers made it difficult to accurately scan the entire tomato genome for QTLs, preventing precise localization. This limitation was later addressed by the development of DNA-based genetic markers, starting with restriction fragment length polymorphisms (RFLPs) (Grandillo and Cammareri 2016).

In 1988, Paterson and colleagues conducted pioneering research by creating a detailed RFLP linkage map using 63 RFLPs and applying suitable statistical methods to a tomato BC_1 interspecific population (Paterson et al. 1988). This method enabled them to map and study QTLs, demonstrating that complex traits could be broken down into single Mendelian factors. Since then, the number of RFLP markers used in tomato genetics has grown to approximately 1,000 (Tanksley et al. 1992). Various segregating populations and mapping strategies have been used for this purpose. A crucial factor for QTL mapping populations is sufficient polymorphism at marker loci and in genes linked to the trait(s) in question. Due to multiple genetic bottlenecks in tomato domestication and breeding, as well as in other self-pollinating crops, the genotypic diversity within cultivated germplasm remains limited (Miller and Tanksley 1990; Blanca et al. 2012). This restriction has prompted tomato geneticists and breeders to utilize the abundant genetic diversity found in unadapted germplasm for creating mapping populations and advancing breeding efforts (Rick 1982; Bai and Lindhout 2007; Grandillo and Cammareri 2016).

A decade ago, after the tomato genome was sequenced, the advent of next-generation sequencing and high-throughput genotyping platforms made more marker sources available, thereby making QTL mapping easier and more efficient. Moreover, Single Nucleotide Polymorphisms (SNPs) arrays enable large-scale parallel genotyping of entire genomes (Sim et al. 2012b; Víquez-Zamora et al. 2013). Similarly, advanced high-throughput genotyping methods have enabled the creation of high-density linkage maps, allowing for the identification of QTLs associated with complex traits. New methodological approaches have enabled successful applications in dissecting complex traits and mapping QTLs in tomatoes (Sauvage et al. 2014; Pascual et al. 2015). These developments facilitate the more effective use of *S. lycopersicum* germplasm in breeding efforts. However, avoiding the linkage drag associated with functional loci remains a significant challenge for tomato breeders and geneticists. To date, several reviews have discussed the current state of QTL mapping in tomatoes (Foolad et al. 2007; Grandillo et al. 2011, 2013; Grandillo and Cammareri 2016). Therefore, this chapter focuses on the overall status of QTL mapping, marker development, and their application in tomato crop improvement briefly under recent advances.

4.2 BASICS OF QTL MAPPING

Quantitative Trait Loci (QTLs) are specific regions in the genome that influence variation in quantitative traits, which are controlled by multiple genes and affected by environmental factors, such as crop yield, plant height, drought tolerance, or disease resistance. Unlike qualitative characteristics, which follow simple Mendelian inheritance, quantitative traits exhibit continuous variation and are typically controlled by multiple loci with minor to moderate effects. Identifying QTLs enables researchers and breeders to understand the genetic basis of complex traits and pinpoint the genomic regions associated with phenotypic differences. This is essential for marker-assisted selection (MAS) and genomic selection, which accelerate crop improvement by enabling the selection of beneficial alleles without relying solely on phenotype (Collard et al. 2005). Additionally, QTL mapping enhances our understanding of gene-environment interactions and the mechanisms underlying stress tolerance, thereby supporting the development of resilient crops in the face of climate change (Xu and Crouch 2008). As a result, QTLs are a vital link between traditional breeding and modern genomics, serving a key role in plant genetics and breeding efforts.

QTL mapping in plants is a genetic technique used to identify chromosomal regions associated with variations in quantitative traits. It involves correlating phenotypic data (trait measurements) with genotypic data (molecular marker profiles) in a segregating population such as F_2, backcross, recombinant inbred lines (RILs), or doubled haploids. The process begins with developing a mapping population from parents who differ in the interest trait. This population is then genotyped using DNA markers, such as SSRs or SNPs, and phenotyped under relevant environmental conditions. Statistical methods, such as interval mapping or composite interval mapping, are used to identify QTLs by scanning for markers significantly associated with the trait, estimating their locations, effects, and modes of gene action (additive or dominant) (Lander and Botstein 1989). QTL mapping relies on recombination events within the population to connect phenotype and genotype. Its accuracy depends on population size, trait heritability, marker density, and environmental stability. In plants, QTL mapping has been crucial for understanding the genetic basis of complex traits such as flowering time, disease resistance, and stress tolerance, aiding marker-assisted breeding and gene cloning efforts (Collard et al. 2005; Xu and Crouch 2008). Figure 4.1 depicts the overall concept of QTL mapping and its applications in tomato crop improvement.

QTL mapping in plants depends on well-designed populations that show genetic recombination and segregation for the traits of interest. Each common population type has its benefits and limitations. F_2 populations are created by selfing the F_1 hybrid of two contrasting parents. They are quick, easy to produce, and ideal for initial QTL detection, but they are genetically diverse and mainly suitable for short-term studies. Backcross (BC) populations are generated by crossing the F_1 with a parent, which helps identify dominant QTLs and is often used in fine mapping. Recombinant Inbred Lines (RILs), produced through repeated selfing or single seed descent over many generations starting from F_2, become homozygous and can be tested across different environments, making them highly reliable for QTL and gene-environment interaction studies (Broman 2005). Near-isogenic lines (NILs) carry a small genomic segment from a donor parent within the background of a recurrent parent, created through repeated backcrossing. NILs are particularly useful for validating QTL effects and precise mapping because they reduce background variation from other loci (Eshed and Zamir 1995). Additional populations, such as doubled haploids (DHs) and advanced backcrosses, are also employed for high-resolution mapping and breeding. The choice of population depends

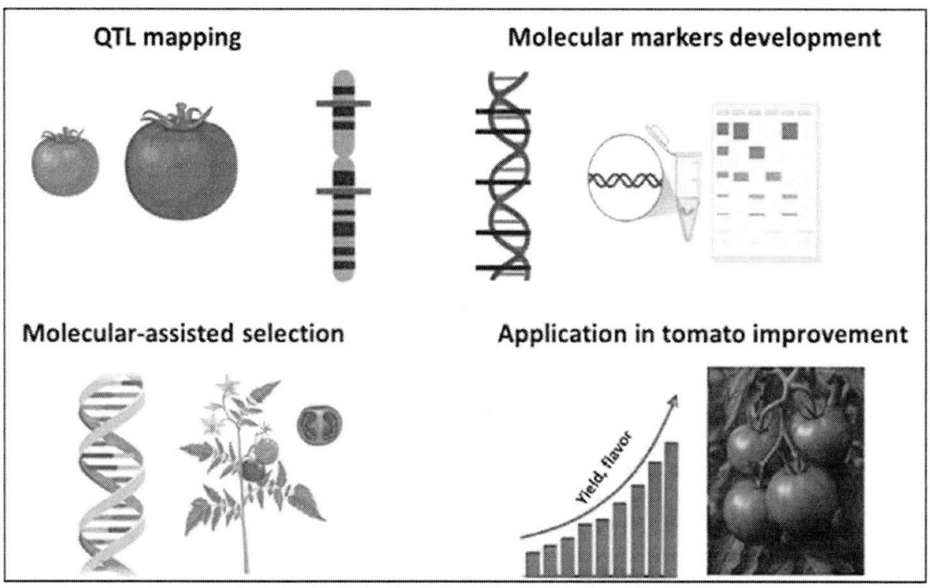

FIGURE 4.1 QTL mapping, molecular marker development, and their application in tomato crop improvement.

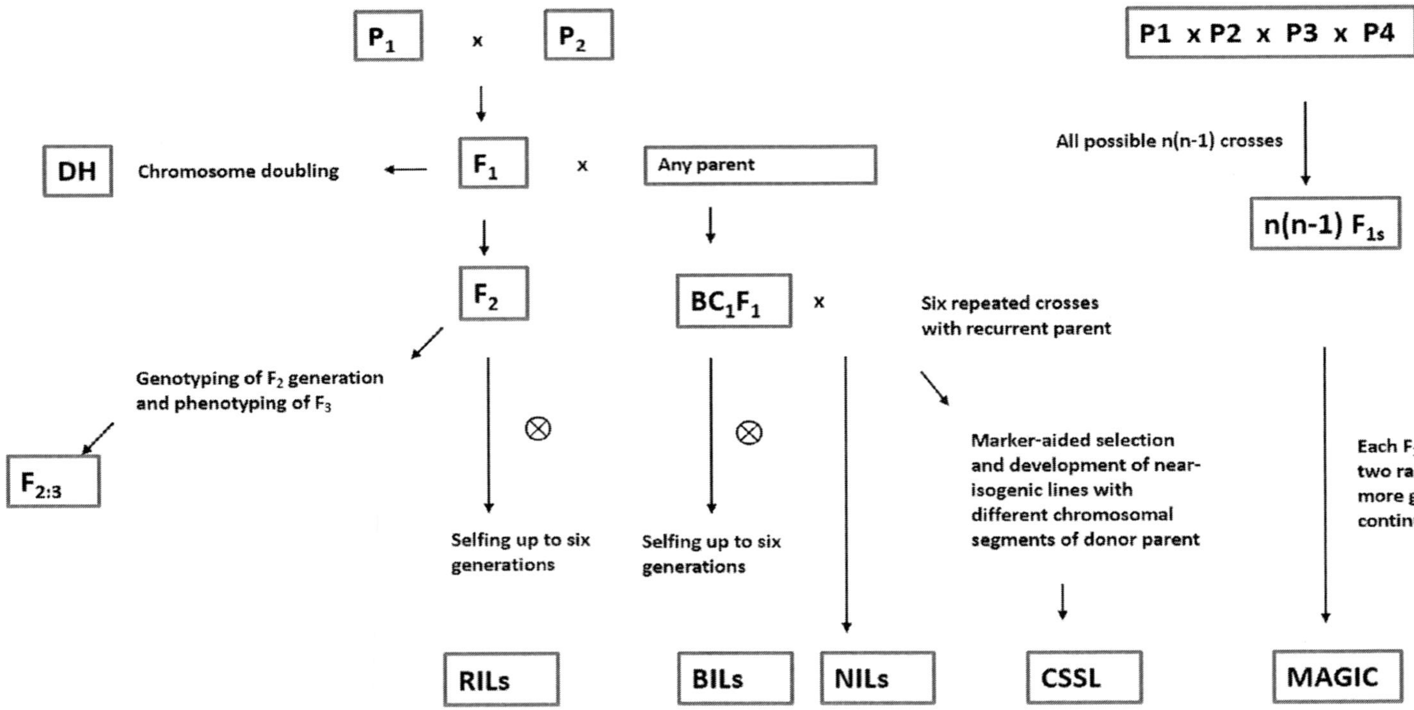

FIGURE 4.2 Mapping populations for QTL mapping in plant breeding. Different kinds of bi-parental mapping populations and their scheme of development. (RILs recombinant inbred lines; NILs near isogenic lines; BILs backcross inbred lines; DH doubled haploid; CSSLs chromosome segment substitution lines; MAGIC Multi-parent Advanced Generation Inter-Cross population)

on the genetic complexity of the trait, the required resolution, and the availability of resources. Figure 4.2 shows an overview of the mapping population and the development process used in QTL mapping.

Linkage mapping, also known as classical QTL mapping, is one of the earliest and most common methods used to identify genomic regions associated with plant quantitative traits. It relies on the idea that genes or markers located close together on a chromosome tend to be inherited together because of limited recombination. In QTL linkage mapping, a bi-parental population (such as F_2, backcross, RILs, or NILs) is genotyped with molecular markers like SSRs, SNPs, or RFLPs. The same population is also phenotyped for the trait of interest. A linkage map is created using JoinMap, MapMaker, or R/qtl, which calculates recombination frequencies and arranges markers along chromosomes. After the map is constructed, statistical techniques like interval mapping, composite interval mapping (CIM), or MQM scan the genome and identify significant marker–trait links.

Statistical methods are crucial in QTL mapping, as they quantify the relationship between genetic markers and trait variation within a population. The earliest method, single-marker analysis, tests each marker separately for trait association but doesn't consider recombination distances and can produce biased results (Xu 2013). Interval mapping (IM), introduced by Lander and Botstein (1989), offers a more precise approach by estimating the likelihood of a QTL at different positions between markers, using models of recombination and trait variance via maximum likelihood or regression. Composite interval mapping (CIM) builds on IM by including background marker cofactors, improving resolution, and reducing confounding effects (Zeng 1994). Likewise, multiple QTL mapping (MQM) and Bayesian methods enable the simultaneous analysis of various QTLs, revealing interactions (epistasis) and complex trait architectures (Kao et al. 1999; Yi et al. 2003). These models are implemented in widely used tools, such as R/qtl (Broman et al. 2003), Windows QTL Cartographer (Wang 2006), MapQTL, and QGene, which facilitate LOD analysis, effect size estimation, and confidence interval determination. Such tools are vital for understanding the genetic basis of quantitative traits and form the core of effective QTL mapping in plant studies. Table 4.1 presents the commonly used QTL tools for bi-parental QTL analysis.

TABLE 4.1
Overview of the Tools Used to Perform QTL Mapping

Tool Name	Reference
R/qtl	Broman et al. (2003)
R/qtl2	Broman et al. (2019)
MapQTL	Van Ooijen (2004)
QTL IciMapping (ICIM)	Meng et al. (2015)
QTL Cartographer	Zeng (1994)
WinQTL Cartographer	Wang (2006)
PLABQTL	Utz and Melchinger (1996)
QGene	Nelson (1997)
MQTL (Meta-QTL software)	Veyrieras et al. (2007)
TASSEL	Bradbury et al. (2007)
QTLNetwork/QTXNetwork	Yang et al. (2008)
QTLbase	Zheng et al. (2020)
QTLViewer	Vincent et al. (2022)
Multimapper	Sillanpaa (1998)
The QTL Café	Manly and Olson (1999)
Epistat	Chase et al. (1997)

Linkage mapping's strength is establishing a high-confidence link between a QTL and a specific chromosomal region within controlled, structured populations. Nonetheless, its resolution is often constrained by the number of recombination events in the population, leading to broad QTL intervals. This limitation can be alleviated by increasing the population size or utilizing advanced lines, such as RILs or NILs. Standard tools such as Windows QTL Cartographer, MapQTL, and QGene are used for statistical analysis and visualizing LOD (logarithm of odds) scores across the genome. Despite the rise of genome-wide association studies (GWAS), linkage mapping remains essential, particularly for species with limited natural diversity or traits that are heavily influenced by environmental factors. It provides a solid foundation for marker-assisted selection (MAS) and facilitates the identification of candidate genes through fine-mapping and positional cloning.

4.3 MOLECULAR MARKERS IN TOMATO

Molecular markers are vital tools in contemporary plant genetics and breeding. They allow researchers to detect genetic variation, build genetic maps, identify genes or QTLs, and speed up breeding via marker-assisted selection (MAS). In tomato (*Solanum lycopersicum*), molecular markers have played a crucial role in elucidating the genetic basis of key traits, including disease resistance, fruit quality, and tolerance to abiotic stress. Sequencing the tomato genome and related resources has further improved the use of molecular markers for functional genomics and precise breeding. This overview covers the types of molecular markers used in tomato, their applications, and the available high-throughput genotyping platforms and databases.

The initial wave of molecular markers in tomato consisted of Restriction Fragment Length Polymorphisms (RFLPs), which identify genetic variation through differences in restriction enzyme cleavage patterns. RFLPs were among the first markers used to construct the tomato genetic linkage map, helping to locate QTLs associated with yield, ripening, and disease resistance (Tanksley et al. 1992). Despite their usefulness, RFLPs are labor-intensive, demanding large DNA quantities and radioactive labels.

Random Amplified Polymorphic DNA (RAPD) markers, introduced later, utilize short, arbitrary primers in PCR to amplify random parts of the genome. Although quick and cost-effective, RAPDs face issues with low reproducibility and being dominant markers, which restricts their use in detailed genetic research. On the other hand, Amplified Fragment Length Polymorphisms (AFLPs) offer improved reproducibility and higher resolution by combining restriction enzyme digestion with selective PCR amplification. AFLPs have been extensively used in tomato for genetic diversity assessments, mapping, and cultivar identification (Zhang and Stommel 2001), but they remain more technically challenging than PCR-based markers.

Simple Sequence Repeats (SSRs), or microsatellites, are short tandem repeats of 1 to 6 base pairs. They are highly polymorphic and exhibit codominance. These features offer advantages over earlier marker methods, such as improved reproducibility, greater abundance, and increased variability. SSRs have been extensively used in tomato genetic mapping, analyzing population structure, and marker-assisted selection (MAS). The discovery of thousands of SSR markers from expressed sequence tags (ESTs) and genomic libraries has broadened their application in cultivated and wild tomato species (Areshchenkova and Ganal 2002). These markers are particularly valuable in breeding programs for tracking introgressed genes from wild relatives such as *Solanum pimpinellifolium* and *S. pennellii*.

Insertion-deletion polymorphisms (Indels) are another genetic variation that can be easily identified through simple PCR tests. These Indels are common throughout the genome and are easily detectable via gel electrophoresis. Indels have been successfully employed in tomatoes to differentiate closely related lines and verify candidate gene polymorphisms (Shirasawa et al. 2010). Other PCR-based markers like Cleaved Amplified Polymorphic Sequences (CAPS) and Derived Cleaved Amplified Polymorphic Sequences (dCAPS) effectively identify known SNPs through restriction

TABLE 4.2
Overview of Molecular Markers and Their Characteristics Used in QTL Mapping

Marker System	Co-Dominant/ Dominant	Reproducibility	Polymorphism Level	Required DNA Quality	Required DNA Quantity	Marker Index	Genome Abundance	Cost	Sequencing	PCR Requirement	Visualization	Required DNA (ng)
RFLP	Co-Dominant	High	Medium	High	High	Low	High	High	Yes	No	Radioactive	10,000
RAPD	Dominant	High	Very High	High	Medium	High	Very High	Less	No	Yes	Agarose gel	20
AFLP	Dominant	Intermediate	High	High	Low	Medium	Very High	High	No	Yes	Agarose gel	500–1000
ISSR	Dominant	Medium–High	High	Low	Low	Medium	Medium	High	No	Yes	Agarose gel	50
SSR	Co-Dominant	High	High	Low	Low	Medium	Medium	High	Yes	Yes	Agarose gel	50
DArT	Dominant	High	High	High	Low	High	Very High	Cheapest	Yes	No	Microarray	50–100
SNP	Co-Dominant	High	High	High	Low	High	Very High	Variable	Yes	Yes	SNP-Plate Form	50

enzyme digestion. These markers are particularly valuable when high-throughput SNP data are translated into inexpensive assays for routine genetic analysis.

Single-nucleotide polymorphisms (SNPs) are the most common and informative molecular markers in the tomato genome. The reference tomato genome (Tomato Genome Consortium 2012) has enabled the identification of millions of SNPs through resequencing efforts. These SNPs are codominant, highly abundant, and suitable for automation and high-throughput genotyping. Similar to the SolCAP Tomato SNP array, which features 7,720 markers, SNP arrays have been developed for genome-wide association studies (GWAS), genomic selection, and high-resolution mapping (Sim et al. 2012a). Furthermore, genotyping-by-sequencing (GBS) and whole-genome resequencing enable genome-wide SNP detection across diverse tomato germplasm. SNP markers have transformed tomato genetics by enabling fine mapping, allele mining, and rapid trait-based selection. Table 4.2 shows the markers and their features in QTL mapping.

Recent progress in sequencing technologies has led to the development of high-throughput genotyping platforms in tomato. Approaches like Genotyping-by-Sequencing (GBS) and Restriction-Site Associated DNA sequencing (RAD-seq) are popular next-generation sequencing (NGS) methods that enable simultaneous SNP discovery and genotyping across the entire genome. These techniques provide dense marker coverage, allowing for high-resolution QTL detection in biparental populations and diversity panels (Rani et al. 2023). Additional high-throughput options include KASP (Kompetitive Allele-Specific PCR) assays, cost-effective, flexible, and scalable solutions for genotyping SNPs validated from NGS data (Semagn et al. 2014). These technologies empower breeders to genotype thousands of samples quickly and affordably, accelerating genetic improvements in tomato breeding programs.

Various databases and genomic platforms have been developed to advance molecular marker research in tomatoes. The Sol Genomics Network (SGN) (https://solgenomics.net) is the most comprehensive resource, offering access to the reference genome, marker details, gene models, phenotype data, and breeding tools. It features tools for BLAST searches, primer design, gene expression analysis, and QTL visualization. The Tomato Genome Re-sequencing Project and the Tomato Functional Genomics Database expand these resources by providing genome-wide SNP data, gene annotations, and transcriptomic datasets across multiple tomato accessions (Hirakawa et al. 2013). Moreover, Ensembl Plants and the NCBI Genome Database supply complete genomic assemblies and variant datasets for cultivated and wild tomato species. These platforms enable researchers to combine molecular marker data with functional genomics, facilitating exceptional trait analysis and genome-assisted breeding accuracy.

Molecular markers have revolutionized tomato research and breeding by enabling detailed analysis of genetic variation and trait inheritance. Marker technologies have become more efficient and informative, from early RFLPs and RAPDs to advanced SNP arrays and sequencing-based platforms. SSRs, Indels, and SNPs remain the most versatile tools, with high-throughput platforms like GBS and KASP supporting large-scale genotyping efforts. Alongside extensive genomic databases, such as SGN, molecular markers continue to accelerate gene discovery, QTL mapping, and precision breeding in tomato. As marker technologies advance, they will continue to play a crucial role in addressing challenges such as climate resilience, nutritional enhancement, and yield improvement in tomato breeding programs.

4.4 QTLS IDENTIFIED IN TOMATO FOR KEY TRAITS

Tomato (*Solanum lycopersicum*) is among the most important horticultural crops worldwide. Enhancing its yield, quality, and stress resistance remains a principal objective in plant breeding (Panthee and Gardner 2010). Quantitative trait loci (QTLs) mapping has played a vital role in uncovering the genetic basis of complex traits in tomato. Developing high-density genetic maps, whole-genome resequencing data, and well-designed mapping populations has facilitated the discovery of many QTLs associated with key agronomic traits (Robertson and Labate 2007; Causse and Grandillo 2016). The QTLs identified in tomato related to yield, fruit quality, resistance to biotic and abiotic stresses, and nutritional characteristics, drawing on recent research and notable breakthroughs, have been reviewed by Foolad et al. (2007), Causse and Grandillo (2016), and Grandillo and Cammareri (2016). However, a brief overview of QTLs associated with economically important traits is provided next.

4.4.1 Yield and Fruit Quality Traits

Yield is a complex trait influenced by factors such as fruit size, number, weight, and plant architecture. Various QTLs that affect these traits have been identified using interspecific populations derived from wild species, such as *Solanum pennellii* and *S. pimpinellifolium* (Torgeman and Zamir 2023). The fruit weight QTL *fw2.2* is among the most extensively studied loci, explaining up to 30% of the variation in fruit weight. It was also the first QTL to be cloned in plants, and it encodes a regulator of cell division (Frary et al. 2000). Likewise, *fw3.2*, another significant QTL on chromosome 3, impacts pericarp cell size and has been linked with a cytochrome P450 gene (Chakrabarti et al. 2013).

Fruit quality traits, such as shape, color, firmness, and soluble solids content, are also influenced by multiple genes. QTL analysis identified the *SUN* and *OVATE* genes as being linked to elongated and pear-shaped fruit forms (Xiao et al. 2008). A QTL for soluble solids content (SSC), named *brix9-2-5*, was located on chromosome 9 using *S. pennellii* introgression lines. This QTL increased fruit sweetness without reducing yield (Eshed and Zamir 1995; Tieman et al. 2012). Such QTLs are crucial for breeding programs that aim to optimize productivity and flavor.

4.4.2 Biotic Stress Resistance

Tomato plants face threats from several pathogens, including bacteria, fungi, and viruses. Among soil-borne diseases, a pathogen called *Ralstonia solanacearum* causes bacterial wilt disease, which is crucial. QTL mapping has been instrumental in pinpointing resistance genes sourced from wild species. For bacterial wilt, significant QTLs like *Bwr-12* on chromosome 12 and *Bwr-6* on chromosome 6 have been identified, originating from *S. pimpinellifolium* and *S. habrochaites* (Wang et al. 2013). A recent study identified three QTLs: *qbw-01*, *qbw-02*, and *qbw-03* at chromosomes 1, 2, and 3 (Siddique et al. 2024). Marker-assisted selection targeting these loci has improved resistance traits in breeding programs.

Among fungal pathogens, *Phytophthora infastans*, which causes late blight, is destructive for tomato cultivation. The *Ph-3* gene on chromosome 9 has been extensively used in resistance breeding. It confers broad-spectrum resistance and has been fine-mapped and cloned in resistant accessions of *S. pimpinellifolium* (Zhang et al. 2013). QTL mapping has also revealed additional loci that enhance partial resistance and complement the action of major R-genes. These resistance QTLs are critical for developing cultivars with durable disease resistance under changing pathogen pressures. Similarly, QTLs have been mapped for several other pathogens, which are discussed in detail in Chapters 6, 8, 9, and 10 of this book.

4.4.3 Abiotic Stress Tolerance

Abiotic stress tolerance is becoming increasingly crucial due to climate change and water shortages (Egea et al. 2022). QTL mapping in tomatoes has pinpointed important regions associated with traits such as drought resistance, salinity tolerance, and heat resilience. A study using RILs derived from *S. lycopersicum* × *S. pennellii* identified QTLs on chromosomes 1 and 9 associated with water use efficiency and root architecture (Foolad 2007). In research on salinity stress, QTLs for Na^+ and Cl^- exclusion and fruit yield under salt conditions have been found on chromosomes 6 and 9 (Ali et al. 2021).

Introgression lines derived from wild species, such as *S. cheesmaniae* and *S. chilense*, have introduced drought-resilient alleles. For instance, the gene *SlDREB2*, which responds to dehydration, has been located within a QTL linked to drought tolerance (Zhou et al. 2020). These discoveries provide a genetic foundation for breeding cultivars that sustain yield in suboptimal environmental conditions.

4.4.4 Nutritional and Processing Traits

Nutritional traits, such as carotenoid levels (especially lycopene and β-carotene), vitamin C, and flavonoids, are essential for consumer health and quality processing. QTL mapping has identified several loci linked to these metabolites. The *lcy-e* and *lcy-b* genes affect lycopene content and have been mapped in populations derived from wild species (Gonzali et al. 2009). A key QTL on chromosome 6, called *QTL-Lyc6*, significantly boosts lycopene accumulation.

QTLs linked to ascorbic acid (vitamin C) levels have been identified, especially on chromosomes 2 and 9. Processing traits such as fruit firmness, viscosity, and color retention are affected by loci like *pg2a* (polygalacturonase gene) and *rin* (ripening inhibitor), both of which are widely used in breeding processing tomatoes (Osorio et al. 2012). Nowadays, marker-assisted backcrossing and genomic selection combine these QTLs to enhance quality in processing cultivars.

A key example of a landmark QTL study is the creation of Introgression Lines (ILs) from *S. pennellii* into *S. lycopersicum* by Eshed and Zamir (1995). This work established the groundwork for fine-mapping many QTLs related to yield and fruit quality traits. The IL population has been extensively utilized in QTL mapping for characteristics such as sugar content, flowering time, and metabolite profiles. Recent use of genome-wide association studies (GWAS) and multi-parent advanced generation inter-cross (MAGIC) populations has enabled more precise mapping of complex traits, such as flavor volatiles and disease resistance (Gao et al. 2019). These studies highlight a shift from traditional QTL mapping to more comprehensive genomic approaches, ushering in a new era for tomato genetic enhancement.

4.5 MARKER-ASSISTED SELECTION (MAS) IN TOMATO BREEDING: INTEGRATING TRAIT-LINKED QTLs

Marker-assisted selection (MAS) uses molecular markers closely linked to desirable genes or QTLs to guide breeding choices. It improves accuracy, speed, and efficiency, especially for complex traits that are hard or costly to phenotype, such as disease resistance, abiotic stress tolerance, and fruit quality. In tomato (*Solanum lycopersicum*), MAS has become an essential breeding method due to the availability of high-resolution linkage maps, genome sequences, and extensively studied QTLs (Tiwari et al. 2022).

MAS involves selecting individuals based on specific molecular markers linked to trait loci. It is used for foreground selection (detecting target genes), background selection (reducing donor genome in backcrossing), and pyramiding (combining multiple genes or QTLs) (Hanson et al. 2016). Tomato is especially suitable for MAS because many important QTLs for yield, resistance, and quality have been cloned or precisely mapped, and wild relatives offer a diverse source of advantageous alleles. MAS has demonstrated effectiveness in tomato breeding for traits identified through QTL mapping. For instance, the *fw2.2* QTL, one of the earliest cloned plants, accounts for up to 30% of fruit weight variation and is frequently utilized in breeding programs to enhance fruit size (Frary et al. 2000). Likewise, QTLs such as *fw3.2*, *SUN*, and *OVATE* are targeted via MAS to adjust fruit shape and size, supporting the development of market-specific tomato cultivars (Xiao et al. 2008; Chakrabarti et al. 2013). *Brix9-2-5*, a QTL from *S. pennellii*, increases soluble solids and is commonly used in breeding processing tomatoes to boost flavor and sweetness (Eshed and Zamir 1995b; Tieman et al. 2012). MAS targeting ripening-related genes such as *rin*, *nor*, and *alc* is employed to enhance shelf life and firmness in fresh-market and processing tomatoes. In biotic stress resistance, MAS plays a crucial role in deploying resistance genes, such as Ty-1, Ty-2, and *Ty-3* (which confer resistance to TYLCV), *Ph-3* (for late blight), and *Ve1* (for Verticillium). For example, the *Ty-1* gene was introgressed from *S. chilense* and is commonly identified using gene-specific markers, allowing for durable virus resistance without affecting yield (Ji et al. 2007; Verlaan et al. 2011). Likewise, *Ph-3*, originally from *S. pimpinellifolium*, is now utilized through MAS to provide resistance against late blight (Zhang et al. 2013).

For abiotic stress tolerance, QTLs associated with drought and salinity resistance, such as those on chromosomes 1, 6, and 9 identified from wild species like *S. pennellii* and *S. cheesmaniae*, are incorporated into elite varieties' backgrounds. Markers associated with *SlDREB2*, a transcription factor responsive to drought, facilitate the early identification of stress-tolerant genotypes (Zhou et al. 2020; Ali et al. 2021). These examples demonstrate how the discovery of QTLs directly enhances MAS pipelines, enabling precise trait improvement with reduced linkage drag and faster generation turnover.

4.5.1 Limitations and Future Outlook

Despite its benefits, MAS has several drawbacks and limitations. It depends on closely linked or gene-specific markers, which are often unavailable, especially for polygenic traits with minor effects (Foolad and Panthee 2012). Moreover, the costs of genotyping and the necessary infrastructure can hinder the widespread use of this approach in resource-limited breeding programs. MAS is most effective for major-effect QTLs or single genes. At the same time, traits influenced by multiple loci, such as drought tolerance or flavor, are better suited to genomic selection (GS) approaches (Osei et al. 2018).

Moreover, MAS's success depends on precise phenotypic data and proper marker validation across various genetic backgrounds and environments. Incorrect marker-trait links, caused by linkage disequilibrium or population structure, can lead to selection mistakes (Foolad and Panthee 2012). Nonetheless, ongoing improvements in QTL mapping, marker development methods such as KASP and GBS, and functional genomics enhance the reliability and efficiency of MAS (Tiwari et al. 2022). MAS combined with high-throughput phenotyping, GS, and CRISPR gene editing provides a strong foundation for next-generation tomato breeding. As tomato genetic resources expand, MAS will remain essential for translating genetic insights into tangible breeding improvements (Mishra 2022).

4.6 INTEGRATION OF QTL MAPPING WITH GENOMICS AND BIOINFORMATICS

4.6.1 Use of Genome-Wide Association Studies (GWAS)

Genome-wide association studies (GWAS) revolutionize the identification of loci linked to complex traits by scanning the entire genome for marker-trait associations across diverse germplasm collections (Tibbs Cortes et al. 2021). Unlike traditional QTL mapping, which relies on bi-parental populations, GWAS leverages natural populations or cultivar panels, capturing greater allelic diversity and enabling more precise mapping. In tomato (*Solanum lycopersicum*), GWAS has been extensively applied to study traits such as fruit flavor, yield components, and disease resistance (Harish et al. 2023). For example, a group of researchers conducted a tomato pan-genome GWAS, which revealed new genes and rare alleles affecting fruit flavor and metabolite profiles, thereby providing breeders with access to previously untapped genetic variation (Gao et al. 2019). When combined with high-density SNP genotyping and advanced phenotyping platforms, GWAS allows researchers to pinpoint candidate genes with unprecedented accuracy, facilitating more targeted breeding strategies (Tripodi et al. 2021).

4.6.2 QTL-Seq and Fine Mapping Approaches

QTL-seq is a rapid and cost-effective technique that combines bulked segregant analysis (BSA) with whole-genome resequencing. It has become a valuable tool for accelerating QTL discovery by sequencing two DNA pools representing contrasting phenotypes from segregating populations and identifying genomic regions with significant differences in allele frequencies (Zhang and Panthee 2022). In tomato research, QTL-seq has effectively mapped loci related to disease resistance and fruit quality traits. For example, researchers utilized QTL-seq to efficiently identify candidate regions associated with bacterial wilt resistance. This approach improves upon traditional fine mapping, which narrows QTL regions through recombination analysis with closely spaced markers (Zhao et al. 2021). When combined, QTL-seq and fine mapping accelerate the cloning and functional analysis of causal genes, thereby supporting molecular breeding strategies.

4.6.3 Role of Transcriptomics, Proteomics, and Metabolomics

Integrating QTL mapping with functional genomics technologies like transcriptomics, proteomics, and metabolomics facilitates a comprehensive understanding of traits at the system-level regulation. Transcriptome analysis (RNA-seq) enables the identification of expression quantitative trait loci (eQTLs), which link variations in gene expression to phenotypic traits. For instance, researchers integrated QTL mapping with RNA-seq to pinpoint genes involved in high-temperature stress tolerance in tomato, uncovering candidate transcription factors and stress-related pathways (Wen et al. 2019). Proteomics complements this by analyzing protein levels and post-translational modifications, providing insights into the functional proteins that affect traits. Metabolomics characterizes the biochemical phenotype, helping to identify metabolite QTLs (mQTLs) linked to flavor, nutrition, and stress responses (Khojasteh et al. 2024). These multi-omics strategies provide a robust framework for prioritizing candidate genes within QTL regions and elucidating their biological roles.

4.6.4 CRISPR and Gene Validation of Candidate Loci

Validating candidate genes identified through QTL mapping and omics analyses is crucial for confirming their role and improving traits. *CRISPR-Cas* genome editing has emerged as a precise and highly effective technique for targeted modification of genes in tomatoes (Berman et al. 2025). With CRISPR, researchers can knockout, knock-in, or precisely edit candidate genes within QTL regions to assess their phenotypic impact (Kang et al. 2025). *CRISPR* has validated and modified genes that influence fruit size, ripening, and disease resistance, thereby accelerating functional genomics and breeding projects (Berman et al. 2025). Integrating QTL mapping with genomics and bioinformatics has transformed tomato genetics and breeding (Rajendra 2025). GWAS boosts discovery potential and resolution, while QTL-seq and fine mapping accelerate candidate gene identification. Multi-omics approaches enhance the understanding of the molecular basis of complex traits, and CRISPR gene editing provides unmatched precision for functional validation and trait improvement (Rajendra 2025). Together, these techniques enable more efficient and targeted breeding strategies to develop tomato varieties with higher yields, improved quality, and increased stress tolerance, thereby supporting sustainable agriculture and meeting consumer expectations.

4.7 FUTURE PROSPECTS AND CHALLENGES IN QTL MAPPING AND GENOMICS INTEGRATION FOR TOMATO IMPROVEMENT

4.7.1 Advances in Sequencing and Phenotyping Technologies

The rapid development of next-generation sequencing (NGS) technologies is significantly advancing tomato genetics by facilitating comprehensive genome analysis with reduced costs and shorter timelines (Ullah et al. 2025). Long-read sequencing platforms, such as PacBio and Oxford Nanopore, produce highly contiguous genome assemblies, facilitating the identification of structural variants and complex genomic regions that were previously challenging to resolve (Michael and VanBuren 2020). These innovations enable the construction of a tomato pan-genome, encompassing a broader spectrum of genetic diversity and facilitating the discovery of new QTLs and candidate genes (Benoit et al. 2025).

Along with genomics, high-throughput phenotyping methods that utilize drones, imaging, spectroscopy, and sensor arrays allow for precise, non-destructive, and rapid measurement of complex traits in both field and controlled environments (Araus and Cairns 2014). Integrating phenomics with genomic information enhances the accuracy of QTL identification and genomic prediction by capturing how phenotypes change in response to environmental stresses. In tomato research, automated phenotyping platforms for traits such as fruit size, color, firmness, and metabolite content are becoming more accessible, allowing breeders to analyze complex traits with unmatched detail.

4.7.2 Integrating Multi-Omics Data for Trait Improvement

A future focus in crop breeding and genetics is on integrating multi-omics datasets, including genomics, transcriptomics, proteomics, metabolomics, and epigenomics, to explore the complex molecular networks that control traits (Wang et al. 2024). Combining these datasets helps identify key regulatory genes, pathways, and biomarkers that cannot be detected through QTL mapping alone. For example, integrating metabolomics with transcriptomics has elucidated how flavor volatiles and nutrient levels are regulated in tomato fruits, thereby aiding in selection processes that extend beyond basic phenotypic traits (Rajendra 2025). Furthermore, advancements in machine learning and network biology enhance the integration and analysis of extensive datasets, enabling predictions of gene functions and interaction networks that impact key agricultural traits (van Dijk et al. 2021). Applying single-cell omics and spatial transcriptomics is an emerging area that offers insights into tissue-specific gene regulation during fruit development and stress responses. These data layers help refine QTL intervals and enhance the accuracy of genome editing targets for trait improvement.

4.7.3 Bottlenecks in Translating QTLs to Field-Ready Varieties

Despite technological advances, transforming QTL discoveries into better, field-ready tomato varieties remains challenging (Bonarota et al. 2025). First, many QTLs account for only a small portion of phenotypic variation and interact with the environment (G×E), which complicates their stable expression across different field conditions (Cooper et al. 2023). This limits the success of marker-assisted selection, particularly for complex traits such as drought tolerance or flavor. Second, because most agronomic traits are polygenic, pinpointing all relevant loci with traditional QTL mapping and MAS is challenging. Genomic selection (GS), which utilizes genome-wide markers for prediction, offers an alternative; however, it demands large training populations and high-quality phenotypic data, making it resource-intensive. Third, linkage drag from wild relatives during introgression breeding remains a persistent challenge, frequently introducing undesirable traits and beneficial alleles. Although fine mapping and genome editing can mitigate this, they stay labor-intensive and costly. Finally, challenges in regulatory approval and public acceptance concerning genome editing and GMOs could influence the deployment of edited varieties based on QTL candidate gene validation. Combining QTL mapping with advanced genomics, phenomics, and bioinformatics significantly accelerates the breeding of tomatoes. Progress in sequencing, phenotyping, multi-omics, and machine learning opens new opportunities to understand and improve complex traits. However, tackling challenges such as G×E interactions, polygenic traits, and practical breeding barriers will be crucial to utilize these innovations fully. Continued interdisciplinary collaborations will be essential to turn genomic insights into durable, high-yielding tomato varieties that meet future agricultural and consumer needs.

4.8 CONCLUSION

Looking ahead, the future of tomato breeding depends on seamlessly integrating multi-omics data, high-throughput phenotyping, and advanced genome editing, combined with traditional breeding and molecular techniques such as QTL mapping. By combining these with cutting-edge bioinformatics and machine learning tools, breeders can more accurately predict phenotypes and develop cultivars tailored to specific environments and consumer preferences. Focus will shift toward sustainable breeding methods that enhance yield stability, nutritional quality, and climate resilience. Furthermore, emerging technologies such as single-cell omics and epigenetic profiling will reveal new regulatory layers that influence key traits, opening up fresh avenues for genetic improvement. Ultimately, combining QTL mapping, genomics, and precision breeding promises to accelerate the development of superior tomato varieties, supporting future food security and sustainable agriculture.

REFERENCES

Ali AAM, Romdhane WB, Tarroum M, et al. (2021) Analysis of salinity tolerance in tomato introgression lines based on morpho-physiological and molecular traits. *Plants* 10:2594

Araus JL, Cairns JE (2014) Field high-throughput phenotyping: The new crop breeding frontier. *Trends Plant Sci* 19:52–61. https://doi.org/10.1016/j.tplants.2013.09.008

Areshchenkova T, Ganal MW (2002) Comparative analysis of polymorphism and chromosomal location of tomato microsatellite markers isolated from different sources. *Theor Appl Genet* 104:229–235

Bai Y, Lindhout P (2007) Domestication and breeding of tomatoes: What have we gained and what can we gain in the future? *Ann Bot* 100:1085–1094

Benoit M, Jenike KM, Satterlee JW, et al. (2025) Solanum pan-genetics reveals paralogues as contingencies in crop engineering. *Nature* 640:135–145. https://doi.org/10.1038/s41586-025-08619-6

Berman A, Su N, Li Z, et al. (2025) Construction of multi-targeted CRISPR libraries in tomato to overcome functional redundancy at genome-scale level. *Nat Commun* 16:4111. https://doi.org/10.1038/s41467-025-59280-6

Blanca J, Cañizares J, Cordero L, et al. (2012) Variation revealed by SNP genotyping and morphology provides insight into the origin of the tomato. *PLoS ONE* 7:e48198

Bonarota M-S, Fenstemaker S, Vasquez-Gross H, et al. (2025) Physiology of salt tolerance introgressions from Solanum galapagense in the domesticated tomato. *Front Plant Sci* 16

Bradbury PJ, Zhang Z, Kroon DE, et al. (2007) TASSEL: Software for association mapping of complex traits in diverse samples. *Bioinformatics* 23:2633–2635

Broman KW (2005) The genomes of recombinant inbred lines. *Genetics* 169:1133–1146

Broman KW, Gatti DM, Simecek P, et al. (2019) R/qtl2: Software for mapping quantitative trait loci with high-dimensional data and multiparent populations. *Genetics* 211:495–502

Broman KW, Wu H, Sen Ś, Churchill GA (2003) R/qtl: QTL mapping in experimental crosses. *Bioinformatics* 19:889–890

Causse M, Grandillo S (2016) Gene mapping in tomato. In: Kole C (ed) *The Tomato Genome*. Springer, Berlin Heidelberg, pp. 23–37

Chakrabarti M, Zhang NA, Sauvage C, et al. (2013) A cytochrome P450 regulates a domestication trait in cultivated tomato. *Proc Natl Acad Sci* 110:17125–17130

Chase K, Adler FR, Lark KG (1997) Epistat: A computer program for identifying and testing interactions between pairs of quantitative trait loci. *Theor Appl Genet* 94:724–730

Collard BCY, Jahufer MZZ, Brouwer JB, Pang ECK (2005) An introduction to markers, quantitative trait loci (QTL) mapping and marker-assisted selection for crop improvement: The basic concepts. *Euphytica* 142:169–196

Cooper M, Powell O, Gho C, et al. (2023) Extending the breeder's equation to take aim at the target population of environments. *Front Plant Sci* 14

Du M, Sun C, Deng L, et al. (2025) Molecular breeding of tomato: Advances and challenges. *J Integr Plant Biol* 67:669–721

Edwards MD, Stuber CW, Wendel J (1987) Molecular-marker-facilitated investigations of quantitative-trait loci in maize. I. Numbers, genomic distribution and types of gene action. *Genetics* 116:113–125

Egea I, Estrada Y, Flores FB, Bolarín MC (2022) Improving production and fruit quality of tomato under abiotic stress: Genes for the future of tomato breeding for a sustainable agriculture. *Environ Exp Bot* 204:105086

Eshed Y, Zamir D (1995) An introgression line population of Lycopersicon pennellii in the cultivated tomato enables the identification and fine mapping of yield-associated QTL. *Genetics* 141:1147–1162

Falconer DS (1996) *Introduction to Quantitative Genetics*. Pearson Education India

FAO FAOSTAT (2022) *Crops and Livestock Products—Tomato*. Food and Agriculture Organization of the United Nations

Foolad MR (2007) Genome mapping and molecular breeding of tomato. *Int J Plant Genomics* 2007:64358. https://doi.org/10.1155/2007/64358

Foolad MR, Panthee DR (2012) Marker-assisted selection in tomato breeding. *CRC Crit Rev Plant Sci* 31:93–123

Foolad MR, Subbiah P, Zhang L (2007) Common QTL affect the rate of tomato seed germination under different stress and nonstress conditions. *Int J Plant Genomics* 2007:097386

Frary A, Nesbitt TC, Frary A, et al. (2000) fw2. 2: A quantitative trait locus key to the evolution of tomato fruit size. *Science (1979)* 289:85–88

Gao L, Gonda I, Sun H, et al. (2019) The tomato pan-genome uncovers new genes and a rare allele regulating fruit flavor. *Nat Genet* 51:1044–1051

Geldermann H (1975) Investigations on inheritance of quantitative characters in animals by gene markers I. Methods. *Theor Appl Genet* 46:319–330

Gonzali S, Mazzucato A, Perata P (2009) Purple as a tomato: Towards high anthocyanin tomatoes. *Trends Plant Sci* 14:237–241

Grandillo S, Cammareri M (2016) Molecular mapping of quantitative trait loci in tomato. In: Kole C (ed) *The Tomato Genome*. Springer, Berlin Heidelberg, pp. 39–73

Grandillo S, Chetelat R, Knapp S, et al. (2011) Solanum sect. Lycopersicon. In: Kole C (ed) *Wild Crop Relatives: Genomic and Breeding Resources: Vegetables*. Springer, Berlin Heidelberg, pp. 129–215

Grandillo S, Termolino P, van der Knaap E (2013) Molecular mapping of complex traits in tomato. In: Foolad MR (ed) *Genetics, Genomics, and Breeding of Tomato*. CRC Press, Boca Raton, pp. 150–227

Hanson P, Lu S-F, Wang J-F, et al. (2016) Conventional and molecular marker-assisted selection and pyramiding of genes for multiple disease resistance in tomato. *Sci Hortic* 201:346–354

Harish A, Ramesh Kumar A, Manivannan S, Senthil Kumar S (2023) An outlook on GWAS with a special focus on solanaceous vegetable crops-a review. *Int J Plant Soil Sci* 35:209–220

Hirakawa H, Shirasawa K, Ohyama A, et al. (2013) Genome-wide SNP genotyping to infer the effects on gene functions in tomato. *DNA Res* 20:221–233

Ji Y, Schuster DJ, Scott JW (2007) Ty-3, a begomovirus resistance locus near the Tomato yellow leaf curl virus resistance locus Ty-1 on chromosome 6 of tomato. *Mol Breeding* 20:271–284

Kang GH, Ko Y, Lee JM (2025) Enhancing virus-mediated genome editing for cultivated tomato through low temperature. *Plant Cell Rep* 44:22. https://doi.org/10.1007/s00299-024-03392-8

Kao C-H, Zeng Z-B, Teasdale RD (1999) Multiple interval mapping for quantitative trait loci. *Genetics* 152:1203–1216

Khojasteh M, Darzi Ramandi H, Taghavi SM, et al. (2024) Unraveling the genetic basis of quantitative resistance to diseases in tomato: A meta-QTL analysis and mining of transcript profiles. *Plant Cell Rep* 43:184. https://doi.org/10.1007/s00299-024-03268-x

Lander ES, Botstein D (1989) Mapping Mendelian factors underlying quantitative traits using RFLP linkage maps. *Genetics* 121:185–199

Mackay TFC, Stone EA, Ayroles JF (2009) The genetics of quantitative traits: Challenges and prospects. *Nat Rev Genet* 10:565–577

Manly KF, Olson JM (1999) Overview of QTL mapping software and introduction to Map Manager QT. *Mamm Genome* 10:327–334

Mather K (1941) Variation and selection of polygenic characters. *J Genet* 41:159–193

Mather K, Jinks JL (2013) *Biometrical Genetics: The Study of Continuous Variation*. Springer

Meng L, Li H, Zhang L, Wang J (2015) QTL IciMapping: Integrated software for genetic linkage map construction and quantitative trait locus mapping in biparental populations. *Crop J* 3:269–283

Michael TP, VanBuren R (2020) Building near-complete plant genomes. *Curr Opin Plant Biol* 54:26–33. https://doi.org/10.1016/j.pbi.2019.12.009

Miller JC, Tanksley SD (1990) RFLP analysis of phylogenetic relationships and genetic variation in the genus Lycopersicon. *Theor Appl Genet* 80:437–448

Mishra A (2022) Recent developments in breeding approaches of tomato (*Solanum lycopersicum* L)—a review. *Int J Farm Sci* 12:1–6

Nelson JC (1997) QGENE: Software for marker-based genomic analysis and breeding. *Mol Breeding* 3:239–245

Osei MK, Prempeh R, Adjebeng-Danquah J, et al. (2018) Marker-assisted selection (MAS): A fast-track tool in tomato breeding. In: *Recent Advances in Tomato Breeding and Production*. IntechOpen

Osorio S, Alba R, Nikoloski Z, et al. (2012) Integrative comparative analyses of transcript and metabolite profiles from pepper and tomato ripening and development stages uncovers species-specific patterns of network regulatory behavior. *Plant Physiol* 159:1713–1729

Panthee DR, Gardner RG (2010) 'Mountain Merit': A late blight-resistant large-fruited tomato hybrid. *HortScience* 45:1547–1548

Paran I, Van Der Knaap E (2007) Genetic and molecular regulation of fruit and plant domestication traits in tomato and pepper. *J Exp Bot* 58:3841–3852

Pascual L, Desplat N, Huang BE, et al. (2015) Potential of a tomato MAGIC population to decipher the genetic control of quantitative traits and detect causal variants in the resequencing era. *Plant Biotechnol J* 13:565–577

Paterson AH, Lander ES, Hewitt JD, et al. (1988) Resolution of quantitative traits into Mendelian factors by using a complete linkage map of restriction fragment length polymorphisms. *Nature* 335:721–726

Rajendra V (2025) Marker-assisted selection (MAS) revolutionized: Integrating 'Omics' data into tomato breeding. In: Waseem M, Prakash CS, Faiz S, Li Z (eds) *Omics Approaches for Tomato Yield and Quality Trait Improvement*. Springer Nature Singapore, pp. 273–305

Rani K, Kumar M, Razzaq A, et al. (2023) Recent advances in molecular marker technology for QTL mapping in plants. In: Kole C (ed) *QTL Mapping in Crop Improvement*. Springer, Cham, pp. 1–15

Rick CM (1982) The potential of exotic germplasm for tomato improvement. In: Atherton JG, Rudich J (eds) *The Biology and Breeding of the Tomato*. Croom Helm, London, pp. 1–28

Robertson LD, Labate JA (2007) Genetic resources of tomato (*Lycopersicon esculentum* Mill.) and wild relatives. In: Razdan MK, Matoo AK (eds) *Genetic Improvement of Solanaceous Crops, Vol. 2: Tomato*. Science Publishers, Enfield (NH), pp. 25–75

Sauvage C, Segura V, Bauchet G, et al. (2014) Genome-wide association in tomato reveals 44 candidate loci for fruit metabolic traits. *Plant Physiol* 165:1120–1132

Sax K (1923) The association of size differences with seed-coat pattern and pigmentation in Phaseolus vulgaris. *Genetics* 8:552

Semagn K, Babu R, Hearne S, Olsen M (2014) Single nucleotide polymorphism genotyping using Kompetitive Allele Specific PCR (KASP): Overview of the technology and its application in crop improvement. *Mol Breeding* 33:1–14

Shirasawa K, Asamizu E, Fukuoka H, et al. (2010) An interspecific linkage map of SSR and intronic polymorphism markers in tomato. *Theor Appl Genet* 121:731–739

Siddique MI, Silverman E, Louws F, Panthee DR (2024) Quantitative trait loci mapping for bacterial wilt resistance and plant height in tomatoes. *Plants* 13:876

Sillanpaa MJ (1998) *Multimapper Reference Manual*. Department of Mathematics and Statistics, University of Helsinki, Helsinki, Finland

Sim S-C, Durstewitz G, Plieske J, et al. (2012a) Development of a large SNP genotyping array and generation of high-density genetic maps in tomato. *PLoS ONE* 7:e40563

Sim S-C, Van Deynze A, Stoffel K, et al. (2012b) High-density SNP genotyping of tomato (*Solanum lycopersicum* L.) reveals patterns of genetic variation due to breeding. *PLoS ONE* 7(9):e45520

Tanksley SD (1993) Mapping polygenes. *Annu Rev Genet* 27:205–233

Tanksley SD, Ganal MW, Prince JP, et al. (1992) High density molecular linkage maps of the tomato and potato genomes. *Genetics* 132:1141–1160

Tanksley SD, Medina-Filho H, Rick CM (1982) Use of naturally-occurring enzyme variation to detect and map genes controlling quantitative traits in an interspecific backcross of tomato. *Heredity (Edinb)* 49:11–25

Thoday JM (1961) Location of polygenes. *Nature* 191:368–370

Tibbs Cortes L, Zhang Z, Yu J (2021) Status and prospects of genome-wide association studies in plants. *Plant Genome* 14:e20077

Tieman D, Bliss P, McIntyre LM, et al. (2012) The chemical interactions underlying tomato flavor preferences. *Curr Biol* 22:1035–1039

Tiwari JK, Yerasu SR, Rai N, et al. (2022) Progress in marker-assisted selection to genomics-assisted breeding in tomato. *CRC Crit Rev Plant Sci* 41:321–350. https://doi.org/10.1080/07352689.2022.2130361

Tomato Genome Consortium xx (2012) The tomato genome sequence provides insights into fleshy fruit evolution. *Nature* 485:635

Torgeman S, Zamir D (2023) Epistatic QTLs for yield heterosis in tomato. *Proc Natl Acad Sci* 120:e2205787119

Tripodi P, Soler S, Campanelli G, et al. (2021) Genome wide association mapping for agronomic, fruit quality, and root architectural traits in tomato under organic farming conditions. *BMC Plant Biol* 21:1–22

Ullah Z, Iqbal J, Abbasi BA, et al. (2025) Future horizons: Emerging "omics" technologies and challenges in tomato. In: Waseem M, Prakash CS, Faiz S, Li Z (eds) *Omics Approaches for Tomato Yield and Quality Trait Improvement*. Springer Nature Singapore, pp. 347–394

Utz HF, Melchinger AE (1996) PLABQTL: A program for composite interval mapping of QTL. In *Institute of Plant Breeding, Seed Science, and Population Genetics*. University of Hohenheim, Stuttgart, Germany, pp. 1–6

Vallejos CE, Tanksley SD (1983) Segregation of isozyme markers and cold tolerance in an interspecific backcross of tomato. *Theor Appl Genet* 66:241–247

van Dijk ADJ, Kootstra G, Kruijer W, de Ridder D (2021) Machine learning in plant science and plant breeding. *iScience* 24. https://doi.org/10.1016/j.isci.2020.101890

Van Ooijen JW (2004) *MapQTL® 5, Software for the Mapping of Quantitative Trait Loci in Experimental Populations*. Kyazma BV:63

Verlaan MG, Szinay D, Hutton SF, et al. (2011) Chromosomal rearrangements between tomato and Solanum chilense hamper mapping and breeding of the TYLCV resistance gene Ty-1. *Plant J* 68:1093–1103. https://doi.org/10.1111/j.1365-313X.2011.04762.x

Veyrieras J, Camus-Kulandaivelu L, Gouesnard B, et al. (2007) Bridging genomics and genetic diversity: Linkage disequilibrium structure and association mapping in maize and other cereals. *Crop Sci* 47:S–60

Vincent M, Gerdes Gyuricza I, Keele GR, et al. (2022) QTLViewer: An interactive webtool for genetic analysis in the Collaborative Cross and Diversity Outbred mouse populations. *G3* 12:jkac146

Víquez-Zamora M, Vosman B, van de Geest H, et al. (2013) Tomato breeding in the genomics era: Insights from a SNP array. *BMC Genomics* 14:1–13

Wang J-F, Ho F-I, Truong HTH, et al. (2013) Identification of major QTLs associated with stable resistance of tomato cultivar 'Hawaii 7996' to Ralstonia solanacearum. *Euphytica* 190:241–252

Wang S (2006) *Windows QTL Cartographer 2.5*. Department of Statistics, North Carolina State University, Raleigh, NC

Wang Y, Sun C, Ye Z, et al. (2024) The genomic route to tomato breeding: Past, present, and future. *Plant Physiol* 195:2500–2514. https://doi.org/10.1093/plphys/kiae248

Weller JI, Soller M, Brody T (1988) Linkage analysis of quantitative traits in an interspecific cross of tomato (Lycopersicon esculentum x Lycopersicon pimpinellifolium) by means of genetic markers. *Genetics* 118:329–339

Wen J, Jiang F, Weng Y, et al. (2019) Identification of heat-tolerance QTLs and high-temperature stress-responsive genes through conventional QTL mapping, QTL-seq and RNA-seq in tomato. *BMC Plant Biol* 19:398. https://doi.org/10.1186/s12870-019-2008-3

Xiao H, Jiang N, Schaffner E, et al. (2008) A retrotransposon-mediated gene duplication underlies morphological variation of tomato fruit. *Science* 319:1527–1530

Xu S (2013) *Principles of Statistical Genomics*. Springer

Xu Y, Crouch JH (2008) Marker-assisted selection in plant breeding: From publications to practice. *Crop Sci* 48:391–407

Yang J, Hu C, Hu H, et al. (2008) QTLNetwork: Mapping and visualizing genetic architecture of complex traits in experimental populations. *Bioinformatics* 24:721–723

Yi N, George V, Allison DB (2003) Stochastic search variable selection for identifying multiple quantitative trait loci. *Genetics* 164:1129–1138

Zeng Z-B (1994) Precision mapping of quantitative trait loci. *Genetics* 136:1457–1468

Zhang C, Liu L, Zheng Z, et al. (2013) Fine mapping of the Ph-3 gene conferring resistance to late blight (Phytophthora infestans) in tomato. *Theor Appl Genet* 126:2643–2653

Zhang J, Panthee DR (2022) Next-generation sequencing-based bulked segregant analysis without sequencing the parental genomes. *G3 Genes|Genomes|Genetics* 12:jkab400. https://doi.org/10.1093/g3journal/jkab400

Zhang Y, Stommel JR (2001) Development of SCAR and CAPS markers linked to the Beta gene in tomato. *Crop Sci* 41:1602–1608

Zhao Y, Chen W, Cui Y, et al. (2021) Detection of candidate genes and development of KASP markers for Verticillium wilt resistance by combining genome-wide association study, QTL-seq and transcriptome sequencing in cotton. *Theor Appl Genet* 134:1063–1081. https://doi.org/10.1007/s00122-020-03752-4

Zheng Z, Huang D, Wang J, et al. (2020) QTLbase: An integrative resource for quantitative trait loci across multiple human molecular phenotypes. *Nucleic Acids Res* 48:D983–D991

Zhou R, Yu X, Ottosen C-O, et al. (2020) Unique miRNAs and their targets in tomato leaf responding to combined drought and heat stress. *BMC Plant Biol* 20:1–10

5 BSA-Seq

A Novel Approach for QTL Mapping and Molecular Marker Development

Jianbo Zhang, Muhammad Irfan Siddique, and Dilip R. Panthee

5.1 GENETICS OF BULKED SEGREGANT ANALYSIS

Bulked segregant analysis (BSA) was developed for rapidly identifying genetic markers linked to a trait of interest (Giovannoni et al. 1991; Michelmore et al. 1991). For a particular trait, a certain number of individuals are selected from a segregating population to make two groups that have contrasting phenotypes of the trait. Equal amounts of DNA are pooled from each individual within a group. The bulked DNA samples are then examined via various methods to identify bulk and parent-specific genetic markers. Collectively, the genomic regions controlling the trait should be different, while the other unlinked regions should be the same or very similar between the bulks. Thus, the bulk and parent-specific markers should be tightly linked to the trait-controlling gene.

In bulked segregant analysis, the allele controlling the trait should be enriched in one bulk, and very likely depleted in the other bulk. Allele enrichment caused via bulking varies among different types of traits (Figure 5.1 and Table 5.1). For bulks constructed with an F_2 population where alleles *A*/*a* control the trait expression and *A* is dominant over *a*, bulk 1 consists of both *AA* and *Aa* individuals while bulk 2 consists of only *aa* individuals, and the frequency of the *A* allele is 2/3 (*AA*:*Aa* = 1:2) in bulk 1 and zero in bulk 2 (Figure 5.1a and Table 5.1). These values are different in incomplete dominant traits (Figure 5.1b and Table 5.1), epistatic traits (Figure 5.1c and Table 5.1), or quantitative traits (Figure 5.1d). Allele enrichment varies as well among quantitative trait loci (QTLs) controlling the same trait; a QTL contributing more to the phenotype would be enriched more. The alleles of the other genes on the same chromosome can be enriched too, depending on their distances from *A*/*a*. Due to meiotic homologous recombination, the frequencies of the alleles of these genes should decrease in bulk 1 and increase in bulk 2 as the distance between them and the allele *A*/*a* increases. When the genes are very far apart from the allele *A*/*a*, they will assort independently and the allele frequencies of these genes will reach 1/2 in both bulks.

BSA can work well without any prior information about the trait as long as it segregates well in the population and one or both bulks contain a unique trait-controlling allele. For qualitative traits controlled by a single gene, the allele *A* exists only in bulk 1 in both Figure 5.1a and Figure 5.1b, while the allele *a* exists only in bulk 2 in Figure 5.1b. If the structural polymorphisms specific to the unique allele or alleles tightly linked to it can be detected via polymerase chain reaction (PCR) procedures such as random amplified polymorphic DNA (RAPD), amplified fragment length polymorphism (AFLP), or simple sequence repeat (SSR, also known as microsatellite) (Giovannoni et al. 1991; Michelmore et al. 1991; Cai et al. 2003; Shen et al. 2003; Asnaghi et al. 2004; Hartl et al. 2011; Zhang et al. 2013), one or a few markers can be developed to assist the trait screening.

It is somewhat complicated for qualitative epistatic traits because they are controlled by multiple genes. The phenomenon that two or more genes interact with one another to affect the expression of

DOI: 10.1201/9781003506539-5

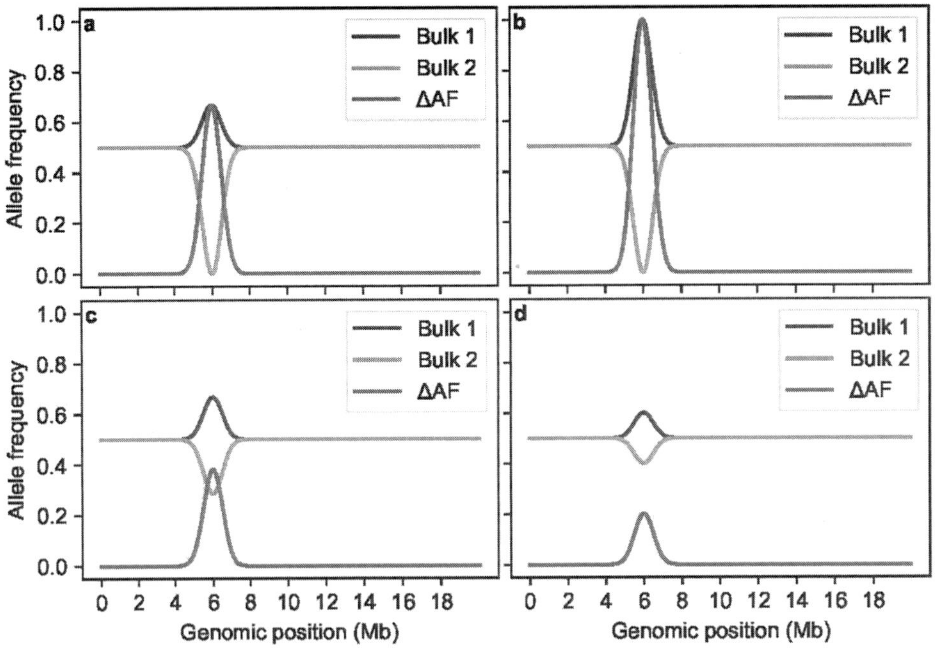

FIGURE 5.1 Allele frequencies of genes on a chromosome in bulks constructed from an F_2 population where alleles *A*/*a* participate in conditioning the trait phenotype. *A*/*a* is located at the position of 6 Mb for all traits. (**a**). A trait controlled by a single dominant gene. (**b**). A trait controlled by a single incomplete dominant gene. (**c**). A trait controlled by two independent genes with an epistatic effect (the chromosome that contains one of the trait-controlling genes is shown here; the other gene is on a different chromosome). (**d**). A trait controlled by multiple QTLs (the chromosome that contains one trait-controlling QTL is shown here; the remaining QTLs are on other chromosomes).

Table 5.1
Allele Enrichment of Different Traits in BSA

	Bulk 1		Bulk 2		
Trait	**Genotype**	**AF1**	**Genotype**	**AF2**	**ΔAF**
A dominant trait controlled by gene *A*	*A_*	2/3	*aa*	0	2/3
An incomplete dominant trait controlled by gene *A*	*AA*	1	*aa*	0	1
An epistatic trait controlled by genes *A* and *B*	*A_B_*	2/3	*aa__* and *__bb*	2/7	8/21

a trait is called epistasis. BSA will work if genes interaction leads to a unique allele in a phenotype group (the chicken comb shape trait is an example (Miko 2008)); otherwise, it will not work (the pea flower color trait is a great example (Miko 2008)). Allele enrichment of pea flower color should be the same as that in Figure 5.1c and Table 5.1.

The simplest and most intuitive epistasis example is phenotypes directly related to metabolic pathways. For a trait controlled by two independently segregated genes (*A* and *B*) that code for enzymes catalyzing two different reactions in a biosynthesis pathway, the *A_B_* (*A_* can be *AA* or *Aa* while *B_* can be *BB* or *Bb*) individuals in an F_2 population have the same phenotype as the F_1 individuals, and can be used to construct bulk 1, while the *aa__* (*__* can be *BB*, *Bb*, or *bb*) and __*bb* (__ can be *AA*, *Aa*, or *aa*) have the recessive phenotype and can be used to construct bulk 2 (Table 5.1).

The *A/B* allele frequency in bulk 1 (*AF1*) is always 2/3. Whereas the *A/B* allele frequency in bulk 2 (*A*F2) can be calculated using equation 5.1, where n (n≥2) is the number of genes (assume all of them segregate independently) involved in epistasis. Thus, the minimal *AF2* value is 2/7 ($n = 2$, Table 5.1), while its maximal value is 1/2 (meaning allele *A* is enriched is bulk 1 but not depleted in bulk 2 when many genes are involved), and the allele difference (ΔAF) of the *A/B* allele between the two bulks is between 1/6 and 8/21 (Figure 5.1c and Table 5.1). The ΔAF gets smaller when more genes are involved in epistasis. Apparently, BSA does not work for this trait because no unique allele exists in either bulk.

$$AF2 = 2\times\frac{4^{n-1}-3^{n-1}}{4^n-3^n} = \frac{1}{2} - \frac{1}{8\times\left(\frac{4}{3}\right)^{n-1}-6} \tag{5.1}$$

For quantitative traits, most likely each bulk contains both of the trait-controlling gene alleles because of quantitative inheritance. Thus, no polymorphism can be identified between the bulks in AFLP/RAPD/SSR assays. Because multiple targets are amplified in a single AFLP/RAPD/SSR reaction, quantitative PCR cannot be used to detect the amount of a specific band in the mixed PCR products. For standard PCR, band intensity ratio varies considerably from experiment to experiment. Often, this is not a reliable metric to judge the enrichment of a specific genomic region in a bulk.

Simulations have shown that QTLs can be identified via BSA using backcross or F_2 populations when the mean difference between bulks is 3/4 to 1 standard deviation. But for most QTLs, the mean difference is much smaller (Wang and Paterson 1994); thus, recombinant inbred lines or doubled haploid lines are needed to identify major QTLs via BSA (Shen et al. 2003). However, constructing populations of such lines is both time and labor consuming.

The structural polymorphisms are detected by PCR procedures and gel electrophoresis in BSA. Thus, BSA has limitations associated with these technologies:

- PCR primers may miss the trait-controlling gene.
- Only SVs leading to size polymorphism of PCR products are informative.
- A few hundred or even thousands of primers or primer pairs must be tested to develop a few tightly linked markers, which is quite labor-intensive.

5.2 NEXT-GENERATION SEQUENCING-BASED BSA (BSA-SEQ)

With the recent dramatic cost reduction, the next-generation sequencing (NGS) technology has been applied to more and more BSA studies (Schneeberger et al. 2009; Ehrenreich et al. 2010; Magwene et al. 2011; Trick et al. 2012; Takagi et al. 2013; Das et al. 2015; Daware et al. 2016; Gu et al. 2017; Chen et al. 2018; Clevenger et al. 2018; Klein et al. 2018; Arikit et al. 2019; Imerovski et al. 2019; Cao et al. 2021); the technology is referred to as BSA-Seq. In BSA-Seq, pooled DNA samples are not subjected to AFLP/RAPD/SSR assays but are directly sequenced instead. Structural variants (SVs) such as single-nucleotide polymorphisms (SNPs) and small insertions/deletions (InDels) across the genome can be detected via structural variant calling. Both SNPs and InDels are so abundant that all trait-controlling genes should contain many such SVs and/or have many such SVs in the vicinity in any reasonably well-designed BSA-Seq studies. Thus, BSA-Seq is capable of detecting all trait-controlling loci that segregate in a population. Moreover, enrichment/depletion of the allele of an SV or the alleles of a stretch of SVs in the bulks can be analyzed using statistical approaches, although most SVs are heterogeneous in allele composition in both bulks. Thus, the presence of a unique allele in a bulk is not a prerequisite anymore; BSA-Seq can work well for both qualitative and quantitative traits using F_2 or backcross populations. Furthermore, only two pooled DNA samples are genotyped in BSA-Seq, which is much more cost- and labor-efficient than traditional genetic mapping, genome-wide association studies (GWAS), and genotyping by sequencing (GBS) that require genotyping every individual/accession in the population/panel. More importantly, the time-consuming and labor-intensive marker development and genetic mapping steps are eliminated in BSA-Seq; thus, BSA-Seq takes much less time than BSA.

5.3 BSA-SEQ DATA ANALYSIS ALGORITHMS

An SV in a BSA-Seq dataset can be biallelic, triallelic, or tetraallelic. Biallelic SVs are the most abundant if the parents for generating a population are highly homozygous, and only this type of SV is used in most BSA-Seq analysis algorithms. For a biallelic SV locus, one of its alleles is termed the reference allele (*REF*), while the other is designated as the alternative allele (*ALT*). When the parental genome sequences are available, the SV alleles from one of the parents serve as *REF* alleles, while the SV alleles from the other parent are *ALT* alleles. Because each bulk contains many individuals, the vast majority of SV loci in a bulk contain both *REF* and *ALT* alleles. The number of reads of the *REF*/*ALT* alleles of each SV in a sequencing dataset is termed allele depth (*AD*). Because of the phenotypic selection via bulking, one of the *REF*/*ALT* alleles of the trait-associated SVs should be enriched in one of the two bulks. However, the *REF*/*ALT* alleles of SVs not associated with the trait would be randomly segregated in both bulks and enriched in neither. Hence, these four *AD* values (a pair of AD_{ref}/AD_{alt} from each bulk as in Table 5.2) can be used to assess how likely an SV is associated with the trait.

The allele frequency of an SV is fixed once bulk construction is complete; the allele frequency of an SV allele linked to a trait-controlling genomic region should be different, while the allele frequency of an SV allele segregating independently should be similar between the bulks. It is costly to sequence the bulked DNA in very high depth, but low sequencing depth would lead to the allele frequencies of some SVs deviating significantly from their real ratios because of random variation, generating noise in the allele frequency (or other parameters) curve. The noise can be reduced if multiple SVs of a genomic region are included in the calculation. Thus, the sliding window technique is used to mitigate noise in most BSA-Seq data analysis algorithms. A sliding window in BSA-Seq is a genomic region of a certain length. Starting from the beginning of a chromosome, the sliding window moves forward at a small incremental step; the allele frequency (or other parameters) of a sliding window is calculated at each step; the sliding window will stop moving if the end of the chromosome is reached.

Statistical analysis is needed to judge if the allele enrichment/depletion of a genomic region between bulks is significant. Quite a few algorithms have been developed to serve this purpose, and the allele frequency method (QTL-seq) (Takagi et al. 2013) and the *G*-statistic method (Magwene et al. 2011) are the most widely used. We have recently developed a significant structural variant method for the same purpose; it has much higher detection power than the allele frequency method (Zhang and Panthee 2020; Zhang and Panthee 2022). We summarized these three algorithms next.

TABLE 5.2
Allele Depths of an SV in a BSA-Seq Dataset

	Bulk 1	Bulk 2	Total
REF	AD_{ref1}	AD_{ref2}	$AD_{ref1} + AD_{ref2}$
ALT	AD_{alt1}	AD_{alt2}	$AD_{alt1} + AD_{alt2}$
Total	$AD_{alt1} + AD_{ref1}$	$AD_{alt2} + AD_{ref2}$	

5.3.1 The *G*-Statistic Method

Bulking leads to allele enrichment/depletion in the trait-associated genomic region. We need to choose a parameter that can reflect such enrichment/depletion and find a value (confidence interval/threshold) of this parameter to judge if the enrichment/depletion is significant. In the *G*-statistic method, the *G*-statistic value of an SV is the chosen parameter, and it can be calculated using the given formula:

$$G - 2\sum_{i=1}^{n} O_i \times ln\frac{O_i}{E_i} \tag{5.2}$$

where O is the observed AD (AD_{ref1}, AD_{alt1}, AD_{ref2}, or AD_{alt2} in Table 5.2), E is the expected AD under the null hypothesis, ln denotes the natural logarithm, and n is equal to 4 (2 × 2 table) in BSA-Seq. For an SV locus, its expected AD values of the *REF*/*ALT* alleles in both bulks under the null hypothesis can be calculated with its four observed AD values using these formulas (Magwene et al. 2011; Mansfeld and Grumet 2018):

$$ADexp_{alt1} = \frac{\left(AD_{alt1} + AD_{alt2}\right)\left(AD_{alt1} + AD_{ref1}\right)}{AD_{alt1} + AD_{ref1} + AD_{alt2} + AD_{ref2}} \tag{5.3}$$

$$ADexp_{ref1} = \frac{\left(AD_{alt1} + AD_{ref1}\right)\left(AD_{ref1} + AD_{ref2}\right)}{AD_{alt1} + AD_{ref1} + AD_{alt2} + AD_{ref2}} \tag{5.4}$$

$$ADexp_{alt2} = \frac{\left(AD_{alt2} + AD_{ref2}\right)\left(AD_{alt1} + AD_{alt2}\right)}{AD_{alt1} + AD_{ref1} + AD_{alt2} + AD_{ref2}} \tag{5.5}$$

$$ADexp_{ref2} = \frac{\left(AD_{alt2} + AD_{ref2}\right)\left(AD_{ref1} + AD_{ref2}\right)}{AD_{alt1} + AD_{ref1} + AD_{alt2} + AD_{ref2}} \tag{5.6}$$

In *G*-test, the sum of the observed *AD* values of an SV should be equal to the sum of the expected *AD* values, and these equations meet this requirement; in addition, all *G*-statistic values are non-negative (https://stats.stackexchange.com/users/5339/guy)). If an SV is associated with a trait-controlling gene, its *REF*/*ALT* allele will be enriched/depleted in the bulks, leading to its observed *AD* values deviating greatly from the expected values under the null hypothesis, and its *G*-statistic value will be high. The *G*-statistic value of a sliding window is the average of the *G*-statistic values of all SVs in it.

Datasets that do not contain trait-associated SVs are needed for confidence interval/threshold estimation in BSA-Seq data analysis. Such a dataset can be experimentally generated via assembling a bulk by random selection (without phenotypic screening), but doing so would significantly increase the project costs, and thus computational approaches were preferred for this purpose (Magwene et al. 2011; Takagi et al. 2013; Zhang and Panthee 2020; Zhang and Panthee 2022). In the *G*-statistic method, such a dataset was obtained by filtering out outliers (trait-associated SVs) in the real dataset using an empirical coefficient, the *ln(G')* (the natural logarithm of the smoothed *G*-statistic) of each SV in the dataset, and the median absolute deviation of each *ln(G')*. Based on the work of Davies and Gather (Davies and Gather 1993), Eq. 5.2 was used as the coefficient in the *G*-statistic method. The G-statistic threshold is estimated using a false discovery rate (FDR) approach with the trimmed dataset without any outliers (Magwene et al. 2011). *G*-statistic values of all sliding windows of the dataset are plotted against their position on each chromosome. Each peak in the curve should represent a potential genetic locus that conditions the trait phenotypes. A peak in the curve would be significant if its *G*-statistic value is greater than the threshold and should contain a trait-controlling gene.

5.3.2 The Allele Frequency Method (QTL-Seq)

The allele frequency method (QTL-Seq) is much more straightforward than the G-statistic method; it uses the allele frequency differences (ΔAF) of either the *REF* or *ALT* allele of an SV between bulks to judge if an SV is likely to be associated with a trait of interest (Takagi et al. 2013). ΔAF of an SV can be calculated using the following formula:

$$\Delta AF = \frac{AD_{alt2}}{AD_{alt2} + AD_{ref2}} - \frac{AD_{alt1}}{AD_{alt1} + AD_{ref1}} \tag{5.7}$$

or

$$\Delta AF = \frac{AD_{alt1}}{AD_{alt1} + AD_{ref1}} - \frac{AD_{alt2}}{AD_{alt2} + AD_{ref2}} \tag{5.8}$$

The ΔAF value reflects the enrichment/depletion of an SV in the bulks, it ranges from −1 to 1. The more the *REF/ALT* allele of an SV is enriched/depleted in the bulks, the greater the absolute value of its ΔAF. The ΔAF value of an SV in a genomic region not associated with and not linked to a trait-controlling gene should be zero. The ΔAF of a sliding window is the mean of the ΔAF values of all SVs in it.

In this method, the confidence interval of each SV in the dataset is estimated via Monte Carlo simulation. If an SV is not associated with the trait of interest, its *REF/ALT* alleles should segregate randomly in both bulks (the null hypothesis), and its *REF/ALT* allele frequency should be 0.5/0.5 in an F_2 population and 0.25/0.75 in a backcross population. Based on its sequencing depths in both bulks ($AD_{ref1} + AD_{alt1}$ in bulk 1 and $AD_{ref2} + AD_{alt2}$ in bulk 2) and the expected REF/ALT allele frequency (binomial distribution), four simulated *AD* values can be obtained using the NumPy software package (Harris et al. 2020), and a ΔAF value can be calculated using these simulated *AD* values. Repeated this process 10,000 times to obtain 10,000 ΔAF values, and the confidence interval can be obtained using either the empirical rule or the percentile method commonly used in bootstrapping. The upper/lower bound of the confidence interval of a sliding window is the average of the upper/lower bounds of all SVs in it. In the *G*-statistic method, only peaks above the threshold represent trait-associated genomic regions. However, in the allele frequency method, both peaks above the *CI* upper bound and valleys below the *CI* lower bound represent genetic loci conditioning the trait.

The confidence interval of the allele frequency method can be assessed using other means. Huang et al. (2020) estimated a genome-wide confidence interval based on the Central Limit Theorem. Using this approach, the estimated confidence interval is much narrower, and the detection power of the allele frequency method is much higher.

5.3.3 The Significant Structural Variant Method

We have recently developed a significant structural variant method for BSA-Seq data analysis. In this method, each SV of a BSA-Seq dataset is assessed with Fisher's exact test using the *AD* values of both bulks (AD_{ref1}, AD_{alt1}, AD_{ref2}, or AD_{alt2}). An SV with a low *p*-value of Fisher's exact test tends to have a high absolute ΔAF and *G*-statistic and is more likely associated with the trait. An SV is considered significant (sSV) if the p-value is lower than a specific type I error rate, e.g., $\alpha = 0.01$, and non-significant otherwise. Since SVs are normally not evenly distributed across chromosomes, we use the ratio of the significant structural variants (*sSV*) to the total structural variants (*totalSV*) within a genomic region to judge if this region is associated with the trait of interest in our method. sSVs associated with a trait-controlling gene should be clustered together, whereas sSVs caused by random variation should be randomly distributed in the genome, and genomic regions not associated with the trait should contain only sporadic sSVs. Thus, a trait-associated genomic region should have a high *sSV/totalSV* ratio.

Monte Carlo simulation is used to estimate the threshold of *sSV/totalSV* in this method. We assume a genomic region (more specifically, a sliding window) does not contain a trait-controlling gene and segregates randomly in both bulks (null hypothesis). The same approach as in the allele frequency method is used to obtain the four simulated *AD* values for each SV in the sliding window we want to test. Fisher's exact test is performed to judge if each SV is significant, and an *sSV/totalSV* ratio is then calculated. We repeat the simulation 10,000 times to obtain 10,000 *sSV/totalSV* ratios, and then the threshold can be estimated with these simulated ratios using the percentile method.

The *sSV/totalSV* ratio is a parameter at the genomic region level, while both the allele frequency and the G-statistic value are parameters at the SV level. Simulation is used for confidence interval/threshold estimation in both the allele frequency method and the significant structural variant method; confidence intervals for the former are estimated at the SV level, while thresholds for the latter are estimated at the genomic region level. A genomic region of the size of a sliding window normally contains hundreds or thousands of SVs, which can potentially translate into a higher statistical power for the latter; therefore, the significant structural variant method should have a higher detection power than that of the allele frequency method. However, it is hard to predict if the significant structural variant method has a higher detection power than that of the G-statistic method, because they use different approaches for threshold estimation.

Genome-wide confidence intervals/thresholds were used in the G-statistic method (Magwene et al. 2011) and the BRM method (a variant of the allele frequency method) (Huang et al. 2020). Each sliding window can have a different number of SVs, and the SVs in it might be sequenced at different depths. If the number of SVs of a sliding window and the sequencing depths of these SVs can affect confidence interval/threshold estimation, a genome-wide value may over- or underestimate the confidence interval/threshold of a specific sliding window; thus, it may not be proper to use it to identify trait-controlling genes.

5.4 A CASE STUDY USING A DATASET OF RICE SEEDLING COLD TOLERANCE

Different parameters are used for data analysis, and different approaches are used for confidence interval/threshold estimation in the algorithms given earlier, and we want to know how they perform in the detection of trait-associated genomic regions. We used a BSA-Seq dataset of rice seedling cold tolerance (Yang et al. 2013) to compare the detection power of these algorithms. Two bulks, 385 of extremely cold-tolerant seedlings and 430 of extremely cold-sensitive seedlings, were assembled from an F_3 population of 10,800 individuals. The pooled DNA samples were sequenced using the Illumina HiSeq 2000 platform, and the accession number of the sequencing data in the NCBI SRA repository (Sayers et al. 2025) is SRP021494. One of the parents of this population was the Japonica rice variety Nipponbare, of which the whole genome sequence is publicly available, and we used it as the reference genome sequence when performing SV calling. Previously, the sequencing data was treated as single-end for SV calling when we developed the significant structural variant method (Zhang and Panthee 2020); here, we treat it as paired-end.

We analyzed the data and estimated the confidence intervals/thresholds as described earlier. However, we encountered the same issue that Mansfeld and Grumet (2018) did when using the G-statistic method; the filtering mechanism, as described earlier, failed to remove any outliers from

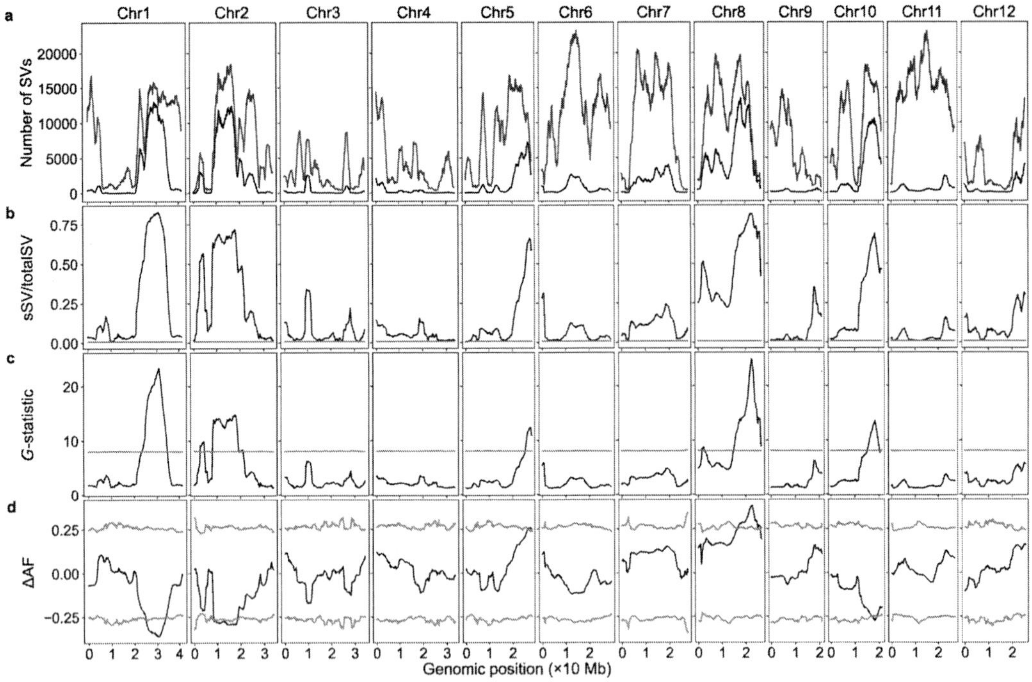

FIGURE 5.2 BSA-Seq data analysis plots. (**a**) Genomic distributions of sSVs and total SVs. Blue curves: total SVs; the black curves: sSVs. (**b**) Genomic distributions of sSV/totalSV ratios. The black curves represent the sSV/totalSV ratios, and the red lines are the thresholds. (**c**) Genomic distributions of *G*-statistic values. The black curves represent the *G*-statistic values, and the magenta curves are the thresholds. (**d**) Genomic distributions of ΔAF values. The black curves represent the ΔAF values, and the magenta curves are the confidence intervals.

the dataset, suggesting that the coefficient Eq. 5.2 is not universally applicable to all datasets. Mansfeld and Grumet (2018) used an arbitrary ΔAF to filter out trait-associated SVs; any SV with its absolute ΔAF greater than 0.1 is considered an outlier. Using this filtering mechanism, the *G*-statistic method had very similar detection power to that of the allele frequency method (Mansfeld and Grumet 2018). However, it is difficult to know if such an arbitrary ΔAF is applicable to all datasets. Thus, we used a Monte Carlo simulation to calculate the *G*-statistic threshold of each SV as in the allele frequency method here. For the significant structural variant method, we used $\alpha = 0.01$ to identify sSVs in both the real dataset and the simulated dataset.

The results indicated that the significant structural variant method has the highest detection power; almost all peaks were significant (Figure 5.2b). Any trait expression is the result of interactions among many cellular processes and biochemical reactions that are controlled by many genes. Thus, it is not surprising that many significant peaks were identified here. However, the minor peaks, although significant, explain a very small portion of the phenotype difference between the bulks. We can safely ignore these peaks and focus on the major ones for practical reasons. The *G*-statistic method has a slightly higher detection power than that of the allele frequency method; the former detected additional peaks on chromosomes 2 and 8 (Figures 5.2c and 5.2d). When estimating *G*-statistic thresholds via simulation, more peaks were detected than those in the study of Mansfeld and Grumet (2018). It is worth noting that the positions of all peaks/valleys are very similar among these three methods.

5.5 MARKER DEVELOPMENT BASED ON THE BSA-SEQ RESULTS

Via BSA-Seq, each trait-controlling gene can be mapped to a peak/valley sliding window. Then we often need to identify the trait-controlling genes and characterize their functions. It would be easier to identify candidate genes by gene function prediction if such a gene is delineated to a narrow genomic region. However, sliding windows flanking this peak/valley have ΔAF, G-statistic, or *sSV/totalSV* values very close to that of the peak/valley, and the value differences among these sliding windows can be caused by random error, so we are not certain which sliding window contains the trait-controlling gene based solely on the BSA-Seq results. Thus, traditional genetic mapping is often performed to overcome this shortcoming of BSA-Seq. The current approach is as follows (Imerovski et al. 2019; Pan et al. 2022; Zhang et al. 2022): (1) design primers based on information obtained via genotyping-by-sequencing or other means and (2) perform PCR and resolve PCR products via gel-based technologies to genotype individuals, and fine-map the trait-controlling gene. This approach is labor-intensive. However, delineating the trait-controlling gene can be done with much less effort by using the approach described next.

After BSA-Seq data analysis, numerous SVs in the vicinity of a peak/valley can be identified, and primers can be designed to amplify DNA fragments containing one or a few such SVs based on the genome sequence. Unlike the work (Imerovski et al. 2019; Pan et al. 2022; Zhang et al. 2022) mentioned earlier, no other assays are needed for this marker development step. We can select a genomic region of a certain length with the peaks/valleys as the center and design a series of primer pairs roughly evenly distributed within this region. PCR analysis can be performed to amplify the DNA fragments, and the PCR products can be resolved using high-resolution melting (HRM) analysis or TaqMan™ SNP Genotyping Assays, not the low-throughput gel-based technologies. These markers can be used to further fine-map the trait-controlling genes and for marker-assisted selection in breeding.

REFERENCES

Arikit S, Wanchana S, Khanthong S, Saensuk C, Thianthavon T, Vanavichit A, Toojinda T. 2019. QTL-seq identifies cooked grain elongation QTLs near soluble starch synthase and starch branching enzymes in rice (Oryza sativa L.). *Sci Rep.* 9(1):1–10. https://doi.org/10.1038/s41598-019-44856-2 [accessed 2019 Dec 17].

Asnaghi C, Roques D, Ruffel S, Kaye C, Hoarau J-Y, Télismart H, Girard JC, Raboin LM, Risterucci AM, Grivet L, et al. 2004. Targeted mapping of a sugarcane rust resistance gene (Bru1) using bulked segregant analysis and AFLP markers. *Theor Appl Genet.* 108(4):759–764. https://doi.org/10.1007/s00122-003-1487-6 [accessed 2022 Feb 2].

Cai H-W, Gao Z-S, Yuyama N, Ogawa N. 2003. Identification of AFLP markers closely linked to the rhm gene for resistance to Southern Corn Leaf Blight in maize by using bulked segregant analysis. *Mol Gen Genomics.* 269(3):299–303. https://doi.org/10.1007/s00438-003-0837-z [accessed 2022 Feb 1].

Cao M, Li S, Deng Q, Wang H, Yang R. 2021. Identification of a major-effect QTL associated with pre-harvest sprouting in cucumber (Cucumis sativus L.) using the QTL-seq method. *BMC Genomics.* 22(1):249. https://doi.org/10.1186/s12864-021-07548-8 [accessed 2021 Jun 29].

Chen Q, Song J, Du W-P, Xu L-Y, Jiang Y, Zhang J, Xiang X-L, Yu G-R. 2018. Identification and genetic mapping for rht-DM, a dominant dwarfing gene in mutant semi-dwarf maize using QTL-seq approach. *Genes Genomics.* 40(10):1091–1099. https://doi.org/10.1007/s13258-018-0716-y [accessed 2023 Jul 9].

Clevenger J, Chu Y, Chavarro C, Botton S, Culbreath A, Isleib TG, Holbrook CC, Ozias-Akins P. 2018. Mapping late leaf spot resistance in peanut (Arachis hypogaea) using QTL-seq reveals markers for marker-assisted selection. *Front Plant Sci.* 9:83. https://doi.org/10.3389/fpls.2018.00083.

Das S, Upadhyaya HD, Bajaj D, Kujur A, Badoni S, Laxmi null, Kumar V, Tripathi S, Gowda CLL, Sharma S, et al. 2015. Deploying QTL-seq for rapid delineation of a potential candidate gene underlying major trait-associated QTL in chickpea. *DNA Res.* 22(3):193–203. https://doi.org/10.1093/dnares/dsv004.

Davies L, Gather U. 1993. The identification of multiple outliers. *J Am Stat Assoc.* 88(423):782–792. https://doi.org/10.2307/2290763 [accessed 2023 Apr 8].

Daware A, Das S, Srivastava R, Badoni S, Singh AK, Agarwal P, Parida SK, Tyagi AK. 2016. An efficient strategy combining SSR markers- and advanced QTL-seq-driven QTL mapping unravels candidate genes regulating grain weight in rice. *Front Plant Sci.* 7:1535. https://doi.org/10.3389/fpls.2016.01535.

Ehrenreich IM, Torabi N, Jia Y, Kent J, Martis S, Shapiro JA, Gresham D, Caudy AA, Kruglyak L. 2010. Dissection of genetically complex traits with extremely large pools of yeast segregants. *Nature.* 464(7291):1039–1042. https://doi.org/10.1038/nature08923 [accessed 2022 May 29].

Giovannoni JJ, Wing RA, Ganal MW, Tanksley SD. 1991. Isolation of molecular markers from specific chromosomal intervals using DNA pools from existing mapping populations. *Nucleic Acids Res.* 19(23):6553–6568. https://doi.org/10.1093/nar/19.23.6553 [accessed 2019 Feb 2].

Gu A, Meng C, Chen Y, Wei L, Dong H, Lu Y, Wang Y, Chen X, Zhao J, Shen S. 2017. Coupling Seq-BSA and RNA-Seq analyses reveal the molecular pathway and genes associated with heading type in Chinese cabbage. *Front Genet.* 8. https://doi.org/10.3389/fgene.2017.00176 [accessed 2021 Feb 27].

Guy (https://stats.stackexchange.com/users/5339/guy). Negative values in G test. https://stats.stackexchange.com/q/372071.

Harris CR, Millman KJ, van der Walt SJ, Gommers R, Virtanen P, Cournapeau D, Wieser E, Taylor J, Berg S, Smith NJ, et al. 2020. Array programming with NumPy. *Nature.* 585(7825):357–362. https://doi.org/10.1038/s41586-020-2649-2 [accessed 2025 Jun 25].

Hartl L, Mohler V, Zeller FJ, Hsam SL, Schweizer G. 2011. Identification of AFLP markers closely linked to the powdery mildew resistance genes Pm1c and Pm4a in common wheat (Triticum aestivum L.). *Genome.* https://doi.org/10.1139/g98-129 [accessed 2022 Feb 1].

Huang L, Tang W, Bu S, Wu W. 2020. BRM: A statistical method for QTL mapping based on bulked segregant analysis by deep sequencing. *Bioinformatics.* 36(7):2150–2156. https://doi.org/10.1093/bioinformatics/btz861 [accessed 2020 Dec 3].

Imerovski I, Dedić B, Cvejić S, Miladinović D, Jocić S, Owens GL, Tubić NK, Rieseberg LH. 2019. BSA-seq mapping reveals major QTL for broomrape resistance in four sunflower lines. *Mol Breeding.* 39(3):41. https://doi.org/10.1007/s11032-019-0948-9 [accessed 2019 Nov 5].

Klein H, Xiao Y, Conklin PA, Govindarajulu R, Kelly JA, Scanlon MJ, Whipple CJ, Bartlett M. 2018. Bulked-segregant analysis coupled to whole genome sequencing (BSA-Seq) for rapid gene cloning in maize. *G3 (Bethesda).* 8(11):3583–3592. https://doi.org/10.1534/g3.118.200499.

Magwene PM, Willis JH, Kelly JK. 2011. The statistics of bulk segregant analysis using next generation sequencing. *PLoS Comput Biol.* 7(11):e1002255. https://doi.org/10.1371/journal.pcbi.1002255 [accessed 2019 Feb 2].

Mansfeld BN, Grumet R. 2018. QTLseqr: An R package for bulk segregant analysis with next-generation sequencing. *Plant Genome.* 11(2). https://doi.org/10.3835/plantgenome2018.01.0006.

Michelmore RW, Paran I, Kesseli RV. 1991. Identification of markers linked to disease-resistance genes by bulked segregant analysis: A rapid method to detect markers in specific genomic regions by using segregating populations. *PNAS.* 88(21):9828–9832. https://doi.org/10.1073/pnas.88.21.9828 [accessed 2021 Feb 27].

Miko I. 2008. Epistasis: Gene interaction and phenotype effects. *Nat Educ.* 1(1):197. http://www.nature.com/scitable/topicpage/epistasis-gene-interaction-and-phenotype-effects-460 [accessed 2024 Nov 20].

Pan J, Zhou X, Ahmad N, Zhang K, Tang R, Zhao H, Jiang J, Tian M, Li C, Li A, et al. 2022. BSA-seq and genetic mapping identified candidate genes for branching habit in peanut. *Theor Appl Genet.* 135(12):4457–4468. https://doi.org/10.1007/s00122-022-04231-8 [accessed 2023 Oct 3].

Sayers EW, Beck J, Bolton EE, Brister JR, Chan J, Connor R, Feldgarden M, Fine AM, Funk K, Hoffman J, et al. 2025. Database resources of the National Center for Biotechnology Information in 2025. *Nucleic Acids Res.* 53(D1):D20–D29. https://doi.org/10.1093/nar/gkae979 [accessed 2025 Jun 26].

Schneeberger K, Ossowski S, Lanz C, Juul T, Petersen AH, Nielsen KL, Jørgensen J-E, Weigel D, Andersen SU. 2009. SHOREmap: Simultaneous mapping and mutation identification by deep sequencing. *Nat Methods.* 6(8):550–551. https://doi.org/10.1038/nmeth0809-550 [accessed 2020 Jan 14].

Shen X, Zhou M, Lu W, Ohm H. 2003. Detection of Fusarium head blight resistance QTL in a wheat population using bulked segregant analysis. *Theor Appl Genet.* 106(6):1041–1047. https://doi.org/10.1007/s00122-002-1133-8 [accessed 2022 Feb 12].

Takagi H, Abe A, Yoshida K, Kosugi S, Natsume S, Mitsuoka C, Uemura A, Utsushi H, Tamiru M, Takuno S, et al. 2013. QTL-seq: Rapid mapping of quantitative trait loci in rice by whole genome resequencing of DNA from two bulked populations. *Plant J.* 74(1):174–183. https://doi.org/10.1111/tpj.12105.

Trick M, Adamski NM, Mugford SG, Jiang C-C, Febrer M, Uauy C. 2012. Combining SNP discovery from next-generation sequencing data with bulked segregant analysis (BSA) to fine-map genes in polyploid wheat. *BMC Plant Biol.* 12(1):14. https://doi.org/10.1186/1471-2229-12-14 [accessed 2020 Feb 19].

Wang GL, Paterson AH. 1994. Assessment of DNA pooling strategies for mapping of QTLs. *Theoret Appl Genetics.* 88(3):355–361. https://doi.org/10.1007/BF00223645 [accessed 2022 Feb 12].

Yang Z, Huang D, Tang W, Zheng Y, Liang K, Cutler AJ, Wu W. 2013. Mapping of quantitative trait loci underlying cold tolerance in rice seedlings via high-throughput sequencing of pooled extremes. *PLoS ONE.* 8(7):e68433. https://doi.org/10.1371/journal.pone.0068433 [accessed 2019 Feb 2].

Zhang J, Panthee DR. 2020. PyBSASeq: A simple and effective algorithm for bulked segregant analysis with whole-genome sequencing data. *BMC Bioinformatics.* 21(1):99. https://doi.org/10.1186/s12859-020-3435-8 [accessed 2020 Mar 6].

Zhang J, Panthee DR. 2022. Next-generation sequencing-based bulked segregant analysis without sequencing the parental genomes. *G3 Genes|Genomes|Genetics.* 12(2):jkab400. https://doi.org/10.1093/g3journal/jkab400 [accessed 2022 Jun 1].

Zhang K, Yuan M, Xia H, He L, Ma J, Wang M, Zhao H, Hou L, Zhao S, Li P, et al. 2022. BSA-seq and genetic mapping reveals AhRt2 as a candidate gene responsible for red testa of peanut. *Theor Appl Genet.* 135(5):1529–1540. https://doi.org/10.1007/s00122-022-04051-w [accessed 2023 Oct 3].

Zhang Y, Guo L, Shu Z, Sun Y, Chen Y, Liang Z, Guo H. 2013. Identification of amplified fragment length polymorphism (AFLP) markers tightly associated with drought stress gene in male sterile and fertile Salvia miltiorrhiza bunge. *Int J Mol Sci.* 14(3):6518–6528. https://doi.org/10.3390/ijms14036518 [accessed 2022 Feb 1].

6 Allele Mining for the Improvement of Tomatoes

Gograj Singh Jat, Suman Lata, Hanuman Ram, Chet Ram, Reshav Naik, Avinash Tomer, and Ramesh Kumar Yadav

6.1 INTRODUCTION

In the era of climate change, it is of utmost importance to develop the tomato genetic stocks or varieties tolerant to adverse climatic conditions over several growing locations and years. The change in the climatic parameters like rainfall, temperatures, and CO_2 concentrations in growing environments alters the plants' circadian rhythm, leading to severe crop losses (Gobu et al., 2020). Tomato (*Solanum lycopersicum*) is one of the most widely cultivated and consumed vegetable crops worldwide, valued for its nutritional content, culinary versatility, and industrial applications. The consumption of tomatoes, both fresh and processed, is still increasing, having been produced about 192 million tons in 2023 worldwide (FAO, 2024). More than half of the total production comes from Asia (63%), followed by the Americas and Europe. Tomato plant is known to be more sensitive to climate change; hence, tomato breeders face a difficult task in improving the genetic base of new varieties that can adapt well to changing climates (Lata et al., 2024). The conventional breeding techniques used to develop new varieties contributed significantly to human beings, but the resources utilized and time consumed in this process are quite high because of long-selection procedure involved (Patra et al., 2020; Chakraborti et al., 2021). With the growing global demand and the need to improve traits such as yield, disease resistance, pest tolerance, abiotic stress tolerance, and fruit quality, advanced genomic tools have become essential in modern breeding programs (Varshney et al., 2016).

Several yield traits are known to be inherited in a complex pattern in tomato, which makes selection only based on phenotypes difficult. With the advances in molecular breeding and marker development, several genomic regions or quantitative trait loci (QTLs) for many economic traits have been identified and utilized in several breeding programs (Causse et al., 2004). Moreover, the development of data science provided the foundation for breeding science's decision-making in the selection of desirable progenies for developing superior varieties. In tomato, several QTLs have been identified for economic traits like plant height (*qPH2a and qPH3a*) with the major objectives to reduce plant height, which is highly beneficial for mechanical harvesting and space optimization. The QTLs have also been identified for fruit weight, variation in fruit size and influencing the yield potential, soluble solids content (*qSSC7a* and *qSSC9a*), locule number (*ln1.1 and ln2.1*), heat tolerance (*qPV11*), and bacterial wilt tolerance, adding-aiding in development of resistant cultivars (Liabeuf et al., 2018) and giving breeders better choices to introgress into an elite background (Celik et al., 2017). Genome-wide association study (GWAS) has found promising results in the identification of genomic regions responsible for multiple stress tolerance in tomato (Kim et al., 2021). Moreover, progeny selection is also possible based on the genotypic values estimated from the genome-wide distribution of marker effects on a specific trait (Kim et al., 2021). Genomic selection has ample scope in tomato breeding; however, only limited studies have been conducted for a few major traits (Kim et al., 2021). Therefore, it is very difficult to dissect the individual gene and study the effect and variation of the gene on the trait's expression. To comprehend the quantitative traits, it is important to dissect all of the variant alleles that contribute to the trait expression as well as diversity (Bhattarai et al., 2016).

 DOI: 10.1201/9781003506539-6

Allele mining, the process of identifying and characterizing allelic variants of genes associated with desirable traits, is a promising approach to dissecting natural variant alleles of a candidate gene that are known to contribute to the expression of a trait (Niu et al., 2021; Selvakumar et al., 2023). By exploring natural genetic diversity in cultivated and wild tomatoes, breeders can uncover novel alleles that are responsible for stress tolerance, improved fruit quality, and other economically important traits (Kim et al., 2021; Selvakumar et al., 2023). The recent advances in genomics studies, including genome editing, high-throughput sequencing, and bioinformatics, have significantly enhanced the efficiency and scope of allele mining in tomato improvement (Osei et al., 2024). To generate the important alleles in a gene of interest, the Targeting Induced Local Lesions IN Genomes (TILLING) combined with chemical mutagenesis is an alternative approach to allele mining (Selvakumar et al., 2023). This approach was applied to generate and identify the variations related to agro-economical traits in tomato. The genome sequencing by next-generation sequencing (NGS) platforms is another important strategy to identify and map genes/QTLs regulating important traits in crop plants. The resequencing clubbed with genotyping and phenotyping of tomato accessions resulted in allelic variability in several traits, including biotic stress, abiotic stress, and fruit quality (soluble solids content and fruit color) (Panthee et al., 2012). These technologies allow the targeted exploration of functional genes and regulatory regions across diverse tomato germplasm collections, which may enable the identification of rare and valuable alleles that may have been lost during domestication and modern breeding (Osei et al., 2024; Selvakumar et al., 2023). The major objectives of allele mining are to identify a small number of genotypes that carry diverse alleles in a large population for a quantitative trait. As a result in the era of genomic-assisted breeding, the mining of alleles for economically important traits in tomato is gaining traction (Bhardwaj et al., 2024). Traditionally, the QTLs and alleles were detected using biparental populations for trait expression. Further, attention was given to using the multiparent populations to harness more alleles for a trait, but GWAS on natural population was found to be equally effective (Kim et al., 2021; Selvakumar et al., 2023). GWAS on natural population identifies a wide array of alleles for economic traits; however, the composition of multi-parent advanced generations inter-cross (MAGIC) population was able to identify only those alleles that exist in the parental populations (Ariizumi et al., 2013; Kim et al., 2021). In early GWAS studies, candidate genes and alleles have rarely been identified, considering the low depth of genotyping. With the development of deep sequencing platforms, it is possible to detect nucleotide variations in all the genes and identify the putative functions of all the alleles (Kim et al., 2021).

In tomato, many studies have been reported to identify candidate genes with putative alleles for several important traits. The availability of genome data provides new opportunities to reduce the number of genotypes and identify desirable genes for a phenotype (Kim et al., 2021; Selvakumar et al., 2023; Osei et al., 2024; Bhardwaj et al., 2024). Known gene sequences can also be searched in the database available to identify the unique genotypes carrying targeted alleles for a phenotype (Selvakumar et al., 2023). Hence, this chapter focuses on the current status and future prospects of allele mining for many traits in tomato, including mining of alleles for yield and its related traits, quality traits, biotic stress tolerance, abiotic stress tolerance, and other economic traits contributing to tomato improvement program.

6.2 USE OF GENETIC DIVERSITY AND TOMATO GENE BANK

Genetic diversity is the foundation for any crop improvement program to sustain population growth and ensure food and nutritional security. Traditional varieties or landraces of tomato harbor a large store of valuable genes that are explored to develop new varieties for enhanced yield, nutritional quality, and tolerance to future climate. Enhancing the genetic makeup of tomatoes has long been a priority for tomato breeders, with major focus on improving yield, disease resistance, fruit quality, and environmental adaptability (Fadhilah et al., 2022; Cecílio et al., 2014; Muthukumar et al., 2024). These modern-day varieties are the result of intensive plant breeding techniques used since the beginning of the 20th century, and the presence of the natural biodiversity of wild tomato species has been key in this success (Mata-Nicolás et al., 2020). All types of disease resistance in contemporary tomato varieties originated from their wild species (Ric & Chetelat, 1995). Breeding efforts often involved the incorporation of genetic material from wild tomato species, particularly for disease resistance (Palanisamy et al., 2023). The incorporation of disease resistance genes from wild species such as *Solanum pimpinellifolium* and *Solanum habrochaites* has been a significant focus (Ric & Chetelat, 1995). *Solanum pimpinellifolium* is the closest wild relative to cultivated tomatoes, *Solanum lycopersicum* var. *lycopersicum* L., and *Solanum lycopersicum* var. *cerasiforme. Solanum pimpinellifolium* hybridized with cultivated tomato make it a valuable species for desirable traits. It has been commercially utilized for the improvement of quality traits such as soluble solid content, fruit firmness, fruit color (Stevens & Rick, 1986), volatile compounds (Capel et al., 2015), resistance against biotic stresses, including tomato leaf curl virus (Banerjee & Kalloo, 1987), *Alternaria solani*, *Fusarium oxysporum*, and *Phytophthora infestans* (Alexander & Hoover, 1955). The species *Solanum lycopersicum* var. *cerasiforme* is a valuable genetic source for moisture tolerance (Nuez & Diez, 2008) and resistance to root rot caused by *Phytophthora* (Rodríguez et al., 2013). The modern breeding programs focused on resistance, yield, and quality such as firmness, color, or texture (Foolad, 2007), plant habit, and mechanized harvesting in processing cultivars or fresh market (Foolad, 2007). However, nowadays the major objective is the sustainable production and adaptation to unfavorable environmental conditions due to climate change and requirement of improved nutritional quality (Panthee et al., 2013b). The germplasm resources, including wild species such as *Solanum pimpinellifolium, Solanum chilense, Solanum habrochaites, Solanum peruvianum, and Solanum pennellii*, have been used for the creation of genetic variation and to bypass the limited genetic diversity available within *Solanum lycopersicum* var. *lycopersicum* (Brekke et al., 2019; Yan et al., 2018). ToLCV-resistant genes have been introgressed into commercial tomato cultivars from three wild species, *S. chilense*, *S. habrochaites*, and *S. peruvianum* (Yan et al., 2018). Resistance to Begomovirus has been another notable example in tomato improvement for successfully introgressing from resistant accessions of the wild species of tomato into cultivated varieties/lines (Ji et al., 2007). To date, six resistant genes (*Ty-1, Ty-2, Ty-3, Ty-4, ty-5,* and *Ty-6*) for tomato leaf curl virus have been transferred from wild accessions to the cultivated lines (Lata et al., 2019; Palanisamy et al., 2023). *Ty-1*, a partial dominant major resistant gene, was introgressed from *S. chilense* accession LA1969 (Zamir et al., 1994); *Ty-2*, a dominant resistant gene, was introgressed from a *Solanum habrochaites* accession into a tomato line H24 (Hanson et al., 2006); *Ty-3*, a partial dominant gene, was introgressed from a *S. chilense* accession, LA2779 and LA1932 (Ji et al., 2007); *Ty-4* was introgressed from *S. chilense* and showed a low level of resistance compared to *Ty-3* (Ji et al., 2009); whereas, *ty-5* is a recessive resistant gene found in tomato cultivar "Tyking" (Hutton et al., 2012). *Ty-6*, a major resistant gene, is found in tomato line FLA456 (Kadirvel et al., 2013). The present-day tomato cultivars carrying *Ty* genes have shown a range of resistance levels and variable responses depending upon strain, isolate, and species, inoculum level, and other factors (Ji et al., 2007, 2009; Shahid et al., 2015; Elbaz et al., 2016). In tomato, several inbred lines (ILs) have been developed using wild species that carry a single introgressed region, which are more useful in QTL identification. These include *S. pennellii* (Eshed & Zamir, 1995), *S. habrochaites* (Monforte & Tanksley, 2000), *S. lycopersicoides,* and *S. sitiens* (Canady et al., 2005, 2006). Wild tomato species have been well characterized for several bioactive compounds, chiefly high lycopene content (Foolad & Panthee, 2012). The wild species *S. pimpinellifolium* contains five times higher lycopene than the cultivated tomato varieties. These large collections of genetic resources have wide genetic variability for several quality traits, including high lycopene, β-carotene, acidity, and TSS, which could be utilized in breeding programs for tomato improvement for quality traits (Singh et al., 2018). Several recombinant inbred lines (RILs) have been developed in tomato for both biotic and abiotic stresses and used extensively. These RILs were developed using different accessions between *S. lycopersicum* × *S. cheesmaniae* and *S. lycopersicum* × *S. lycopersicum* var. *cerasiformae* for desirable traits. The development of the resistant inbred lines are the successful examples in tomatoes for several biotic and abiotic stresses, including CLN1466EA (bacterial wilt and TMV resistance, heat tolerance, and large size fruit suitable for fresh market), CLN5915-2-6-3-3-0-4 (Heat tolerance), CLN399-19-6-13-12-4 (heat tolerance and resistance to bacterial wilt and TMV (*Tm2a* gene), CRA84-26-1-3 (bacterial wilt resistance and large fruit size), NC 84173 (Fusarium and Verticillium wilt resistance, superior fruit quality and large fruit size) (Siddique et al., 2024), NC84173 (multiple disease resistance, fresh market tomato inbred line), LA 0722 (superior fruit quality, disease resistance and abiotic stress tolerance), H7996 (resistance to bacterial wilt), UC 204B (suitable for processing), and LA 2093 (high fruit quality and disease resistance) (Foolad, 2015). The accessions of wild species, *S. chilense* (LA1938, LA2879, LA1969, LA2748), carry novel alleles resistant to ToLCV and are a potential source for introgression of ToLCV-resistance genes into commercial tomato cultivars (Palanisamy et al., 2023).

Hence, maintenance and characterization of available resources are crucial to achieve the present-day tomato breeding objectives. Moreover, the creation of gene bank collections characterized at genotypic and phenotypic levels is the main objective of sustainable breeding. It is estimated that more than 80 thousand accessions are maintained *ex-situ* across gene banks located around the globe (Ebert & Chou, 2015). The World Vegetable Centre (WorldVeg, Taiwan) holds the largest collection comprising 8,200 cultivated tomatoes, mostly cultivars and landraces, and several hundred accessions of wild species. Another major collection is maintained at the Plant Genetic Resources Unit of the USDA (USDA-PGRU, NY, USA), which consists of more than 6,600 accessions. The Tomato Genetic Resource Centre at the University of Davis (TGRC, CA, USA) also maintains over 1,000 wild accessions. In addition to this, TGRC, CA, USA, also holds several pre-breeding genetic stocks consisting of interspecific mapping populations, introgression lines, backcrosses, and recombinant inbred lines. Other main gene banks of tomato are IPK, Leibniz Institute of Plant Genetics and Crop Plant Research, Gatersleben, Germany; CGN, Centre for Genetic Resources, Netherlands; INRAE-GAFL, French National Institute for Agricultural Research Genetics and improvement of Fruit and Vegetables, Avignon, France; ICAR-National Bureau of Plant Genetic Resources, New Delhi, India, holds thousands of individuals, including scientific core collections, re-sequenced genotypes, intra- and inter-specific populations, and multi-parental populations.

6.3 SIGNIFICANCE OF TOMATO REFERENCE GENOME

Tomato is a model vegetable crop to study fleshy fruit traits. It has some intriguing developmental traits, viz., compound leaves, meaty fruits, and sympodial habit of shoot branching (Townsley & Sinha, 2012; Manning et al., 2006; Schmitz & Theres, 1999). It is easy to maintain due to a short generation cycle, routine transformation technology, and simple diploid genetics. All these factors combine to make tomatoes a great plant model species for studying basic and applied plant research. Many genetic and genomic resources have already been developed in tomato before commencement of the tomato genome sequencing effort. The 12 linkage groups that correspond to the cytologically visible chromosomes were reported by tomato geneticists using a variety of morphological, isozyme, and molecular markers to create a genetic map of the tomato. Because of the availability of comprehensive molecular linkage map, breeders can find several quantitative trait loci (QTLs), which help them to comprehend the genetic basis of several agronomical quantitative traits (Paterson et al., 1988). Comprehensive information on tomato genetic and genomic resources is provided by the Solanaceae Genomics Network website (Sol Genomics Network). Moreover, self/hybrid incompatibility studies have also been conducted using tomato and its wild relatives as models (Li & Chetelat, 2010; Moyle & Nakazato, 2012).

A high-quality reference genome is a critical foundation for functional genomics research and crop genetic improvement. In 2005, 14 countries began a worldwide project to sequence the tomato genome. High-coverage, shorter SOLiD and Illumina GAIIx reads were used in combination with longer Sanger and 454/Roche GS FLX reads to sequence the genome of the domesticated tomato cultivar Heinz 1706. The sequences were grouped into 91 scaffolds that covered 760 Mb of the roughly 900 Mb genome and were aligned to the 12 tomato chromosomes with 34,727 putative protein-coding genes. Of them, 33,840 genes (97.4%) were correctly mapped to chromosomes (The Tomato Genome Consortium, 2012). Later improvement to the Heinz 1706 genome sequence project have greatly enhanced the information. 801.8 Mb of sequencing data is included in the most recent SL5.0 genome assembly, featuring fewer gaps (31 in SL5.0 vs. 259 in SL4.0) and a higher contig N50 size (41.7 Mb in SL5.0 vs. 5.5 Mb in SL4.0).

The Heinz sequence was also matched to a draft sequence of *S. pimpinellifolium*. There is only 0.6% nucleotide divergence between these two genomes, indicating their extreme similarity at the evolutionary level. Between domesticated and wild tomatoes, 60% of the genes are the same or only have synonymous changes, while 40% have non-synonymous changes, including changes in stop codons that may have an impact on gene functionality. The tomato and *S. pimpinellifolium* genomes exhibit nucleotide divergence of over 8% when compared to the potato genome. Furthermore, the genomes of other economically significant members of the Solanaceae family, such as pepper and eggplant, possess strong syntenic relationship with tomato genomes.

Besides the Heinz 1706 reference genome, high-quality genome assemblies have been achieved for 24 other tomato cultivars, including wild species of tomato. *Solanum pimpinellifolium* was the first species of wild tomato whose genome was sequenced (Wang et al., 2024). Additionally, the genomes of 10 other wild tomato species, such as *S. habrochaites* (LYC4, LA0407, and LA1777), *S. pennellii* (LA0716 and LA5240), *S. galapagense* (LA0436 and LA0317), *S. chilense* (LA1969 and LA3111), *S. arcanum* (LA2157), *S. peruvianum* (LA0446), *S. corneliomulleri* (LA1331), *S. neorickii* (LA0247), *S. chmielewskii* (LA1028), and *S. lycopersicoides* (LA2951) are now available (Yu et al., 2022; Li et al., 2023a; Wang et al., 2024). The application of genomics in tomato breeding is accelerated by these tomato genome sequencing initiatives, which lay a strong basis for the later identification of functional genes and study of the evolutionary transition from wild to cultivated species.

The incapacity of a single genome sequence to capture all the genetic variety within a species is one of its limitations. When compared to conventional single-reference genomes, a pan-genome, a collective sum of all known genomic information about a species, significantly increases the inclusivity of genetic diversity and successfully reduces the biases caused by a single-reference genome when used to detect genetic variations. The present large-sized tomato evolved from a small-fruited cherry tomato through artificial selection. For better understanding of evolution and domestication, and genes involved in tomato domestication and trait improvement, pan-genomes have been sequenced. By combining more diverse tomato genetic resources, such as 10 wild species and one cultivated tomato accession, a "super-pangenome" was created very recently (Li et al., 2023a). In addition to covering a wide variety of genetic variability within the Solanaceae family, this genomic resource discovered 9,320 new genes that had not been found in previous genome sequencing investigations.

Domestication and improvement in tomatoes significantly impacted traits, including disease resistance, yield, quality, and environmental adaptation. By combining cutting-edge methods like genome resequencing and genome-wide association studies (GWAS), scientists have successfully identified several important genes and genetic loci that control important horticultural traits in tomato. They have also located multiple genomic regions that were subjected to artificial and natural selection during the domestication and improvement processes (Lin et al., 2014). These important discoveries not only shed light on the genetic basis of fruit traits but also established a strong framework for upcoming tomato molecular breeding initiatives. Due to their exceptional genetic and trait richness, domesticated tomatoes and comparable wild species are ideal for research on domestication and evolution (Larry & Joanne, 2007; Nakazato et al., 2010). The discovery of the tomato genome sequence and the pan-genomes of many wild species facilitate the search for new genes/QTLs and trait-linked markers for marker-assisted breeding, as well as new genetic approaches such as genome editing in tomato. This has sped up the process of breeding tomatoes using genome sequencing.

6.4 ALLELE MINING FOR BIOTIC STRESS TOLERANCE GENES IN TOMATO

6.4.1 Resistance (R) Genes in Tomato

Tomato is served as a model species for studying fruit development, plant–pathogen interactions, and metabolic processes. It is affected by a range of pathogens that impact yield and economic quality of the produce (Panno et al., 2021). This disease vulnerability of the tomato crop has resulted in a rich pool of Resistance (R) genes that have evolved or been introgressed from wild species to regulate defense response through effector-triggered immunity (ETI) (Andolfo et al., 2021; He et al., 2025). R genes in plants play a critical role in the innate immune system by encoding specialized proteins. These proteins detect pathogen-associated molecules (effectors), triggering the plant defense mechanisms (Bent & Mackey, 2007). ETI is a robust and specialized defense response that can lead to a hypersensitive response (HR), characterized by localized cell death and activation of systemic acquired resistance (Spoel & Dong, 2012). Similar to R genes in plants, the pathogens possess *Avr* (Avirulence) genes that produce the effector proteins. These genes are critical in providing race-specific resistance against pathogens through gene-for-gene interactions, where each *R* gene corresponds to a specific pathogen avirulence (*Avr*) gene (Liu et al., 2007). The cloning of *Pto* gene that confers resistance to *Pseudomonas syringae* pv. *Tomato* causing bacterial speck was the first disease-resistance gene discovered in plants that abides by the "gene-for-gene" hypothesis (Martin et al., 1993; Zhang & Panthee, 2021).

R genes are grouped into different categories according to their domain architecture, such as coiled coil (CC), leucine-rich repeats (LRRs), Toll/Interleukin-1 receptor (TIR), the nucleotide-binding site (NBS), receptor, and kinase domains (Andolfo et al., 2013). According to the proteins encoded by the cloned R genes, they are categorized under five typical classes (Table 6.1). Furthermore, Kourelis and Hoorn (2018) classified the R genes according to their mechanism into nine classes. The nucleotide binding site-leucine rich repeats proteins (NLB-LRR or NLR) gene family represents majority of the R genes (Michelmore et al., 2013) and their ETI triggers a hypersensitive response, inducing programmed cell death (Lee & Yeom, 2015). The NBS-LRR family is generally subdivided into TIR-(TNL) and non-TIR, typically coiled-coil (CNL) groups.

TABLE 6.1
Classes of R Genes According to Protein Family

Protein Family	Domain Structure	Example Gene	Pathogen/Function	References
TIR-NBS-LRR (TNL)	Toll/Interleukin-1 Receptor (TIR), Nucleotide-Binding Site (NBS), Leucine-Rich Repeat (LRR)	*N* gene	Tobacco mosaic virus resistance	Erickson et al. (1999)
CC-NBS-LRR (CNL)	Coiled-Coil (CC), NBS, LRR	*I-2*	Fusarium wilt resistance in tomato	Simons et al. (1998)
Receptor-Like Kinase (RLK)	Extracellular LRR/lectin domain, transmembrane region, intracellular kinase domain	*SlFls2*	Flagellin sensing, bacterial defense	Robatzek et al. (2007)
Receptor-Like Protein (RLP)	Extracellular LRR domain, transmembrane region (lacks kinase domain)	*Cf4*	Fungal (*Cladosporium fulvum*) resistance in tomato	Thomas et al. (1997)
Kinase-Like Protein	Serine/threonine kinase domain (lacks extracellular domain)	*Pto*	Bacterial speck disease resistance in tomato	Martin et al. (1993)

6.4.2 R Genes Mining in Tomato

Tomato production is challenged by a wide range of diseases, including bacterial infections like bacterial speck, wilt, and spot; fungal diseases such as Fusarium crown and root rot, Fusarium wilt, stem rot, Verticillium wilt, gray leaf spot, early and late blight, leaf mold, gray mold, and powdery mildew; and viral diseases including tomato yellow leaf curl, tobacco mosaic virus, tomato spotted wilt, tomato chlorosis virus, tomato leaf curl New Delhi virus, and tomato brown rugose fruit virus (Shi et al., 2011; Panthee et al., 2017; Shi & Panthee, 2020). Breeding for disease resistance has historically been challenging due to the polygenic nature of many resistance traits and the narrow genetic base of cultivated tomato. However, its wild relatives possess extensive genetic variation and harbor novel alleles that confer resistance to numerous diseases, making them indispensable for breeding programs, which are aimed at improving disease resistance and overall crop resilience. Allele mining involves the systematic amplification and sequencing of candidate resistance gene regions from cultivated varieties, landraces, wild relatives, and inbred lines to uncover novel or superior alleles that impart disease resistance (Vossen et al., 2014). In the current genomic era, the conventional breeding techniques have been supplemented with molecular marker-assisted selection (MAS), genomics-led strategies that employ next-generation sequencing (NGS), and bioinformatics pipelines (Wei et al., 2020). R gene mining in tomato is therefore an essential strategy not only for understanding the genetic basis of resistance but also for facilitating rapid deployment of robust resistance traits into elite cultivars. The majority of characterized disease resistance genes in tomato encode nucleotide-binding site leucine-rich repeat (NBS-LRR) proteins (Du et al., 2025). The list of well-characterized R genes of tomato is given in Table 6.2.

6.4.3 Bacterial Diseases

The *Pto* gene, identified in *Solanum pimpinellifolium* L., confers resistance to bacterial speck (*Pseudomonas syringae* pv. *Tomato*), is one of the most studied R genes. It is closely linked to *Prf*, an NBS-LRR gene essential for downstream signaling (Salmeron et al., 1996). Quantitative trait loci (QTLs) associated with resistance to bacterial wilt caused by *Ralstonia solanacearum*, designated as *Bwr-6* and *Bwr-12*, were initially identified through linkage mapping in a biparental segregating population (Carmeille et al., 2006). Subsequently, genome-wide SNP analysis of resistant and susceptible bulks revealed that *Bwr-12* encodes a gene belonging to the NBS-LRR family (Kim et al., 2018). Ke et al. (2023) have discovered that tomatoes have evolved an unknown receptor-like kinase to recognize extracellular polygalacturonase (PehC) secreted by the *Ralstonia solanacearum* and trigger an immune response to limit the spread of the pathogen in the plant. Bacterial canker caused by *Clavibacter michiganensis subsp. michiganensis* (Cmm) is a serious disease affecting both field and protected cultivation of tomato. Pereyra-Bistraín et al. (2021) performed a comparative RNA-seq analysis between a susceptible *S. lycopersicum* cv. Alisa Craig and resistant *S. arcanum* LA2157 during the first 24 hrs post-inoculation with Cmm to identify potential resistance genes for bacterial canker of tomato. Bacterial spot of tomato is caused by multiple *Xanthomonas* spp., and identifying an R gene that is effective against all the evolved races of this pathogen is challenging. Four resistance genes (*Rx1*, *Rx2*, *Rx3*, and *Rx4*) have been identified that trigger a hypersensitive response against bacterial spot, but none provides effective resistance against all pathogen races (Adhikari et al., 2020; Adhikari et al., 2023).

6.4.4 Fungal Diseases

Fusarium wilt is a destructive fungal disease of tomato caused by *Fusarium oxysporum* f. sp *lycopersici* (Fol). The *I* and *I-2* genes, derived from *Solanum pimpinellifolium*, provide resistance to Fol races 1 and 2, respectively, while the *I-3* and *I-7* genes, sourced from *Solanum pennellii*, confer resistance to race 3 (Simons et al., 1998; Catanzariti et al., 2015; Gonzalez-Cendales et al., 2016).

TABLE 6.2
List of R Genes in Tomato, Their Source Wild Species, Protein Class, and Corresponding Chromosome Locations

Gene	Chromosome No.	Pathogen	Type/Protein class	Source	References
I-2	11	*Fusarium oxysporum*	CC-NBS-LRR	*S. pimpinellifolium*	Simons et al. (1998)
Mi-1.2	6	*Meloidogyne* spp.	CC-NBS-LRR	*S. peruvianum*	Milligan et al. (1998)
Mi-9	6	*Meloidogyne* spp.	CC-NBS-LRR	*S. arcanum*	Jablonska et al. (2007)
Sw-5	9	*Tospovirus* (Tomato Spotted wilt virus)	CC-NBS-LRR	*S. peruvianum*	Brommonschenkel et al. (2000)
Hero	4	*Globodera rostochiensis*	NBS-LRR	*S. pimpinellifolium*	Ernst et al. (2002)
Pto	5	*Pseudomonas syringae*	Ser/Thr protein kinase	*S. pimpinellifolium*	Martin et al. (1993)
Cf genes	All chromosome except 10 and 12	*Cladosporium fulvum*	RLPs	*S. pimpinellifolium*, *S. lycopersicum* var. *cerasiforme*, *S. habrochaites*, *S. peruvianum*	Kanwar et al. (1980), Zhao et al. (2022)
Ve1	9	*Verticillium dahliae*	RLPs	*S. peruvianum*	Fradin et al. (2009)
Sm	11	*Stemphylium lycopersici*	NBS-LRR	*S. pimpinellifolium*	Yang et al. (2022)
Bwr-12	12	*Bacterial wilt*	NBS-LRR	*S. pimpinellifolium*	Kim et al., (2018)
Sl5R-1	5	TSWV	NBS-LRR	*S. lycopersicum* var. *cerasiforme*	Qi et al. (2022)
I-3	7	*Fusarium oxysporum*	S-receptor like kinase	*S. pennellii*	Catanzariti et al. (2015)
I-7	8	*Fusarium oxysporum*	LRR-RLP	*S. pennellii*	Gonzalez-Cendales et al. (2016)
Prf	5	*Pseudomonas syringae*	NBS-LRR	*S. pimpinellifolium*	Salmeron et al. (1996)
Rx4	11	*Xanthomonas perforans* Race 3	NBS-LRR	*S. pimpinellifolium*	Pei et al. (2012)
Ph-3	9	*Phytophthora infestans*	CC-NBS-LRR	*S. pennellii*	Zhang et al. (2014)
EB 5 and *EB 9*	5. 9	*Alternaria solani*	QTLs	*S. pimpinellifolium*	Anderson et al. (2021)
Tm-22	9	ToMV	CC-NBS-LRR	*S. peruvianum*	Lanfermeijer et al. (2003)
Ty1	6	TLCV	RNA-dependent polymerase	*S. chilense*	Zamir et al. (1994), Verlaan et al. (2013)
Ty3	6		RNA-dependent polymerase		Ji et al. (2007),Verlaan et al. (2013)
Ty4	3				Ji et al. (2009)
Ty6	10				Hutton and Scott (2014)
Ty2	11		NBS-LRR	*S. habrochaites*	Hanson et al. (2006), Shen et al. (2020)
ty5	4		*Pelota* (messenger RNA surveillance factor)	*Tyking (S. lycopersicum)/S. peruvianum*	Lapidot et al. (2015)

The *I-7* gene provides broad-spectrum resistance to multiple Fol races, and its effectiveness is maintained even in the presence of pathogen effectors that can suppress other R gene-mediated defenses, therefore offering a promising approach for durable resistance breeding (Gonzalez-Cendales et al., 2016). A single dominant gene *Frl* that provides resistance against the Fusarium crown and root rot caused by *Fusarium oxysporum* f. sp. *radicis-lycopersici* was introgressed from *S. peruvianum* and fine-mapped on chromosome number 9 (Fazio et al., 1999; Devran et al., 2018). Tomato grey leaf spot caused by *Stemphylium lycopersici* has only a single known R gene (*Sm*) identified from *S. pimpinellifolium* that imparts strong resistance against it (Yang et al., 2022). Verticillium wilt of tomato, caused by the soil-borne fungus *Verticillium dahliae*, is countered by two resistance genes, *Ve1* and *Ve2*, which act independently to provide resistance against race 1 of the pathogen (Kawchuk et al., 2001). A total of five resistance genes have been identified against the devastating late blight disease of tomato caused by *Phytopthora infestans. Solanum pimpinellifolium* serves as a major source, contributing four of these genes—*Ph-1*, *Ph-2*, *Ph-3*, and *Ph-5* (Foolad et al., 2008). The *Ph-4* resistance gene was identified in an introgression line derived from *Solanum habrochaites* (Kole et al., 2006; Muthukumar et al., 2024). Introgressing multiple R genes into tomato cultivars enhances the level and durability of resistance against late blight. Numerous QTLs associated with resistance to early blight in tomato, caused by *Alternaria solani*, have been identified from wild relatives. However, these QTLs typically display polygenic inheritance, which contributes only modestly to individual resistance and is often linked to unfavorable horticultural traits (Adhikari et al., 2017). *Cladosporium fulvum*, the causal agent of tomato leaf mold, impairs leaf photosynthesis and exhibits high genetic variability, complicating disease management. There are a number of R genes (*cf1* to *cf24*) against leaf mold discovered from various wild relatives of tomato (Zhao et al., 2022). However, with the development of new resistant varieties, there is always a possibility of the emergence of new *C. fulvum* races capable of evading *Cf*-mediated immunity (Du et al., 2025). *Oidium neolycopersici*, the causative agent of tomato powdery mildew, impairs photosynthesis by infecting leaves. Six dominant (*Ol-1*, *Ol-3* to *Ol-6*) and one recessive (*ol-2*) resistance genes have been identified from wild species such as *S. chilense*, *S. habrochaites*, and *S. peruvianum* (Bai et al., 2008).

6.4.5 Viral Diseases

Tomato leaf curl virus (TLCV) is the most serious viral disease of tomato that can cause huge crop losses. To date, six independently inherited resistance genes (*Ty-1*, *Ty-2*, *Ty-3*, *Ty-4*, *ty-5*, and *Ty-6*) and a few QTLs, exhibiting differential levels of resistance to tomato yellow leaf curl virus, have been identified from wild tomato germplasm and introgressed into cultivated varieties (Palanisamy et al., 2023). The *Ty-2* gene originated from *S. habrochaites*; *ty-5* is presumed to be derived from *S. peruvianum*; and *Ty-1*, *Ty-3*, *Ty-4*, and *Ty-6* have been sourced from various accessions of *S. chilense* (Dhaliwal et al., 2020). Gene pyramiding for TLCV resistance often involves *Ty-1* and *Ty-3*, two allelic genes derived from *S. chilense*, which are among the most thoroughly characterized and widely deployed resistance loci (Verlaan et al., 2013). Resistance to tomato mosaic virus (ToMV) is governed by three dominant genes: *Tm-1*, *Tm-2*, and *Tm-2²* (also referred to as *Tm-2a*) discovered from the wild species *S. habrochaites* and *S. peruvianum* (Panthee et al., 2013a; Du et al., 2025).

6.4.6 Nematodes

Root knot nematode (RKN), *Meloidogyne* Spp., causes large-scale losses to tomato production and is classified as one of the most destructive plant pathogens (Jones et al., 2013). There are a total of 10 genes (*Mi-1, Mi-2, Mi-3, Mi-4, Mi-5, Mi-6, Mi-7, Mi-8, Mi-9, Mi-HT*) that are identified to provide resistance against the RKN, and only 5 of these genes have been mapped (El-sappah et al., 2019). While most root-knot nematode resistance genes have been identified from *Solanum peruvianum*, the Mi-9 gene was discovered in *S. arcanum*. The potato cyst nematode *Globodera rostochiensis* parasitizes the roots, affecting the crop production. The *Hero* gene provides broad-spectrum resistance to potato cyst nematodes, conferring high resistance against all *Globodera rostochiensis* pathotypes and partial resistance against *Globodera pallida* (Ernst et al., 2002).

6.4.7 microRNAs (miRNAs) for Biotic Stress Resistance

miRNA is a type of non-coding RNAs (ncRNAs) that plays an important role in gene regulation, development process, cell signaling, and response to various biotic and abiotic stresses (Sunkar et al., 2012). miRNAs are of 20–22 nucleotides in length, and sometimes long miRNAs of 24 nucleotides are there (Axtell & Bowman, 2008; Chávez Montes et al., 2014). Plant miRNAs are transcribed by RNA polymerase II from MICRORNA (MIR) genes as primary miRNAs (pri-miRNAs) (Lee et al., 2004; Jones-Rhoades et al., 2006). pri-miRNAs are cleaved by Dicer-like RNase III endonuclease 1 (DCL1) into shorter miRNA precursor or pre-miRNA (Kurihara & Watanabe, 2004). Further, pre-miRNA is processed by DCL1 as miRNA/miRNA* (miRNA star) duplex (Kurihara & Watanabe, 2004). miRNA/miRNA duplex is transported to the cytoplasm through HASTY (HST), and the selected guide strand is loaded on the Argonautel (AGO1), where miRNA-induced silencing complex (miRISC) is assembled.

miRNAs play an important role in defense against biotic stresses by regulating resistance (R) gene expression. miRNAs, diseases involved, and their putative targets are listed in Table 6.3. Olmo et al. (2024) used high-throughput RNA sequencing to examine miRNA profiles in 30-day-old tomato plants whose seeds were treated with the fungus *Trichoderma atroviride*. Compared to controls, 10 miRNAs showed differential expression (five up, five down) in the *Trichoderma*-treated plants, seven of which were known and three novel. These findings suggest that *Trichoderma* influences tomato defense responses through miRNA modulation, with potentially long-term effects, by targeting genes involved in pathogen interaction, membrane trafficking, and protein kinases.

Many miRNAs are reported to act as negative regulators of defense genes, for example, miR6024, miR396a-5p–3p, miR1916, sly-miR482f, and sly-miR5300. Negative correlation was revealed between expression of miR6024 (solanaceae-specific) and its target nucleotide-binding leucine-rich repeat (NLR) genes during infection with necrotrophic fungus *Alternaria solani* (Dey et al., 2022). Further infection of *Phytophthora infestans* and *Botrytis cinerea* showed downregulation of miR396a-5p and -3p in tomato. However, overexpression of miR396a-5p and -3p increased the susceptibility to *Phytophthora infestans* and *Botrytis cinerea* infection due to inhibition of R gene expression (Chen et al., 2017). In another study, levels of sly-miR1916 decreased in the defense against *Phytophthora infestans* (which causes late blight) and Botrytis cinerea (which causes grey mold) (Chen et al., 2019). On the other hand, sly-miR1916 overexpression transgenics became more vulnerable to both *P. infestans* and *B. cinerea* infections and also showed an increased tendency to produce harmful reactive oxygen species. However, silencing of sly-miR1916 leads to increased expression of defense genes like Strictosidine synthase (STR-2), MYB transcription factor (MYB12), disease resistance protein RPP13-like, and UDP-glycosyltransferases (UGTs). Ultimately, this increased gene expression, particularly of *STR-2*, *UGTs*, and MYB12, resulted in the accumulation of two important protective compounds in tomato plants: α-tomatine and anthocyanins. These findings suggest that sly-miR1916 plays a crucial role in regulating tomato's susceptibility to these fungal diseases by influencing its immune responses and the production of defensive compounds. Fusarium wilt, caused by the soil-borne fungus *Fusarium oxysporum*, is a major threat to tomato production worldwide. Resistance of certain tomato cultivars to this disease is linked to the carefully orchestrated expression of two miRNAs: sly-miR482f and sly-miR5300 (Ouyang et al., 2014). In resistant tomato plants, the levels of these two miRNAs are significantly suppressed upon infection. This downregulation allows for the increased expression of their target genes, which are predominantly from the Nucleotide-Binding Site-Leucine-Rich Repeat (NBS-LRR) family. NBS-LRR proteins are crucial components of the plant's innate immune system, acting as receptors that detect pathogen-derived molecules and trigger a defense response. Conversely, in susceptible tomato varieties, the levels of sly-miR482f and sly-miR5300 remain high, leading to the continued suppression of these vital NBS-LRR resistance genes. This effectively dampens the plant's ability to recognize and combat the invading fungus, resulting in disease development. This intricate interplay highlights how the precise regulation of miRNAs can be the deciding factor between a plant's survival and succumbing to a pathogen.

The fight against another formidable foe, the oomycete *Phytophthora infestans*, which causes late blight, involves a different set of miRNA players. Overexpressing miR172a and miR172b in tomato plants significantly enhances their resistance to this disease. These miRNAs achieve this by targeting and suppressing an AP2/ERF transcription factor, a type of protein that can act as a negative regulator of defense responses. By silencing this repressor, miR172 unleashes a more robust defense against the pathogen (Luan et al., 2018). The role of miR166 in tomato's defense is more multifaceted. Some studies indicate that, similar to miR172, its overexpression can enhance resistance to late blight. However, in the case of viral diseases, the story can be different. For instance, miR166 has been implicated as a potential susceptibility factor for the *Tomato Leaf Curl New Delhi Virus* (ToLCNDV), suggesting that its function can be pathogen-specific and context-dependent. This complexity underscores the need for further research to fully understand the diverse roles of individual miRNAs (Prasad et al., 2022).

TABLE 6.3
miRNAs, Disease Involved, and Their Putative Targets

S. No	miRNAs and lncRNA	Biotic Stresses	Targets and Functions	References
1.	miR6026	Potato virus X (PVX) and tobacco mosaic virus (TMV)	DCL2	Wang et al. (2018)
2.	miR482/2118 family miRNAs	Bacterial and oomycete pathogens	nucleotide-binding leucine-rich repeat (NLR) genes	Canto-Pastor et al. (2019)
3.	slmiR482f and slmiR5300	*Fusarium oxysporum*	Full or partial nucleotide-binding (NB) domains	Ouyang et al. (2014)
4.	miR6024	tomato fusarium wilt	*I2*	Wei et al. (2014)
5.	miR172	*Phytophthora infestans*	AP2/ERF	Luan et al. (2018)
6.	miR1916	*Phytophthora infestans* and *Botrytis cinerea*	Strictosidine synthase (STR-2), UDP-glycosyltransferases (UGTs), and MYB transcription factor (MYB12)	Chen et al. (2019)
7.	miR396a-5p and -3p	*Phytophthora infestans* and *Botrytis cinerea*	nucleotide-binding site-leucine-rich repeat (NBS-LRR), salicylic acid carboxyl methyltransferase (SAMT), growth-regulating factor1 (GRF1), and glycosyl hydrolases (GH)	Chen et al. (2017)
8.	miRNA_1767, miRNA_257, miRNA_275	*Trichoderma atroviride*	Genes involved in plant pathogen interactions	Olmo et al. (2024)
9.	miR398 and 398a-3p	Potato Spindle Tuber Viroid	Cu/Zn-superoxide dismutase 1 and 2 (SOD1 and SOD2) and the copper chaperon for SOD (CCS1) mRNAs	Suzuki et al. (2019)
10.	miR6024	*A. solani*	Nucleotide-binding leucine-rich repeat (NLR) genes	Dey et al. (2022)
11.	mIR166 *HB*	*Tomato Leaf Curl New Delhi Virus*		Prasad et al. (2022)

(*Source:* Adapted and modified from Li et al., 2023b)

6.5 VARIATIONS OF GENES/ALLELES RESPONSIBLE FOR ABIOTIC STRESS TOLERANCE IN TOMATO

Allelic variation is frequently found in crop wild relatives (CWRs) that might help to overcome biotic and abiotic stresses, as they are often found and bred in marginal domains (Ortiz, 2015; Bohra et al., 2021). *Solanum* section *Lycopersicon* (Mill.) Wettst. consists of cultivated tomato (*S. lycopersicum*) and 12 wild relatives (Ramírez-Ojeda et al., 2021). The natural resistance to various diseases and insect pests is found in tomato wild relatives' germplasm. Despite the diversity in wild tomatoes, they are not frequently utilized in breeding programs due to linkage drag, progeny sterility, and self-incompatibility (Zhang et al., 2017). There are many methods that can be used to increase or broaden the genetic base of tomatoes. These are advanced backcross QTL analysis (Tanksley & Nelson, 1996), chromosome segments substitution lines (CSSL) (Ali et al., 2010; Bessho-Uehara et al., 2017), backcross inbred lines (BIL), and introgression lines (ILs), which may increase the chances of breaking linkage drag by producing lines harboring tiny fragments of donor parent chromosomes. The generation of superior lines, which can be promoted in the variety release pipeline, is combined with QTL analysis in these procedures. Using wild relatives, multiple sets of introgression lines have already been developed for tomatoes. (e.g., *Solanum habrochaites*, *Solanum pennellii*, *S. pimpinellifolium*). Although QTL linked to tomato performance under heat stress were reported by Grilli et al. (2007) and Lin et al. (2010), their applicability for breeding was limited since the QTLs were not mapped to the chromosomes. QTL linked to heat tolerance traits were also found by Xu et al. (2017), Driedonks (2018), and Wen et al. (2019). Using an interspecific mapping population derived from *S. lycopersicum* and *S. habrochaites*, Bernacchi and Tanksley (1997) discovered the first major QTL on chromosome 2, which they named se2.1. In this complicated region, the *style2.1* gene (which regulates style length) was found among total five closely linked genes. Wen et al. (2019) produced an F_2 population from the susceptible tomato genotype LA1698 and the heat-tolerant wild tomato accession LA2093 (*S. pimpinellifolium*), which was subjected to QTL mapping, QTL-seq, and RNA-seq analysis. Five QTLs were found to be associated with CC, REC, and the heat injury index. QTL analysis of heat tolerance has recently been used in advanced populations. Gonzalo et al. (2020) also identified tomato heat tolerance QTLs using 167 RILs and 56 ILs taken from Moneymaker (*S. lycopersicum*) and TO-937 (*S. pimpinellifolium*). Eight additional QTLs pertaining to pollen viability and 22 QTLs pertaining to reproductive features were also discovered by multi-environment QTL analysis.

The ability to withstand abiotic stress conditions in tomato is a complicated trait due to the involvement of many genes. These genes regulate various physiological processes such as transpiration and ion transport, and gene expression, as well as metabolism of biomolecules, including carbohydrates. The knowledge of the molecular basis of heat and drought tolerance mechanisms in tomato has significantly increased, which resulted in the discovery of numerous abiotic stress-tolerant genes through reverse genetic techniques, such as overexpression of the gene, RNA interference (RNAi), virus-induced gene silencing (VIGS), and genome editing (CRISPR/Cas9). Endogenous ABA and gene transcript levels in the anthers are regulated by *SlNCED1* (Wang et al., 2021). According to Mao et al. (2020), *SlDREBA4* is linked to electrolyte leakage, a rise in MDA, and a decrease in antioxidant enzymes. Similarly, Hu et al. (2021) reported that *CPK28* is responsible for raising of ROS and decreasing the *APX* activity during heat stress. According to Ding et al. (2021), *SlCML39* is linked to the downregulation of genes related to the ABA signaling pathway and stress response. *SlMPK1* alters ROS, MDA, and antioxidant enzyme activity (Ding et al., 2018). According to Yu et al. (2019), *SlMAPK3* is linked to upregulation of genes for antioxidant enzymes and downregulation of *SlRBOH1*. *Hsp70* is a candidate gene for preserving internal cell stability under stressful circumstances, while *Hsp20* is engaged in protein processing in the endoplasmic reticulum (Usman et al., 2017). *SlDREBA4* gene is linked to electrolyte leakage, elevated MDA, and decreased antioxidant enzymes. Several HSFs, including *HSFA1*, *HSFA2*, *HSFA7*, *HSFB1*, *HSFB4a*, *HSP17-CI*, and *HSP104* were found responsive due to heat stress in tomato (Mishra et al., 2002; Rao et al., 2022).

The *SlVQ6* gene is responsible for downregulating many stress-responsive genes (Ding et al., 2018). The *SlGSNOR1* gene is associated with accumulation of S-nitrosothiol and inhibition of *RBOH* activity (Song et al., 2022). According to Wang et al. (2020), *SlSNAT* elevates the transcription of HSFs and HSPs while preserving the balance of ROS and enzyme activity. Pessoa et al. (2023) reported some genes responsible for drought stress tolerance, viz., *AHG2* (involved in the post-transcriptional gene regulation of salinity and drought responses), *At1g55840* (plays a role in the lipid regulation of the Rho-mediated signaling pathway, which is involved in water stress tolerance), *SAP5* (a positive regulator of stress responses), *PRXQ* (plays a role in the cellular response to oxidative stress caused by peroxide decomposition), *CFS1* (can interact with some transcription factors to regulate their transcriptional activities, resulting in improved abiotic and biotic stress tolerance in several species), *LCD* (regulation of stomatal closure in plant responses to drought stress), and *CCD1*(encodes 9-cis-epoxycarotenoid dioxygenase, which is involved in the response to water deprivation. Liu et al. (2023) reported nine stress-related genes (*AREB1, WSIP-3, ASR4, TAS14, APX*, glutathione peroxidase, *GSH-Px, GSH-Px-1*, and dehydrin), seven of which were closely related to drought resistance responses in tomato. They found that *AREB1* could directly bind to the promoter region of *Hsp20* and *Hsp70*. These two genes were significantly enriched in stress-related pathways, such as response to stress and those involved in protein processing in the endoplasmic reticulum. The QTLs, genes, and transcription factor promoters associated with heat/drought tolerance are presented in Table 6.4.

TABLE 6.4
The QTLs, Genes, and Transcription Factor Promoters Associated with Heat/Drought Tolerance

Gene	Gene ID	Effects	References
		Heat tolerance	
CPK28	Solyc02g083850	Decrease in APX activity and increase in ROS	Hu et al. (2021)
SlCML39	Solyc11g071740	Downregulation of genes involved in stress response and ABA signaling pathway	Ding et al. (2021)
SlMPK1	Solyc12g019460	Change of antioxidant enzyme activity, MDA, and ROS	Ding et al. (2018)
SlMAPK3	Solyc06g005170	Downregulation of *SlRBOH1*, upregulation of antioxidant enzyme genes	Yu et al. (2019)
SlHK2	Solyc07g047770	Membrane stabilized, cuticle thickened, increase in water content, and osmo protectants	Mushtaq et al. (2022)
SlBAG9	Solyc01g095320	Elimination of ROS and protection of photosynthesis	Huang et al. (2022)
SlNCED1	Solyc03g121880	Regulates endogenous ABA and gene transcript levels in the anthers	Wang et al. (2021)
Hsp20	Solyc04g014480	Involved in protein processing in the endoplasmic reticulum	
Hsp70	Solyc04g011440	Maintain internal cell stability	Usman et al. (2017)
SlDREBA4	Solyc06g066540	Decrease of antioxidant enzymes, increase of MDA, and electrolyte leakage	Mao et al. (2020)
BZR1	Solyc04g079980	Regulation of *RBOH1* production	Yin et al. (2018)
HSFA1	Solyc08g005170	Increase of *HSFA2, HSFB1, HSP17-CI*, and *HSP104*	Mishra et al. (2002)
HSFA7	Solyc09g065660	Changes in survival rate and ROS accumulation	Rao et al. (2022)
HSFB4a	Solyc04g078770	Changes in survival rate and ROS accumulation	Rao et al. (2022)
SlHSBP1	Solyc12g099570	Regulation of HSFs	Marko et al. (2019)
SlNAC1	Solyc04g009440	Decrease in antioxidant activity, proline, and HSP; increase in membrane lipid peroxidation; and PSII damage	Liang et al. (2015)
SlWRKY3	Solyc02g088340	Regulation of ROS-scavenging genes, such as *SlGRXS1* gene cluster	Wang et al. (2022)
LeAN2	Solyc10g086250	Upregulation of anthocyanin biosynthesis genes, reduction of ROS, photoinhibition, and membrane damage of PSII	Meng et al. (2015)

(Continued)

Table 6.4 (*Continued*)
The QTLs, Genes, and Transcription Factor Promoters Associated with Heat/Drought Tolerance

Gene	Gene ID	Effects	References
SlBBX17	Solyc07g052620	Decrease of ion leakage, MDA and ROS contents, increase of HSFs, HSPs, and the activity of antioxidant enzymes	Xu et al. (2022)
SlWHY1	Solyc05g007100	Activation of *SlHSP21.5A* transcription, increase of soluble sugar, decrease of ROS	Zhuang et al. (2020)
SlVQ6	Solyc02g078030	Downregulation of many stress-responsive genes	Ding et al. (2018)
GMPase	Solyc03g096730	Increase of ASA and antioxidant capacity, reduction of ion leakage	Wang et al. (2011)
LeUCP	Solyc09g011920	Increase in ASA content, decrease in MDA, and ROS	Chen et al. (2013)
SlGSNOR1	Solyc09g064370	Accumulation of S-nitrosothiol, inhibition of RBOH activity	Song et al. (2022)
SlSNAT	Solyc10g074910	Enhanced transcription of HSFs and HSPs, maintaining ROS balance, and enzyme activity	Wang et al. (2020)
SlSPS	Solyc07g007790	Change of soluble sugar content and SPS enzyme activity, regulation of ROS damage	Zhang et al. (2022)
SlEGY2	Solyc06g019200	Impaired ROS scavenging, inhibition of PSII	Song et al. (2022)
LeCDJ1	Solyc03g117590	Alleviating photoinhibition of PSII, protecting chloroplast antioxidant enzyme	Kong et al. (2014)
		Drought Tolerance	
AREB1	Solyc04g078840	Plays an important role in regulating drought responses by activating the expression of downstream stress-related target genes	Zhu (2016), Liu et al. (2023)
WSIP-3	Solyc04g071610	*water-stress inducible protein 3*	Liu et al. (2023)
ASR4	Solyc04g071615	closely related to drought resistance responses	
TAS14	Solyc02g084850	closely related to stress resistance responses	
APX	Solyc09g007270	*ascorbate peroxidase*	
glutathione peroxidase	Solyc08g006720	closely related to drought resistance responses	
GSH-Px	Solyc08g006720	closely related to drought resistance responses	
GSH-Px-1	Solyc08g080940	closely related to drought resistance responses	
dehydrin	Solyc04g082200	closely related to drought resistance responses	
AHG2	–	Involved in the post-transcriptional gene regulation of salinity and drought responses	Pessoa et al. (2023)
At1g55840	–	Role in the lipid regulation of the Rho-mediated signaling pathway, which is involved in water stress tolerance	
PRXIIF	–	Oxidative stress response	
SAP5	–	Positive regulator of stress responses	
REF4-RELATED1	–	Regulation of the phenylpropanoid metabolic process	
PRXQ	–	Role in the cellular response to oxidative stress caused by peroxide decomposition	
CFS1	–	Can interact with some transcription factors to regulate their transcriptional activities, resulting in improved abiotic and biotic stress tolerance in several species	
LCD	–	Regulation of stomatal closure in plant responses to drought stress	
CCD1	–	Encodes 9-cis-epoxycarotenoid dioxygenase, which is involved in the response to water deprivation.	
SCS	–	Encodes SnRK2 described as a crucial regulator of plant responses to abiotic stresses. regulator of ABA signaling during water stress at the vegetative and reproductive stages.	

6.6 ALLELE MINING FOR YIELD COMPONENT TRAITS IN TOMATO

The interplay of the environment has a significant impact on tomato yield, which is a complex quantitative trait. Many of these genes/QTLs may not have been utilized in recent varietal development programs and may still be present in the wild gene pool, leading to the beginning of the search for genes/QTLs governed by polygenes. Several genes/QTLs have been identified with favorable alleles for yield and its component traits following different approaches in tomato. Marker-assisted breeding has identified several genes/OTLs that can be stacked in existing high-yielding varieties for increasing genetic gain. In tomato, many genes influencing the fruit size and shape were identified, such as *SUN, OVATE*, cell number regulator (*CNR*)/*FW2.2, SlKLUH/FW3.2, FASCIATED* (*FAS*), and *LOCULE NUMBER* (*LC*) (Chakrabarti et al., 2013; Cong et al., 2008; Frary et al., 2000; Liu et al., 2002). *fw2.2* was the first QTL controlling fruit weight to be cloned from any fruit or vegetable crops. The large fruit alleles of *Fw11.3* (fruit cell size regulator QTL) encode a shorter protein with 194 fewer amino acids than the small fruit alleles due to a 1.4kb deletion, and it results in an increase in the size of fruit cells and, thus, fruit size (Mu et al., 2017). A great increase in fruit weight was found in accessions that carried the derived allele of *fw11.3*, most of which were *Solanum lycopersicum* var. *lycopersicum* (Pereira et al., 2021b). The loci *fas* and *lc* affect the number of locules and fruit weight (Van der Knaap et al., 2014). *LC* encodes *SlWUS*, and two SNPs 1080bp downstream of this gene affect the expression of *SlWUS* (Chu et al., 2019; Munos et al., 2011). FAS encodes *SlCLV3*, and a 294kb inversion downregulates the expression of *SlCLV3*, resulting in an increase in the number of locules (Xu et al., 2015). The *globe* locus affects various dimensions of fruit, including fruit length. Pereira et al. (2021b) identified six QTLs (*fw2.3, fw7.1, fw1.1, fw11.2, fw3.3, fw8.1*) for fruit weight in three different bi-parental populations and evaluated the presence of these QTLs in additional F_2 populations. The percentage of variation explained for fruit weight by these QTLs ranged from 10 to 25%. Hence, tomato fruit weight is important for the breeding program. The major QTL *fw2.3* was also validated using progeny testing in the F_3 population, which accounted for 25% fruit weight variation. Nearly 30 loci affecting tomato fruit weight have been identified (Illa-Berenguer et al., 2015; Grandillo et al., 1999; Paterson et al., 1998; Paterson et al., 1990; Liu et al., 2002), and six of these have been mapped and cloned (Frary et al., 2000; Chakrabarti et al., 2013; Mu et al., 2017; Munos et al., 2011; Sierra-Orozco et al., 2021; Ning et al., 2023). Gene (s)/QTL (s) governing yield and contributing traits in tomato are presented in Table 6.5.

TABLE 6.5
Gene (s)/QTL (s) Governing Yield and Contributing Traits in Tomato

Trait	Gene (s)/QTL (s)	References
Fruit shape	*fs8.1*	Grandilio et al. (1996), Ku et al. (2000), Clevenger (2012)
Obovoid and elongated shape	*sov1*	Rodríguez et al. (2013)
Elongated fruit shape	*sov2*	
Fruit shape triangle	*tri*	Brewer et al. (2006), Adhikari et al. (2020)
Pear shape	*Yr2_Ob, Yr1_Ob*	Adhikari et al. (2020)
Ovate fruit shape	*SlOFP16*	Huang et al. (2013)
Fruit size	*ovate, fw2.2,* and *lcn2.1*	Lin et al. (2010)
Fruit weight	*fw6.3*	Ning et al. (2023)
	fw2.2	Frary et al. (2000)
	fw3.2	Chakrabarti et al. (2013)
	fw11.3	Mu et al. (2017)
	fas	Cong et al. (2008), Xu et al. (2015)
	lc	Chu et al. (2019), Munos et al. (2011)
	globe	Sierra-Orozco et al. (2021)

(Continued)

Table 6.5 *(Continued)*
Gene (s)/QTL (s) Governing Yield and Contributing Traits in Tomato

	10 candidate genes	Kim et al. (2021)
	fw2.3	Pereira et al. (2021b)
	fw7.1	
	fw8.1	
	fw1.1	
	fw11.2	
	fw3.3	
	fw8.1	

6.7 ALLELE MINING EXPLORED THE ALLELIC DIVERSITY FOR NUTRITIONAL TRAITS IN TOMATO

Tomato is an excellent source of phytochemicals like lycopene, carotenoids, vitamins (vitamin K and C), minerals (potassium, calcium, iron), and folates. These phytochemicals have great potential for human health and nutritional benefits. Tomato-based diet is strongly evidenced to reduce devastating diseases like cancer, inflammation, hypertension, coronary heart disease, diabetes, and obesity (Canene-Adams et al., 2005; Raiola et al., 2014; Martí et al., 2016). Tomato is considered a nutritional powerhouse; thus, enhancement of the nutritional constituents with quality traits is one of the major objectives of the present-day tomato breeding program. The synthesis of nutritional compounds in tomato fruit is governed by the action of plenty of genes/QTLs of primary and secondary metabolic pathways (Quinet et al., 2019). Thus, identification of key genes/regulators or QTLs, which regulate the process of nutritional quality traits, is very necessary in tomato breeding. Several traditional breeding approaches, like selection and cross-breeding methods, have been adopted to enhance the nutritional quality of tomato. With the advancement of molecular approaches, many nutritional biosynthesis pathway genes have been identified and validated through transgenic approaches (Zhu et al., 2007; Datir & Regan, 2023; Scarano et al., 2024). However, the transgenic tomato plants, which ensure nutritional quality, cannot be accepted due to the ethical issues in some countries. Therefore, the regulation-free molecular breeding techniques may be the alternative approach to mine the alleles linked with nutritional quality traits in tomato. The technique of allele mining for identification of genes/QTLs associated with nutritional traits is composed of genome-wide association mapping, TILLING/Eco-TILLING, and next-generation sequencing (NGS) approaches. In these approaches, the natural genetic resources like land races, wild relatives, and germplasm were exploited for mining alleles of interest associated with nutritional quality traits in tomato, such as high lycopene, carotenoids, vitamins, and mineral contents (Ruggieri et al., 2014; Orchard et al., 2021; Burgos et al., 2021; Petrozza et al., 2023). The wild tomato *Solanum pennellii* is the potential genetic resource for introgression of important genes/QTLs related to agro-economical traits, including nutritional traits in tomato (Liu et al., 2016). For instance, mapping populations of a wild species, *S. pennellii*, was utilized to map several genes/QTLs associated with fruit quality and nutritional traits in tomato (Rousseaux et al., 2005; Stevens et al., 2006). Similarly, the populations of *S. pennellii* were also utilized to map two QTLs on chromosome 7 and chromosome 12 associated with increased phenolics and ascorbic acid content in tomato fruits (Di Matteo et al., 2010, 2013). The characterized alleles and alternative genes/QTLs are being utilized as potential genomic resources for enhancing nutritional quality in tomato for better human health and dietary benefits enhancement. Genome-wide association studies (GWAS), TILLING, and Next-generation sequencing (NGS) methods have been successfully utilized to identify genes/QTLs associated with polygenic traits in tomato. Tomato flavor and taste are the key preferences for consumers and tomato breeders. Breeders have significantly improved the tomato production under adverse environmental conditions. However, the end users always prefer the fruit quality of tomatoes in terms of flavor and texture. Therefore, several experiments were carried out by adopting QTL mapping and GWAS for identification of alleles responsible for fruit flavor and texture (Table 6.6).

Tomato flavor is mainly ascertained by the balanced amount of acid, sugars, and volatile compounds in the ripened fruit. To get insights into the genetic makeup of flavor quality trait in tomato, several genes responsible of this trait were exploited for allelic variations using association studies in a set of 166 tomato germplasm. The association mapping of 166 Varitome panel of tomato revealed that five candidate genes, like *ALMT9, AAT1, CXE1, LIN5*, and *LoxC*, are responsible for controlling the flavor content in tomato fruit (Pereira et al., 2021a). Besides, several other genes have also been reported for controlling metabolic regulation pathways related to flavor content in tomato (Pereira et al., 2021a). Sauvage et al. (2014) have performed GWAS in 163 tomato accessions and identified 44 candidate loci associated with fruit metabolic traits; 19 were related to sucrose, ascorbate, malate, and citrate metabolism. Another report also evidenced about mining of major loci and genomic regions by conducting GWAS on 300 tomato accessions and implicating flavor improvement by controlling acid and volatile content in tomato fruit (Bauchet et al., 2017). The study clearly identified two close associations related to fruit quality, such as acidity and phenylpropanoid volatile compound production, and identified two associated candidate alleles (Hypothetical protein-Solyc06g072930.2; *Glycosyltransferase* Solyc04g064490.2.1) controlling this pathway. Similarly, Li et al. (2020) performed metabolite-based GWAS and identified a lipase gene associated with the accumulation of short-chain FA-VOCs in tomato fruit, which is involved in short-chain fatty acid-derived flavor content in tomato fruits. The in vitro assay of *SlLIP8* protein suggested the cleavage of glycerolipids into 18:2 and 18:3 acyl groups. The CRISPR/Cas9-mediated editing of these genes resulted in lower content of multiple fruit short-chain FA-VOCs and exhibited the importance in the lipase metabolic pathway. In addition, genetic control of flavor quality of tomato fruit was depicted by meta-analysis of GWAS using 775 accessions and revealed 305 most significant associations pertain to flavor-related volatiles, acids, sugar, and amino acids. The findings suggested that the lipoxygenase gene (*LoxC*, Solyc01g006540), which encodes an enzyme for C6 and C5 fatty acid-derived volatile synthesis, was found with candidate association for flavor-related traits in tomato (Zhao et al., 2019). Zhao et al. (2019) also identified many candidate genes that showed significant association with amino acid content in tomato. For instance, a locus on chromosome 3 showed a significant association with serine content. The locus was annotated into a serine biosynthesis pathway gene encoding threonine synthase. Similarly, the proline content was also revealed to have a significant association with a locus, Solyc03g117770. Additionally, the content of four amino acids, like asparagine, GABA, glutamine, and threonine, was closely associated with a locus Solyc12g089350 annotating GDSL esterase/lipase, which is regulating the biosynthesis pathway of these amino acid content in tomato.

The tocochromanols are essential components of the human diet that are constituted by the different forms of vitamin E. A combined interval mapping and GWAS approach using MAGIC populations unveiled the genetic determinants of tocochromanol accumulation in tomato fruits (Burgos et al., 2021). This study revealed that the allelic variation in the sequences of four candidate genes, namely *TyrA2* (Solyc09g011870), *fbt* (Solyc06g068560), *ADCL* (Solyc11g071280), and *DHQS* (Solyc02g083590), involves regulation of the shikimate pathway and catalyzing reactions for tocochromanol accumulation in tomato fruits.

The cholesterol-derived biomolecules, such as steroidal glycoalkaloids (SGAs), are synthesized in the green part of solanaceous crops and act as a deterrent against herbivores and plant pathogens. However, they are considered as anti-nutritional factors that disturb the digestive system of humans (Jared et al., 2016). To reduce such a detrimental effect on human health, the mining of alleles regulating this trait must be dissected at the genetic level. One-base allelic variation in *GAME9* gene regulated the steroidal glycoalkaloid biosynthesis in tomato by altering the binding capacity in cis-regulatory sequence of *GAME17* promoter (Yu et al., 2020). Similarly, GWAS was performed

TABLE 6.6
Important Alleles Associated with Important Nutritional Quality Traits in Tomato

S. No.	Nutritional Trait	Genes/Alleles	Pathway	Reference
1.	beta-carotene	*Cyc-B* (promoter)	Carotenoid biosynthesis pathway	Orchard et al. (2021)
2.	beta-carotene	Solyc01g087600.2.1 Solyc01g087670.2.1 Solyc01g087880.2.1	Carotenoid biosynthesis pathway	Ruggieri et al. (2014)
3.	Antioxidant properties	*DET1* *COP10* *COP1like* *DDB1*	Light signaling pathway	Jones et al. (2012)
4.	Fruit flavor	*LIN5* *ALMT9* *AAT1* *CXE1* *LoxC*	Metabolite regulation pathway	Pereira et al. (2021a)
5.	Fruit sugar	*SlINVINH1* *SlVPE5*	Sugar metabolic pathway	Zhao et al. (2016)
6.	Fruit metabolic traits	Cell wall invertase (Solyc09g010080.2) *MDHAR* (Solyc09g009390.2) Nicotinate phosphoribosyl transferase protein (Solyc02g093290.2) Aluminum-activated malate transporter-like (Solyc06g072910.2) Sugar transporter (Solyc04g076920.2)	Sugar metabolism Ascorbate	Sauvage et al. (2014)
7.	Acid and volatile content	Hypothetical protein (Solyc06g072930.2) Glycosyltransferase (Solyc04g064490.2.1)	Malate Phenylacetaldehyde and phenylethanol pathway	Bauchet et al. (2017)
8.	Fruit flavor	*SlLIP8*	Lipase metabolic pathway	Li et al. (2020)
9.	Fruit flavor	*LoxC* (Solyc01g006540)	Fatty acid biosynthetic pathway	Zhao et al. (2019)
10.	Amino acids	Threonine synthase Proline synthase (Solyc03g117770) *GDSL* esterase/lipase (Solyc12g089350)	Amino acid biosynthesis pathway	Zhao et al. (2019)
11.	Tocochromanol (Vit. E)	*TyrA2* (Solyc09g011870) fbt (Solyc06g068560) *ADCL* (Solyc11g071280) *DHQS* (Solyc02g083590)	Shikimate pathway	Burgos et al. (2021)
12.	Ascorbate biosynthesis	*SlbHLH59*	D-mannose/L-galactose pathway	Ye et al. (2019)
13.	Lycopene	Solyc03g031480.2.1 Solyc03g031820.2.1	Trans-lycopene biosynthesis	Ruggieri et al. (2014)
14.	Steroidal alkaloid	*GAME9* *GAME17*	Steroidal glycoalkaloids biosynthesis	Yu et al. (2020)
15.	Steroidal alkaloid	*SlGAME5*-like glycosyltransferase (Solyc10g085240) SlDOG1 (Solyc10g085210)	Steroidal alkaloids (SAs) and steroidal glycoalkaloids (SGAs) biosynthesis	Zhao et al. (2023)

to explain the genetic basis of Steroidal alkaloids (SAs) and steroidal glycoalkaloids (SGAs) biosynthesis and regulation in tomato fruits. In association mapping study, two candidate genes, namely *SlGAME5*-like glycosyltransferase (Solyc10g085240), and a transcription factor, *SlDOG1* (Solyc10g085210), showed significant association with steroidal alkaloid deposition in tomato fruit (Zhao et al., 2023).

Lycopene is abundantly found in tomato. It is responsible for red color development in tomato fruits and is also considered a strong anti-oxidant for cardiovascular disease prevention (Gupta et al., 2024). It contributes about 80% of total lycopene in the human diet (Landrier et al., 2023). Thus, it is very much essential to know the biosynthetic pathways and controlling genes/alleles of lycopene synthesis in tomato (Duduit et al., 2022). Ruggieri et al. (2014) have conducted an association study using 96 genotypes of tomato and revealed two genes, namely Solyc03g031480.2.1 and Solyc03g031820.2.1, significantly associated with lycopene content in tomato. In addition to this, they have also mined the alleles related to ascorbic acid, β-carotene, titratable acidity, and pH level in fruit. Similarly, tomato fruits are also a vital source of carotenoids, especially beta-carotene, which is a precursor of provitamin A for human health. As compared to lycopene content, beta-carotene is found in smaller quantities in tomatoes. Being beneficial to human health, it is necessary to identify the alleles and enhance the beta-carotene content in tomato fruits. It is reported that the allelic variation (SNP) in the promoter region of *Cyc-B* gene is responsible for enhancing beta-carotene content in tomato fruits (Orchard et al., 2021). A set of 96 genotypes was characterized phenotypically for higher content of beta-carotene by Ruggieri et al. (2014). By association analysis between phenotype and genotype, they have identified QTL harboring three loci (Solyc01g087600.2.1, Solyc01g087670.2.1, and Solyc01g087880.2.1) associated with beta-carotene content trait in tomato. The carotenoid and phenylpropanoid phytonutrients in ripe tomato fruits were induced by generating a point mutation in *DEETIOLATED1* gene using TILLING method (Jones et al., 2012). The resequencing clubbed with genotyping and phenotyping of tomato core set, comprised of 122 accessions, resulted in allelic variability in fruit quality traits, including soluble solids content and fruit color (Roohanitaziani et al., 2020). The important alleles associated with important nutritional quality traits in tomato are presented in Table 6.6.

6.8 CONVERTING LANDRACE DIVERSITY TO BREEDING RESOURCES

The process of allele mining suggests that it is possible to identify new genotypes for targeted phenotyping. Once the genotypes are validated phenotypically, these can be used as donors to incorporate new diversity into elite cultivars. Traditionally, pedigree, hybrid development, and backcrossing are the standard breeding procedures for introducing diversity into elite germplasms, where bookcrossing of selected genotypes to elite cultivars and creating a segregating bi-parental population have been common practices. Plants carrying the targeted genes/traits are further backcrossed to elite genetic backgrounds. These conventional breeding approaches require extensive selections and observations of the best progenies for many generations and are, therefore, time-consuming and laborious. However, given the large number of genotypes expected from allele mining, an alternate breeding procedure should be considered in the conversion of landrace diversity into genetic resources to harness multiple favorable alleles. The development of MAGIC (Multiparent Advanced Generation Inter-Cross) population is an alternative to conventional bi-parental and advanced backcross methods to combine the favorable alleles, which involves several cycles of intermating among multiple parental lines (Cavanagh et al., 2008). In MAGIC population, lines are inter-crossed in a half-diallel design and F_1 plants are then inter-crossed to produce new F_1s with four different genomes (four-way cross). The developed F_1 is crossed with other F_1s in eight-way crosses. These F_1 plants from eight-way crosses are selfed by single-seed descent for several generations to develop advanced inbred lines. This breeding methodology generates population with higher recombination and generates pre-breeding materials with new genotypic diversity (Bandillo et al., 2013). While MAGIC population offers a number of advantages for the extraction of new alleles, other breeding methods should be considered depending on breeding objectives. In the present genomic era, selection of specific genotypes can be assisted by molecular markers (Collard and Mackill, 2008). For the introgression of a single trait, marker-assisted backcrossing (MABC) has been widely used to improve a single genotype by adding a favorable donor allele, while for multiple traits, marker-assisted recurrent selection (MARS) is an improvement over MABC. Recently, new plant breeding technologies like clustered regularly interspaced short palindromic repeats (CRISPR)/CRISPR associated proteins (Cas)-based approaches have emerged as the superior precision plant breeding technologies for crop genetic improvement and bringing a new hope for future of agriculture (Belhaj et al., 2013; Chen et al., 2019). This technology has been successfully adopted for tomato genome editing and the subsequent precision breeding of new varieties without linkage drag, which has helped in accelerating the introgression of novel traits from tomato wild relatives into their elite cultivars and made it feasible with *de novo* domestication of wild tomato (Zsogon et al., 2018; Vu et al., 2020). The resulting plants carried improved traits such as better growth, larger fruits, and higher yield and quality but still retained important wild traits, including stress tolerance and other traits determined by multiple alleles (Zsogon et al., 2018). Hence, the CRISPR approach would help to save time in developing super tomato cultivars resilient to climate change.

6.9 CONCLUSION

The development of climate-resilient tomato varieties is more important in the future as extreme climatic conditions will become more frequent across the globe and will certainly present new environmental stresses that cause yield losses. Both genotypic and allelic diversity are important to meet the demand of futuristic tomato varieties that will tolerate the adverse climatic conditions, along with high yield and nutritional quality. Breeding efforts often involved the incorporation of genetic material from wild tomato species, particularly for disease resistance. The incorporation of disease resistance genes from wild species such as *Solanum pimpinellifolium* and *Solanum habrochaites* has been a significant focus. Since the release of high-quality reference genome of tomato cultivar Heinz 1706, later improvements to the Heinz 1706 genome sequence and most recent SL5.0 genome assembly, new advances in genome sequencing technologies have made it feasible to reveal the genome diversity of many genetic populations. The resequencing clubbed with genotyping and phenotyping of tomato accessions resulted in allelic variability in several traits, including biotic stress, abiotic stress, and fruit quality traits in tomato. The known gene sequences can also be searched in the database available to identify the unique genotypes carrying targeted alleles for a phenotype. Improvement in data science and the availability of genomics and bioinformatics tools may assist in mining the most valuable alleles for nutritional and stress-tolerant traits. Therefore, tomato, with its high species and cultivated genotype diversity, can benefit from combining allele mining and novel mating designs to develop new genetic resources for breeding.

REFERENCES

Adhikari P, Adhikari TB, Louws FJ, Panthee DR. 2020. Advances and challenges in bacterial spot resistance breeding in tomato (*Solanum lycopersicum* L.). *Int J Mol Sci.* 21(5):1734. https://doi.org/10.3390/ijms21051734.

Adhikari P, Oh Y, Panthee DR. 2017. Current status of early blight resistance in tomato: An update. *Int J Mol Sci.* 18(10):2019.

Adhikari P, Siddique MI, Louws FJ, Panthee DR. 2023. Identification of quantitative trait loci associated with bacterial spot race T4 resistance in intra-specific populations of tomato (*Solanum lycopersicum* L.). *PLoS ONE.* 18(12):e0295551.

Alexander L, Hoover M. 1955. Disease resistance in wild species of tomato: Report of the National Screening Committee. *Agric Exp Stn Res Bull.* 752:1–76.

Ali ML, Sanchez PL, Yu S, Lorieux M, Eizenga GC. 2010. Chromosome segment substitution lines: A powerful tool for the introgression of valuable genes from Oryza wild species into cultivated rice (*O. sativa*). *Rice*. 3:218–234.

Anderson TA, Zitter SM, De Jong DM, Francis D, Mutschler MA. 2021. Cryptic introgressions contribute to transgressive segregation for early blight resistance in tomato. *Theor Appl Genet*. 134:2561–2575. https://doi.org/10.1007/s00122-021-03842-x

Andolfo G, D'Agostino N, Frusciante L, Ercolano MR. 2021. The tomato interspecific NB-LRR gene arsenal and its impact on breeding strategies. *Genes*. 12(2):184.

Andolfo G, Sanseverino W, Rombauts S, Van de Peer Y, Bradeen JM, Carputo D, Frusciante L, Ercolano MR. 2013. Overview of tomato (*Solanum lycopersicum*) candidate pathogen recognition genes reveals important Solanum R locus dynamics. *New Phytol*. 197:223–237.

Ariizumi T, Shinozaki Y, Ezura H. 2013. Genes that influence yield in tomato. *Breed Sci*. 63:3–13.

Axtell MJ, Bowman JL. 2008. Evolution of plant MicroRNAs and their targets. *Trends Plant Sci*. 13:343–349.

Bai Y, Pavan S, Zheng Z, Zappel NF, Reinstädler A, Lotti C, De Giovanni C, Ricciardi L, Lindhout P, Visser R, Theres K, Panstruga R. 2008. Naturally occurring broad-spectrum powdery mildew resistance in a Central American tomato accession is caused by loss of mol function. *Mol Plant Microbe Interact*. 21(1):30–39. https://doi.org/10.1094/MPMI-21-1-0030. PMID: 18052880.

Bandillo N, Raghavan C, Muyco PA, et al. 2013. Multi-parent advanced generation inter-cross (MAGIC) populations in rice: Progress and potential for genetics research and breeding. *Rice*. 6(11). https://doi.org/10.1186/1939-8433-6-11.

Banerjee MK, Kalloo MK. 1987. Sources and inheritance of resistance to leaf curl virus in Lycopersicon. *Theor Appl Genet*. 73:707–710. https://doi.org/10.1007/BF00260780.

Bauchet G, Grenier S, Samson N, Segura V, Kende A, Beekwilder J, Cankar K, Gallois JL, Gricourt J, Bonnet J, Baxter C, Grivet L, Causse M. 2017. Identification of major loci Mathilde Causse et al. 13 and genomic regions controlling acid and volatile content in tomato fruit: Implications for flavor improvement. *New Phytol*. 215:624–641.

Belhaj K, Chaparro-Garcia A, Kamoun S, Nekrasov V. 2013. Plant genome editing made easy: Targeted mutagenesis in model and crop plants using the CRISPR/Cas system. *Plant Methods*. 9:39. https://doi.org/10.1186/1746-4811-9-39.

Bent AF, Mackey D. 2007. Elicitors, effectors, and R genes: The new paradigm and a lifetime supply of questions. *Annu Rev Phytopathol*. 45(1):399–436.

Bernacchi D, Tanksley SD. 1997. An interspecific backcross of *Lycopersicon esculentum*× L. *hirsutum*: Linkage analysis and a QTL study of sexual compatibility factors and floral traits. *Genetics*. 147(2):861–877.

Bessho-Uehara K, Furuta T, Masuda K, Yamada S, Angeles-Shim RB, Ashikari M, Takashi T. 2017. Construction of rice chromosome segment substitution lines harboring Oryza genome and evaluation of yield-related traits. *Breed Sci*. 67:408–415.

Bhardwaj A, Prasad P, Kumar S. 2024. High-throughput genomics for allele mining in tomato. In: C Kole, TK Behera, P Kaushik, eds. *Allele Mining for Genomic Designing of Vegetable Crops*:300. https://doi.org/10.1016/B978-0-10033764-39.

Bhattarai K, Frank JL, Williamson JD, Panthee DR. 2016. Diversity analysis of tomato genotypes based on morphological traits with commercial breeding significance for fresh market production in eastern USA. *Aust J Crop Sci*. 10(8):1098–1103.

Bohra A, Kilian B, Sivasankar S, Caccamo M, Mba C, Mccouch SR, Varshney RK. 2021. Reap the crop wild relatives for breeding future crops. *Trends Biotechnol*. 40:412–431. https://doi.org/10.1016/j.tibtech.2021.08.009.

Brekke TD, Stroud JA, Shaw DS, Crawford S, Steele KA. 2019. QTL mapping in salad tomatoes. *Euphytica*. 215:115.

Brewer MT, Lang LX, Fujimura K, Dujmovic N, Gray S, van der Knaap E. 2006. Development of a controlled vocabulary and software application to analyze fruit shape variation in tomato and other plant species. *Plant Physiol*. 141:15–25. https://doi.org/10.1104/pp.106.077867.

Brommonschenkel SH, Frary A, Tanksley SD. 2000. The broad-spectrum Tospovirus resistance gene Sw-5 of tomato is a homolog of the root-knot nematode resistance gene Mi. *Mol Plant Microbe Interact*. 13:1130–1138.

Burgos E, Maria MB, Diouf I, deHaro LA, Albert E, et al. 2021. Validated MAGIC and GWAS population mapping reveals the link between vitamin E content and natural variation in chorismate metabolism in tomato. *Plant J*. 105:907–923.

Canady MA, Ji Y, Chetelat RT. 2006. Homeologous recombination in *Solanum lycopersicoides* introgression lines of cultivated tomato. *Genetics*. 174:1775–1788.

Canady MA, Meglic V, Chetelat RT. 2005. A library of *Solanum lycopersicoides* introgression lines in cultivated tomato. *Genome*. 48:685–697.

Canene-Adams K, Campbell JK, Zaripheh S, Jeffery EH, Erdman JW, Jr. 2005. The tomato as a functional food. *J Nutr*. 135(5):1226–1230. https://doi.org/10.1093/jn/135.5.1226.

Canto-Pastor A, Santos, BAMC, Valli A, Summers W, Schornack S, Baulcombe DC. 2019. Enhanced resistance to bacterial and oomycete pathogens by short tandem target mimic RNAs in tomato. *Proc Natl Acad Sci U S A*. 116(7):2755–2760.

Capel C, et al. 2015. Wide-genome QTL mapping of fruit quality traits in a tomato RIL population derived from the wild-relative species *Solanum pimpinellifolium* L. *Theor Appl Genet*. 128:2019–2035. https://doi.org/10.1007/s00122-015-2563-4.

Carmeille A, Caranta C, Dintinger J, Prior P, Luisetti J, Besse P. 2006. Identification of QTLs for *Ralstonia solanacearum* race 3-phylotype II resistance in tomato. *Theor Appl Genet*. 113:110–121. https://doi.org/10.1007/s00122-006-0277-3.

Catanzariti AM, Lim GT, Jones DA. 2015. The tomato *I-3* gene: A novel gene for resistance to Fusarium wilt disease. *New Phytol*. 207:106–118.

Causse MP, Duffe MC, Gomez M, Buret R, Damidaux D, Zamir D, et al. 2004. A genetic map of candidate genes and QTLs involved in tomato fruit size and composition. *J Exp Bot*. 55:1671–1685.

Cavanagh C, Morell M, Mackay I, Powell W. 2008. From mutations to MAGIC: Resources for gene discovery, validation and delivery in crop plants. *Curr Opin Plant Biol*. 11:215–221. https://doi.org/10.1016/j.pbi.2008.01.002.

Cecílio Filho A, Maia M, Mendoza-Cortez J, Rodrigues M, Nowaki R. 2014. Growing seasons and fractional fertilization for arugula. *Comun Sci*. 5:252–258. [Google Scholar]

Celik I, Gurbuz N, Uncu AT, Frary A, Doganlar S. 2017. Genome-wide SNP discovery and QTL mapping for fruit quality traits in inbred backcross lines (IBLs) of solanum pimpinellifolium using genotyping by sequencing. *BMC Genomics*. 18:1.

CGN Centre for genetic resources. https://www.wur.nl/en/Research-Results/Statutory-research-tasks/Centre-for-Genetic-Resources-the-Netherlands-1/Plant-Genetic Resources/Genebank.htm [Last Access 09 November 2022].

Chakrabarti M, Zhang N, Sauvage C, Munos S, Blanca J, Canizares J, Diez MJ, Schneider R, Mazourek M, McClead J, et al. 2013. A cytochrome P450 regulates a domestication trait in cultivated tomato. *Proc Natl Acad Sci USA*. 110:17125–17130. https://doi.org/10.1073/pnas.1307313110.

Chakraborti M, Anilkumar C, Verma RL, Abdul Fiyaz R, Reshmi Raj KR, Patra BC, Balakrishnan D, Sarkar S, Mondal P, Kar M, Meher J, Sundaram RM, Subba Rao LV. 2021. Rice breeding in India: Eight decades of journey towards enhancing the genetic gain for yield, nutritional quality, and commodity value. *Oryza*. 58:69–88. https://doi.org/10.35709/ory.2021.58.spl.2.

Chávez Montes RA, Rosas-Cárdenas DF, De Paoli E, Accerbi M, Rymarquis LA, Mahalingam G, Marsch-Martínez N, Meyers BC, Green PJ, de Folter S. 2014. Sample sequencing of vascular plants demonstrates widespread conservation and divergence of MicroRNAs. *Nat Commun*. 5:3722.

Chen L, Meng J, Zhai J, Luan Y. 2017 MicroRNA396a-5p and -3p induce tomato disease susceptibility by suppressing target genes and upregulating salicylic acid. *Plant Sci*. 265:177–187.

Chen L, Meng J, Zhang M, Luan YS. 2019. *Solanum lycopersicum* microRNA1916 targets multiple target genes and negatively regulates the immune response in tomato. *Plant Cell Environ*. 42(4):1393–1407.

Chen SC, Liu AR, Zhang SJ, Li C, Chang R, Liu DL, Ahammed GJ, Lin XM. 2013. Overexpression of mitochondrial uncoupling protein conferred resistance to heat stress and Botrytis cinerea infection in tomato. *Plant Physiol Biochem*. 73:245–253.

Chu YH, Jang JC, Huang Z, van der Knaap E. 2019. Tomato locule number and fruit size controlled by natural alleles of LC and FAS. *Plant Direct*. 3(7):e00142. https://doi.org/10.1002/pld3.142. PMID: 31312784; PMCID: PMC6607973.

Clevenger J. 2012. *Metabolic and Genomic Analysis of Elongated Fruit Shape in Tomato (Solanum lycopersicum)*. Master's Thesis. Ohio State University, Wooster, OH.

Collard BC, Mackill DJ. 2008. Marker-assisted selection: An approach for precision plant breeding in the twenty-first century. *Phil Trans R Soc Lond Biol Sci*. 363:557–572. https://doi.org/10.1098/rstb.2007.2170.

Cong B, Barrero LS, Tanksley SD. 2008. Regulatory change in YABBY-like transcription factor led to evolution of extreme fruit size during tomato domestication. *Nat Genet*. 40:800–804. https://doi.org/10.1038/ng.144.

Datir SS, Regan S. 2023. *Transgenic Approaches for Nutritional Enhancement of Potato*. Intech Open. https://doi.org/10.5772/intechopen.106898.

Devran Z, Kahveci E, Hong YG, Studholme DJ, Tör M. 2018. Identifying molecular markers suitable for Frl selection in tomato breeding. *Theor Appl Genet*. 131:2099–2105.

Dey S, Sarkar A, Chowdhury S, Singh R, Mukherjee A, Ghosh Z, Kundu P. 2022 Heightened miR6024-NLR interactions facilitate necrotrophic pathogenesis in tomato. *Plant Mol Biol*. 109(6):717–739.

Dhaliwal MS, Jindal SK, Sharma A, Prasanna HC. 2020. Tomato yellow leaf curl virus disease of tomato and its management through resistance breeding: A review. *J Hortic Sci Biotechnol*. 95(4):425–444.

Di Matteo A, Ruggieri V, Sacco A, Rigano MM, Carriero F, Bolger A, Fernie AR, Frusciante L, Barone A. 2013. Identification of candidate genes for phenolics accumulation in tomato fruit. *Plant Sci*. 205–206:87–96.

Di Matteo A, Sacco A, Anacleria M, Pezzetti M, Delledonne M, Ferrarini A, Frusciante L, Barone A. 2010. The ascorbic acid content of tomato fruits is associated with the expression of genes involved in pectin degradation. *BMC Plant Biol*. 10:163.

Ding HD, He J, Wu Y, Wu XX, Ge CL, Wang YJ, Zhong SL, Peiter E, Liang J S, Xu WF. 2018. The tomato mitogen-activated protein kinase SlMPK1 is as a negative regulator of the high-temperature stress response. *Plant Physiol*. 177:633–651.

Ding HD, Qian Y, Fang YF, Ji YR, Sheng JR, Ge CL. 2021. Characteristics of SlCML39, a tomato calmodulin-like gene, and its negative role in high temperature tolerance of Arabidopsis thaliana during germination and seedling growth. *Int J Mol Sci*. 22:11475.

Driedonks N. 2018. *From Flower to Fruit in the Heat Reproductive Thermotolerance in Tomato and Its Wild Relatives*. Ph.D. Thesis, Radboud University, Nijmegen, The Netherlands.

Du M, Sun C, Deng L, Zhou M, Li J, Du Y, Ye Z, Huang S, Li T, Yu J, Li CB. 2025. Molecular breeding of tomato: Advances and challenges. *J Integr Plant Biol*. 67(3):669–721.

Duduit JR, Kosentka PZ, Miller MA, Blanco-Ulate B, Lenucci MS, Panthee DR, Perkins-Veazie P, Liu W. 2022. Coordinated transcriptional regulation of the carotenoid biosynthesis contributes to fruit lycopene content in high-lycopene tomato genotypes. *Hortic Res*. 9:uhac084.

Ebert AW, Chou YY. 2015. The tomato collection maintained by AVRDC—The World Vegetable Center: Composition, germplasm dissemination and use in breeding. *Acta Hortic*. 1101:169–176.

Elbaz M, Hanson P, Fgaier S, Laarif A. 2016. Evaluation of tomato entries with different combinations of resistance genes to tomato yellow leaf curl disease in Tunisia. *Plant Breed*. 135:525–530.

El-Sappah AH, El-Awady H, Yan S, Qi S, Liu J, Cheng GT, Liang Y. 2019. Tomato natural resistance genes in controlling the root-knot nematode. *Genes*. 10(11):925. https://doi.org/10.3390/genes10110925.

Erickson FL, Holzberg S, Calderon-Urrea A, Handley V, Axtell M, Corr C, Baker B. 1999. The helicase domain of the TMV replicase proteins induces the N-mediated defence response in tobacco. *Plant J*. 18:67–75.

Ernst K, Kumar A, Kriseleit D, Kloos DU, Phillips MS, Ganal MW. 2002. The broad-spectrum potato cyst nematode resistance gene (*Hero*) from tomato is the only member of a large gene family of NBS-LRR genes with an unusual amino acid repeat in the LRR region. *Plant J*. 31:127–136.

Eshed H, Zamir D. 1995. An introgression line population of *Lycopersicon pennellii* in the cultivated tomato enables the identification and fine mapping of yield-associated QTL. *Genetics*. 141:1147–1162.

Fadhilah A, Farid M, Ridwan I, Anshori M, Yassi A. 2022. Genetic parameters and selection index of high-yielding tomato F2 populations. *SABRAO J Breed Genet*. 54:1026–1036. [Google Scholar] [CrossRef]

FAO. 2024. *FAOSTAT: Production: Crops and Livestock Products*. https://www.fao.org/faostat/en/#data/QCL.

Fazio G, Stevens MR, Scott JW. 1999. Identification of RAPD markers linked to Fusarium crown and root rot resistance (*Frl*) in tomato. *Euphytica*. 105:205–210. https://doi.org/10.1023/A:1003497719 705.

Foolad MR. 2007. Genome mapping and molecular breeding of tomato. *Int J Plant Genomics*. 64358. https://doi.org/10.1155/2007/64358.

Foolad MR. 2015. A recombinant inbred line population of tomato and its genetic map constructed based on a *Solanum lycopersicum* × *S. pimpinellifolium* cross. *Adv Stud Bio*. 7(11):441–471.

Foolad MR, Merk HL, Ashrafi H. 2008. Genetics, genomics and breeding of late blight and early blight resistance in tomato. *Crit Rev Plant Sci*. 27(2):75–107.

Foolad MR, Panthee DR. 2012. Marker-assisted selection in tomato breeding. *Crit Rev Plant Sci*. 31:93–123.

Fradin EF, Zhang Z, Juarez Ayala JC, Castroverde CD, Nazar RN, Robb J, Liu CM, Thomma BP. 2009. Genetic dissection of Verticillium wilt resistance mediated by tomato *Ve1*. *Plant Physiol*. 150(1):320–332.

Frary A, Nesbitt TC, Grandillo S, van der Knaap E, Cong B, Liu JP, Meller J, Elber R, Alpert KB, Tanksley SD. 2000. fw2.2: A quantitative trait locus key to the evolution of tomato fruit size. *Science*. 289:85–88. https://doi.org/10.1126/science.289.5476.85.

Gobu R, Shiv A, Anilkumar C, Basavaraj PS, Harish D, Adhikari S, Ramtekey V, Hudedamani U, Mulpuri S. 2020. Accelerated crop breeding towards development of climate resilient varieties. In: CH Srinivasarao, T Srinivas, RVS Rao, N Srinivasarao, SS Vinayagam, P Krishnan, eds. *Climate Change and Indian Agriculture: Challenges and Adaptation Strategies*. ICAR-National Academy of Agricultural Research Management, Hyderabad, Telangana:59–79.

Gonzalez-Cendales Y, Catanzariti AM, Baker B, Mcgrath DJ, Jones DA. 2016. Identification of *I-7* expands the repertoire of genes for resistance to Fusarium wilt in tomato to three resistance gene classes. *Mol Plant Pathol.* 17(3):448–463.

Gonzalo MJ, Li YC, Chen KY, Gil D, Montoro T, Nájera I, et al. 2020. Genetic control of reproductive traits in tomatoes under high temperature. *Front Plant Sci.* 11:326. https://doi.org/10.3389/fpls.2020.00326.

Grandillo S, Ku HM, Tanksley SD. 1996. Characterization of fs8.1, a major QTL influencing fruit shape in tomato. *Mol Breed.* 2:251–260. https://doi.org/10.1007/BF00564202.

Grandillo S, Ku HM, Tanksley SD. 1999. Identifying the loci responsible for natural variation in fruit size and shape in tomato. *Theor Appl Genet.* 99:978–987.

Grilli GVG, Braz LT, Lemos EGM. 2007. QTL identification for tolerance to fruit set in tomato by FAFLP markers. *Crop Breed Appl Biotechnol.* 7:234–241.

Gupta E, Mishra P, Singh P, Mishra N. 2024. Evidence and prospects of lycopene as powerful red superfood: Modern approach to food science. *Curr Tradit Med.* 10(3):50–60.

Hanson P, Green SK, Kuo G. 2006. Ty2, a gene on chromosome 11 conditioning geminivirus resistance in tomato. *Tomato Genet Coop Rep.* 56:17–18.

He W, Li X, Zhang Z, Chen Y, Zhang J, Panthee DR, Meadows I, Xiang L. 2025. High-throughput robotic phenotyping for quantifying tomato disease severity enabled by synthetic data and domain-adaptive semantic segmentation. *J Field Robot.* 42(3):657–678.

Hu ZJ, Li JX, Ding ST, Cheng F, Li X, Jiang YP, Yu JQ, Foyer CH, Shi K. 2021. The protein kinase CPK28 phosphorylates ascorbate peroxidase and enhances thermotolerance in tomato. *Plant Physiol.* 186:1302–1317.

Huang HM, Liu CX, Yang C, Kanwar MK, Shao SJ, Qi ZY, Zhou J. 2022. BAG9 confers thermotolerance by regulating cellular redox homeostasis and the stability of heat shock proteins in *Solanum lycopersicum*. *Antioxidants.* 11:1467.

Huang ZJ, Van Houten J, Gonzalez G, Xiao H, van der Knaap E. 2013. Genome-wide identification, phylogeny and expression analysis of SUN, OFP and YABBY gene family in tomato. *Mol Genet Genomics.* 288:111–129. https://doi.org/10.1007/s00438-013-0733-0.

Hutton SF, Scott JW. 2014. *Ty-6*, a major begomovirus resistance gene located on chromosome 10. *Rept Tomato Genet Coop.* 64:14–18.

Hutton SF, Scott JW, Schuster DJ. 2012. Recessive resistance to tomato yellow leaf curl virus from the tomato cultivar tyking is located in the same region as *Ty-5* on chromosome 4. *HortScience.* 47(3):324–327.

Illa-Berenguer E, Van Houten J, Huang Z, Van der Knaap E. 2015. Rapid and reliable identification of tomato fruit weight and locule number loci by QTL-seq. *Theor Appl Genet.* 128:1329–1342.

IPK the Leibniz Institute of Plant Genetics and Crop Plant Research. 2022. https://www.ipk-gatersleben.de/en/infrastructure/gene-bank.

Jablonska B, Ammiraju JSS, Bhattarai KK, Mantelin S, de Ilarduya OM, Roberts PA, Kaloshian I. 2007. The *Mi-9* gene from Solanum arcanum conferring heat-stable resistance to root-knot nematodes is a homolog of *Mi-1*. *Plant Physiol.* 143:1044.

Jared JJ, Murungi LK, Wesonga J, Torto B. 2016. Steroidal glycoalkaloids: Chemical defence of edible African nightshades against the tomato red spider mite, Tetranychus evansi (Acari: Tetranychidae). *Pest Manag Sci.* 72(4):828–836. https://doi.org/10.1002/ps.4100.

Ji Y, Schuster DJ, Scott JW. 2007. *Ty-3*, a begomovirus resistance locus near the Tomato yellow leaf curl virus resistance locus *Ty-1* on chromosome 6 of tomato. *Mol Breed.* 20:271–284.

Ji Y, Scott JW, Schuster DJ, Maxwell DP. 2009. Molecular mapping of *Ty-4*, a new Tomato yellow leaf curl virus resistance locus on chromosome 3 of tomato. *J Am Soc Hortic Sci.* 134:281–288.

Jones JT, Haegeman A, Danchin EGJ, Gaur HS, Helder J, Jones MGK, Kikuchi T, Manzanilla-López R, Palomares-Rius JE, Wesemael WML, Perry RN. 2013. Top 10 plant-parasitic nematodes. *Mol Plant Pathol.* 14:946–961. https://doi.org/10.1111/mpp.12057.

Jones MO, Piron-Prunier F, Marcel F, Piednoir-Barbeau E, Alsadon AA. 2012. Characterization of alleles of tomato light signalling genes generated by TILLING. *Phytochemistry.* 79:78–86. https://doi.org/10.1016/j.phytochem.2012.04.005.

Jones-Rhoades MW, Bartel DP, Bartel B. 2006. MicroRNAs and their regulatory roles in plants. *Annu Rev Plant Biol*. 57(1):19–53. https://doi.org/10.1146/annurev.arplant.57.032905.105218.

Kadirvel P, de la Peña R, Schafleitner R, Huang S, Geethanjali S, Kenyon L, Tsai W, Hanson P. 2013. Mapping of QTLs in tomato line FLA456 associated with resistance to a virus causing tomato yellow leaf curl disease. *Euphytica*. 190(2):297–308.

Kanwar JS, Harney PM, Kerr EA. 1980. Allelic relationships of genes for resistance to tomato leaf mold, *Cladosporium fulvum* Cke. *Hortscience*. 15:418–428.

Kawchuk LM, Hachey J, Lynch DR, Kulcsar F, van Rooijen G, Waterer DR, Robertson A, Kokko E, Byers R, Howard RJ. 2001. Tomato *Ve* disease resistance genes encode cell surface-like receptors. *Proc Natl Acad Sci USA*. 98:6511–6515.

Ke JJ, Zhu WT, Yuan Y, Du XY, Xu A, Zhang D, Cao S, Chen W, Lin Y, Xie JT, et al. 2023. Duality of immune recognition by tomato and virulence activity of the *Ralstonia solanacearum*exo-polygalacturonase PehC. *Plant Cell*. 35:2552–2569.

Kim B, Hwang IS, Lee HJ, Lee JM, Seo E, Choi D, Oh CS. 2018. Identification of a molecular marker tightly linked to bacterial wilt resistance in tomato by genome-wide SNP analysis. *Theor Appl Genet*. 131:1017–1030. https://doi.org/10.1007/s00122-018-3054-1.

Kim M, Nguyen TTP, Ahn JH, et al. 2021. Genome-wide association study identifies QTL for eight fruit traits in cultivated tomato (*Solanum lycopersicum* L.). *Hortic Res*. 8:203. https://doi.org/10.1038/s41438-021-00638-4.

Kole C, Ashrafi H, Lin G, Foolad M. 2006. Identification and molecular mapping of a new R gene, Ph-4, conferring resistance to late blight in tomato. In *Proceedings of Solanaceae Conference*. University of Wisconsin, Madison, Abstract 449.

Kong FY, Deng YS, Wang GD, Wang JR, Liang XQ, Meng QW. 2014. LeCDJ1, a chloroplast DnaJ protein, facilitates heat tolerance in transgenic tomatoes. *J Integr Plant Biol*. 56:63–74.

Kourelis J, Van Der Hoorn RA. 2018. Defended to the nines: 25 years of resistance gene cloning identifies nine mechanisms for R protein function. *Plant Cell*. 30(2):285–299.

Ku HM, Grandillo S, Tanksley SD. 2000. A major QTL, sets the pattern of tomato carpel shape well before anthesis. *Theor Appl Genet*. 101:873–878. https://doi.org/10.1007/s001220051555.

Kurihara Y, Watanabe Y. 2004. Arabidopsis micro-RNA biogenesis through dicer-like 1 protein functions. *Proc Natl Acad Sci USA*. 101:12753–12758.

Landrier JF, Breniere T, Sani L, Desmarchelier C, Mounien L, Borel P. 2023. Effect of tomato, tomato-derived products and lycopene on metabolic inflammation: From epidemiological data to molecular mechanisms. *Nutr Res Rev*. Online ahead of print: 1–43. https://doi.org/10.1017/S095442242300029X.

Lanfermeijer FC, Dijkhuis J, Sturre MJ, de Haan P, Hille J. 2003. Cloning and characterization of the durable tomato mosaic virus resistance gene Tm-2^2 from *Lycopersicon esculentum*. *Plant Mol Biol*. 52:1037–1049.

Lapidot M, Karniel U, Gelbart D, Fogel D, Evenor D, Kutsher Y, Makhbash Z, Nahon S, Shlomo H, Chen L, Reuveni M. 2015. A novel route controlling begomovirus resistance by the messenger RNA surveillance factor pelota. *PLoS Genetics*. 11(10):1005538.

Larry R, Joanne L. 2007. Genetic resources of tomato. In: MK Razdan, AK Mattoo, eds. *Genetic Improvement of Solanaceous Crops. Volume 2*. Science Publishers, Enfield, NH.

Lata S, Hussain Z, Mangal M, Yadav RK, Jat GS, Gosavi G, Kumar P, Perveen S, Tomar B. 2019. QPCR analysis of Ty-2 and Ty-3 gene pyramided lines of tomato for resistance to tomato leaf curl New Delhi virus (ToLCNDV). *Indian J Agric Sci*. 89(10):1719–1722.

Lata S, Hussain Z, Yadav RK, Jat GS, Kumar P, Tomar BS. 2024. Insights into the genetic improvement of tomato. In: S Tiwari, B Koul, eds. *Genetic Engineering of Crop Plants for Food and Health Security*. Springer, Singapore. https://doi.org/10.1007/978-981-97-3119-0_7.

Lee HA, Yeom SI. 2015. Plant NB-LRR proteins: Tightly regulated sensors in a complex manner. *Brief Funct Genomics*. 14:233–242.

Lee Y, Kim M, Han J, Yeom K, Lee S, Baek SH, et al. 2004. MicroRNA genes are transcribed by RNA polymerase II. *EMBO J*. 23(20):4051–4060. https://doi.org/10.1038/sj.emboj.7600385.

Li N, He Q, Wang J, Wang BK, Zhao JT, Huang SY, Yang T, Tang YP, Yang SB, Aisimutuola P, et al. 2023a. Superpangenome analyses highlight genomic diversity and structural variation across wild and cultivated tomato species. *Nat Genet*. 55:852–860.

Li Q, Shen H, Yuan S, Dai X, Yang C. 2023b. miRNAs and lncRNAs in tomato: Roles in biotic and abiotic stress responses. *Front Plant Sci*. 13:1094459.

Li W, Chetelat RT. 2010. A pollen factor linking inter- and intraspecific pollen rejection in tomato. *Science*. 330:1827–1830.
Li X, Tieman D, Liu Z, Chen K, Klee HJ. 2020. Identification of a lipase gene with a role in tomato fruit short-chain fatty acid-derived flavor volatiles by genome wide association. *Plant J*. 104:631–644.
Liabeuf D, Sim SC, Francis DM. 2018. Comparison of marker-based genomic estimated breeding values and phenotypic evaluation for selection of bacterial spot resistance in tomato. *Phytopathology*. 108:392–401.
Liang XQ, Ma NN, Wang GD, Meng X, Ai XZ, Meng QW. 2015. Suppression of SlNAC1 reduces heat resistance in tomato plants. *Biol Plant*. 59:92–98.
Lin KH, Yeh WL, Chen HM, Lo HF. 2010. Quantitative trait loci influencing fruit-related characteristics of tomato grown in high-temperature conditions. *Euphytica*. 174:119–135.
Lin T, Zhu GT, Zhang JH, Xu XY, Yu QH, Zheng Z, Zhang ZH, Lun YY, Li S, Wang XX, et al. 2014. Genomic analyses provide insights into the history of tomato breeding. *Nat Genet*. 46:1220–1226. https://doi.org/10.1038/ng.3117.
Liu J, Liu X, Dai L, Wang G. 2007. Recent progress in elucidating the structure, function and evolution of disease resistance genes in plants. *J Genet Genomics*. 34(9):765–776. https://doi.org/10.1016/s1673-8527(07)60087-3.
Liu J, Van Eck J, Cong B, Tanksley SD. 2002. A new class of regulatory genes underlying the cause of pear-shaped tomato fruit. *Proc Natl Acad Sci USA*. 99:13302–13306.
Liu M, Zhao G, Huang X, Pan T, Chen W, Qu M, Ouyang B, Yu M, Shabala S. 2023. Candidate regulators of drought stress in tomato revealed by comparative transcriptomic and proteomic analyses. *Front Plant Sci*. 14:1282718. https://doi.org/10.3389/fpls.2023.1282718. PMID: 37936934; PMCID: PMC10627169.
Liu Z, Alseekh S, Brotman Y, Zheng Y, Fei Z, Tieman DM, Giovannoni JJ, Fernie AR, Klee HJ. 2016. Identification of a Solanum pennellii chromosome 4 fruit flavor and nutritional quality-associated metabolite QTL. *Front Plant Sci*. 7:1671. https://doi.org/10.3389/fpls.2016.01671.
Luan Y, Cui J, Li J, Jiang N, Liu P, Meng J. 2018. Effective enhancement of resistance to Phytophthora infestans by overexpression of miR172a and b in *Solanum lycopersicum*. *Planta*. 247(1):127–138.
Manning K, Tor M, Poole M, Hong Y, Thompson AJ, King GJ, Giovannoni JJ, Seymour GB. 2006. A naturally occurring epigenetic mutation in a gene encoding an SBP-box transcription factor inhibits tomato fruit ripening. *Nat Genet*. 38:948–952.
Mao L, Deng M, Jiang S, Zhu H, Yang Z, Yue Y, Zhao K. 2020. Characterization of the DREBA4-type transcription factor (SlDREBA4), which contributes to heat tolerance in tomatoes. *Front Plant Sci*. 11:554520.
Marko D, El-Shershaby A, Carriero F, Summerer S, Petrozza A, Iannacone R, Schleiff E, Fragkostefanakis S. 2019. Identification and characterization of a thermotolerant TILLING allele of heat shock binding protein 1 in tomato. *Genes*. 10:516.
Martí R, Roselló S, Cebolla-Cornejo J. 2016. Tomato as a source of carotenoids and polyphenols targeted to cancer prevention. *Cancers (Basel)*. 8:E58. https://doi.org/10.3390/cancers8060058.
Martin GB, Brommonschenkel SH, Chunwongse J, Frary A, Ganal MW, Spivey R, Wu T, Earle ED, Tanksley SD. 1993. Map-based cloning of a protein kinase gene conferring disease resistance in tomato. *Science*. 262:1432–1436.
Mata-Nicolás E, Montero-Pau J, Gimeno-Paez E, et al. 2020. Exploiting the diversity of tomato: The development of a phenotypically and genetically detailed germplasm collection. *Hortic Res*. 7:66. https://doi.org/10.1038/s41438-020-0291-7.
Meng X, Wang JR, Wang GD, Liang XQ, Li XD, Meng QW. 2015. An R2R3-MYB gene, LeAN2, positively regulated the thermo-tolerance in transgenic tomato. *J Plant Physiol*. 175:1–8.
Michelmore RW, Christopoulou M, Caldwell KS. 2013. Impacts of resistance gene genetics, function, and evolution on a durable future. *Annu Rev Phytopathol*. 51:291–319.
Milligan SB, Bodeau J, Yaghoobi J, Kaloshian I, Zabel P, Williamson VM. 1998. The root knot nematode resistance gene *Mi* from tomato is a member of the leucine zipper, nucleotide binding, leucine-rich repeat family of plant genes. *Plant Cell*. 10(8):1307–1319. https://doi.org/10.1105/tpc.10.8.1307.
Mishra SK, Tripp J, Winkelhaus S, Tschiersch B, Theres K, Nover L, Scharf KD. 2002. In the complex family of heat stress transcription factors, HsfA1 has a unique role as master regulator of thermotolerance in tomato. *Gene Dev*. 16:1555–1567.
Monforte AJ, Tanksley SD. 2000. Fine mapping of a quantitative trait locus (QTL) from *Lycopersicon hirsutum* chromosome 1 affecting fruit characteristics and agronomic traits: Breaking linkage among QTLs affecting different traits and dissection of heterosis for yield. *Theor Appl Genet*. 100:471–479.

Moyle LC, Nakazato T. 2012. Hybrid incompatibility “snowballs” between Solanum species. *Science*. 329:1521–1523.
Mu QI, Huang Z, Chakrabarti M, Illa-Berenguer E, Liu X, Wang Y, Ramos A, van der Knaap E. 2017. Fruit weight is controlled by Cell Size Regulator encoding a novel protein that is expressed in maturing tomato fruits. *PLoS Genet*. 13(8):e1006930.
Munos S, et al. 2011. Increase in tomato locule number is controlled by two single nucleotide polymorphisms located near WUSCHEL. *Plant Physiol*. 156:2244–2254.
Mushtaq N, Wang Y, Fan J, Li Y, Ding J. 2022. Down-regulation of cytokinin receptor gene SlHK2 improves plant tolerance to drought, heat, and combined stresses in tomato. *Plants*. 11:154.
Muthukumar P, Jat GS, Kalia P, et al. 2024. Morphological characterization and screening of *Solanum habrochaites* accessions for late blight (*Phytophthora infestans*) disease resistance. *Genet Resour Crop Evol*. 71:1369–1377. https://doi.org/10.1007/s10722-023-01700-x.
Nakazato T, Warren DL, Moyle LC. 2010. Ecological and geographic modes of species divergence in wild tomatoes. *Am J Bot*. 97:680–693.
Niu Y, Chen T, Wang C, Chen K, Shen C, Chen H, Zhu S, Wu Z, Zheng T, Zhang F, Xu J. 2021. Identification and allele mining of new candidate genes underlying rice grain weight and grain shape by genome-wide association study. *BMC Genomics*. 22(1):1–4. https://doi.org/10.1186/s12-864-021-07901-x.
Nuez F, Diez MJ. 2008. Tomato. In: J Prohens, F Nuez, eds. *Vegetables II. Handbook of Plant Breeding*:249–323. https://doi.org/10.1007/978-0-387-74110-9_7.
Olmo R, Quijada NM, Morán-Diez ME, Hermosa R, Monte E. 2024. Identification of tomato microRNAs in late response to *Trichoderma atroviride*. *Int J Mol Sci*. 25(3):1617. https://doi.org/10.3390/ijms25031617.
Orchard CJ, Cooperstone JL, Gas-Pascual E, Andrade MC, Abud G, Schwartz SJ, Francis DM. 2021. Identification and assessment of alleles in the promoter of the Cyc-B gene that modulate levels of β-carotene in ripe tomato fruit. *Plant Genome*. 14:e20085.
Ortiz R. 2015. The importance of crop wild relatives, diversity, and genetic potential for adaptation to abiotic stress-prone environments. In: R Redden, SS Yadav, N Maxted, ME Dulloo, L Guarino, P Smith, eds. *Crop Wild Relatives and Climate Change*. John Wiley & Sons, Inc, Hoboken, NJ:80–87.
Osei MK, Adjebeng-Danquah J, Amankwaah VA, Awuku FJ, Zohoungbogbo HPF, Melomey LD, Agyare RY, Puozaa DK, Mensah INB, Osei-Bonsu I. 2024. Allele mining in tomato: Strategies and applications in gene discovery for food and nutritional security. In: C Kole, TK Behera, P Kaushik, eds. *Allele Mining for Genomic Designing of Vegetable Crops*, vol. 28:300. https://doi.org/10.1201/9781003376439.
Ouyang S, Park G, Atamian HS, Han CS, Stajich JE, Kaloshian I, Borkovich KA. 2014. MicroRNAs suppress NB domain genes in tomato that confer resistance to Fusarium oxysporum. *PLoS Pathog* 10(10):e1004464.
Palanisamy M, Jat GS, Kalia P, et al. 2023 Screening and validation of Ty-1, Ty-3 and Ty-3a specific markers in Solanum chilense accessions for resistance to tomato leaf curl virus. *Genet Resour Crop Evol*. 70:2567–2574. https://doi.org/10.1007/s10722-023-01586-9.
Panno S, Davino S, Caruso AG, Bertacca S, Crnogorac A, Mandi AA. 2021. Review of the most common and economically important diseases that undermine the cultivation of tomato crop in the Mediterranean basin. *Agronomy*. 11:2188.
Panthee DR, Brown AF, Yousef GG, Ibrahem R, Anderson C. 2013a. Novel molecular marker associated with T m2a gene conferring resistance to tomato mosaic virus in tomato. *Plant Breeding*. 132(4):413–416.
Panthee DR, Cao CX, Debenport SJ, Rodríguez GR, Labate JA, Robertson LD, Breksa AP III, van der Knapp E, Gardener BBM. 2012. Magnitude of genotype × environment interactions affecting tomato fruit quality. *HortScience*. 47(6):721–726.
Panthee DR, Labate JA, McGrath MT, Breksa AP, Robertson LD. 2013b. Genotype and environmental interaction for fruit quality traits in vintage tomato varieties. *Euphytica*. 193:169–182.
Panthee DR, Piotrowski A, Ibrahem R. 2017. Mapping quantitative trait loci (QTL) for resistance to late blight in tomato. *Int J Mol Sci*. 18(7):1589.
Paterson AH, Deverna JW, Lanini B, Tansksley SD. 1990. Fine mapping of quantitative trait loci using selected overlapping recombinant chromosomes in an interspecies cross of tomato. Genetics. 28:379–392.
Paterson AH, Lander ES, Hewitt JD, Peterson S, Lincoln SE, Tanksley SD. 1988. Resolution of quantitative traits into Mendelian factors by using a complete linkage map of restriction fragment length polymorphisms. *Nature*. 335:721–726.
Patra BC, Anilkumar C, Chakraborti M. 2020. Rice breeding in India: A journey from phenotype based pureline selection to genomics assisted breeding. *Agril Res J*. 57(6):816–825.

Pei CC, Wang H, Zhang JY, Wang YY, Francis DM, Yang WC. 2012. Fine mapping and analysis of a candidate gene in tomato accession PI128216 conferring hypersensitive resistance to bacterial spot race T3. *Appl Genet*. 124:533–542.

Pereira L, Sapkota M, Alonge M, Zheng Y, Zhang Y, Razifard H, Taitano NK, Schatz MC, Fernie AR, Wang Y, Fei Z, Caicedo AL, Tieman DM and van der Knaap E. 2021a. Natural genetic diversity in tomato flavor genes. *Front Plant Sci*. 12:642828. https://doi.org/10.3389/fpls.2021.642828.

Pereira L, Zhang L, Sapkota M, Ramos A, Razifard H, Caicedo AL, van der Knaap E. 2021b. Unraveling the genetics of tomato fruit weight during crop domestication and diversification. *Theor Appl Genet*. 134(10):3363–3378. https://doi.org/10.1007/s00122-021-03902-2. PMID: 34283260; PMCID: PMC8440300.

Pereyra-Bistraín LI, Ovando-Vázquez C, Rougon-Cardoso A, Alpuche-Solís ÁG. 2021. Comparative RNA-Seq analysis reveals potentially resistance-related genes in response to bacterial canker of tomato. *Genes*. 12(11):1745.

Pessoa HP, Dariva FD, Copati MGF, de Paula RG, de Oliveira Dias F, Gomes CN. 2023. Uncovering tomato candidate genes associated with drought tolerance using *Solanum pennellii* introgression lines. *PLoS ONE*. 18(6):e0287178.

Petrozza A, Summerer S, Melfi D, Mango T, Vurro F, Bettelli M, Janni M, Cellini F, Carriero F. 2023. A lycopene ε-cyclase TILLING allele enhances lycopene and carotenoid content in fruit and improves drought stress tolerance in tomato plants. *Genes*. 14(6):1284. https://doi.org/10.3390/genes14061284.

Prasad A, Sharma N, Chirom O, et al. 2022 The sly-miR166-SlyHB module acts as a susceptibility factor during ToLCNDV infection. *Theor Appl Genet*. 135:233–242.

Qi S, Shen Y, Wang X, Zhang S, Li Y, Islam MM, Wang J, Zhao P, Zhan X, Zhang F, Liang Y. 2022. A new NLR gene for resistance to Tomato spotted wilt virus in tomato (*Solanum lycopersicum*). *Theor Appl Genet*. 135(5):1493–1509. https://doi.org/10.1007/s00122-022-04049-4. Epub 2022 Feb 18. PMID: 35179614.

Quinet M, Angosto T, Yuste-Lisbona FJ, Blanchard-Gros R, Bigot S, Martinez JP, Lutts S. 2019. Tomato fruit development and metabolism. *Front Plant Sci*. 10:1554. https://doi.org/10.3389/fpls.2019.01554.

Raiola A, Rigano MM, Calafiore R, Frusciante L, Barone A. 2014. Enhancing the health-promoting effects of tomato fruit for biofortified food. *Mediators Inflamm*. 2014:139873. https://doi.org/10.1155/2014/139873.

Ramírez-Ojeda G, Peralta IE, Rodríguez-Guzmán E, Sahagún-Castellanos J, Chávez-Servia JL, Medina-Hinostroza TC, Rijalba-Vela JR, Vásquez-Núñez LP, Rodríguez-Pérez JE. 2021. Edaphoclimatic descriptors of wild tomato species (*Solanum* Sect. *Lycopersicon*) and closely related species (*Solanum* Sect. *Juglandifolia* and Sect. *Lycopersicoides*) in South America. *Front Genet*. 12:748979. https://doi.org/10.3389/fgene.2021.748979.

Rao S, Das JR, Balyan S, Verma R, Mathur S. 2022. Cultivar-biased regulation of HSFA7 and HSFB4a govern high-temperature tolerance in tomato. *Planta*. 255(2):31.

Ric C, Chetelat R. 1995. Utilization of related wild species for tomato improvement. In: *Proceedings of the International Symposium on Solanacea for Fresh Market. Acta Horticulturae 412*. Málaga, Spain. 28–31:21–38. [Google Scholar]

Robatzek S, Bittel P, Chinchilla D, Kochner P, Felix G, Shiu SH, Boller T. 2007. Molecular identification and characterization of the tomato flagellin receptor *LeFLS2*, an orthologue of *Arabidopsis FLS2* exhibiting characteristically different perception specificities. *Plant Mol Biol*. 64:539–547.

Rodríguez, LJA, Guzmán, ER, Parra, JR, Ramírez, JL, Saldaña, HL, Martínez, JS, Ildefonso, RL. 2013. Evaluation of the resistance against *Phytophthora infestans* of wild populations of *Solanum lycopersicum* var. *cerasiforme*. *Rev Mex Cienc Agr*. 4:753–766.

Rodríguez GR, Kim HJ, van der Knaap E. 2013. Mapping of two suppressors of OVATE (sov) loci in tomato. *Heredity*. 111:256–264. https://doi.org/10.1038/hdy.2013.45.

Roohanitaziani R, de Maagd RA, Lammers M, Moltho J, Meijer-Dekens F, et al. 2020. Exploration of a resequenced tomato core collection for phenotypic and genotypic variation in plant growth and fruit quality traits. *Genes*. 11:1278. https://doi.org/10.3390/genes11111278.

Rousseaux MC, Jones CM, Adams D, Chetelat R, Bennett A, Powell A. 2005. QTL analysis of fruit antioxidants in tomato using *Lycopersicon pennellii* introgression lines. *Theor Appl Genet*. 111:1396–1408.

Ruggieri V, Francese G, Sacco A, D'Alessandro A, Rigano MM, Parisi M, Milone M, Cardi T, Mennella G, Barone A. 2014. An association mapping approach to identify favourable alleles for tomato fruit quality breeding. *BMC Plant Biol*. 14:337. https://doi.org/10.1186/s12870-014-0337-9.

Salmeron JM, Oldroyd GE, Rommens CM, Scofield SR, Kim HS, Lavelle DT, Dahlbeck D, Staskawicz BJ. 1996. Tomato Prf is a member of the leucine-rich repeat class of plant disease resistance genes and lies embedded within the Pto kinase gene cluster. *Cell.* 86:123–133.

Sauvage C, Segura V, Bauchet G, Stevens R, Do PT, Nikoloski Z, Fernie AR, Causse M. 2014. Genome-wide association in tomato reveals 44 candidate loci for fruit metabolic traits. *Plant Physiol.* 165:1120–1132.

Scarano A, Santino A, Ricroch A, Eriksson D, Miladinović D, Sweet J, Van Laere K, Woźniak-Gientka E. 2024. Engineering phytonutrient content in tomato by genome editing technologies. In: *A Roadmap for Plant Genome Editing*. Springer, Cham. https://doi.org/10.1007/978-3-031-46150-7_22.

Schmitz G, Theres K. 1999. Genetic control of branching in *Arabidopsis* and tomato. *Curr Opin Plant Biol.* 2:51–55.

Selvakumar R, Jat GS, Manjunathagowda DC. 2023. Allele mining through TILLING and EcoTILLING approaches in vegetable crops. *Planta.* 258:15. https://doi.org/10.1007/s00425-023-04176-2.

Shahid MS, Kimbara J, Onozato A, Natsuaki KT, Ikegami M. 2015. Comparative analysis of gene expression of Ty-1 hybrid and non-hybrid tomatoes exposed to tomato yellow leaf curl virus strains. *Aust J Crop Sci.* 9(9):819–825.

Shen X, Yan Z, Wang X, Wang Y, Arens, M, Du Y, Visser, RGF, Kormelink R, Wolters AMA. 2020. The NLR protein encoded by the resistance gene *Ty-2* is triggered by the replication-associated protein *Rep/C1* of Tomato yellow leaf curl virus. *Front Plant Sci.* 11:1384.

Shi A, Vierling R, Grazzini R, Chen P, Caton H, Panthee D. 2011. Identification of molecular markers for Sw-5 gene of tomato spotted wilt virus resistance. *Am J Biotechnol Mol Sci.* 1(1):8–16.

Shi R, Panthee DR. 2020. Transcriptome-based analysis of tomato genotypes resistant to bacterial spot (*Xanthomonas perforans*) race t4. *Int J Mol Sci.* 21(11):4070.

Siddique MI, Silverman E, Louws F, Panthee DR. 2024. Quantitative trait loci mapping for bacterial wilt resistance and plant height in tomatoes. *Plants (Basel).* 13(6):876. https://doi.org/10.3390/plants13060876. PMID: 38592886; PMCID: PMC10976105.

Sierra-Orozco E, Shekasteband R, Illa-Berenguer E, Snouffer A, Van der Knaap E, Lee TG, Hutton SF. 2021. Identification and characterization of GLOBE, a major gene controlling fruit shape and impacting fruit size and marketability in tomato. *Hortic Res.* 8:138.

Simons G, Groenendijk J, Wijbrandi J, Reijans M, Groenen J, Diergaarde P, Van der Lee T, Bleeker M, Onstenk J, de Both M, et al. 1998. Dissection of the *Fusarium* I2 gene cluster in tomato reveals six homologs and one active gene copy. *Plant Cell.* 10:1055–1068.

Singh S, Selvakumar R, Mangal M, Kalia P. 2018. Breeding and genomic investigations for quality and nutraceutical traits in vegetable crops-a review. *Indian J Hortic.* 77(1):1–40.

Song X, Wang T, Zhang Y, Yu JQ, Xia XJ. 2022. S-Nitrosoglutathione reductase contributes to thermotolerance by modulating high temperature-induced apoplastic H2O2 in *Solanum lycopersicum. Front Plant Sci.* 13:862649.

Spoel S, Dong X. 2012. How do plants achieve immunity? Defence without specialized immune cells. *Nat Rev Immunol.* 12:89–100. https://doi.org/10.1038/nri3141.

Stevens M, Rick CM. 1986. Genetics and breeding. In: Atherton JG, Rudich J, eds. *The Tomato Crop. A Scientific Basis for Improvement.* Chapman & Hall:35–109.

Stevens R, Buret M, Garchery C, Carretero Y, Causse M. 2006. Technique for rapid small-scale analysis of vitamin C levels in fruit and application to a tomato mutant collection. *J Agric Food Chem.* 54:6159–6165.

Sunkar R, Li YF, Jagadeeswaran G. 2012. Functions of MicroRNAs in plant stress responses. *Trends Plant Sci.* 17:196–203.

Suzuki T, Ikeda S, Kasai A, Taneda A, Fujibayashi M, Sugawara K, Sano T. 2019. RNAi-mediated down-regulation of dicer-like 2 and 4 changes the response of 'Moneymaker' tomato to potato spindle tuber viroid infection from tolerance to lethal systemic necrosis, accompanied by up-regulation of miR398, 398a-3p and production of excessive amount of reactive oxygen species. *Viruses.* 11(4):344.

Tanksley SD, Nelson JC. 1996. Advanced backcross QTL analysis: A method for the simultaneous discovery and transfer of valuable QTLs from unadapted germplasm into elite breeding lines. *Theor Appl Genet.* 92:191–203.

Thomas CM, Jones DA, Parniske M, Harrison K, Balint-Kurti PJ, Hatzixanthis K, Jones JD. 1997. Characterization of the tomato *Cf-4* gene for resistance to *Cladosporium fulvum* identifies sequences that determine recognitional specificity in *Cf-4* and *Cf-9. Plant Cell.* 9:2209–2224.

The Tomato Genome Consortium. 2012. The tomato genome sequence provides insights into fleshy fruit evolution. *Nature.* 485:635–641.

Townsley BT, Sinha NR. 2012. A new development: Evolving concepts in leaf ontogeny. *Annu Rev Plant Biol.* 63:535–562.

USDA-PGRU Plant Genetic Resources Unit. https://www.ars.usda.gov/northeast-area/geneva-ny/plant-genetic-resources-unit-pgru/docs/tomato-collection/ [Last Access 09 November 2022].

Usman MG, Rafii MY, Martini MY, Yusuff OA, Ismail MR, Miah G. 2017. Molecular analysis of Hsp70 mechanisms in plants and their function in response to stress. *Biotechnol Genet Eng Rev.* 33(1):26–39. https://doi.org/10.1080/02648725.2017.1340546. Epub Jun 25. PMID: 28649918

Van der Knaap E, Chakrabarti M, Chu YH, Clevenger JP, Illa-Berenguer E, Huang Z, Keyhaninejad N, Mu Q, Sun L, Wang Y. et al. 2014. What lies beyond the eye: The molecular mechanisms regulating tomato fruit weight and shape. *Front Plant Sci.* 5:227.

Varshney RK, Singh VK, Hickey JM, Xun X, Marshall DF, Wang J, Edwards D, Ribaut JM. 2016. Analytical and decision support tools for genomics-assisted breeding. *Trends Plant Sci.* 21(4):354–363. https://doi.org/10.1016/j.tplants.2015.10.018.

Verlaan MG, Hutton SF, Ibrahem RM, Kormelink R, Visser RG, Scott JW, Edwards JD, Bai Y. 2013. The tomato yellow leaf curl virus resistance genes *Ty-1* and *Ty-3* are allelic and code for DFDGD-class RNA-dependent RNA polymerases. *PLoS Genet.* 9:e1003399.

Vossen JH, Jo KR, Vosman B. 2014. Mining the genus Solanum for increasing disease resistance. In: Tuberosa, R., Graner, A., Frison, E. (eds) *Genomics of Plant Genetic Resources.* Springer, Dordrecht. https://doi.org/10.1007/978-94-007-7575-6_2

Vu TV, Das S, Tran MT, Hong JC, Kim JY. 2020. Precision genome engineering for the breeding of tomatoes: Recent progress and future perspectives. *Front Genome Ed.* 2:612137. https://doi.org/10.3389/fgeed.2020.612137.

Wang HS, Yu C, Zhu ZJ, Yu XC. 2011. Overexpression in tobacco of a tomato GMPase gene improves tolerance to both low and high temperature stress by enhancing antioxidation capacity. *Plant Cell Rep.* 30:1029–1040.

Wang X, Xu N, Dong K, Li H, Shi S, Liu Z, Cao X, Cui X, Dai S. 2021. SlNCED1 affects pollen maturation in tomato by regulating the expression of anther-specific genes. *Plant Growth Regul.* 95:191–205. https://doi.org/10.1007/s10725-021-00732-6.

Wang XY, Zhang HJ, Xie Q, Liu Y, Lv HM, Bai RY, Ma R, Li XD, Zhang XC, Guo YD, Zhang N. 2020. SlSNAT interacts with HSP40, a molecular chaperone, to regulate melatonin biosynthesis and promote thermotolerance in tomato. *Plant Cell Physiol.* 61:909–921.

Wang Y, Gai W, Yuan L, Shang L, Li F, Gong Z, Ge P, Wang Y, Tao J, Zhang X, Dong H, Zhang Y. 2022. Heat-inducible SlWRKY3 confers thermotolerance by activating the SlGRXS1 gene cluster in tomato. *Hortic Plant J.* 10(2):515–531.

Wang Y, Sun C, Ye Z, Li C, Huang S, Lin T. 2024. The genomic route to tomato breeding: Past, present, and future. *Plant Physiol.* 195:2500–2514.

Wang Z, Hardcastle TJ, Canto Pastor A, Yip WH, Tang S, Baulcombe DC. 2018. A novel DCL2-dependent miRNA pathway in tomato affects susceptibility to RNA viruses. *Genes Dev.* 32(17–18):1155–1160.

Wei C, Kuang H, Li F, Chen J. 2014. The I2 resistance gene homologues in Solanum have complex evolutionary patterns and are targeted by miRNAs. *BMC Genomics.* 15:1–14.

Wei H, Liu J, Guo Q, Pan L, Chai S, Cheng Y, Ruan M, Ye Q, Wang R, Yao Z, Zhou G. 2020. Genomic organization and comparative phylogenic analysis of NBS-LRR resistance gene family in *Solanum pimpinellifolium* and *Arabidopsis thaliana. Evol Bioinform.* 16:1176934320911055.

Wen J, Jiang F, Weng Y, Sun M, Shi X, Zhou Y, Yu L, Wu Z. 2019. Identification of heat-tolerance QTLs and high-temperature stress-responsive genes through conventional QTL mapping, QTL-seq and RNA-seq in tomato. *BMC Plant Biol.* 19:1–17.

The World Vegetable Center. https://avrdc.org/our-work/managing-germplasm/ [Last Access 09 November 2022].

Xu C, Liberatore KL, MacAlister CA, Huang Z, Chu Y-H, Jiang K, Lippman ZB. 2015. A cascade of arabinosyl transferases controls shoot meristem size in tomato. *Nat Genet.* 47:784–792. https://doi.org/10.1038/ng.3309.

Xu J, Driedonks N, Rutten MJM, Vriezen WH, de Boer G-J, Rieu I. 2017. Mapping quantitative trait loci for heat tolerance of reproductive traits in tomato (*Solanum lycopersicum*). *Mol Breed.* 37:58.

Xu X, Wang Q, Li WQ, Hu TX, Wang QQ, Yin Y, Liu XH, He S, Zhang MK, Liang Y, Zhu JH, Zhan XQ. 2022. Overexpression of SlBBX17 affects plant growth and enhances heat tolerance in tomato. *Int J Biol Macromol.* 206:799–811.

Yan Z, Pérez-De-Castro A, Diez MJ, Hutton SF, Visser RG F, Wolters A.-M.-A., Bai Y, Li J. 2018. Resistance to tomato yellow leaf curl virus in tomato germplasm. *Front Plant Sci.* 9:1198. https://doi.org/10.3389/fpls.2018.01198.

Yang H, Wang H, Jiang J, Liu M, Liu Z, Tan Y, Zhao T, Zhang H, Chen X, Li J, Wang A, Du M, Xu X. 2022. The Sm gene conferring resistance to gray leaf spot disease encodes an NBS-LRR (nucleotide-binding site-leucine-rich repeat) plant resistance protein in tomato. *Theor Appl Genet.* 135(5):1467–1476.

Ye J, Li W, Ai G, Li C, Liu G, Chen W, Wang B, Wang W, Lu Y, Zhang J, Li H, Ouyang B, Zhang H, Fei Z, Giovannoni JJ, Ye Z, Zhang Y. 2019. Genome-wide association analysis identifies a natural variation in basic helix-loop-helix transcription factor regulating ascorbate biosynthesis via D-mannose/L-galactose pathway in tomato. *PLoS Genet.* 15(5):e1008149. https://doi.org/10.1371/journal.pgen.1008149.

Yin YL, Qin KZ, Song XW, Zhang QH, Zhou YH, Xia XJ, Yu JQ. 2018. BZR1 transcription factor regulates heat stress tolerance through FERONIA receptor-like kinase-mediated reactive oxygen species signaling in tomato. *Plant Cell Physiol.* 59:2239–2254.

Yu G, Li C, Zhang L, Zhu G, Munir S, et al. 2020. An allelic variant of GAME9 determines its binding capacity with the GAME17 promoter in the regulation of steroidal glycoalkaloid biosynthesis in tomato. *J Exp Bot.* 71(9):2527–2536.

Yu N, Wei K, Li S, Zhang L, Chen Z, Lu F, Yang P, Yang M, Liu X, Liu X, et al. 2023. Fine mapping of fw6.3, a major-effect quantitative trait locus that controls fruit weight in tomato. *Plants.* 12(11):2065. https://doi.org/10.3390/plants12112065.

Yu WQ, Wang L, Zhao RR, Sheng JP, Zhang SJ, Li R, Shen L. 2019. Knockout of SlMAPK3 enhances tolerance to heat stress involving ROS homeostasis in tomato plants. *BMC Plant Biol.* 19:354.

Yu X, Qu M, Shi Y, Hao C, Guo S, Fei Z, Gao L. 2022. Chromosome-scale genome assemblies of wild tomato relatives *Solanum habrochaites* and *Solanum galapagense* reveal structural variants associated with stress tolerance and terpene biosynthesis. *Hortic Res.* 9:uhac139.

Zamir D, Ekstein-Michelson I, Zakay Y, Navot N, Zeidan M, Sarfatti M, Eshed Y, Harel E, Pleban T, Van-Oss H, Kedar N, Rabinowitch HD, Czosnek H. 1994. Mapping and introgression of a tomato yellow leaf curl virus tolerance gene, TY-1. *Theor Appl Genet.* 88:141–146.

Zhang C, Liu L, Wang X, Vossen J, Li G, Li T, Zheng Z, Gao J, Guo Y, Visser RG, Li J, Bai Y, Du Y. 2014. The *Ph-3* gene from *Solanum pimpinellifolium* encodes CC-NBS-LRR protein conferring resistance to *Phytophthora infestans. Theor Appl Genet.* 127(6):1353–1364. https://doi.org/10.1007/s00122-014-2303-1.

Zhang H, Mittal N, Leamy LJ, Barazani O, Song BH. 2017. Back into the wild—Apply untapped genetic diversity of wild relatives for crop improvement. *Evol Appl.* 10:5–24.

Zhang H, Zhou JF, Kan Y, Shan JX, Ye WW, Dong NQ, Guo T, Xiang YH, Yang YB, Li YC, Zhao HY, Yu HX, Lu ZQ, Guo SQ, Lei JJ, Liao B, Mu XR, Cao YJ, Yu JJ, Lin YS, Lin HX. 2022. A genetic module at one locus in rice protects chloroplasts to enhance thermotolerance. *Science.* 376:1293–1300.

Zhang J, Panthee DR. 2021. Development of codominant SCAR markers to detect the Pto, Tm22, I3 and Sw5 genes in tomato (*Solanum lycopersicum*). *Plant Breed.* 140(2):342–348.

Zhao J, Sauvage C, Zhao J, Bitton F, Bauchet G, Liu D, Huang S, Tieman DM, Klee HJ, Causse M. 2019. Meta-analysis of genome-wide association studies provides insights into genetic control of tomato flavor. *Nat Commun.* 10:1534. https://doi.org/10.1038/s41467-019-09462-w.

Zhao J, Xu Y, Ding Q, Huang X, Zhang Y, Zou Z, Li M, Cui L and Zhang J.2016. Association mapping of main tomato fruit sugars and organic acids. *Front Plant Sci.* 7:1286. https://doi.org/10.3389/fpls.2016.01286.

Zhao T, Pei T, Jiang J, Yang H, Zhang H, Li J, Xu X. 2022. Understanding the mechanisms of resistance to tomato leaf mold: A review. *Hortic Plant J.* 8(6):667–675.

Zhao X, Zhang Y, Lai J, Deng Y, Hao Y, Wang S, Yang J. 2023–24. The SlDOG1 affect biosynthesis of steroidal glycoalkaloids by regulating GAME expression in tomato. *Int J Mol Sci.* 3360. https://doi.org/10.3390/ijms24043360.

Zhu CS, Naqvi S, Gomez G, Pelacho AM, et al. 2007. Transgenic strategies for the nutritional enhancement of plants. *Trends Plant Sci.* 12(12):548–555. https://doi.org/10.1016/j.tplants.2007.09.007.

Zhu JK. 2016. Abiotic stress signaling and responses in plants. *Cell.* 167(2):313–324.

Zhuang KY, Gao YY, Liu ZB, Diao PF, Sui N, Meng QW, Meng C, Kong FY. 2020. WHIRLY1 regulates HSP21.5A expression to promote thermotolerance in tomato. *Plant Cell Physiol.* 61:169–177.

Zsogon A, Cermak T, Naves ER, Notini MM, Edel KH, Weinl S, et al. 2018. *De novo* domestication of wild tomato using genome editing. *Nat Biotechnol.* 36:1211–1216. https://doi.org/10.1038/nbt.4272.

7 Genetic Transformation and Genome Editing

Its Applications in Tomato Improvement

Kedong Da, Anna Nelson, and Dilip R. Panthee

7.1 INTRODUCTION

With the world population increasing and climate change, global agricultural production faces significant challenges. Though the extreme poverty population has fallen dramatically since 1990, ~700 million people still live in extreme poverty globally, 40–60% of children in sub-Saharan Africa and 50% of pregnant women in South Asian countries are living undernourished (FAO et al., 2023). Environmental stresses, such as temperature fluctuations, irregular precipitation patterns, and increasing soil salinization, significantly hinder the successful cultivation of existing crop varieties. These changing conditions disrupt plant growth, reduce yields, and threaten the overall stability of agricultural systems, making it increasingly challenging to maintain food production under current climatic scenarios. Cultivable land is shrinking, especially per person, due to a combination of environmental degradation, urban expansion, and desertification (World Bank, 2023). In addition to international collaboration to reduce greenhouse gas emissions and support low-carbon development, sustainable infrastructure, clean energy investments, and efforts to equip developing countries with tools for adaptation and mitigation, and community capacity building, developing new tools to breed climate-resilient crop varieties for food and nutrients is essential to mitigating the adverse effects of ongoing climate change. Plant tissue culture regeneration based genetic transformation and genome editing, plays a pivotal role in developing new crop varieties that can help combat climate change and alleviate poverty. With the launch of the first genome-edited tomato into the market (Ezura, 2022), these advanced biotechnological tools enable precise modification of plant genomes to enhance traits such as drought tolerance, heat resistance, salinity tolerance, and pest or disease resistance—traits that are increasingly essential under changing environmental conditions. By improving crop resilience and productivity, genetic transformation and genome editing contribute to food security, reducing the need for chemical inputs, and supporting sustainable agriculture, particularly in resource-limited regions most vulnerable to climate impacts. Ultimately, these innovations hold significant potential to support both environmental adaptation and socioeconomic development. Tomato (*Solanum lycopersicum*), the world's number one vegetable crop, is cultivated globally and plays a critical role in global food security and human nutrition. Tomatoes are among the most versatile vegetables and contribute to a healthy, well-balanced diet due to their unique nutritional value. It contains considerable amounts of vitamins (B and C), minerals, essential amino acids, sugars, iron, phosphorus, potassium, and dietary fiber. Red tomatoes are rich in lycopene, an antioxidant that may help protect against carcinogenic substances.

Lycopene also serves as an important quality indicator in the tomato processing industry. With a relatively small genome, a short life cycle, and the availability of genetic/sequence resources (Tomato Genome Consortium, 2012), tomato serves as a prime candidate for molecular breeding

DOI: 10.1201/9781003506539-7

and functional genomics studies. Biotechnological tools such as *Agrobacterium*-mediated T-DNA transformation and the clustered regularly interspaced short palindromic repeat (CRISPR) and CRISPR-associated 9 protein (CRISPR/Cas9)-based genome editing have revolutionized the pace and precision of tomato improvement (Tiwari et al., 2023). Tomatoes have an advantage over other crops in their potential to use new breeding tools to develop climate-resilient, productive varieties that help fight climate change.

Plant transformation has long been a cornerstone of crop biotechnology, enabling the integration of foreign genes to improve traits such as pest resistance, herbicide tolerance, and nutritional content. Traditionally, *Agrobacterium*-mediated transformation and biolistic (gene gun) methods have been employed to introduce transgenes into plant genomes. However, the advent of genome editing technologies has revolutionized the landscape, shifting the focus from random transgene integration to precise, targeted modifications. Presently, plant transformation acts as a gateway to genome editing. While genome editing tools like CRISPR/Cas9 offer unprecedented precision, delivering editing reagents into plant cells and regenerating whole plants remain challenging for recalcitrant or elite cultivars. Plant transformation protocols are essential and serve as the primary method for introducing genome-editing machinery (e.g., Cas nucleases and guide RNAs) into plant cells.

7.2 GENETIC TRANSFORMATION (GT) IN TOMATO

Tomato (*Solanum lycopersicum*) is a model horticultural plant for functional genomics due to its small genome, short life cycle, and established transformation systems. The first transgenic tomato plant was reported by McCormick et al. (1986). Genetic transformation techniques, particularly those coupled with efficient tissue culture systems, are instrumental in developing transgenic tomato lines with enhanced traits such as disease resistance, abiotic stress tolerance, extended shelf life, and improved nutrition. The *Agrobacterium*-mediated transformation is the most common method for producing transgenic tomatoes. Particle bombardment is occasionally used for tomato transformation.

7.2.1 Tissue Culture Regeneration in Tomato

A robust tissue culture system is essential for successful genetic transformation and genome editing. Regeneration responses differ substantially among tomato cultivars. Table 7.1 provides an overview

TABLE 7.1
Tomato Varieties Successfully Reported for Tissue Culture Regeneration

Cultivar	Regeneration potential	Notes	References
Ailsa Craig	Moderate	Reported in early literature, requires optimization	Hanson (1981)
Better Boy, Starfire, UC 134-1-2	High	Early report for regeneration	Kurtz and Lineberger (1983)
UC82B/UC82	Good	Common in tissue culture studies	Hamza and Chupeau (1993)
Micro-Tom	Very high	Model cultivar, regenerates easily from cotyledon and hypocotyl explants	Sun et al. (2006)
Rio Grande	Moderate to high	Good response in multiple studies, used in CRISPR-related works	Yasmeen (2009)
M82	High	Used in functional genomics, improved with specific hormone combinations	Brooks et al. (2014)

(*Continued*)

TABLE 7.1 ***(Continued)***
Tomato Varieties Successfully Reported for Tissue Culture Regeneration

Cultivar	Regeneration potential	Notes	References
Pusa Ruby	Moderate to good	Used in hot/humid climates, responds well to MS + BAP/Zeatin	Sarker et al. (2009)
Heinz 1706	Moderate	Reference genome published, responds to TDZ and BAP	Kassahun et al. (2021)
Money Maker	High	Common cultivar, good regeneration with BA + IAA	Yaroshko et al. (2023)

of popular cultivars and their respective regeneration potential. Tomato regeneration protocols rely mainly on organogenesis using explants such as cotyledons, hypocotyls, or leaf discs, shoot tips, or internodes. Cotyledons have been regarded as the most responsive explant and are more commonly used in genetic transformation and genome editing research. A protocol for cotyledon regeneration, developed in the author's lab and combined with new advancements, is described next.

Protocol:

A. Explant preparation: Tomato seeds were pre-soaked for 1 hour in sterile distilled water. The seeds were soaked in 70% ethanol for 45 seconds, then rinsed with sterile distilled water. The seeds were transferred to a 20% commercial bleach solution containing 25 μL of Tween-20™ per 100 ml. The seeds were stirred continuously in the bleach solution for 30 min. The bleach-treated seeds were then rinsed four times with sterile distilled water and germinated on half-strength Murashige and Skoog medium (MS) (Murashige & Skoog, 1962).

B. Organogenesis: Cotyledons from seven-day-old seedlings are used as explants. It is very important that the first true leaves have not enlarged or opened. The first true leaves should only be barely visible at the best time for explant harvest. A popular medium used for tomato seedling explant regeneration is composed of MS basal salts supplemented with Gamborg B5 vitamins (Gamborg et al., 1968), 2 mg/L BAP, and 0.5 mg/L IAA. Different varieties may vary in BAP concentration in the medium. The plates with the explants were incubated at 25 ± 2°C under a 16/8 h light/dark photoperiod. The explants were subcultured every two weeks onto fresh regeneration medium. Organogenesis occurs in 20 days. Shoot regeneration will be observed in 30 days. Regenerated shoots are transferred to an elongation medium when they reach 12.5 mm in length.

C. Shoot elongation: The regenerated shoots are excised from the source material and transferred onto a shoot elongation medium consisting of full-strength MS basal salts and vitamins without the addition of any plant growth regulators. The cultures were maintained at 25 ± 2°C under a 16/8 h light/dark photoperiod for 30 days, during which the shoots elongated and developed further, preparing them for subsequent rooting.

D. Root induction: Elongated shoots are carefully trimmed to remove senescent or older leaves at the basal region. The cleaned shoots are then transferred to a rooting medium composed of full-strength MS medium supplemented with 1.0 mg/L indole-3-acetic acid (IAA). Adventitious root formation will be observed within 20 days.

E. Acclimatization: To harden the plantlets, the culture vessels were partially opened for seven days to reduce humidity gradually. The rooted plantlets were then transferred to 24-cell seed-starting trays filled with seed germination mix for acclimatization. A clear plastic dome covered the tray to keep moisture for one week. The dome was gradually opened and completely removed over the next two weeks to acclimate the plants.

TABLE 7.2
Frequently Reported Varieties in Tomato Genetic Transformation

Cultivar	Features	Common Use	References
Beef master, Better boy, Perfect Peel	Standard cultivar, Kanamycin resistance, Southern blot positive	*Agrobacterium*-mediated transformation	McCormick et al. (1986)
Ailsa Craig	Standard cultivar	Gene function studies	Lipp and Brown (1993)
Money Maker	Widely grown	Genetic studies and transformation	Van Roekel et al. (1993)
IPA-5, IPA-6	Processing type	Transformation studies	Costa et al. (2000)
UC82	Processing type	Transformation studies	Pandolfini et al. (2002)
Micro-Tom	Dwarf, fast life cycle	Model system in labs; high transformation efficiency	Sun et al. (2006)
Rio Grande	High transformation rate	Used in CRISPR studies	Yasmeen (2009)
Heinz 1706	Genome reference	Reference genome line	Brooks et al. (2014)
Pusa Ruby	Adapted to South Asia	Transformation research	Madhu and Savithramma (2014)
M82	Processing type	Functional genomics	Gupta et al. (2016)

7.2.2 *Agrobacterium*-Mediated Transformation in Tomato

Agrobacterium tumefaciens-mediated transformation is the most widely used and efficient method for tomato genetic transformation and genome editing due to its simplicity, cost-effectiveness, and high transformation frequency. Tomato varieties exhibit substantial variation in their transformation responses. An overview of commonly utilized cultivars for *Agrobacterium*-mediated transformation is presented in Table 7.2. A protocol for *Agrobacterium*-mediated transformation, as performed in the author's lab, is described next.

Protocol:

A. Explant preparation:
   a. Tomato seeds were pre-soaked for 1 h in sterile distilled water. The seeds were then rinsed well with sterile distilled water, soaked in 70% ethanol for 45 seconds, and rinsed again with sterile distilled water. The seeds were then soaked in a 20% commercial bleach solution containing 25 µL of Tween-20™ per 100 ml. The seeds were stirred continuously in the bleach solution for 30 min. The bleach-treated seeds were then rinsed four times with sterile distilled water and germinated on dishes with half-strength MS medium supplemented with 7 g/L agar and 30 g/L sucrose. The pH was adjusted to 5.8 before autoclaving.
   b. One day before inoculation, the cotyledons were cut from six-day-old seedlings. Remove the cotyledon from the seedling by cutting at the petiole and removing the cotyledonary apex using #10 scalpel blades. Place 50 explants between two moistened sterile Whatman® #3 filter papers in a petri dish. Seal with Parafilm® Lab Film and incubate at 25 ± 2°C with a 16/8 h light/dark photoperiod.

B. *Agrobacterium* preparation
   a. Streak *Agrobacterium* onto a solid LB dish containing the appropriate antibiotics. Incubate at 28°C for 48 hours until well-formed colonies have developed.
   b. Select four, single, well-formed colonies from the plate and transfer to 10 ml of liquid LB with appropriate antibiotics. Culture overnight in an orbital shaker incubator at 220 rpm and 28°C.

c. Use 50 ul of the bacterial suspension to inoculate 50 ml of LB medium containing the appropriate antibiotic in a flask. Culture overnight in an orbital shaker at 220 rpm and 28°C.
d. Check the OD600 with a NanoDrop-2000C microvolume spectrophotometer. The optimal OD600 = 0.8–1.0.
e. Centrifuge the bacterial solution at 4,000 rpm for 10 minutes at 20°C.
f. Discard the supernatant. Add 30 ml of sterile, full-strength MS0 liquid medium containing 30 g/L sucrose, with a pH of 5.8.
g. Resuspend the pellet by using a vortex mixer. Regulate the OD600 to 0.1 by diluting the suspension with MS0.

C. Transformation
a. Incubate explants in *Agrobacterium* culture resuspended in MS0. Pipette 25 ml of *Agrobacterium* culture into a sterile 50 ml tube. Add 0.1mM acetosyringone. Transfer 100 explants to the sterile tube containing 25 ml of inoculum. Sonicate for 30 seconds. Incubate for 20 min on an orbital shaker set to 120 rpm. Remove the explants and put them between two sterile, moistened Whatman® #3 filter papers, with the abaxial side down. Seal plates with Parafilm ® Lab Film.
b. Co-cultivate explants in the dark at 28°C for 72 hrs.
c. Rest culture: Transfer 25 explants to each plate of Transgenic Selection Media #1 (TSM1), which contains full-strength MS basal salts supplemented with Gamborg B5 vitamins, 2 mg/L BAP, 0.5 mg/L IAA, and 300 mg/L Timentin, with the explants adaxial side up. Seal plates with Parafilm ® Lab Film. Culture at 25°C, 16/8 h light/dark photoperiod, for two weeks. Refresh the medium every five days.
d. Transgenic selection: After two weeks, transfer explants to Transgenic Selection Medium #2 (TSM2), which is full-strength MS basal salts supplemented with Gamborg B5 vitamins, 2 mg/L BAP, 0.5 mg/L IAA, the selective agent, and 300 mg/L Timentin. Each petri dish should contain no more than 15 explants. Transfer explants to fresh TSM2 selection medium plates every five days. When shoots reach 12.5 mm in length, transfer the shoots to Selection Elongation Medium (SEM), which contains full-strength MS medium with 30 g/L sucrose, 7 g/L agar, pH 5.8, and the appropriate selective agent, plus 300 mg/L Timentin, in Magenta™ boxes for elongation (5 explants per box).

D. Rooting and acclimatization
a. Elongated shoots were placed in Magenta™ boxes containing tomato rooting media (TRM) composed of full-strength MS medium supplemented with 1.0 mg/L indole-3-acetic acid (IAA), the appropriate selective agent, and 300 mg/L Timentin. Place four shoot cuttings per box. Roots begin to appear in five days.
b. Rooted plantlets were acclimated in the growth chamber and then transferred to the greenhouse under the same conditions as those used in the regeneration protocol.

7.2.3 Biolistic (Particle Bombardment) Transformation in Tomato

Biolistic transformation uses high-velocity metal particles (usually gold or tungsten) coated with DNA, RNA, or RNPs to physically deliver genes into plant cells. Common target tissues include embryogenic callus or leaf discs. Biolistic methods have been infrequently used in tomato transformation and genome editing because of the simplicity, efficiency, and economics of *Agrobacterium*-mediated cotyledon transformation protocols. Biolistic methods have an advantage over *Agrobacterium*-mediated transformation for studying transient gene expression and transgene-free genome editing.

Protocol:

A. Preparation of explants (same as *Agrobacterium* transformation)
One day prior to biolistic transformation, cut the cotyledons from six-day-old seedlings. Remove the cotyledon from the seedling by cutting at the petiole and removing the

cotyledonary apex. Place 50 explants between two moistened sterile Whatman® #3 filter papers in a petri dish. Seal with Parafilm® Lab Film and incubate at 25 ± 2°C with a 16/8 h light/dark photoperiod.

B. Plasmid DNA preparation (modified from Sambrook et al., 1989)
   1. Streak *E. coli* onto LB plates with appropriate antibiotics. Incubate for 48 hours at 37°C until well-formed colonies have developed.
   2. Select one single, well-formed colony from the plate and transfer it to a 50 ml tube containing 5 ml of LB with appropriate antibiotics. Culture in an orbital incubator set to 250 rpm at 37°C overnight.
   3. Use 50 μL of bacterial suspension to inoculate 50 ml of LB medium containing appropriate antibiotics in a 125 ml flask. Culture overnight in an orbital incubator at 37°C and 250 rpm.
   4. Check the OD600 with a microvolume spectrophotometer. The optimal OD600 = 0.8–1.0.
   5. Centrifuge the bacterial solution at 5,000 rpm for 10 minutes at 20°C.
   6. Discard the supernatant. Resuspend bacteria in a 50 ml sterile tube with 5 mL of Solution I (50 mM glucose, 25 mM Tris.HCl (pH 8.0), 10 mM EDTA (pH 8.0)) by vortexing.
   7. Add newly made 10 ml Solution II (0.2 N NaOH (freshly diluted from a 10 N stock), 1% SDS), gently mix by inverting, and incubate at room temperature for 3 min.
   8. Add 7.5 ml Solution III (3 M potassium acetate by mixing 60 ml 5 M potassium acetate, 11.5 ml glacial acetic acid, and 28.5 ml distilled water). Mix immediately by inverting four to six times. Proceed directly to the next step.
   9. Add 15 mL of chloroform to the tube and invert four to six times.
   10. Centrifuge at 5,000 rpm for 5 minutes at 20°C.
   11. Transfer the supernatant to a new tube.
   12. Precipitate the plasmid DNA with two volumes of ethanol at room temperature.
   13. Allow the mixture to stand at room temperature for 5 minutes.
   14. Centrifuge at 8,000 rpm for 10 minutes at 4°C
   15. Remove the supernatant with a pipette. Stand the tube in an inverted position on a paper towel to allow all the fluid to drain away.
   16. Rinse the DNA pellet with 1 mL of 70% ethanol.
   17. Allow the DNA pellet to dry in the air overnight.
   18. Resuspend the DNA in 2 ml Tris-EDTA solution (1x TE, pH 8.0), then add 2 ul DNAse-free RNase. Mix by pipetting briefly.
   19. Store the DNA at −20°C for long-term use.

C. Plasmid DNA coating to gold particle (modified from Bio-Rad Laboratories (2006))
   1. Move 6 mg (50ul) of pretreated gold particles (microcarriers) to a 1.5 ml microcentrifuge tube. Vortex to mix vigorously.
   2. Add in order:
      a. 10 ul DNA (1ug/ul)
      b. 50 ul 2.5 M $CaCl_2$
      c. 20 ul 0.1 M spermidine
   3. Continue vortexing for 2–3 min. Allow to settle for 1 minute. Pellet microcarriers by spinning for 2 seconds in a microcentrifuge. Remove the liquid and discard. Add 140 μL of 70% ethanol (HPLC grade) and spin the tube in the centrifuge for 2 seconds. Remove the liquid and discard. Add 140 μL of 100% ethanol and spin the tube in the centrifuge for 2 seconds. Remove the liquid and discard. Add 48 μL of 100% ethanol. Gently tap the tube several times, then vortex at low speed for 2–3 seconds. The particle solution is ready to be spread on the macrocarrier holder.

D. Bombardment operation (special training is needed for operating PDS1000/He Biolistic Particle Delivery System).

Notes: It is recommended to bombard each sample twice for better results. A visual marker (GFP, DsRed, etc.) gene is recommended for a transient expression study. The GUS or Luciferase gene can also be used in conjunction with histochemical reactions.

The success of tomato genetic transformation depends heavily on a well-optimized tissue culture regeneration system and the choice of gene-delivery method. *Agrobacterium*-mediated transformation remains the preferred method due to its efficiency, reliability, and simplicity, while biolistic transformation offers an alternative for genotypes or applications where *Agrobacterium*-mediated transformation is less effective. Together, these tools support advanced research and breeding strategies to improve tomato traits for climate resilience, productivity, and nutritional quality.

7.3 GENOME EDITING (GE) IN TOMATO

Agrobacterium-mediated transfer DNA (T-DNA) genetic transformation (GT) and CRISPR-Cas9 (Clustered Regularly Interspaced Short Palindromic Repeats and CRISPR-associated protein 9)-based genome editing (GE) are two common methods for introducing genetic modifications. Tomato is one of the most tractable crop species for transformation and editing, and has been the first plant species to benefit from GT and GE technologies. Various genome-editing technologies existed prior to CRISPR, such as Zinc Finger Nucleases (ZFNs) and Transcriptional Activator-like Effector Nucleases (TALENs). The latter two were successfully applied to tomato genome editing (Hilioti et al., 2016; Lor et al., 2014), but subsequent research used CRISPR due to higher editing efficiency. GT and GE differ significantly in their mechanisms and outcomes. T-DNA transformation, associated with *Agrobacterium*-mediated gene transfer, involves integrating a piece of T-DNA into the plant genome, potentially introducing foreign genes and creating genetically modified organisms (GMOs). CRISPR-Cas9 is a more precise gene-editing tool that can be used to make targeted changes to a plant's existing DNA sequence without introducing foreign DNA or creating GMOs. CRISPR-based GE technology is rapidly advancing toward the development of transgene-free, non-GMO organisms. Modifications such as CRISPR/Cas12a (Cpf1), Cas9 nickase, CRISPR-Cas13, and base editors expand the toolbox for precise editing.

7.3.1 CRISPR/Cas Systems

CRISPR and its associated locus encoding Cas proteins were initially discovered in bacterial and archaeal genomes. This system acts like an immune defense, helping these microbes fight off viruses (bacteriophages) and foreign DNA, such as plasmids. When a virus infects a bacterium for the first time, pieces of the viral DNA are inserted into the bacterium's CRISPR region. Upon a new viral infection, the CRISPR region is transcribed into CRISPR RNA (crRNA), which carries sequences (called spacers) that match past viral invaders. These spacers are separated by repetitive endogenous sequences, known as direct repeats. The Cas proteins, which have cutting (nuclease) and unwinding (helicase) abilities, use the crRNA to recognize and cut matching foreign DNA. This process also needs another small RNA called tracrRNA, which helps form a complex with crRNA and Cas proteins. This complex targets and destroys invading DNA that matches the crRNA sequence (Figure 7.1c). For the Cas protein to cut the DNA, a short specific sequence next to the target—called the protospacer-adjacent motif (PAM)—must also be present. Without the PAM, even a perfect match to the crRNA will not be cut. After discovering how CRISPR works, scientists quickly succeeded in editing tomato genes using the CRISPR/Cas system. A key breakthrough was the development of single-guide RNA (sgRNA)—a fusion of two RNA types (crRNA and tracrRNA)—which helps guide the system to the specific gene to be edited. By changing the sgRNA sequence, different parts of the DNA can be targeted. There are three main types of CRISPR-Cas systems: Type I, Type II, and Type III. These types are distinguished by the signature Cas protein they utilize: Cas3 for Type I, Cas9 for Type II, and Cas10 for Type III. Each type also has subtypes, further differentiating them based on their effector complexes and mechanisms.

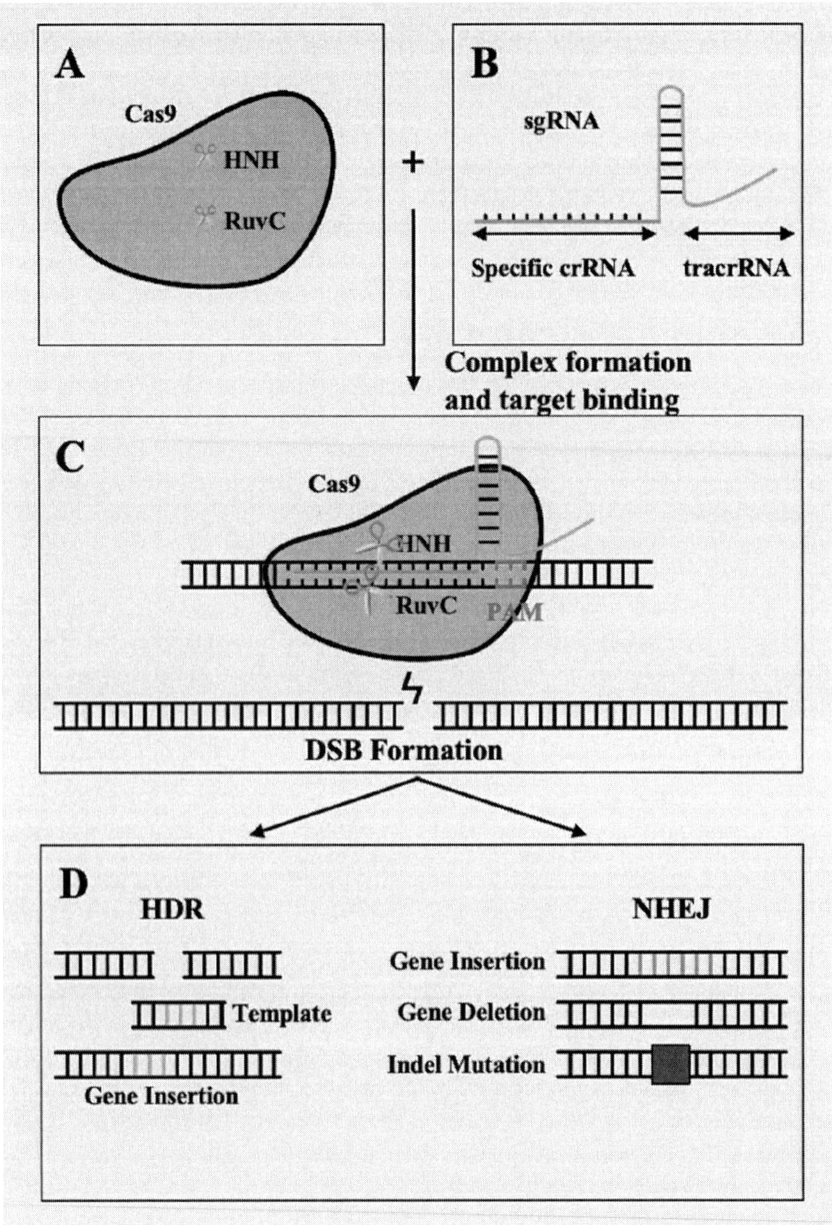

FIGURE 7.1 Schematic diagram of the CRISPR/Cas9 system molecular mechanism. The CRISPR/Cas9 system consists of sgRNA and Cas9. The Cas9 protein contains two nuclease domains, the RuvC domain, which cleaves non-complementary DNA strands, and the HNH domain, which cleaves complementary DNA strands (**A**). The sgRNA includes the trans-activating CRISPR RNA (tracrRNA), and the CRISPR RNA (crRNA). The crRNA comprises a 20 nt protospacer element and a few additional nucleotides, which are complementary to the tracrRNA (**B**). The tracrRNA hybridizes to the crRNA and binds to the Cas9 protein, forming the CRISPR/Cas9/sgRNA complex to edit genomic sequences. The Cas9–sgRNA complex unwinds dsDNA, and the complementary sequence in the sgRNA anneals to one of the DNA strands. Upon binding, the endonuclease domains cleave both DNA strands three bases upstream of the protospacer-adjacent motif (PAM) sequence (**C**). A double-strand break (DSB) in DNA forms and is then repaired either by homology-directed repair (HDR), if an appropriate donor is present, or by non-homologous end joining (NHEJ). HDR repair allows precise genome editing at the target site, and NHEJ introduces short insertions, deletions, or indels (**D**).

Researchers have studied many types of CRISPR/Cas systems, but the most commonly used ones are Cas9 from *Streptococcus pyogenes* and Cas12a (also called Cpf1) from *Acidaminococcus* and *Lachnospiraceae* species.

7.3.2 Cas9

CRISPR/Cas9 is considered a third-generation genome-editing tool (Figure 7.1; Janik et al., 2020). It was first used in plants in 2013 and has since become the most widely used method for gene editing. Originally discovered as a bacterial immune system that defends against viruses and plasmids, the CRISPR/Cas9 system used in research is based on the type II CRISPR system. This system includes three key components: the Cas9 protein and two RNA molecules—tracrRNA and pre-crRNA. Cas9 is an enzyme that cuts DNA and plays a role in processing and guiding the RNA. The tracrRNA pairs with pre-crRNA to form a duplex, which is essential for targeting and cutting DNA. The pre-crRNA is derived from repeat-spacer arrays in the genome. Repeats are short sequences that often form secondary structures, while spacers are unique sequences copied from past viral or plasmid infections. Together, tracrRNA and pre-crRNA form a complex that guides Cas9 to specific DNA targets. Cas9 then cuts both strands of the DNA, creating a double-strand break (DSB) at a specific site next to a short sequence called the PAM (protospacer adjacent motif). In engineered systems, tracrRNA and crRNA are fused into a single guide RNA (sgRNA) to simplify the process. The sgRNA directs Cas9 to the target DNA, requiring only a 20-base sequence for recognition. This makes CRISPR/Cas9 easier to design and more efficient than older tools like ZFNs and TALENs. CRISPR/Cas9 can also be used for multiplex editing by introducing multiple sgRNAs targeting different sequences. Its simplicity and effectiveness have made it widely accessible for gene editing in many laboratories. Since 2013, CRISPR/Cas9 has been successfully used in a wide range of plant species, including rice, wheat, tobacco, Arabidopsis, maize, tomato, potato, and soybean.

7.3.3 Cas12a

Cas12a and its variants are generally smaller than Cas9 because they lack a helicase domain. Some Cas12a proteins act alone to cut target DNA, using shorter crRNAs (~42 nucleotides) compared to Cas9's longer crRNAs (~100 nucleotides). Cas12a can process its own pre-crRNA into a mature crRNA without the need for tracrRNA. The substrate specificity for the protospacer adjacent motif (PAM) sites differs significantly between Cas12a and Cas9 nucleases. Cas12a preferentially recognizes thymidine-rich PAM sequences, while Cas9 favors guanine-rich ones. After cutting DNA, Cas12a creates a 5′ overhang, which improves the precision of editing in AT-rich regions—areas where Cas9 struggles. Although both Cas12a and Cas9 can tolerate mismatches in their target sequences, Cas12a tends to have fewer off-target effects. The first genome editing of tomato using a Cpf1/Cas12a system (derived from *Francisella novicida*) was reported by Bernabe-Orts et al. (2019).

7.3.4 Base Editing

Base editing is a type of genome editing that directly converts one DNA base into another, such as changing a C to a T, without creating a double-strand break. It uses a CRISPR-Cas9 system, but the Cas9 protein is modified to be catalytically inactive (dCas9) and fused to a deaminase enzyme. The gRNA directs the dCas9-deaminase complex to a specific DNA sequence, where the deaminase then chemically modifies the target base. The gRNA is designed to be complementary to the target DNA sequence, allowing it to bind to its target. After binding to the target sequences, the deaminase enzyme is positioned at the target base by the dCas9 protein. The deaminase then chemically modifies the target nucleotide, converting it to a different base. For example, a cytosine base editor

(CBE) converts cytosine to uracil, which is then read as thymine during DNA replication or repair. An adenine base editor (ABE) converts adenine to guanine.

7.4 DELIVERY APPROACHES

7.4.1 Methods for CRISPR/Cas Delivery

To induce DSBs and activate the endogenous DNA repair mechanism, the Cas protein and sgRNA(s) components of the CRISPR/Cas system must reach the plant cell nucleus. This can be achieved by conventional delivery of binary vector-encoded CRISPR/Cas-gRNA components via *Agrobacterium*-mediated transformation. Biolistic bombardment can also be used to deliver CRISPR components, either as preassembled gRNA/Cas ribonucleoprotein (RNP) complexes or as plasmid DNA encoding the CRISPR-Cas-gRNA. The RNP complexes target and cleave genomic DNA at their sites immediately after delivery and are rapidly degraded in those cells. Transient DNA or RNA expression systems select for genome-edited cells lacking stable integration of genetic material into the plant genome, thereby generating transgene-free GEs. Next, you will find a brief overview of these methods, with an emphasis on recent developments for CRISPR/Cas delivery in tomatoes (Table 7.3).

TABLE 7.3
Implementation of Genome Editing for Tomato Improvement

Variety	Editing Tool	Target Genes	Type of Edit	Traits	References
Biotic stress resistance					
FL8000	CRISPR/Cas9	*SlDMR6–1*	Gene knockout	*P. syringae, P. capsici,* and *Xanthomonas* spp resistance	Thomazella et al. (2021)
Moneymaker	CRISPR/Cas9	*SlMlo1* to *SlMlo16*	Deletion	Powdery mildew resistance	Nekrasov et al. (2017)
Rio Grande	CRISPR/Cas9	*Solyc08g075770*	Gene knockout	*Fusarium oxysporum* resistance	Prihatna et al. (2018)
Ailsa Craig	CRISPR/Cas9	*MAPK3*	Gene knockout	Botrytis cinerea	Zhang et al. (2018)
Moneymaker	CRISPR/Cas9	*CP* and *Rep gene*	Gene knockout	TYLCV	Tashkandi et al. (2018)
Ailsa Craig	CRISPR/Cas9	*DCL2b*	Gene knockout	ToMV	Wang et al. (2018)
Moneymaker	CRISPR/Cas9	*SlJAZ2*	Gene knockdown	Bacterial speck resistant	Ortigosa et al. (2019)
Moneymaker	CRISPR/Cas9	*PMR4*	Gene knockout	Powdery mildew resistance	Martínez et al. (2020)
Moneymaker	CRISPR/Cas9	*rgsCaM promoter*	Gene knockout	TYLCV	Faal et al. (2020)
Micro-Tom	CRISPR/Cas9	*SlMYC2*	Gene knockout	*B. cinerea*	Shu et al. (2020)
Moneymaker	CRISPR/Cas9	*rgsCaM promoter*	Gene knockout	TYLCV	Faal et al. (2020)
Inbred line S8	CRISPR/Cas9	*eIF4E1*	Gene knockout	PVY and CMV	Atarashi et al. (2020)
Zaofen no. 2	CRISPR/Cas9	*miR482b* and *miR482c*	Gene knockout	Late blight resistance	Hong et al. (2021)
BN-86	CRISPR/Cas9	*SlPelo* and *SlMlo1, SlPelo*	Gene knockout	TYLCV and Oidium sp. resistance	Pramanik et al. (2021)

(*Continued*)

TABLE 7.3 ***(Continued)***
Implementation of Genome Editing for Tomato Improvement

Variety	Editing Tool	Target Genes	Type of Edit	Traits	References
Alisa Craig	CRISPR/Cas9	*SlMYBS2*	Gene knockout	Late blight	Liu et al. (2021a)
Hongkwang 15T01	CRISPR/Cas9	*SlHyPRP1*	Gene knock out	Enhanced resistance to *Pto DC3000*, Enhanced susceptibility to *FOL* and *B.cinerea*	Tran et al. (2023)
Abiotic stress tolerance					
M82	RNA interference	*SlHyPRP1*	RNAi	Oxidative stress, dehydration, and salinity tolerance	Li et al. (2016)
Ailsa Craig	CRISPR/Cas9	*SlMAPK3, SlLOX, SlGST,* and *SlDREB*	Gene knockout	Drought stress	Wang et al. (2017)
Condine Red	CRISPR/Cas9	*BZR1*	Gene knockout	Heat tolerant	Yin et al. (2018)
Micro-Tom	CRISPR/Cas9	*SlSBPASE*	Gene knockout	C-N metabolism	Ding et al. (2018)
Ailsa Craig	CRISPR/Cas9	*SlCBF1*	Gene knockout	Cold tolerance	Li et al. (2018e)
Ailsa Craig	CRISPR/Cas9	*SlNPR1*	Gene knockout	Drought tolerance	Li et al. (2019)
WVA106	CRISPR/Cas9 base editing	*ALS*	Gene knockout	Herbicide resistance	Veillet et al. (2019)
Micro-Tom	CRISPR/Cas9	*PHO1;1, PHO1;2, PHO1;3, PHO1;4, PHO1;5, PHO1;6*	Gene knockout	Phosphate transporter	Zhao et al. (2019)
	CRISPR/Cas9	*ALS1, ALS2, ALS3*	Gene knockin	Herbicide resistance	Danilo et al. (2019)
Micro Tom	CRISPR/Cas9	*SlLBD40*	Gene knockout	Drought tolerance	Liu et al. (2020a)
Ailsa Craig	CRISPR/Cas9	*SlUVR8*	Gene knockout	UV stress	Liu et al. (2020b)
Hongkwang	CRISPR/Cas9	*HKT1;2 RAD51/54*	Gene knockout	Salt tolerance	Vu et al. (2020b)
Micro-Tom	CRISPR/Cas9	*SlARF4*	Gene knockout	Tolerance to salinity and osmotic stress	Bouzroud et al. (2020)
Rubion	CRISPR/Cas9	*GRXS14/15/16/17*	Gene knockout	Multiple stresses	Kakeshpour et al. (2021)
Condine Red	CRISPR/Cas9	*ALD1, FMO1*	Gene knockout	Drought tolerance	Wang et al. (2021b)
Condine Red	CRISPR/Cas9	*CPK28, APX2*	Gene knockout	Heat tolerance	Hu et al. (2021)
Hongkwang and 15T01	CRISPR/Cas9	*SlHyPRP1*	Gene knockout	Salt tolerant	Tran et al. (2021)
S. lycopersicum, S. pimpinellifolium	CRISPR/Cas9	*SlSOS1*	Gene knockout	Salt tolerance	Wang et al. (2021c)
Micro-Tom	CRISPR/Cas9	*PDS, ALS, EPSPS*	Gene knockout	Herbicide resistance	Yang et al. (2022)
Ailsa Craig	CRISPR/Cas9	*TAGL1*	Gene knockout	Increase firmness	Jeon et al. (2024)
Plant architecture					
Zhongshu 6 (ZS6)	RNA interference	*SlSGR1*	RNAi	Color formation, fruit shelf life	Luo et al. (2013)

(Continued)

TABLE 7.3 *(Continued)*
Implementation of Genome Editing for Tomato Improvement

Variety	Editing Tool	Target Genes	Type of Edit	Traits	References
M82	TALEN	*Solyc11g011260*	Targeted mutagenesis	Gibberellin signaling	Lor et al. (2014)
M82	CRISPR/Cas9	*AGO7*	Gene knockout	Leaf shape	Brooks et al. (2014)
M82	CRISPR/Cas9	*SHORT-ROOT SCARECROW*	Gene knockout	Root size	Ron et al. (2014)
Micro-Tom	CRISPR/Cas9, TALEN	*ANT1*	Precise editing	Anthocyanin	C'ermak et al. (2015)
Heinz 1706	ZFN	*LEC1-LIKE4*	Targeted mutagenesis	Plant height, number of florets	Hilioti et al. (2016)
Micro-Tom	CRISPR/Cas9	*SlPDS* and *SlPIF4*	Gene knockout	PDS and PIF	Pan et al. (2016)
M82	CRISPR/Cas9	*PSY1*	Gene knockin	Yellow fruit	Hayut et al. (2017)
Micro-Tom, Ailsa Craig	CRISPR/Cas9	*SlIAA9*	Gene knockout	Parthenocarpy	Ueta et al. (2017)
M82	CRISPR/Cas9	*SlAGL6*	Gene knockout	Parthenocarpy	Klap et al. (2017)
M82	CRISPR/Cas9	*ALC*	Gene knockin	Shelf life	Yu et al. (2017)
Ailsa Craig	CRISPR/Cas9	*RIN locus*	Gene knockout	Fruit ripening	Ito et al. (2017)
Micro-Tom	CRISPR/Cas9	*SlGAD2* and *SlGAD3*	Gene knockout	γ-Aminobutyric acid (GABA)	Nonaka et al. (2017)
S. lycopersicum, S. pimpinellifolium, S.l. var *cerasiforme*	CRISPR/Cas9	*TFM6, ALMT*	Gene knockout	Sour taste, aluminium tolerance	Ye et al. (2017)
LA1589, M82	CRISPR/Cas9	*SlWUS*	Gene knockout	Fruit size and inflorescence architecture	Rodriguez-Leal et al. (2017)
Micro-Tom	CRISPR/Cas9 (base editing)	*DELLA, ETR1*	Gene knockout	Leaf shape	Shimatani et al. (2017)
M82	CRISPR/Cas9	*SlMBP21 MADS-box gene*	Gene knockout	Jointless fruit	Roldan et al. (2017)
Micro-Tom, Ailsa Craig	CRISPR/Cas9	*SlIAA9*	Gene knockout	Parthenocarpy	Ueta et al. (2017)
M82	CRISPR/Cas9	*SlAGL6*	Gene knockout	Parthenocarpy	Klap et al. (2017)
M82	CRISPR/Cas9	*ALC*	Gene knockin	Shelf life	Yu et al. (2017)
Ailsa Craig	CRISPR/Cas9	*RIN locus*	Gene knockout	Fruit ripening	Ito et al. (2017)
Micro-Tom	CRISPR/Cas9	*SlGAD2* and *SlGAD3*	Gene knockout	γ-Aminobutyric acid (GABA)	Nonaka et al. (2017)
Micro-Tom	CRISPR/Cas9	*CRISTO, PSY1*	Gene knockin	Orange and yellow fruit	Dahan-Meir et al. (2018)
Red Setter	CRISPR/Cas9	*Psy1, CrtR-b2*	Gene knockout	Yellow fruit	D'Ambrosio et al. (2018)
Ailsa Craig	CRISPR/Cas9	*lncRNA1459*	Gene knockout	Fruit ripening	Li et al. (2018a)
Micro-Tom, Ailsa Craig	CRISPR/Cas9	*PDS* *GABA1–6*	Gene knockout	Plant phenotype GABA	Li et al. (2018b)
LA1589, LA1547, LA1357	CRISPR/Cas9	*SP, SP5G, SlCLV3, SlWUS*	Gene knockout	Fruit size, fruit ripening, shoot architecture, and lycopene content	Li et al. (2018c)

(Continued)

TABLE 7.3 ***(Continued)***
Implementation of Genome Editing for Tomato Improvement

Variety	Editing Tool	Target Genes	Type of Edit	Traits	References
Ailsa Craig	CRISPR/Cas9	*SGR1, LCY-E, Blc, LCY-B1, LCY-B2*	Gene knockout	Lycopene	Li et al. (2018d)
TB0240, TB0216	CRISPR/Cas9	*SlMYB12*	Gene knockout	Pink fruit	Deng et al. (2018)
LA1589	CRISPR/Cas9	*SP, O, FW2.2, Cycb*	Gene knockout	Fruit size and number, and lycopene content	Zsogon et al. (2018)
M82	CRISPR/Cas9	*Slddm1a Slddm1b*	Gene knockout	Vegetative and reproductive phenotypes	Corem et al. (2018)
LA1589	CRISPR/Cas9	*SP, O, FW2.2, Cycb*	Gene knockout	Fruit size and number, and lycopene content	Zsogon et al. (2018)
M82	CRISPR/Cas9	*Slddm1a Slddm1b*	Gene knockout	Vegetative and reproductive phenotypes	Corem et al. (2018)
S. lycopersicum, S. pimpinellifolium	CRISPR/Cas9	*TRM3/4/5, OFP*	Gene knockout	Fruit size	Wu et al. (2018)
IndigoRose	CRISPR/Cas9	*HY5*	Gene knockout	Anthocyanin	Qiu et al. (2019)
Moneymaker	CRISPR/Cas9	*PRO*	Gene knockout	Dwarf	Tomlinson et al. (2019)
Ailsa Craig	CRISPR/Cas9	*PL, PG2a,* and *TBG4*	Gene knockout	Fruit ripening	Wang et al. (2019a)
Moneyberg, AilsaCraig	CRISPR/Cas9	*FUL1, FUL2*	Gene knockout	Fruit ripening	Wang et al. (2019a)
Micro-Tom	CRISPR/Cas9	*DDB1, DET1, CYC-B*	Base editing	Carotenoid, lycopene,	Hunziker et al. (2020)
F6(RILs)	CRISPR/Cas9	*FLORAL4*	Gene knockout	Fruit flavor	Tikunov et al. (2020)
KS-13	CRISPR/Cas9	*SlMS10*	Gene knockout	Male sterility	Jung et ak. (2020)
Indigo Rose	CRISPR/Cas9	*SlAN2*	Gene knockout	Fruit color	Zhi et al. (2020)
Ailsa Craig	CRISPR/Cas9	*YABBY2b, GH3.8*	Gene knockout	Tomato plants' height, flower, and fruit size	Sun et al. (2020)
IndigoRose	CRISPR/Cas9	*R2R3-MYB TF AN2-like/ Anthocyanin fruit (Aft)*	Gene knockout	Anthocyanin	Yan et al. (2020)
P73	CRISPR/Cas9	*ENO*	Gene knockout	Fruit size	Yuste-Lisbona et al. (2020)
Ailsa Craig	CRISPR/Cas9	*NAM1*	Gene knockout	Fruit ripening	Gao et al. (2021)
Periyakulam1 (PKM1)	CRISPR/Cas9	*CRTISO*	Gene knockout	Carotenoid	Jayaraj et al. (2021)
Dwarf Tomato	CRISPR/Cas9	*SlMAPK6*	Gene knockout	Axillary buds	Li et al. (2024)
Micro-Tom	CRISPR/Cas9	*Ms1035· GSTAA*	Gene knockout	Male sterility	Liu et al. (2021b)
M82	CRISPR/Cas9	*INVINH1, VPE5*	Gene knockout	Glucose, fructose, and TSS	Wang et al. (2021a)
Suzukoma	CRISPR/Cas9	*INVINH1*	Gene knockout	Sugar	Kawaguchi et al. (2021)

(Continued)

TABLE 7.3 *(Continued)*
Implementation of Genome Editing for Tomato Improvement

Variety	Editing Tool	Target Genes	Type of Edit	Traits	References
Micro-Tom	CRISPR/Cas9	*MIR164A*	Gene knockout	Fruit ripening	Lin et al. (2022b)
Ailsa Craig	CRISPR/Cas9	*LeRBOH, LeRBOHE*	Gene knockout	Male sterility	Dai et al. (2022)
Ailsa Craig	CRISPR/Cas9	*SlAMS*	Gene knockout	Pollen viability	Bao et al. (2022)
Ailsa Craig	CRISPR/Cas9	*SlCMT4*	Gene knockout	Pollen development	Guo et al. (2022)
LA1589	CRISPR/Cas9	*SlKLUH*	Gene knockout	Fruit weight	Li et al. (2022b)
Ailsa Craig	CRISPR/Cas9	*SUN, BZR1.7*	Gene knockout	Fruit shape	Yu et al. (2022)
Moneymaker	CRISPR/Cas9	*7-DR2*	Gene knockout	Vitamin D	Li et al. (2022a)
Micro-Tom	CRISPR/Cas9	*APX4*	Gene knockout	Vitamin C	Do et al. (2022)
Micro-Tom	CRISPR/Cas9	*SlPG*	Gene knockout	Fruit softening	Nie et al. (2022)
RedSetter, AilsaCraig, M82, Moneymaker	CRISPR/Cas9 (RNP)	*SP, SP5G*	Gene knockout	Protoplast regeneration	Liu et al. (2022)
Moneymaker, SanMarzano	CRISPR/Cas9	*GF*	Gene knockout	Greenflesh	Gianoglio et al. (2022)
MM and M82	CRISPR/Cas9	*SlCDPK27 SlCDPK27*	Gene knockout	Sugar content	Zhang et al. (2024)
Ailsa Craig	CRISPR/Cas9	*SlMTC*	Gene knockout	Small seeds, sensitivity to salt	Gao et al. (2024)
Miro-Tom	CRISPR/Cas9	*SlWRKY29*	Gene knockout	Somatic embryogenesis	Valencia-Lozano et al. (2024)
Micro-Tom	RNA virus/ CRISPR/Cas9	*SGR1*	Gene knockout	Fruit color	Uranga et al. (2024)
Miro-Tom Sicilian Rouge CF	CRISPR/Cas9	*DWARF (D) and SELF-PRUNING (SP)*	Gene knockout	Dwarf	Nagamine and Ezura (2024)

Note: 7-DR2, 7-Dehydrocholesterol reductase; AGL6, Gamous-like 6; AGO7, Argonaute7; ALC, Alcobaca; ALD1, AGD2-like defense response protein; ALMT, Al-activated malate transporter; ALS1, Acetolactate synthase 1; AMS, Aborted microspores; AN2, Anthocyanin2; ANT1, Anthocyanin 1; APX2/4, Ascorbate peroxidase2/4; Blc, Beta-lycopene cyclase; BZR1, Brassinazole-resistant 1; CRTISO, Carotenoid isomerise; CBF1, C-repeat/dehydration responsive element binding factor1; CLV3, Clavata3, CMT4 Chromomethylase; CPK28, Calcium-dependent protein kinase28; CP, Coat protein; CrtR-b2, Beta-carotene hydroxylase 2; CYC-B/CycB, Lycopene beta cyclase; DDB1, DNA damage UV binding protein 1; DELLA, Aspartic acid–glutamic acid–leucine–leucine–alanine; DCL2b, Dicer-like 2b; DET1, Deetiolated1; DMR6, Downy mildew resistance 6; ENO, Excessive number of floral organs; EPSPS, 5-Enolpyruvylshikimate-3-phosphate synthase; ETR1, Ethylene receptor 1; FMO1, Flavin-dependent monooxygenase; FUL1/2, Fruitfull; GAD2, Glutamate decarboxylase 2; GAD3, Glutamate decarboxylase 3; GF, Greenflesh/Staygreen; GRXS, CGFS-type glutaredoxin; GSTAA, Glutathione S-transferase; HKT1/2, High-affinity potassium transporter 1/2; HY5, Elongated hypocotyl 5; HyPRP1, Hybrid proline-rich protein 1; IAA9, Auxin-induced 9; INVINH1, Invertase inhibitor 1; JAZ2, Jasmonate zim domain; LBD40, Lateral organ boundaries domain40; LCY B1, Lycopene bcyclase1; LCY-B2, Lycopene b-cyclase 2; LCY-E, Lycopene e-cyclase; MAPK3, Mitogen activated protein kinase 3; MIR164A, MicroRNA164A; Mlo1, Mildew resistance locus O 1; MS10, Male sterile 10; Ms1035, Male sterile 1035; MYBS2, MYB transcription factor S2; MYC2, Basic helix–loop–helix transcription factor; NAC, NAM-ATAF-CUC; NAM1/2/3, No apical meristem1/2/3; NPR1, Nonexpressor of pathogenesis-related gene 1; PDS, Phytoene desaturase; OFP, OVATE family protein; PG, Polygalacturonase; PG2a, Polygalacturonase 2a; PHO1, Phosphate 1; PIF4, Phytochrome interacting factor 4; PMR4, Powdery mildew resistance 4; PRO, Procera; PSY1, Phytoene synthase 1; RAD51/54, DNA repair and recombination protein51/54; RIN, Ripening inhibitor; S, Compound inflorescence; SGR1, Stay-green 1; SOS1, Salt overly sensitive 1; SP, Self pruning; SP5G, Self pruning 5G; TBG4, B-galactanase; ToMV, Tomato Mosaic virus; TRM3/4/5, TONNEAU1 Recruiting Motif3/4/5; VPE5, Vacuolar processing enzyme 5; WUS, Wuschel; Pto, *Pseudomonas syringae pv. Tomato*; FOL, *Fusarium oxysporum* f. sp. *Lycopersici*.

7.4.2 Agrobacterium-Mediated Delivery of CRISPR/Cas Binary Vectors to Tomato

The method is the same as traditional *Agrobacterium*-mediated T-DNA transformation. A binary CRISPR/Cas-gRNA vector is constructed and used for *Agrobacterium*-mediated transformation. CRISPR/Cas editing elements may not be incorporated into the plant genome. Stable component incorporation results in GMO plants. Transient expression of the CRISPR/Cas components permits gene editing while preventing the stable incorporation of foreign nucleic acids. The resulting transgene-free status of these gene-edited plants is beneficial for biotechnological applications.

7.4.3 Biolistic Bombardment of Editing Elements

The biolistic method delivers editing elements into plant nuclei using accelerated, nanosized particles coated with CRISPR DNA, RNA, or RNP complexes. DNA delivery may produce transgene-free plants via a transient expression transformation system or GMO plants with stable component incorporation. Both RNA and RNP delivery only produce transgene-free plants. Transgene-free genome editing has the advantage over conventional T-DNA insertion methods. Transgene-free editing often exempts the edited plant from GMO regulations and provides a simplified pathway for swift commercial deployment. For this reason, biolistic methods are widely used in plant genome editing. There are limiting factors to this approach. The cargo DNA may be randomly integrated at multiple genomic sites. The equipment and reagents can be prohibitively expensive. Finally, regenerating whole plantlets from transformed explants is a time-consuming process that is also dependent on the species and the genotype.

7.5 GENOME EDITING APPLICATIONS IN TOMATO IMPROVEMENT

7.5.1 Biotic Stress Tolerance

Pathogen susceptibility to bacterial, fungal, and viral infections poses a major threat to global tomato production; infections often lead to severe yield decreases and economic losses. The use of disease-resistant crop varieties can be a sustainable approach to meeting the world's growing food demand. While traditional breeding and genetic engineering have been used to develop disease-resistant varieties, these approaches are often slow, labor-intensive, and limited in scope. CRISPR/Cas9-mediated genome editing has emerged as a powerful tool for rapidly introducing precise genetic modifications that enhance tomato disease resistance. New materials/lines have been generated through gene-targeted genome editing with resistance or higher tolerance to bacteria, fungi, viruses, and parasitic weeds. CRISPR-based gene editing has significantly enhanced disease resistance in tomatoes by targeting specific genes involved in pathogen defense mechanisms. A 48-bp deletion in *Mlo1* provided durable resistance to powdery mildew disease (Nekrasov et al., 2017), knockout of *Solyc08g075770* reduced susceptibility to *Fusarium oxysporum* (Prihatna et al., 2018), mutating *MAPK3* enhanced resistance to *Botrytis cinerea* by increasing ROS accumulation (Zhang et al., 2018), and mutations in *dcl2b* influenced susceptibility to *Tobacco mosaic virus* (TMV) and *Potato virus X* (PVX) (Wang et al., 2018). Disrupting *JAZ2* in the jasmonic acid (JA) pathway improved resistance to *Pseudomonas syringae* by preventing stomatal reopening (Ortigosa et al., 2019), while editing DMR6–1 conferred broad-spectrum resistance against *P. syringae* and *Phytophthora capsici* without growth penalties (Thomazella et al., 2021). Late blight (*Phytophthora infestans*) resistant mutants were also developed by CRISPR/Cas9 editing of the MYBS2 (MYB transcription factor S2) gene (Liu et al., 2021a), while targeting viral genes (coat protein and replicase) or disrupting *SlPelo*-restricted *Tomato yellow leaf curl virus* (TYLCV) infection (Pramanik et al., 2021). Editing strigolactone biosynthesis genes (*CCD7*, *CCD8*, *MAX1*) enhanced resistance to the parasitic weed *Phelipanche aegyptiaca*

(Bari et al., 2021). These advancements demonstrate CRISPR's potential in developing robust, multi-pathogen-resistant tomato varieties. Knockout of the HyPRP1 gene enhanced resistance to *Pseudomonas syringae pv. Tomato DC3000* (Tran et al., 2023).

7.5.2 Abiotic Stress Tolerance

Drought stress severely impacts tomato cultivation. Mitogen-activated protein kinases (MAPKs) play a key role in stress response. *MAPK3*, a critical signaling enzyme, was studied using CRISPR/Cas9 knockout mutants. These mutants showed improved drought resistance, with lower H_2O_2 levels, reduced membrane damage, and higher antioxidant enzyme activity (Kong et al., 2012). The *SlHyPRP1* gene, a cell wall glycoprotein, was found to regulate the stress response negatively (Li et al., 2016). Altered expression of *LOX*, *GST*, and *DREB* genes suggests their role in drought adaptation (Wang et al., 2017). CRISPR-edited *NPR1* mutants exhibited better stress resilience (Li et al., 2019). The *SlLBD40* gene, highly expressed in roots and fruits, negatively regulates jasmonic acid (JA) signaling. CRISPR knockout of *SlLBD40* improved drought tolerance, while overexpression had the opposite effect (Liu et al., 2020a). *SlMAPK6* knockout affected axillary bud growth by altering gibberellin and strigolactone biosynthesis, influencing genes such as *CCD7*, *CCD8*, *GA20ox3*, and *GA3ox1* (Li et al., 2024). High salinity disrupts Na^+/K^+ balance, affecting tomato growth. CRISPR-generated SOS1 mutants showed increased Na^+/K^+ ratios and greater salt sensitivity when compared to wild-type plants (Wang et al., 2021c). CRISPR-mediated disruption of *HyPRP1* enhanced salt tolerance (>150 mM NaCl) without altering plant growth, making it a promising target for improving salinity resilience (Tran et al., 2021). *TAGL1*-KO mutants showed reduced ethylene biosynthesis, increased firmness, and delayed onset of ripening (Jeon et al., 2024).

7.5.3 Plant Cultivation and Fruit Improvement

CRISPR/Cas9 gene editing has been used to enhance important tomato traits. Fruit softening affects flavor, shelf life, and transport. CRISPR edits revealed that Pectate Lyase (PL) is crucial for maintaining firmness. Two more genes, *PG2a* and *TBG4*, primarily influence fruit color and weight, not softening (Wang et al., 2019). Anthocyanins create red/purple/blue colors and act as antioxidants. Editing the *SlAN2* gene increases anthocyanin production, leading to deeper color (Yan et al., 2020). The *SlMYB12* gene controls the red-to-pink shift. Editing of *SlMYB12* resulted in non-GMO pink tomatoes being produced much faster than through traditional breeding (Deng et al., 2018). Editing the *ARF4* gene (involved in auxin response) impacts plant structure (roots, stems, stomata) and photosynthesis more effectively than older methods (Bouzroud et al., 2020). "Multiplex" CRISPR (editing multiple genes at once) can quickly domesticate wild tomato relatives (Zsogon et al., 2018). Editing genes like *SlORRM4* (which delays ripening) and *DELLA* (which regulates growth hormones) improves traits such as plant shape, lycopene content, disease/salt resistance, and growth (Tomlinson et al., 2019). Targeted disruption of the *SlMTC* gene in tomato, using CRISPR/Cas9 technology, resulted in smaller seeds, stunted seedling development, and heightened sensitivity to salt stress (Gao et al., 2024). Editing DNA methylation genes (*DDM1*) reduces methylation, altering gene expression and affecting plant development (Corem et al., 2018). Editing *YABBY2b* influences plant height, fruit size, and flowers (Sun et al., 2020). Mutations in genes like *SlPDS/SlPIF4* show high success rates and stable inheritance across generations (Pan et al., 2016). Editing the flowering repressor gene SP5G accelerates flowering and boosts early yields by manipulating day-length sensitivity (Soyk et al., 2017). Gene-edited Sl*CDPK27* and Sl*CDPK26* knockouts increased glucose and fructose contents by up to 30%, enhancing perceived sweetness without fruit weight or yield penalty (Zhang et al., 2024). Gene-edited tomatoes targeting *DWARF* and *SELF-PRUNING* genes displayed a dwarf phenotype while retaining normal fruit characteristics (Nagamine & Ezura, 2024).

7.6 FUTURE PERSPECTIVES

Future perspectives for plant transformation and genome editing are broad and transformative, with potential to revolutionize agriculture, sustainability, and food security. Tissue culture regeneration-based transformation will still be the major method for plant transformation and genome editing. Novel methods to enhance editing efficiency are still being developed (Kang et al., 2025). The primary goal for tomato plant transformation is to develop a genotype-independent transformation protocol. Advancements in molecular biology and plant transformation methodologies offer the possibility of achieving this goal. Morphogenic gene-assisted transformation and in planta transformation (Yasmeen et al., 2009) are two areas of interest that provide promising results. Nanoparticle-mediated delivery and transgene-free transformation systems are two other focus areas. Improved protoplast regeneration systems across the broad plant spectrum also enhance editing efficiency in tomatoes. (Liu et al., 2022; Lin et al., 2022a). The development of next-generation tools such as prime editing, base editing, and CRISPR/Cas variants (e.g., Cas12a, Cas13) enables more precise and efficient genome modifications without creating double-stranded breaks. There is also an expansion of editing to non-model and recalcitrant plant species due to improved delivery methods (e.g., nanoparticles, in planta delivery, viral vectors, DNA-free systems). Genome editing is rapidly integrating with synthetic biology design principles to build novel biosynthetic pathways for pharmaceuticals, biofuels, or industrial enzymes in plants. Gene expression can also be more easily regulated by editing regulatory regions. Future tools may allow modification of epigenetic markers (e.g., methylation) to regulate gene expression without altering the DNA sequence. Targeted RNA editing can transiently modulate gene function without altering the DNA sequence. All of these new methods mean that researchers can accelerate breeding timelines for genotypes with high climate resilience, drought tolerance, disease resistance, and nutrient-use efficiency, as well as nutritional quality (e.g., higher vitamin, mineral, or protein content). It is also easier to improve yield potential and post-harvest shelf life. In the future, tomatoes may also be the first plant species used to introduce a gene drive to spread a beneficial trait throughout the population rapidly.

Genetic transformation and genome editing offer unprecedented opportunities to improve tomato cultivars for diverse traits. Continued innovation in delivery systems, regulatory alignment, and public engagement will be key to fully harnessing these technologies for global food security and agricultural sustainability.

REFERENCES

Atarashi, H., Jayasinghe, W. H., Kwon, J., Kim, H., Taninaka, Y., Igarashi, M., et al. (2020). Artificially edited alleles of the eukaryotic translation initiation factor 4E1 gene differentially reduce susceptibility to cucumber mosaic virus and potato virus Y in tomato. *Frontiers in Microbiology*, *11*, 564310. https://doi.org/10.3389/fmicb.2020.564310

Bao, H., Ding, Y., Yang, F., Zhang, J., Xie, J., Zhao, C., et al. (2022). Gene silencing, knockout and overexpression of a transcription factor ABORTED MICROSPORES (SlAMS) strongly affects pollen viability in tomato (*Solanum lycopersicum*). *BMC Genomics*, *23*(Suppl 1), 346. https://doi.org/10.1186/s12864-022-08549-x

Bari, V. K., Nassar, J. A., & Aly, R. (2021). CRISPR/Cas9 mediated mutagenesis of MORE AXILLARY GROWTH 1 in tomato confers resistance to root parasitic weed *Phelipanche aegyptiaca*. *Scientific Reports*, *11*, 3905. https://doi.org/10.1038/s41598-021-82897-8

Bernabé-Orts, J. M., Casas-Rodrigo, I., Minguet, E. G., Landolfi, V., García-Carpintero, V., Gianoglio, S., Vázquez-Vilar, M., Granell, A., & Orzaez, D. (2019). Assessment of Cas12a-mediated gene editing efficiency in plants. *Plant Biotechnology Journal*, *17*(10), 1971–1984. https://doi.org/10.1111/pbi.13113

Bio-Rad Laboratories. (2006). *PDS-1000/He Biolistic Particle Delivery System User Manual* (Revision C). Bio-Rad.

Bouzroud, S., Gasparini, K., Hu, G., Barbosa, M., Rosa, B. L., Fahr, M., et al. (2020). Down regulation and loss of auxin response factor 4 function using CRISPR/Cas9 alters tomato plants growth, stomatal function and improves tomato tolerance to salinity and osmotic stress. *Genes*, *11*, 272. https://doi.org/10.3390/genes11030272

Brooks, C., Nekrasov, V., Lippman, Z. B., & Van Eck, J. (2014). Efficient gene editing in tomato in the first generation using the clustered regularly interspaced short palindromic repeats/CRISPR-associated9 system. *Plant Physiology*, *166*, 1292–1297. https://doi.org/10.1104/pp.114.247577

C'ermak, T., Baltes, N. J., Cegan, R., Zhang, Y., & Voytas, D. F. (2015). High-frequency, precise modification of the tomato genome. *Genome Biology*, *16*, 232. https://doi.org/10.1186/s13059-015-0796-9

Corem, S., Doron-Faigenboim, A., Jouffroy, O., Maumus, F., Arazi, T., & Bouché, N. (2018). Redistribution of CHH methylation and small interfering RNAs across the genome of tomato ddm1 mutants. *The Plant Cell*, *30*, 1628–1644. https://doi.org/10.1105/tpc.18.00167

Costa, M., Nogueira, F., Figueira, M., Otoni, W., Brommonschenkel, S., & Cecon, P. (2000). Influence of the antibiotic timentin on plant regeneration of tomato (*Lycopersicon esculentum* Mill.) cultivars. *Plant Cell Reports*, *19*, 327–332. https://doi.org/10.1007/s002990050021

Dahan-Meir, T., Filler-Hayut, S., Melamed-Bessudo, C., Bocobza, S., Czosnek, H., Aharoni, A., et al. (2018). Efficient in planta gene targeting in tomato using geminiviral replicons and the CRISPR/Cas9 system. *The Plant Journal*, *95*, 5–16. https://doi.org/10.1111/tpj.13932

Dai, X., Han, H., Huang, W., Zhao, L., Song, M., Cao, X., et al. (2022). Generating novel male sterile tomatoes by editing respiratory burst oxidase homolog genes. *Frontiers in Plant Science*, *12*, 817101. https://doi.org/10.3389/fpls.2021.817101

D'Ambrosio, C., Stigliani, A. L., & Giorio, G. (2018). CRISPR/Cas9 editing of carotenoid genes in tomato. *Transgenic Research*, *27*, 367–378. https://doi.org/10.1007/s11248-018-0079-9

Danilo, B., Perrot, L., Mara, K., Botton, E., Nogue, F., & Mazier, M. (2019). Efficient and transgene-free gene targeting using Agrobacterium-mediated delivery of the CRISPR/Cas9 system in tomato. *Plant Cell Reports*, *38*, 459–462. https://doi.org/10.1007/s00299-019-02373-6

Deng, L., Wang, H., Sun, C., Li, Q., Jiang, H., Du, M., et al. (2018). Efficient generation of pink-fruited tomatoes using CRISPR/Cas9 system. *Journal of Genetics and Genomics*, *45*, 51–54. https://doi.org/10.1016/j.jgg.2017.10.002

Ding, F., Hu, Q., Wang, M., & Zhang, S. (2018). Knockout of SlSBPASE suppresses carbon assimilation and alters nitrogen metabolism in tomato plants. *International Journal of Molecular Sciences*, *19*, 4046. https://doi.org/10.3390/ijms19124046

Do, J. H., Park, S. Y., Park, S. H., Kim, H. M., Ma, S. H., & Mai, T. D. (2022). Development of a genome-edited tomato with high ascorbate content during later stage of fruit ripening through mutation of SlAPX4. *Frontiers in Plant Science*, *13*, 836916. https://doi.org/10.3389/fpls.2022.836916

Ezura, H. (2022). The world's first CRISPR tomato launched to a Japanese market: The social-economic impact of its implementation on crop genome editing. *Plant and Cell Physiology*, *63*, 731–733. https://doi.org/10.1093/pcp/pcac048

Faal, P. G., Farsi, M., Seifi, A., & Mirshamsi Kakhki, A. (2020). Virus-induced CRISPR-Cas9 system improved resistance against tomato yellow leaf curl virus. *Molecular Biology Reports*, *47*, 3369–3376. https://doi.org/10.1007/s11033-020-05409-3

FAO, IFAD, UNICEF, WFP, & WHO. (2023). *The State of Food Security and Nutrition in the World 2023: Urbanization, Agrifood Systems Transformation and Healthy Diets Across the Rural-Urban Continuum*. FAO. https://doi.org/10.4060/cc3017en

Gamborg, O. L., Miller, R. A., & Ojima, K. (1968). Nutrient requirements of suspension cultures of soybean root cells. *Experimental Cell Research*, *50*, 151–158. https://doi.org/10.1016/0014-4827(68)90403-5

Gao, Y., Fan, Z. Q., Zhang, Q., Li, H. L., Liu, G. S., Jing, Y., et al. (2021). A tomato NAC transcription factor, SlNAM1, positively regulates ethylene biosynthesis and the onset of tomato fruit ripening. *The Plant Journal*, *108*, 1317–1331. https://doi.org/10.1111/tpj.15512

Gao, Z., Yang, Q., Shen, H., Guo, P., Xie, Q., Chen, G., et al. (2024). The knockout of SlMTC impacts tomato seed size and reduces resistance to salt stress in tomato. *Plant Science*, *349*, 112228. https://doi.org/10.1016/j.plantsci.2024.112228

Gianoglio, S., Comino, C., Moglia, A., Acquadro, A., García-Carpintero, V., Diretto, G., et al. (2022). In-depth characterization of greenflesh tomato mutants obtained by CRISPR/Cas9 editing: A case study with implications for breeding and regulation. *Frontiers in Plant Science*, *13*, 936089. https://doi.org/10.3389/fpls.2022.936089

Guo, X., Zhao, J., Chen, Z., Qiao, J., Zhang, Y., Shen, H., et al. (2022). CRISPR/Cas9-targeted mutagenesis of SlCMT4 causes changes in plant architecture and reproductive organs in tomato. *Horticulture Research, 9*, uhac081. https://doi.org/10.1093/hr/uhac081

Gupta, S., & Van Eck, J. (2016). Modification of plant regeneration medium decreases the time for recovery of *Solanum lycopersicum* cultivar M82 stable transgenic lines. *Plant Cell, Tissue and Organ Culture, 127*, 417–423. https://doi.org/10.1007/s11240-016-1063-9

Hamza, S., & Chupeau, Y. (1993). Re-evaluation of conditions for plant regeneration and Agrobacterium-mediated transformation from tomato (*Lycopersicon esculentum*). *Journal of Experimental Botany, 44*, 1837–1845. https://doi.org/10.1093/jxb/44.12.1837

Hanson, M. R. (1981). Regeneration of tomato plants in vitro from leaf and hypocotyl explants. *Tomato Genetics Cooperative Report, 32*, 15–16.

Hayut, S. F., Bessudo, C. M., & Levy, A. A. (2017). Targeted recombination between homologous chromosomes for precise breeding in tomato. *Nature Communications, 8*, 15605. https://doi.org/10.1038/ncomms15605

Hilioti, Z., Ganopoulos, I., Ajith, S., Bossis, I., & Tsaftaris, A. (2016). A novel arrangement of Zinc Finger Nuclease system for in vivo targeted genome engineering: The tomato LEC1-LIKE4 gene case. *Plant Cell Reports, 35*, 2241–2255. https://doi.org/10.1007/s00299-016-2031-x

Hong, Y., Meng, J., He, X., Zhang, Y., Liu, Y., Zhang, C., et al. (2021). Editing miR482b and miR482c simultaneously by CRISPR/Cas9 enhanced tomato resistance to *Phytophthora infestans*. *Phytopathology*. https://doi.org/10.1094/PHYTO-08-20-0360-R

Hu, Z., Li, J., & Ding, S. (2021). The protein kinase CPK28 phosphorylates ascorbate peroxidase and enhances thermotolerance in tomato. *Plant Physiology, 186*, 1302–1317. https://doi.org/10.1093/plphys/kiab120

Hunziker, J., Nishida, K., Kondo, A., Kishimoto, S., Ariizumi, T., & Ezura, H. (2020). Multiple gene substitution by target-AID base-editing technology in tomato. *Scientific Reports, 10*, 20471. https://doi.org/10.1038/s41598-020-77379-2

Ito, Y., Nishizawa-Yokoi, A., Endo, M., Mikami, M., Shima, Y., Nakamura, N., et al. (2017). Re-evaluation of the rin mutation and the role of RIN in the induction of tomato ripening. *Nature Plants, 3*, 866–874. https://doi.org/10.1038/s41477-017-0041-5

Janik, E., Niemcewicz, M., Ceremuga, M., Krzowski, L., Saluk-Bijak, J., & Bijak, M. (2020). Various aspects of a gene editing system—CRISPR–Cas9. *International Journal of Molecular Sciences, 21*, 9604. https://doi.org/10.3390/ijms21249604

Jayaraj, K. L., Thulasidharan, N., Antony, A., John, M., Augustine, R., Chakravartty, N., et al. (2021). Targeted editing of tomato carotenoid isomerase reveals the role of 5' UTR region in gene expression regulation. *Plant Cell Reports, 40*, 621–635. https://doi.org/10.1007/s00299-020-02659-0

Jeon, C., Chung, M. Y., & Lee, J. M. (2024). Reassessing the contribution of TOMATO AGAMOUS-LIKE1 to fruit ripening by CRISPR/Cas9 mutagenesis. *Plant Cell Reports, 43*, 41. https://doi.org/10.1007/s00299-023-03105-7

Jung, Y. J., Kim, D. H., Lee, H. J., Nam, K. H., Bae, S., Nou, I. S., et al. (2020). Knockout of SlMS10 gene (Solyc02g079810) encoding bHLH transcription factor using CRISPR/Cas9 system confers male sterility phenotype in tomato. *Plants, 9*, 1189. https://doi.org/10.3390/plants9091189

Kakeshpour, T., Tamang, T. M., Motolai, G., Fleming, Z. W., Park, J. E., Wu, Q., et al. (2021). CGFS-type glutaredoxin mutations reduce tolerance to multiple abiotic stresses in tomato. *Physiologia Plantarum, 173*, 1263–1279. https://doi.org/10.1111/ppl.13522

Kang, G. H., Ko, Y., & Lee, J. M. (2025). Enhancing virus-mediated genome editing for cultivated tomato through low temperature. *Plant Cell Reports, 44*, 22. https://doi.org/10.1007/s00299-024-03392-8

Kassahun, B. M., Kang, B. C., Su-Ji, B., Nam, Y. J., Mundo, G. F., Kang, G. H., et al. (2021). Rapid delivery of Cas9 gene into the tomato cv. 'Heinz 1706' through an optimized Agrobacterium-mediated transformation procedure. *Biocell, 45*, 199–215. https://doi.org/10.32604/biocell.2021.012353

Kawaguchi, K., Takei-Hoshi, R., Yoshikawa, I., Nishida, K., Kobayashi, M., Kusano, M., et al. (2021). Functional disruption of cell wall invertase inhibitor by genome editing increases sugar content of tomato fruit without decrease fruit weight. *Scientific Reports, 11*, 21534. https://doi.org/10.1038/s41598-021-00966-4

Klap, C., Yeshayahou, E., Bolger, A. M., Arazi, T., Gupta, S. K., Shabtai, S., et al. (2017). Tomato facultative parthenocarpy results from SlAGAMOUS-LIKE 6 loss of function. *Plant Biotechnology Journal, 15*, 634–647. https://doi.org/10.1111/pbi.12662

Kong, F. L., Wang, J., Cheng, L., Liu, S. Y., Wu, J., Peng, Z., & Lu, G. (2012). Genome-wide analysis of the mitogen-activated protein kinase gene family in Solanum lycopersicum. *Gene*, *499*, 108–120. https://doi.org/10.1016/j.gene.2012.01.048

Kurtz, S. M., & Lineberger, R. D. (1983). Genotypic differences in morphogenic capacity of cultured leaf explants of tomato. *Journal of the American Society for Horticultural Science*, *108*, 710–714. https://doi.org/10.21273/JASHS.108.5.710

Li, J., Ouyang, B., Wang, T., Luo, Z., Yang, C., Li, H., et al. (2016). HyPRP1 gene suppressed by multiple stresses plays a negative role in abiotic stress tolerance in tomato. *Frontiers in Plant Science*, *7*, 967. https://doi.org/10.3389/fpls.2016.00967

Li, J., Scarano, A., Gonzalez, N. M., D'Orso, F., Yue, Y., Nemeth, K., et al. (2022a). Biofortified tomatoes provide a new route to vitamin D sufficiency. *Nature Plants*, *8*, 611–616. https://doi.org/10.1038/s41477-022-01154-6

Li, Q., Feng, Q., Snouffer, A., Zhang, B., Rodríguez, G. R., & van der Knaap, E. (2022b). Increasing fruit weight by editing a cis-regulatory element in tomato KLUH promoter using CRISPR/Cas9. *Frontiers in Plant Science*, *13*, 879642. https://doi.org/10.3389/fpls.2022.879642

Li, R., Fu, D., Zhu, B., Luo, Y., & Zhu, H. (2018a). CRISPR/Cas9-mediated mutagenesis of lncRNA1459 alters tomato fruit ripening. *The Plant Journal*, *94*, 513–524. https://doi.org/10.1111/tpj.13872

Li, R., Li, R., Li, X., Fu, D., Zhu, B., & Tian, H. (2018b). Multiplexed CRISPR/Cas9-mediated metabolic engineering of γ-aminobutyric acid levels in *Solanum lycopersicum*. *Plant Biotechnology Journal*, *16*, 415–427. https://doi.org/10.1111/pbi.12781

Li, R., Liu, C., Zhao, R., Wang, L., Chen, L., Yu, W., et al. (2019). CRISPR/Cas9-mediated SlNPR1 mutagenesis reduces tomato plant drought tolerance. *BMC Plant Biology*, *19*, 38. https://doi.org/10.1186/s12870-018-1627-4

Li, R., Zhang, L., Wang, L., Chen, L., Zhao, R., Sheng, J., et al. (2018e). Reduction of tomato-plant chilling tolerance by CRISPR/Cas9-mediated SlCBF1 mutagenesis. *Journal of Agricultural and Food Chemistry*, *66*, 9042–9051. https://doi.org/10.1021/acs.jafc.8b02177

Li, T., Yang, X., Yu, Y., Si, X., Zhai, X., Zhang, H., et al. (2018c). Domestication of wild tomato is accelerated by genome editing. *Nature Biotechnology*, *36*, 1160–1163. https://doi.org/10.1038/nbt.4273

Li, X., Wang, Y., Chen, S., Tian, H., Fu, D., Zhu, B., et al. (2018d). Lycopene is enriched in tomato fruit by CRISPR/Cas9-mediated multiplex genome editing. *Frontiers in Plant Science*, *9*, 559. https://doi.org/10.3389/fpls.2018.00559

Li, Y., Yue, N., Basit, A., Li, Y., Zhang, D., Qin, L., et al. (2024). Targeted editing of SlMAPK6 using CRISPR/Cas9 technology to promote the development of axillary buds in tomato plants. *Journal of Agricultural Science*, *13*, 1–11. https://doi.org/10.5539/jas.v13n2p11

Lin, C. S., Hsu, C. T., Yuan, Y. H., Zheng, P. X., Wu, F. H., Cheng, Q. W., et al. (2022a). DNA-free CRISPR-Cas9 gene editing of wild tetraploid tomato *Solanum peruvianum* using protoplast regeneration. *Plant Physiology*, *188*, 1917–1930. https://doi.org/10.1093/plphys/kiac022

Lin, D., Zhu, X., Qi, B., Gao, Z., Tian, P., Li, Z., et al. (2022b). SlMIR164A regulates fruit ripening and quality by controlling SlNAM2 and SlNAM3 in tomato. *Plant Biotechnology Journal*, *20*, 1456–1469. https://doi.org/10.1111/pbi.13824

Lipp, J. K. H., & Brown, T. A. (1993). Enhanced transformation of tomato co-cultivated with *Agrobacterium tumefaciens* C58C1Rifr: pGSFR1161 in the presence of acetosyringone. *Plant Cell Reports*, *12*, 422–425. https://doi.org/10.1007/BF00234705

Liu, C., Zhang, Y., Tan, Y., Zhao, T., Xu, X., Yang, H., et al. (2021a). CRISPR/Cas9-mediated SlMYBS2 mutagenesis reduces tomato resistance to *Phytophthora infestans*. *International Journal of Molecular Sciences*, *22*, 11423. https://doi.org/10.3390/ijms222111423

Liu, J., Wang, S., Wang, H., Luo, B., Cai, Y., Li, X., et al. (2021b). Rapid generation of tomato male-sterile lines with a marker use for hybrid seed production by CRISPR/Cas9 system. *Molecular Breeding*, *41*, 25. https://doi.org/10.1007/s11032-021-01215-2

Liu, L., Zhang, J., Xu, J., Li, Y., Guo, L., Wang, Z., et al. (2020a). CRISPR/Cas9 targeted mutagenesis of SlLBD40, a lateral organ boundaries domain transcription factor, enhances drought tolerance in tomato. *Plant Science*, *301*, 110683. https://doi.org/10.1016/j.plantsci.2020.110683

Liu, X., Zhang, Q., Yang, G., Zhang, C., Dong, H., Liu, Y., et al. (2020b). Pivotal roles of Tomato photoreceptor SlUVR8 in seedling development and UV-B stress tolerance. *Biochemical and Biophysical Research Communications*, *522*, 177–183. https://doi.org/10.1016/j.bbrc.2019.11.073

Liu, Y., Andersson, M., Granell, A., Cardi, T., Hofvander, P., & Nicolia, A. (2022). Establishment of a DNA-free genome editing and protoplast regeneration method in cultivated tomato (*Solanum lycopersicum*). *Plant Cell Reports*, *41*, 1843–1852. https://doi.org/10.1007/s00299-022-02893-8

Lor, V. S., Starker, C. G., & Voytas, D. F. (2014). Targeted mutagenesis of the tomato PROCERA gene using transcription activator-like effector nucleases. *Plant Physiology*, *166*, 1288–1291. https://doi.org/10.1104/pp.114.247593

Luo, Z., Zhang, J., Li, J., Yang, C., Wang, T., Ouyang, B., et al. (2013). A STAY-GREEN protein SlSGR1 regulates lycopene and β-carotene accumulation by interacting directly with SlPSY1 during ripening processes in tomato. *New Phytologist*, *198*, 442–452. https://doi.org/10.1111/nph.12175

Madhu, S., & Savithramma, D. (2014). Standardization of high efficient and rapid regeneration protocol for Agrobacterium mediated transformation of tomato (*Solanum lycopersicum* L.) cv. Pusa Ruby, Vaibhav and Arka Meghali. *International Journal of Agronomy and Agricultural Research*, *5*, 161–169.

Martínez, M. I. S., Bracuto, V., Koseoglou, E., Appiano, M., Jacobsen, E., Visser, R., et al. (2020). CRISPR/Cas9-targeted mutagenesis of the tomato susceptibility gene PMR4 for resistance against powdery mildew. *BMC Plant Biology*, *20*, 284. https://doi.org/10.1186/s12870-020-02497-y

McCormick, S., Niedermeyer, J., Fry, J., Barnason, A., Horsch, R., & Fraley, R. (1986). Leaf disc transformation of cultivated tomato (*L. esculentum*) using *Agrobacterium tumefaciens*. *Plant Cell Reports*, *5*, 81–84. https://doi.org/10.1007/BF00269239

Murashige, T., & Skoog, F. (1962). A revised medium for rapid growth and bioassays with tobacco tissue cultures. *Physiologia Plantarum*, *15*, 473–497. https://doi.org/10.1111/j.1399-3054.1962.tb08052.x

Nagamine, A., & Ezura, H. (2024). Genome editing of DWARF and SELF-PRUNING rapidly confers traits suitable for plant factories while retaining useful traits in tomato. *Breeding Science*, *74*, 59–72. https://doi.org/10.1270/jsbbs.23063

Nekrasov, V., Wang, C., Win, J., Lanz, C., Weigel, D., & Kamoun, S. (2017). Rapid generation of a transgene-free powdery mildew resistant tomato by genome deletion. *Scientific Reports*, *7*, 482. https://doi.org/10.1038/s41598-017-00578-x

Nie, H., Shi, Y., Geng, X., & Xing, G. (2022). CRISPR/Cas9-mediated targeted mutagenesis of tomato polygalacturonase gene (SlPG) delays fruit softening. *Frontiers in Plant Science*, *13*, 729128. https://doi.org/10.3389/fpls.2022.729128

Nonaka, S., Arai, C., Takayama, M., Matsukura, C., & Ezura, H. (2017). Efficient increase of γ-aminobutyric acid (GABA) content in tomato fruits by targeted mutagenesis. *Scientific Reports*, *7*, 7057. https://doi.org/10.1038/s41598-017-06400-y

Ortigosa, A., Gimenez-Ibanez, S., Leonhardt, N., & Solano, R. (2019). Design of a bacterial speck-resistant tomato by CRISPR/Cas9-mediated editing of SlJAZ2. *Biotechnology Journal*, *17*, 665–673. https://doi.org/10.1111/pbi.13006

Pan, C., Ye, L., Qin, L., Liu, X., He, Y., Wang, J., et al. (2016). CRISPR/Cas9-mediated efficient and heritable targeted mutagenesis in tomato plants in the first and later generations. *Scientific Reports*, *6*, 24765. https://doi.org/10.1038/srep24765

Pandolfini, T., Rotino, G. L., Camerini, S., Defez, R., & Spena, A. (2002). Optimisation of transgene action at the post-transcriptional level: High quality parthenocarpic fruits in industrial tomatoes. *BMC Biotechnology*, *2*, 1. https://doi.org/10.1186/1472-6750-2-1

Pramanik, D., Shelake, R. M., & Kim, M. J., Kim, J.-Y. (2021). CRISPR-mediated engineering across the central dogma in plant biology for basic research and crop improvement. *Molecular Plant*, *14*, 127–150. https://doi.org/10.1016/j.molp.2020.11.002

Prihatna, C., Barbetti, M. J., & Barker, S. J. (2018). A novel tomato Fusarium wilt tolerance gene. *Frontiers in Microbiology*, *9*, 1226. https://doi.org/10.3389/fmicb.2018.01226

Qiu, Z., Wang, H., Li, D., Yu, B., Hui, Q., Yan, S., et al. (2019). Identification of candidate HY5-dependent and -independent regulators of anthocyanin biosynthesis in tomato. *Plant Cell Physiology*, *60*, 643–656. https://doi.org/10.1093/pcp/pcy236

Rodriguez-Leal, D., Lemmon, Z. H., Man, J., Bartlett, M. E., & Lippman, Z. B. (2017). Engineering quantitative trait variation for crop improvement by genome editing. *Cell*, *171*, 470–480.e8. https://doi.org/10.1016/j.cell.2017.08.030

Roldan, M. V. G., Perilleux, C., Morin, H., Huerga-Fernandez, S., Latrasse, D., Benhamed, M., et al. (2017). Natural and induced loss of function mutations in SlMBP21 MADS-box gene led to jointless-2 phenotype in tomato. *Scientific Reports*, *7*, 4402. https://doi.org/10.1038/s41598-017-04556-1

Ron, M., Kajala, K., & Pauluzzi, G. (2014). Hairy root transformation using *Agrobacterium rhizogenes* as a tool for exploring cell type-specific gene expression and function using tomato as a model. *Plant Physiology*, *166*, 455–469. https://doi.org/10.1104/pp.114.239392

Sambrook, J., Fritsch, E. F., & Maniatis, T. (1989). *Molecular Cloning: A Laboratory Manual* (2nd ed.). Cold Spring Harbor Laboratory Press.

Sarker, R. H., Islam, K., & Hoque, M. I. (2009). In vitro regeneration and Agrobacterium-mediated genetic transformation of tomato (*Lycopersicon esculentum* Mill.). *Plant Tissue Culture and Biotechnology*, *19*, 101–111. https://doi.org/10.3329/ptcb.v19i1.5004

Shimatani, Z., Kashojiya, S., Takayama, M., Terada, R., Arazoe, T., Ishii, H., et al. (2017). Targeted base editing in rice and tomato using a CRISPR-Cas9 cytidine deaminase fusion. *Nature Biotechnology*, *35*, 441–443. https://doi.org/10.1038/nbt.3833

Shu, P., Li, Z., Min, D., Zhang, X., Ai, W., Li, J., et al. (2020). CRISPR/Cas9-mediated SlMYC2 mutagenesis adverse to tomato plant growth and MeJA-induced fruit resistance to *Botrytis cinerea*. *Journal of Agricultural and Food Chemistry*, *68*, 5529–5538. https://doi.org/10.1021/acs.jafc.9b08069

Soyk, S., Müller, N. A., Park, S. J., Schmalenbach, I., Jiang, K., Hayama, R., et al. (2017). Variation in the flowering gene SELF PRUNING 5G promotes day-neutrality and early yield in tomato. *Nature Genetics*, *49*, 162–168. https://doi.org/10.1038/ng.3733

Sun, H. J., Uchii, S., Watanabe, S., & Ezura, H. (2006). A highly efficient transformation protocol for Micro-Tom, a model cultivar for tomato functional genomics. *Plant Cell Physiology*, *47*, 426–431. https://doi.org/10.1093/pcp/pci251

Sun, M., Li, H., Li, Y., Xiang, H., Liu, Y., He, Y., et al. (2020). Tomato YABBY2b controls plant height through regulating indole-3-acetic acid-amido synthetase (GH3.8) expression. *Plant Science*, *297*, 110530. https://doi.org/10.1016/j.plantsci.2020.110530

Tashkandi, M., Ali, Z., Aljedaani, F., Shami, A., & Mahfouz, M. M. (2018). Engineering resistance against tomato yellow leaf curl virus via the CRISPR/Cas9 system in tomato. *Plant Signaling & Behavior*, *13*, e1525996. https://doi.org/10.1080/15592324.2018.1525996

Thomazella, D., Seong, K., Mackelprang, R., Dahlbeck, D., Geng, Y., Gill, U. S., et al. (2021). Loss of function of a DMR6 ortholog in tomato confers broad-spectrum disease resistance. *Proceedings of the National Academy of Sciences of the United States of America*, *118*(27), e2026152118. https://doi.org/10.1073/pnas.2026152118

Tikunov, Y. M., Roohanitaziani, R., Meijer-Dekens, F., Molthoff, J., Paulo, J., Finkers, R., et al. (2020). The genetic and functional analysis of flavor in commercial tomato: The FLORAL4 gene underlies a QTL for floral aroma volatiles in tomato fruit. *The Plant Journal*, *103*, 1189–1204. https://doi.org/10.1111/tpj.14795

Tiwari, J. K., Singh, A. K., & Behera, T. K. (2023). CRISPR/Cas genome editing in tomato improvement: Advances and applications. *Frontiers in Plant Science*, *14*, 1121209. https://doi.org/10.3389/fpls.2023.1121209

Tomato Genome Consortium. (2012). The tomato genome sequence provides insights into fleshy fruit evolution. *Nature*, *485*, 635–641. https://doi.org/10.1038/nature11119

Tomlinson, L., Yang, Y., Emenecker, R., Smoker, M., Taylor, J., Perkins, S., et al. (2019). Using CRISPR/Cas9 genome editing in tomato to create a gibberellin-responsive dominant dwarf DELLA allele. *Plant Biotechnology Journal*, *17*, 132–140. https://doi.org/10.1111/pbi.12952

Tran, M. T., Doan, D. T. H., Kim, J., Song, Y. J., Sung, Y. W., Das, S., et al. (2021). CRISPR/Cas9-based precise excision of SlHyPRP1 domain(s) to obtain salt stress-tolerant tomato. *Plant Cell Reports*, *40*, 999–1011. https://doi.org/10.1007/s00299-020-02622-z

Tran, M. T., Son, G. H., Song, Y. J., Nguyen, N. T., Park, S., Thach, T. V., et al. (2023). CRISPR-Cas9-based precise engineering of SlHyPRP1 protein towards multi-stress tolerance in tomato. *Frontiers in Plant Science*, *14*, 1186932. https://doi.org/10.3389/fpls.2023.1186932

Ueta, R., Abe, C., Watanabe, T., Sugano, S. S., Ishihara, R., Ezura, H., et al. (2017). Rapid breeding of parthenocarpic tomato plants using CRISPR/Cas9. *Scientific Reports*, *7*, 507. https://doi.org/10.1038/s41598-017-00501-4

Uranga, M., Aragonés, V., García, A., Mirabel, S., Gianoglio, S., Presa, S., et al. (2024). RNA virus-mediated gene editing for tomato trait breeding. *Horticulture Research*, *11*, uhad279. https://doi.org/10.1093/hr/uhad279

Valencia-Lozano, E., Cabrera-Ponce, J. L., Barraza, A., López-Calleja, A. C., García-Vázquez, E., Rivera-Toro, D. M., et al. (2024). Editing of SlWRKY29 by CRISPR-activation promotes somatic embryogenesis in

Solanum lycopersicum cv. Micro-Tom. *PLoS ONE, 19*, e0301169. https://doi.org/10.1371/journal.pone.0301169

Van Roekel, J. S., Damm, B., Melchers, L. S., & Hoekema, A. (1993). Factors influencing transformation frequency of tomato (*Lycopersicon esculentum*). *Plant Cell Reports, 12*, 644–647. https://doi.org/10.1007/BF00232816

Veillet, F., Perrot, L., Chauvin, L., Kermarrec, M. P., Guyon-Debast, A., Chauvin, J. E., et al. (2019). Transgene-free genome editing in tomato and potato plants using Agrobacterium-mediated delivery of a CRISPR/Cas9 cytidine base editor. *International Journal of Molecular Sciences, 20*, 402. https://doi.org/10.3390/ijms20020402

Vu, T. V., Sivankalyani, V., Kim, E. J., Doan, D., & Tran, M. T., & Kim, J. (2020b). Highly efficient homology-directed repair using CRISPR/Cpf1-geminiviral replicon in tomato. *Plant Biotechnology Journal, 18*, 2133–2143. https://doi.org/10.1111/pbi.13373

Wang, B., Li, N., Huang, S., Hu, J., Wang, Q., Tang, Y., et al. (2021a). Enhanced soluble sugar content in tomato fruit using CRISPR/Cas9-mediated SlINVINH1 and SlVPE5 gene editing. *PeerJ, 9*, e12478. https://doi.org/10.7717/peerj.12478

Wang, L., Chen, L., Li, R., Zhao, R., Yang, M., Sheng, J., et al. (2017). Reduced drought tolerance by CRISPR/Cas9-mediated SlMAPK3 mutagenesis in tomato plants. *Journal of Agricultural and Food Chemistry, 65*, 8674–8682. https://doi.org/10.1021/acs.jafc.7b02745

Wang, P., Luo, Q., Yang, W., Ahammed, G. J., Ding, S., Chen, X., et al. (2021b). A novel role of pipecolic acid biosynthetic pathway in drought tolerance through the antioxidant system in tomato. *Antioxidants, 10*, 1923. https://doi.org/10.3390/antiox10121923

Wang, R., Tavano, E., Lammers, M., Martinelli, A. P., Angenent, G. C., & de Maagd, R. A. (2019). Re-evaluation of transcription factor function in tomato fruit development and ripening with CRISPR/Cas9-mutagenesis. *Scientific Reports, 9*, 1696. https://doi.org/10.1038/s41598-018-38170-6

Wang, T., Deng, Z., Zhang, X., Wang, H., Wang, Y., Liu, X., et al. (2018). Tomato DCL2b is required for the biosynthesis of 22-nt small RNAs, the resulting secondary siRNAs, and the host defense against ToMV. *Horticulture Research, 5*, 62. https://doi.org/10.1038/s41438-018-0073-7

Wang, Z., Hong, Y., Li, Y., Shi, H., Yao, J., Liu, X., et al. (2021c). Natural variations in SlSOS1 contribute to the loss of salt tolerance during tomato domestication. *Plant Biotechnology Journal, 19*, 20–22. https://doi.org/10.1111/pbi.1344

World Bank. (2023). *Poverty and inequality*. World Bank Open Data. https://data.worldbank.org/topic/poverty

Wu, S., Zhang, B., Keyhaninejad, N., Liu, S., Cheng, Y., Han, Y., et al. (2018). A common genetic mechanism underlies morphological diversity in fruits and other plant organs. *Nature Communications, 9*, 4734. https://doi.org/10.1038/s41467-018-07216-8

Yan, S., Chen, N., Huang, Z., Li, D., Zhi, J., & Yu, B. (2020). Anthocyanin fruit encodes an R2R3-MYB transcription factor, SlAN2-like, activating the transcription of SlMYBATV to fine-tune anthocyanin content in tomato fruit. *New Phytologist, 225*, 2048–2063. https://doi.org/10.1111/nph.16272

Yang, S. H., Kim, E., Park, H., & Koo, Y. (2022). Selection of the high efficient sgRNA for CRISPR/Cas9 to edit herbicide related genes, PDS, ALS, and EPSPS in tomato. *Applied Biological Chemistry, 65*, 13. https://doi.org/10.1186/s13765-022-00679-w

Yaroshko, O., Pasternak, T., Larriba, E., & Pérez-Pérez, J. M. (2023). Optimization of callus induction and shoot regeneration from tomato cotyledon explants. *Plants, 12*, 2942. https://doi.org/10.3390/plants12162942

Yasmeen, A. (2009). An improved protocol for the regeneration and transformation of tomato (cv Rio Grande). *Acta Physiologiae Plantarum, 31*, 1271–1277. https://doi.org/10.1007/s11738-009-0364-6

Yasmeen, A., Mirza, B., Inayatullah, S., Safdar, N., Jamil, M., Ali, S., et al. (2009). In planta transformation of tomato. *Plant Molecular Biology Reporter, 27*, 20–28. https://doi.org/10.1007/s11105-008-0044-5

Ye, J., Wang, X., Hu, T., Zhang, F., Wang, B., Li, C., et al. (2017). An InDel in the promoter of Al-ACTIVATED MALATE TRANSPORTER9 selected during tomato domestication determines fruit malate contents and aluminium tolerance. *The Plant Cell, 29*, 2249–2268. https://doi.org/10.1105/tpc.17.00211

Yin, Y., Qin, K., Song, X., Zhang, Q., Zhou, Y., Xia, X., et al. (2018). BZR1 transcription factor regulates heat stress tolerance through FERONIA receptor-like kinase-mediated reactive oxygen species signaling in tomato. *Plant Cell Physiology, 59*, 2239–2254. https://doi.org/10.1093/pcp/pcy146

Yu, Q. H., Wang, B., Li, N., Tang, Y., Yang, S., Yang, T., Xu, J., et al. (2017). CRISPR/Cas9-induced targeted mutagenesis and gene replacement to generate long-shelf life tomato lines. *Scientific Reports, 7*, 11874. https://doi.org/10.1038/s41598-017-12262-1

Yu, T., Ai, G., Xie, Q., Wang, W., Song, J., Wang, J., et al. (2022). Regulation of tomato fruit elongation by transcription factor BZR1.7 through promotion of SUN gene expression. *Horticulture Research*, *9*, uhac121. https://doi.org/10.1093/hr/uhac121

Yuste-Lisbona, F. J., Fernández-Lozano, A., Pineda, B., Bretones, S., Ortiz-Atienza, A., García-Sogo, B., et al. (2020). ENO regulates tomato fruit size through the floral meristem development network. *Proceedings of the National Academy of Sciences of the USA*, *117*, 8187–8195. https://doi.org/10.1073/pnas.1913688117

Zhang, J., Lyu, H., Chen, J., Cao, X., Du, R., Ma, L., et al. (2024). Releasing a sugar brake generates sweeter tomato without yield penalty. *Nature*, *635*, 647–656. http://doi.org/10.1038/s41586-024-08186-2

Zhang, S., Wang, L., Zhao, R., Yu, W., Li, R., Li, Y., et al. (2018). Knockout of SlMAPK3 reduced disease resistance to *Botrytis cinerea* in tomato plants. *Journal of Agricultural and Food Chemistry*, *66*, 8949–8956. http://doi.org/10.1021/acs.jafc.8b02191

Zhao, P., You, Q., & Lei, M. (2019). A CRISPR/Cas9 deletion into the phosphate transporter SlPHO1;1 reveals its role in phosphate nutrition of tomato seedlings. *Physiologia Plantarum*, *167*, 556–563. https://doi.org/10.1111/ppl.12897

Zhi, J., Liu, X., Li, D., Huang, Y., Yan, S., Cao, B., et al. (2020). CRISPR/Cas9-mediated SlAN2 mutants reveal various regulatory models of anthocyanin biosynthesis in tomato plant. *Plant Cell Reports*, *39*, 799–809. https://doi.org/10.1007/s00299-020-02531-1

Zsogon, A., Cermak, T., Naves, E. R., Notini, M. M., Edel, K. H., Weinl, S., et al. (2018). De novo domestication of wild tomato using genome editing. *Nature Biotechnology*, *36*, 1211–1216. https://doi.org/10.1038/nbt.4272

8 Fruit Quality in Tomato
Current Status, Genetic Insights, and Future Breeding Opportunities

Manoj Sapkota and Yasin Topcu

8.1 INTRODUCTION

Tomato (*Solanum lycopersicum* L.) is one of the world's most widely cultivated and consumed vegetable crops. It serves as both a dietary staple and a high-value commodity. Tomatoes are cultivated across nearly all agroecological zones, consumed fresh in distinct market classes (cherry, plum, beefsteak), and widely processed into paste, sauces, and juices. Tomato production contributes substantially to local economies and international trade everywhere, while also serving as a major source of nutrients globally. This exhibits its global economic and social importance. Tomatoes are also among the most widely consumed vegetables worldwide on a per capita basis, representing a dietary source of lycopene and vitamin C in many regions, thereby linking fruit quality directly to human nutrition and health.

In addition to its commercial value, the tomato has become a model for fleshy fruit biology, offering key insights into fruit development, ripening, and the genetics of quality traits like weight, shape, and size. Its short life cycle, ease of cultivation, and transformation have made it a highly preferred model crop for research on fruit development, ripening, and quality traits. Advances in genomic resources—from the completion of the reference genome to transcriptomic, metabolomic, and pangenomic datasets—have accelerated the mapping and cloning of quantitative trait loci (QTLs) and genes underlying fruit quality traits. These tools are now central to both basic research and applied breeding, enabling integration of marker-assisted selection, genomic selection, and genome editing into breeding and improvement programs. Heirlooms, landraces, and wild relatives harbor rich genetic diversity, providing valuable alleles for improving fruit quality.

Historically, breeding efforts that prioritized yield and post-harvest traits have come at the expense of sensory quality, leading to consumer dissatisfaction with modern commercial varieties compared with heirloom tomatoes. This highlights the challenge and the opportunity of developing cultivars that simultaneously meet consumer expectations, supply chain requirements, and production constraints. In this chapter, we first briefly explore the genetic and physiological basis of fruit quality, review recent genomic advances, and highlight future opportunities to develop cultivars that combine sensory appeal, nutritional value, and resilience.

8.2 DISSECTING KEY FRUIT QUALITY TRAITS

Tomato fruit quality is shaped by physical, sensory, and nutritional traits that together define consumer perception as well as market value. These traits do not operate separately; rather, they are integrated outcomes of developmental physiology, metabolic regulation, and genetic control. They are further modulated by environmental and postharvest conditions. For breeders, understanding the biological basis of these traits is critical not only for meeting consumer expectations but also for combining them with other important traits that are highly desirable for production, distribution, and resilience goals. In this section, we examine the major fruit trait categories that define tomato

DOI: 10.1201/9781003506539-8

fruit quality and briefly highlight the physiological and molecular mechanisms underlying these traits.

The first fruit quality trait is the physical attributes. This comprises size, texture, and firmness. Fruit weight, shape, and size are among the first qualities perceived by consumers and are tightly linked to market class definitions (e.g., cherry, plum, beefsteak). Size and weight reflect cell division and expansion, controlled by a network of QTLs and regulatory genes (van der Knaap et al. 2014). Firmness and texture, meanwhile, are central to both sensory perception and postharvest performance. They are determined by cuticle structure, cell wall remodeling, and turgor maintenance. Genes involved in pectin modification, cuticle biosynthesis, and hormonal signaling contribute to the balance between juiciness and shelf-life durability (Uluisik et al. 2016).

The second quality parameter is the flavor. Flavor perception in tomato arises from the integration of soluble sugars, organic acids, and a complex of volatile organic compounds (VOCs). Glucose and fructose drive sweetness, while citric and malic acids provide sourness or acidity. The sugar-acid ratio largely defines taste intensity. A high sugar-to-acid ratio generally enhances perceived sweetness and overall flavor quality, whereas balanced acidity contributes to the characteristic tangy taste associated with fresh, vine-ripened tomatoes. Several QTLs, including those for soluble solids (°Brix), have been mapped (Xu et al. 2025; Kayikci et al. 2025; Fulton et al. 2002). However, most have small additive effects, making taste improvement challenging. Beyond taste, aroma plays an equally important role in flavor perception and is largely determined by VOCs produced through lipid, amino acid, and carotenoid metabolism (Tieman et al. 2017). Although more than 400 VOCs have been identified (Baldwin et al. 2000), only 20–30 strongly influence perception due to low odor thresholds and synergistic interactions (Tieman et al. 2017; Klee and Tieman 2013; Rambla et al. 2017). Notable VOCs include cis-3-hexenal (green, fresh aroma), β-ionone (floral), and 6-methyl-5-hepten-2-one (fruity) (Rambla et al. 2013; Klee and Tieman 2018; Distefano et al. 2022). Compounds such as cis-3-hexenal (fresh, green), phenylacetaldehyde (floral), and β-ionone (fruity) also play major roles in consumer liking despite being present in trace amounts (Tieman et al. 2012).

The next parameter is the color and pigmentation of the fruits. Color is both a primary trait for consumer acceptance and an indicator of nutritional content. Color perception plays a decisive role in shaping consumer expectations for ripeness, flavor, and nutritional value. Color often serves as a proxy for internal quality traits such as lycopene content and antioxidant potential (Stommel et al. 2005). In tomato, color development is primarily driven by carotenoid accumulation and ranges from red, pink, yellow, orange, green, white, and purple, over their entire surface or in stripes. Lycopene imparts the deep red hue, β-carotene contributes orange tones, and chlorophyll degradation during ripening reveals these pigments (Liu et al. 2015; Martí et al. 2016; Sun et al. 2018). The red coloration of most cultivated tomatoes is due to lycopene, a carotenoid with antioxidant properties.

The final key quality trait is the nutrition, that is, nutritional and health-related compounds in the fruit. Beyond immediate sensory qualities, tomato serves as a major dietary source of several nutritional metabolites like vitamins, minerals, and bioactive compounds. vitamin C (ascorbic acid) and vitamin E (tocopherols) are abundant antioxidants with broad health benefits. Polyphenols (e.g., flavonoids, phenolic acids) and carotenoids contribute to anti-inflammatory and cardioprotective effects (Giovannucci 1999).

Although these traits are discussed separately, they are deeply interconnected. Pigments influence both color and flavor. Sugars affect postharvest physiology as well as taste, and many volatiles are derived from carotenoid or lipid pathways. The challenge for breeders lies in leveraging this interrelationship through multi-omics and precision breeding approaches that can efficiently optimize all or the majority of the quality profiles in parallel. While individual traits such as firmness, flavor, pigmentation, and nutritional compounds can be dissected and improved, the real challenge lies in integrating them within cultivars that also deliver high yield, disease resistance, and resilience to environmental stress. The genetic and physiological bases of these quality traits illustrate both the opportunities and challenges of breeding for consumer satisfaction, nutritional enhancement, and supply chain performance. This complexity requires breeding strategies that go beyond considering single-trait improvement. It requires harnessing advanced genomic tools, high-throughput phenotyping, and precision approaches to balance multiple quality dimensions simultaneously.

8.3 GENETIC AND GENOMIC ADVANCES IN FRUIT QUALITY

The complexity of tomato fruit quality requires tools beyond conventional phenotypic selection. Over the past two decades, advances in molecular genetics, genomics, and functional biology have transformed tomato into a model crop for dissecting quality-related traits and applying this knowledge to breeding. These advances reveal the genes, molecular mechanisms, and biochemical pathways underlying key traits. These advances also offer practical strategies to accelerate cultivar development that combine sensory appeal, nutritional value, and agronomic performance.

8.3.1 Fruit Weight, Shape, and Size

Fruit weight (FW) has undergone dramatic changes during domestication, from the small 2–3 g fruits of the wild *S. pimpinellifolium* to modern cultivars that can exceed 1 kg per fruit. Size is determined by a combination of cell number, cell size, and locule number influenced by a complex network of genes and QTL, including *CNR/FW2.2*, *SlKLUH/FW3.2*, *CSR/FW11.3*, *SlWUS/LC* (locule number), and *SlCLV3/FAS* (fasciated), *SUN*, *OVATE*, *FS8.1*, *SlOFP20*, *GLOBE*, which influence cell division, cell expansion, and fruit morphology (Frary et al. 2000; Xiao et al. 2008; Muños et al. 2011; Chakrabarti et al. 2013; Sun et al. 2015; Xu et al. 2015; Mu et al. 2017; Wu et al. 2018; Sierra-Orozco et al. 2021; Zhu et al. 2023). Early quantitative genetic studies identified several major QTLs. Tomato was the first crop where a fruit weight QTL was cloned, *FW2.2*, encoding a membrane-associated regulator of cell division being the cloned QTL (Frary et al. 2000). Cultivated alleles of *FW2.2* are expressed at lower levels, permitting increased cell proliferation. Other major loci have been identified through advances in high-resolution mapping and genome sequencing. For example, *FW3.2* (*SlKLUH*, a cytochrome P450) increases fruit weight by increasing cell number, and a structural variant (a 50-kb tandem duplication) was found to increase the weight (Alonge et al. 2020). In contrast, *FW11.3* enlarges fruit primarily by increasing pericarp cell size (Mu et al. 2017). Modern pangenomic approaches have revealed that many size-related QTLs are underpinned by copy number variants, inversions, or promoter indels that would have been missed by single-reference mapping (Alonge et al. 2020).

Regarding fruit shape, locule number strongly influences fruit shape and size. Wild tomatoes typically have two locules, whereas beefsteak varieties may have more than eight. Three major genes—*LC*, *FAS*, and *ENO*—have been cloned and shown to affect floral meristem size via the CLAVATA–WUSCHEL pathway (Muños et al. 2011; Xu et al. 2015; Yuste-Lisbona et al. 2020). Mutations such as SNPs near *SlWUS*, a 294 kb inversion affecting *SlCLV3*, and promoter deletions in *ENO* each contribute to higher locule numbers. Domestication has also produced remarkable variation in fruit shape—from round wild types to pear-shaped, elongated, or flattened fruits in cultivated forms. Four key loci (*OVATE*, *SUN*, *FS8.1*, *GLOBE*) explain much of this variation. Mutations in *OVATE* produce pear-shaped fruits (Liu et al. 2002), while *SUN* alleles derived from retrotransposon-mediated duplications cause fruit elongation (Xiao et al. 2008). In addition to these major players, other minor regulators like *TRMs* and *OFPs* have also been studied and identified to modulate the shape of the fruit (Zhang et al. 2023). Recently, *FS8.1* and *GLOBE* have been identified as regulators of rectangular and round/flat-round shapes, respectively (Zhu et al. 2023; Sierra-Orozco et al. 2021). These discoveries illustrate how both classic mutational events and modern structural variant mapping contribute to our understanding of fruit morphology. Even minor morphological features, such as the pointed tip in cherry tomatoes, have genetic bases now elucidated by genomics. The *POINTED TIP* (*PT*) gene encodes a zinc-finger protein that suppresses tip formation via regulation of *FUL2* and auxin transport. Loss-of-function alleles, selected during domestication, increased the prevalence of blossom-end tips in cultivated types (Song et al. 2022).

8.3.2 Fruit Texture

Texture is a crucial determinant of tomato quality, affecting consumer perception, shelf life, and postharvest handling. Genetic and genomic approaches have revealed the molecular basis of fruit softening and cuticle development, enabling precise manipulation of texture traits. Softening is largely mediated by cell wall–modifying enzymes, including polygalacturonase (PG), pectin lyase (PL), pectin methylesterase (PME), tomato β-galactosidase (TBG), and xyloglucan endotransglycosylase (XET). Functional genomics has shown that simultaneous knockout of *SLEXP1* (*expansin*) and *SLCEL2* (*endo-β-1,4-glucanase*) increases firmness without compromising quality, while CRISPR/Cas9-mediated *PL* disruption similarly enhances fruit firmness (Brummell et al. 1999; Su et al. 2024; Uluisik et al. 2016). Ethylene signaling genes such as *EIN4* also modulate firmness, providing targets for breeding programs for developing varieties with extended shelf life (Zhang et al. 2024b). The cuticle, composed of cutin and wax, influences both texture and resistance to water loss and pathogens. Genes controlling cuticle biosynthesis, including *CD1/GDSL1*, *CD2*, *CD3*, and *GDSL2*, determine cuticle thickness, composition, and surface properties. Allelic variants such as the *Sticky peel* (*pe*) mutation reduce peel-flesh separation, improving sensory texture while maintaining durability (Yeats et al. 2012; Nadakuduti et al. 2012; Shi et al. 2013; Petit et al. 2014). Advances such as the identification of *qFIS1*, a QTL enhancing firmness without compromising taste, highlight how targeted genetic interventions can mitigate trade-offs. *qFIS1* (*GA2-oxidase*) integrates hormonal regulation and cuticle biosynthesis, controlling bioactive gibberellins, cutin/wax accumulation, firmness, and shelf life (Li et al. 2020). By combining QTL mapping with genome editing and functional genomics, breeders can optimize firmness and texture, balancing postharvest resilience with consumer-preferred mouthfeel.

8.3.3 Fruit Color

Fruit color, a key visual and nutritional quality trait, emerges from tightly coordinated ripening processes involving pigment accumulation, cell wall remodeling, and metabolite fluxes. Tomato fruits can display a wide range of colors, including red, pink, yellow, orange, green, white, and purple, and in striped pattern as well. In tomato, carotenoids, chlorophylls, and flavonoids define pericarp and peel pigmentation, with lycopene accounting for up to 90% of carotenoids in red fruit and β-carotene contributing an additional 5–10% (Ronen et al. 1999). Genetic studies have identified key biosynthetic and regulatory loci, including *PSY1*, *CRTISO*, *BETA*, *DELTA*, and transcription factors such as *RIN*, *NOR*, and *GLK2*, which control pigment accumulation, chloroplast-to-chromoplast transitions, and spatial color uniformity (Fray and Grierson 1993; Isaacson et al. 2002; Ronen et al. 1999; Ronen et al. 2000; Vrebalov et al. 2002; Ng and Tigchelaar 1977; Nguyen et al. 2014). Engineering transcription factors like *AtMYB12* has successfully boosted flavonoid and anthocyanin accumulation, yielding high-pigment lines with dual nutritional and aesthetic benefits (Zhang et al. 2015; Meng et al. 2020). Modern genome editing has enabled combinatorial manipulation of carotenoid, flavonoid, and chlorophyll pathways to generate novel fruit colors, from yellow and orange to purple anthocyanin-enriched fruits (Song et al. 2022). Mutations such as *old-gold (ogc)* and introgressions from wild relatives have also been used to enhance carotenoid accumulation (Ronen et al. 2000). Beyond aesthetics, these pigments influence nutritional quality, linking color selection to health-promoting traits. Color-related traits are generally less prone to negative trade-offs with yield and, thus, making them attractive targets for simultaneous nutritional and sensory improvement.

Although anthocyanin accumulation is rare in cultivated tomato, introgressions from wild species or transgenic overexpression can induce purple pigmentation. The *AFT* (*SlAN2-like*) and *ATV* (*SlMYBATV*) loci act together to induce anthocyanin biosynthesis, with the *bHLH* factor *SlAN1* partnering with *SlAN2-like* to activate pathway genes (Sun et al. 2020). Transgenic approaches using *Ros1* and *Del* from *Antirrhinum majus* have successfully produced high-anthocyanin purple tomatoes (Butelli et al. 2008), some of which have achieved regulatory approval and commercial release, expanding the sensory and nutritional diversity of tomato fruit.

8.3.4 Fruit Flavor Quality

Tomato fruit flavor quality is determined by a balance of sweetness, acidity, and aroma, driven by soluble sugars, organic acids, amino acids, and a complex blend of volatile compounds (Pereira et al. 2021). While heirloom tomatoes often exhibit intense, rich flavors, many modern commercial cultivars suffer from diminished taste. This is largely due to linkage drag during introgression of disease resistance loci, breeding for extended shelf life, and/or other traits preferred by growers like yield.

Sweetness in tomato is primarily determined by soluble sugars, mainly glucose and fructose in cultivated varieties, and sucrose in some wild species (*S. peruvianum*, *S. chmielewskii*, *S. habrochaites*). Genetic studies have identified key loci regulating sugar accumulation, including *LIN5* and *SUCR*, which encode invertases responsible for sucrose hydrolysis. Sugar transporters such as *STP1* and *FGR* modulate the distribution of sugars from source tissues to fruits (Fridman et al. 2004; Shammai et al. 2018; Wang et al. 2023). Transcription factors such as *ZAT10-LIKE* and *SlbZIP1* further regulate sugar metabolism, with allelic variants influencing hexose content and perceived sweetness (Sagor et al. 2016; Nguyen et al. 2023). Additionally, *SlCDPK26* and *SlCDPK27* negatively regulate sugar accumulation. Targeted knockouts of these genes increase glucose and fructose without sacrificing fruit weight or yield (Zhang et al. 2024a). Acidity is largely governed by malic and citric acids. Genes such as *TFM6* (*ALMT9*) control malate accumulation, while genome-wide association studies have identified multiple candidate genes regulating citric acid content (Ye et al. 2017; Gai et al. 2023; Gai et al. 2024). Manipulating these loci offers the potential to fine-tune fruit acidity for improved flavor balance.

Tomato aroma arises from volatile compounds derived from lipid, carotenoid, and amino acid pathways. Some key genes among several that have been identified include *TomLoxC* and *HPL* for lipid-derived volatiles, *SlAAT1* for ester synthesis, and *FLORAL4*, which regulates phenylalanine-derived aroma compounds (Goulet et al. 2012; Tieman et al. 2017; Distefano et al. 2022). Apocarotenoid volatiles such as α-ionone are produced by *SlCCD1A* and *SlCCD1B*, while *SlYTH2* influences their biosynthesis (Ilg et al. 2014; Bian et al. 2024). Negative flavor contributors such as guaiacol and methyl salicylate are controlled by genes including *NSGT1* and *Methylesterases*, which have been targets for flavor improvement as well (Tikunov et al. 2013; Frick et al. 2023; Sapkota et al. 2023).

8.3.5 Fruit Nutritional Quality

Tomato nutritional quality is shaped by the accumulation of bioactive compounds such as carotenoids, vitamins, and amino acid derivatives, which confer health benefits and contribute to fruit appeal. Genomic studies have elucidated the genes and regulatory networks controlling these metabolites, enabling targeted breeding strategies. Carotenoids are the dominant pigments in tomato, with lycopene accounting for 80–90% of total carotenoid content. Mutations in carotenoid biosynthesis genes, such as the *Beta* gene in the *og* and *ogc* mutants, increase lycopene accumulation and deepen fruit color (Ronen et al. 1999; Ronen et al. 2000). Modern breeding approaches like gene editing now enable precise introduction of these alleles into fresh-market backgrounds without linkage drag.

Vitamin C accumulation is positively regulated by transcription factors including *SlHZ24* and *bHLH59*, which control ascorbate biosynthesis genes (*SlGMP3*, *GME*, *GGP*), and by the zinc-finger protein *SlZF3*, which modulates light-mediated regulation (Hu et al. 2016; Li et al. 2018; Ye et al.

2019). Similarly, manipulation of vitamin D3 precursors (*7-DR2*) and vitamin E loci (*VTE3*) has opened avenues for biofortified tomatoes with enhanced micronutrient content (Quadrana et al. 2014; Li et al. 2022). For example, RNAi suppression of *APX* genes (ascorbate peroxidases) increased vitamin C accumulation (Zhang et al. 2011), while manipulation of carotenoid biosynthesis pathways has produced tomatoes with enhanced lycopene or β-carotene content (Ronen et al. 1999; Ronen et al. 2000). Tomatine, a bitter glycoalkaloid, is largely reduced in ripe fruit due to domestication and selective breeding. Genes such as *GORKY*, *GAME9*, and *GAME36* regulate tomatine biosynthesis and degradation, and regulatory elements like the *GAME Enhancer 1* (*GE1*) orchestrate the expression of gene clusters controlling this metabolite (Cárdenas et al. 2016; Sonawane et al. 2023; Bai et al. 2024). These insights allow breeders to minimize anti-nutritional factors while maintaining plant defense. Through the integration of genomics, gene editing, and metabolic pathway analysis, tomato breeding now has unprecedented capacity to enhance the nutritional profile of fruits while preserving other quality attributes.

Genetic and genomic advances have transformed tomato breeding from a largely phenotype-driven process to one that can exploit precise molecular knowledge of quality traits. While technical and logistical challenges of integrating multi-trait targets and complex gene–environment interactions remain, the convergence of QTL/GWAS discoveries, pangenomics, genomic selection, and genome editing provides opportunities to develop cultivars that align with both consumer preferences and agronomic demands.

8.4 OPPORTUNITIES AND CHALLENGES IN IMPROVING TOMATO FRUIT QUALITY

Tomato fruit quality is a long-standing objective. It is also a rapidly evolving frontier in crop science. Despite significant advances in breeding, consumer surveys have highlighted dissatisfaction with the flavor and sensory attributes of modern commercial cultivars compared with traditional or heirloom types (Bruhn et al. 1991; Klee and Tieman 2018; Tieman et al. 2017). At the same time, new genomic tools, expanded germplasm resources, and high-throughput phenotyping are providing opportunities to dissect and enhance the complex fruit quality traits. This persistent gap between consumer expectations and current commercial varieties, and the emerging technological capacity to address it, shows both the opportunities and the challenges for breeders working to develop tomatoes that meet the needs of consumers, growers, and supply chains.

One of the most persistent of these challenges is the trade-off between shelf life and flavor, where genetic strategies that enhance longer shelf life can compromise sensory quality. Consumers demand tomatoes with rich flavor, while growers and distributors require high yield and fruits that remain firm during storage, handling, and transport. Historically, breeding programs prioritized yield and shelf life, often at the expense of flavor. Comparative metabolomic studies have shown that modern commercial cultivars contain significantly lower levels of key flavor-associated volatiles and soluble solids than heirloom varieties, reflecting the unintended sensory costs of these strategies (Tieman et al. 2017; Sapkota et al. 2023). A major contributor to this trade-off has been the use of ripening-associated mutations such as *rin* (ripening-inhibitor), *nor* (non-ripening), and *alc* (alcobaca). These alleles extend postharvest life by delaying softening but also suppress ethylene-mediated ripening pathways. This then leads to reduced volatile synthesis, less sugar accumulation, and diminished flavor intensity (Klee and Tieman 2018). However, recent advances suggest that flavor improvements need not come at the expense of firmness or yield. Targeted knockout of *SlCDPK27* and *SlCDPK26*, negative regulators of sucrose synthase, resulted in a 30% increase in glucose and fructose without reducing fruit weight or yield (Zhang et al. 2024a). Similarly, overexpression of *SlCYP90B3*, a brassinosteroid biosynthetic enzyme, enhanced favorable carotenoid-derived aroma volatiles while maintaining desirable agronomic performance (Hu et al. 2020). Another example is the firmness QTL *qFIS1*, encoding a GA2-oxidase, which improves cuticle development and shelf life without compromising taste, color, or ripening (Li et al. 2020). These findings highlight that targeted allele introgression and molecular breeding can provide a blueprint for solving the long-standing quality trade-offs.

Genetic complexity and environment pose a second challenge. Tomato fruit quality traits are genetically complex and strongly influenced by the environment. This complicates efforts to achieve stable improvement across growing regions. Furthermore, these traits are highly polygenic, controlled by multiple QTLs of small to moderate effect, often showing epistasis and strong genotype-by-environment interactions.

Introgression from wild relatives or traditional germplasm has been a powerful tool for incorporating resistance alleles and expanding trait diversity. However, such efforts may introduce pleiotropic effects or linkage drag that negatively impact fruit quality or yield. This illustrates the complexity of balancing resistance and quality but also demonstrates that careful allele selection can minimize trade-offs.

Breeding priorities versus sensory quality is a third hurdle. A recurring challenge in tomato improvement is the disconnect between breeding priorities and consumer expectations. Breeders and growers emphasize yield, uniformity, and shelf life to ensure profitability and supply chain efficiency. Yet consumers often value flavor, aroma, and freshness above all else. This misalignment has contributed to the perception that modern tomatoes lack the taste of heirloom varieties (Klee and Tieman 2013; Tieman et al. 2017). Part of the difficulty lies in the multidimensional nature of flavor as this integrates sugars, acids, and a diverse array of VOCs. Unlike easily measurable traits such as yield or firmness, flavor is challenging to quantify in high-throughput, standardized ways. It requires technical and biochemical knowledge to run and interpret the metabolite data quantification from expensive equipment. This is not readily available with the breeders. Traditional and trained sensory panels remain valuable but are costly and prone to variability in panelist perception (Sinesio et al. 2021). Participatory breeding approaches that integrate consumer panels with molecular profiling have successfully identified alleles linked to enhanced flavor (Tieman et al. 2017). In parallel, analytical advances such as gas chromatography–olfactometry (GC–O) coupled with metabolomic fingerprinting allow precise identification of odor-active compounds and their sensory thresholds. This enables breeders to select for more targeted volatile profiles. By incorporating these chemical–sensory datasets into breeding pipelines, it is increasingly possible to combine sensory improvements with stable yields and postharvest performance. Such integrative approaches provide a pathway toward cultivars that satisfy consumer taste expectations while meeting production and distribution needs.

Quality improvement means balancing what consumers taste with what supply chains require. Trade-offs between shelf life and flavor, the polygenic and environmentally sensitive nature of quality traits, and the historical disconnect between breeding priorities and sensory attributes have shaped modern tomato breeding outcomes. Yet targeted allele introgression, molecular breeding, and integrated sensory omics are beginning to resolve long-standing conflicts.

8.5 TOWARD NEXT-GENERATION BREEDING FOR QUALITY

The main aim of modern tomato breeding is to go beyond incremental improvements in yield or disease resistance and to achieve holistic enhancement of fruit quality, including flavor, nutritional content, texture, and appearance. This progress depends on the following four practical moves:

- **Integrating multi-omics data for trait improvement:** Advances in genomics, transcriptomics, metabolomics, and proteomics have enabled a systems-level understanding of the molecular networks underlying fruit quality. By linking metabolite profiles with gene expression and genetic variation, breeders can identify key regulatory hubs that coordinate flavor, nutrition, and ripening traits.

- **Breeding strategies balancing quality with yield and stress tolerance:** While consumer-preferred quality traits are essential, they must be integrated without compromising agronomic performance. Genomic selection, marker-assisted breeding, and allele pyramiding enable simultaneous selection for flavor, nutritional content, fruit firmness, and resilience to biotic and abiotic stress. Incorporating alleles from wild relatives or landraces by carefully managing the linkage drags offers opportunities to enhance stress tolerance while preserving high-quality sensory traits.
- **Precision breeding and CRISPR/Cas applications:** Genome editing has revolutionized crop breeding, including tomato breeding. This allows direct modification of key genes controlling sugar accumulation, carotenoid content, and volatile profiles. CRISPR/Cas-mediated knockouts or promoter edits can increase sweetness, enhance lycopene accumulation, or restore lost flavor alleles, often without affecting yield or plant vigor. Multiplex editing also facilitates simultaneous manipulation of multiple pathways, offering better control over complex traits.
- **Participatory and consumer-driven breeding models:** Breeding for quality increasingly relies on integrating consumer preferences and sensory evaluation into selection schemes. Participatory breeding approaches engage growers, processors, and consumers in evaluating flavor, texture, and appearance. This will ensure that genomic improvements align with market demands. Combining phenotypic assessment with molecular markers can create a feedback loop that accelerates the development of varieties with both high quality and broad adoption potential.

8.6 FUTURE PERSPECTIVES AND RESEARCH PRIORITIES

The next frontier in tomato quality breeding, we believe, is defined by integration, innovation, and scalability:

- **High-throughput phenotyping for flavor and nutrition:** Advances in imaging, spectroscopy, and automated chemical assays enable rapid, non-destructive quantification of sugar, acid, pigment, and volatile profiles both preharvest and postharvest. Coupled with genomics, these technologies can allow breeders to assess hundreds or thousands of lines efficiently, accelerating selection for superior quality.
- **Leveraging AI/ML for trait prediction and selection:** In the growing era of machine learning and AI, machine learning models trained on multi-omics datasets, combined with real-time supply-chain sensor data (temperature, humidity, ethylene, storage duration), can predict outcomes for flavor, texture, and shelf stability reliably and quickly. Such integrative models can uncover hidden correlations among traits, guide multi-trait selection, and prioritize elite genotypes with higher precision than conventional classical breeding approaches.
- **Expanding pangenomic resources to capture diversity:** Reference genomes alone are insufficient to capture the full spectrum of allelic variation underlying fruit quality. Thus, the use of comprehensive pangenomes, including wild species and landraces, provides a rich catalog of structural variants, regulatory alleles, and rare genes. This can easily be leveraged to improve flavor, nutrition, and stress resilience.
- **Climate resilience and fruit quality interactions:** Environmental factors such as temperature, water availability, and soil salinity highly affect fruit composition, including sugar–acid balance, carotenoids, and volatile production. Future research must elucidate genotype-by-environment interactions to develop varieties that maintain quality under variable climates, ensuring both consumer satisfaction and crop stability.
- **Need for integrated, interdisciplinary breeding pipelines:** The next era of tomato varieties requires coordinated efforts across genomics, plant physiology, biochemistry,

computational biology, sensory science, and food and nutrition science. Integrated pipelines that combine all these multi-omics and interdisciplinary science insights, precision editing, and participatory evaluation will be critical to delivering high-quality, climate-resilient, and consumer-preferred fruits.
- **Postharvest and supply-chain quality modeling:** Fruit quality is not determined solely in the field but continues to evolve throughout storage, transport, and retail. Storage time, temperature, ethylene, and humidity profiles influence sugar retention, pigment stability, firmness, and volatile loss. Predictions derived from sensor-enabled logistics and packaging trials can guide improvements in both breeding targets (e.g., cuticle traits, volatile-retention alleles) and supply-chain management strategies.

The next era of breeding tomato varieties will emerge from approaches that bridge genes, environments, and supply chains. This will be done by embedding advanced phenotyping, machine learning, pangenomic diversity, and interdisciplinary collaboration into breeding pipelines, making it possible to deliver tomatoes that consistently meet consumer expectations for flavor and nutrition while withstanding the challenges of modern agriculture and climate change.

8.7 CONCLUSION

Advances in genetic and genomic tools have transformed our understanding of the molecular and genetic basis of tomato fruit quality. From fruit weight, shape, size, texture, and color to flavor and nutritional content, key traits are now increasingly dissected at the gene, pathway, and regulatory network levels. Multi-omics integration, genome editing, and AI-driven predictive models are enabling breeders to design next-generation varieties that balance consumer preferences, agronomic performance, and resilience. By leveraging pangenomic diversity, high-throughput phenotyping, and participatory selection, the future of tomato breeding is sure to deliver fruits that meet both the sensory expectations of consumers and the practical demands of production systems, setting a model for quality-focused breeding in other horticultural crops.

REFERENCES

Alonge M, Wang X, Benoit M, Soyk S, Pereira L, Zhang L, Suresh H, Ramakrishnan S, Maumus F, Ciren D (2020) Major impacts of widespread structural variation on gene expression and crop improvement in tomato. *Cell* 182 (1):145–161.e123

Bai F, Shu P, Deng H, Wu Y, Chen Y, Wu M, Ma T, Zhang Y, Pirrello J, Li Z (2024) A distal enhancer guides the negative selection of toxic glycoalkaloids during tomato domestication. *Nature Communications* 15 (1):2894

Baldwin EA, Scott JW, Shewmaker CK, Schuch W (2000) Flavor trivia and tomato aroma: Biochemistry and possible mechanisms for control of important aroma components. *HortScience* 35 (6):1013–1022

Bian H, Song P, Gao Y, Deng Z, Huang C, Yu L, Wang H, Ye B, Cai Z, Pan Y (2024) The m6A reader SlYTH2 negatively regulates tomato fruit aroma by impeding the translation process. *Proceedings of the National Academy of Sciences* 121 (28):e2405100121

Bruhn CM, Feldman N, Garlitz C, Harwood J, Ivans E, Marshall M, Riley A, Thurber D, Williamson E (1991) Consumer perceptions of quality: Apricots, cantaloupes, peaches, pears, strawberries, and tomatoes. *Journal of Food Quality* 14 (3):187–195

Brummell DA, Harpster MH, Civello PM, Palys JM, Bennett AB, Dunsmuir P (1999) Modification of expansin protein abundance in tomato fruit alters softening and cell wall polymer metabolism during ripening. *The Plant Cell* 11 (11):2203–2216

Butelli E, Titta L, Giorgio M, Mock H-P, Matros A, Peterek S, Schijlen EGWM, Hall RD, Bovy AG, Luo J, Martin C (2008) Enrichment of tomato fruit with health-promoting anthocyanins by expression of select transcription factors. *Nature Biotechnology* 26 (11):1301–1308. https://doi.org/10.1038/nbt.1506

Cárdenas PD, Sonawane PD, Pollier J, Vanden Bossche R, Dewangan V, Weithorn E, Tal L, Meir S, Rogachev I, Malitsky S (2016) GAME9 regulates the biosynthesis of steroidal alkaloids and upstream isoprenoids in the plant mevalonate pathway. *Nature Communications* 7 (1):10654

Chakrabarti M, Zhang N, Sauvage C, Muños S, Blanca J, Cañizares J, Diez MJ, Schneider R, Mazourek M, McClead J (2013) A cytochrome P450 regulates a domestication trait in cultivated tomato. *Proceedings of the National Academy of Sciences* 110 (42):17125–17130

Distefano M, Mauro RP, Page D, Giuffrida F, Bertin N, Leonardi C (2022) Aroma volatiles in tomato fruits: The role of genetic, preharvest and postharvest factors. *Agronomy* 12 (2):376

Frary A, Nesbitt TC, Frary A, Grandillo S, Knaap Evd, Cong B, Liu J, Meller J, Elber R, Alpert KB (2000) fw2. 2: A quantitative trait locus key to the evolution of tomato fruit size. *Science* 289 (5476):85–88

Fray RG, Grierson D (1993) Identification and genetic analysis of normal and mutant phytoene synthase genes of tomato by sequencing, complementation and co-suppression. *Plant Molecular Biology* 22 (4):589–602

Frick EM, Sapkota M, Pereira L, Wang Y, Hermanns A, Giovannoni JJ, Van Der Knaap E, Tieman DM, Klee HJ (2023) A family of methyl esterases converts methyl salicylate to salicylic acid in ripening tomato fruit. *Plant Physiology* 191 (1):110–124

Fridman E, Carrari F, Liu Y-S, Fernie AR, Zamir D (2004) Zooming in on a quantitative trait for tomato yield using interspecific introgressions. *Science* 305 (5691):1786–1789

Fulton TM, Bucheli P, Voirol E, Lopez J, Petiard V, Tanksley SD (2002) Quantitative trait loci (QTL) affecting sugars, organic acids and other biochemical properties possibly contributing to flavor, identified in four advanced backcross populations of tomato. *Euphytica* 127 (2):163–177

Gai W, Yang F, Yuan L, Ul Haq S, Wang Y, Wang Y, Shang L, Li F, Ge P, Dong H (2023) Multiple-model GWAS identifies optimal allelic combinations of quantitative trait loci for malic acid in tomato. *Horticulture Research* 10 (4):uhad021

Gai W, Yuan L, Yang F, Ahiakpa JK, Li F, Ge P, Zhang X, Tao J, Wang F, Yang Y (2024) Genome-wide variants and optimal allelic combinations for citric acid in tomato. *Horticulture Research* 11 (5):uhae070

Giovannucci E (1999) Tomatoes, tomato-based products, lycopene, and cancer: Review of the epidemiologic literature. *Journal of the National Cancer Institute* 91 (4):317–331

Goulet C, Mageroy MH, Lam NB, Floystad A, Tieman DM, Klee HJ (2012) Role of an esterase in flavor volatile variation within the tomato clade. *Proceedings of the National Academy of Sciences* 109 (46):19009–19014

Hu S, Liu L, Li S, Shao Z, Meng F, Liu H, Duan W, Liang D, Zhu C, Xu T (2020) Regulation of fruit ripening by the brassinosteroid biosynthetic gene SlCYP90B3 via an ethylene-dependent pathway in tomato. *Horticulture Research* 7

Hu T, Ye J, Tao P, Li H, Zhang J, Zhang Y, Ye Z (2016) The tomato HD-Zip I transcription factor Sl HZ 24 modulates ascorbate accumulation through positive regulation of the d-mannose/l-galactose pathway. *The Plant Journal* 85 (1):16–29

Ilg A, Bruno M, Beyer P, Al-Babili S (2014) Tomato carotenoid cleavage dioxygenases 1A and 1B: Relaxed double bond specificity leads to a plenitude of dialdehydes, mono-apocarotenoids and isoprenoid volatiles. *FEBS Open Bio* 4:584–593

Isaacson T, Ronen G, Zamir D, Hirschberg J (2002) Cloning of tangerine from tomato reveals a carotenoid isomerase essential for the production of β-carotene and xanthophylls in plants. *The Plant Cell* 14 (2):333–342

Kayikci HC, Aydin S, Adak A, Dogan A, Sapkota M, Feng Q, Topcu Y (2025) Association mapping of tomato fruit quality for weight, firmness, brix, and color using GWAS. *BMC Plant Biology*. December. https://doi.org/10.1186/ s12870-025-07838-3

Klee HJ, Tieman DM (2013) Genetic challenges of flavor improvement in tomato. *Trends in Genetics* 29 (4):257–262

Klee HJ, Tieman DM (2018) The genetics of fruit flavour preferences. *Nature Reviews Genetics* 19 (6):347–356

Li J, Scarano A, Gonzalez NM, D'Orso F, Yue Y, Nemeth K, Saalbach G, Hill L, de Oliveira Martins C, Moran R (2022) Biofortified tomatoes provide a new route to vitamin D sufficiency. *Nature Plants* 8 (6):611–616

Li R, Sun S, Wang H, Wang K, Yu H, Zhou Z, Xin P, Chu J, Zhao T, Wang H (2020) FIS1 encodes a GA2-oxidase that regulates fruit firmness in tomato. *Nature Communications* 11 (1):5844

Li Y, Chu Z, Luo J, Zhou Y, Cai Y, Lu Y, Xia J, Kuang H, Ye Z, Ouyang B (2018) The C2H2 zinc-finger protein Sl ZF 3 regulates AsA synthesis and salt tolerance by interacting with CSN 5B. *Plant Biotechnology Journal* 16 (6):1201–1213

Liu J, Van Eck J, Cong B, Tanksley SD (2002) A new class of regulatory genes underlying the cause of pear-shaped tomato fruit. *Proceedings of the National Academy of Sciences* 99 (20):13302–13306

Liu L, Shao Z, Zhang M, Wang Q (2015) Regulation of carotenoid metabolism in tomato. *Molecular Plant* 8 (1):28–39

Martí R, Roselló S, Cebolla-Cornejo J (2016) Tomato as a source of carotenoids and polyphenols targeted to cancer prevention. *Cancers* 8 (6):58

Meng X, Zhao X, Ding X, Li Y, Cao G, Chu Z, Su X, Liu Y, Chen X, Guo J (2020) Integrated functional omics analysis of flavonoid-related metabolism in AtMYB12 transcript factor overexpressed tomato. *Journal of Agricultural and Food Chemistry* 68 (24):6776–6787

Mu Q, Huang Z, Chakrabarti M, Illa-Berenguer E, Liu X, Wang Y, Ramos A, van der Knaap E (2017) Fruit weight is controlled by Cell Size Regulator encoding a novel protein that is expressed in maturing tomato fruits. *PLoS Genetics* 13 (8):e1006930

Muños S, Ranc N, Botton E, Bérard A, Rolland S, Duffé P, Carretero Y, Le Paslier M-C, Delalande C, Bouzayen M (2011) Increase in tomato locule number is controlled by two single-nucleotide polymorphisms located near WUSCHEL. *Plant Physiology* 156 (4):2244–2254

Nadakuduti SS, Pollard M, Kosma DK, Allen Jr C, Ohlrogge JB, Barry CS (2012) Pleiotropic phenotypes of the sticky peel mutant provide new insight into the role of CUTIN DEFICIENT2 in epidermal cell function in tomato. *Plant Physiology* 159 (3):945–960

Ng TJ, Tigchelaar E (1977) Action of the non-ripening (nor) mutant on fruit ripening of tomato. *Journal of the American Society for Horticultural Science* 102 (4):504–509

Nguyen CV, Vrebalov JT, Gapper NE, Zheng Y, Zhong S, Fei Z, Giovannoni JJ (2014) Tomato GOLDEN2-LIKE transcription factors reveal molecular gradients that function during fruit development and ripening. *The Plant Cell* 26 (2):585–601

Nguyen NH, Bui TP, Le NT, Nguyen CX, Le MTT, Dao NT, Phan Q, Van Le T, To HMT, Pham NB (2023) Disrupting Sc-uORFs of a transcription factor bZIP1 using CRISPR/Cas9 enhances sugar and amino acid contents in tomato (*Solanum lycopersicum*). *Planta* 257 (3):57

Pereira L, Sapkota M, Alonge M, Zheng Y, Zhang Y, Razifard H, Taitano NK, Schatz MC, Fernie AR, Wang Y (2021) Natural genetic diversity in tomato flavor genes. *Frontiers in Plant Science* 12:642828

Petit J, Bres C, Just D, Garcia V, Mauxion J-P, Marion D, Bakan B, Joubès J, Domergue F, Rothan C (2014) Analyses of tomato fruit brightness mutants uncover both cutin-deficient and cutin-abundant mutants and a new hypomorphic allele of GDSL lipase. *Plant Physiology* 164 (2):888–906

Quadrana L, Almeida J, Asís R, Duffy T, Dominguez PG, Bermúdez L, Conti G, Corrêa da Silva JV, Peralta IE, Colot V (2014) Natural occurring epialleles determine vitamin E accumulation in tomato fruits. *Nature Communications* 5 (1):4027

Rambla JL, Medina A, Fernández-del-Carmen A, Barrantes W, Grandillo S, Cammareri M, López-Casado G, Rodrigo G, Alonso A, García-Martínez S (2017) Identification, introgression, and validation of fruit volatile QTLs from a red-fruited wild tomato species. *Journal of Experimental Botany* 68 (3):429–442

Rambla JL, Tikunov YM, Monforte AJ, Bovy AG, Granell A (2013) The expanded tomato fruit volatile landscape. *Journal of Experimental Botany* 65 (16):4613–4623

Ronen G, Carmel-Goren L, Zamir D, Hirschberg J (2000) An alternative pathway to β-carotene formation in plant chromoplasts discovered by map-based cloning of Beta and old-gold color mutations in tomato. *Proceedings of the National Academy of Sciences* 97 (20):11102–11107

Ronen G, Cohen M, Zamir D, Hirschberg J (1999) Regulation of carotenoid biosynthesis during tomato fruit development: Expression of the gene for lycopene epsilon-cyclase is down-regulated during ripening and is elevated in the mutant Delta. *The Plant Journal* 17 (4):341–351

Sagor G, Berberich T, Tanaka S, Nishiyama M, Kanayama Y, Kojima S, Muramoto K, Kusano T (2016) A novel strategy to produce sweeter tomato fruits with high sugar contents by fruit-specific expression of a single bZIP transcription factor gene. *Plant Biotechnology Journal* 14 (4):1116–1126

Sapkota M, Pereira L, Wang Y, Zhang L, Topcu Y, Tieman D, van der Knaap E (2023) Structural variation underlies functional diversity at methyl salicylate loci in tomato. *PLoS Genetics* 19 (5):e1010751

Shammai A, Petreikov M, Yeselson Y, Faigenboim A, Moy-Komemi M, Cohen S, Cohen D, Besaulov E, Efrati A, Houminer N (2018) Natural genetic variation for expression of a SWEET transporter among wild species of Solanum lycopersicum (tomato) determines the hexose composition of ripening tomato fruit. *The Plant Journal* 96 (2):343–357

Shi JX, Adato A, Alkan N, He Y, Lashbrooke J, Matas AJ, Meir S, Malitsky S, Isaacson T, Prusky D (2013) The tomato S l SHINE 3 transcription factor regulates fruit cuticle formation and epidermal patterning. *New Phytologist* 197 (2):468–480

Sierra-Orozco E, Shekasteband R, Illa-Berenguer E, Snouffer A, van der Knaap E, Lee TG, Hutton SF (2021) Identification and characterization of GLOBE, a major gene controlling fruit shape and impacting fruit size and marketability in tomato. *Horticulture Research* 8

Sinesio F, Cammareri M, Cottet V, Fontanet L, Jost M, Moneta E, Palombieri S, Peparaio M, Romero del Castillo R, Saggia Civitelli E (2021) Sensory traits and consumer's perceived quality of traditional and modern fresh market tomato varieties: A study in three European countries. *Foods* 10 (11):2521

Sonawane PD, Gharat SA, Jozwiak A, Barbole R, Heinicke S, Almekias-Siegl E, Meir S, Rogachev I, Connor SEO, Giri AP (2023) A BAHD-type acyltransferase concludes the biosynthetic pathway of non-bitter glycoalkaloids in ripe tomato fruit. *Nature Communications* 14 (1):4540

Song J, Shang L, Li C, Wang W, Wang X, Zhang C, Ai G, Ye J, Yang C, Li H (2022) Variation in the fruit development gene POINTED TIP regulates protuberance of tomato fruit tip. *Nature Communications* 13 (1):5940

Stommel J, Abbott JA, Saftner RA, Camp MJ (2005) Sensory and objective quality attributes of beta-carotene and lycopene-rich tomato fruit. *Journal of the American Society for Horticultural Science* 130 (2):244–251

Su G, Lin Y, Wang C, Lu J, Liu Z, He Z, Shu X, Chen W, Wu R, Li B (2024) Expansin SlExp1 and endoglucanase SlCel2 synergistically promote fruit softening and cell wall disassembly in tomato. *The Plant Cell* 36 (3):709–726

Sun C, Deng L, Du M, Zhao J, Chen Q, Huang T, Jiang H, Li C-B, Li C (2020) A transcriptional network promotes anthocyanin biosynthesis in tomato flesh. *Molecular Plant* 13 (1):42–58

Sun L, Rodriguez GR, Clevenger JP, Illa-Berenguer E, Lin J, Blakeslee JJ, Liu W, Fei Z, Wijeratne A, Meulia T (2015) Candidate gene selection and detailed morphological evaluations of fs8. 1, a quantitative trait locus controlling tomato fruit shape. *Journal of Experimental Botany* 66 (20):6471–6482

Sun T, Yuan H, Cao H, Yazdani M, Tadmor Y, Li L (2018) Carotenoid metabolism in plants: The role of plastids. *Molecular Plant* 11 (1):58–74

Tieman D, Bliss P, McIntyre LM, Blandon-Ubeda A, Bies D, Odabasi AZ, Rodríguez GR, van der Knaap E, Taylor MG, Goulet C (2012) The chemical interactions underlying tomato flavor preferences. *Current Biology* 22 (11):1035–1039

Tieman D, Zhu G, Resende Jr MF, Lin T, Nguyen C, Bies D, Rambla JL, Beltran KSO, Taylor M, Zhang B (2017) A chemical genetic roadmap to improved tomato flavor. *Science* 355 (6323):391–394

Tikunov YM, Molthoff J, de Vos RC, Beekwilder J, van Houwelingen A, van der Hooft JJ, Nijenhuis-de Vries M, Labrie CW, Verkerke W, van de Geest H (2013) Non-smoky glycosyltransferase1 prevents the release of smoky aroma from tomato fruit. *The Plant Cell* 25 (8):3067–3078

Uluisik S, Chapman NH, Smith R, Poole M, Adams G, Gillis RB, Besong TM, Sheldon J, Stiegelmeyer S, Perez L (2016) Genetic improvement of tomato by targeted control of fruit softening. *Nature Biotechnology* 34 (9):950–952

van der Knaap E, Chakrabarti M, Chu YH, Clevenger JP, Illa-Berenguer E, Huang Z, Keyhaninejad N, Mu Q, Sun L, Wang Y, Wu S (2014) What lies beyond the eye: the molecular mechanisms regulating tomato fruit weight and shape. *Frontiers in Plant Science* 5 (227): 1–13

Vrebalov J, Ruezinsky D, Padmanabhan V, White R, Medrano D, Drake R, Schuch W, Giovannoni J (2002) A MADS-box gene necessary for fruit ripening at the tomato ripening-inhibitor (rin) locus. *Science* 296 (5566):343–346

Wang Y, Shi C, Ge P, Li F, Zhu L, Wang Y, Tao J, Zhang X, Dong H, Gai W (2023) A 21-bp InDel in the promoter of STP1 selected during tomato improvement accounts for soluble solid content in fruits. *Horticulture Research* 10 (3):uhad009

Wu S, Zhang B, Keyhaninejad N, Rodríguez GR, Kim HJ, Chakrabarti M, Illa-Berenguer E, Taitano NK, Gonzalo MJ, Díaz A (2018) A common genetic mechanism underlies morphological diversity in fruits and other plant organs. *Nature Communications* 9 (1):4734

Xiao H, Jiang N, Schaffner E, Stockinger EJ, van der Knaap E (2008) A retrotransposon-mediated gene duplication underlies morphological variation of tomato fruit. *Science* 319 (5869):1527–1530

Xu C, Liberatore KL, MacAlister CA, Huang Z, Chu Y-H, Jiang K, Brooks C, Ogawa-Ohnishi M, Xiong G, Pauly M (2015) A cascade of arabinosyltransferases controls shoot meristem size in tomato. *Nature Genetics* 47 (7):784–792

Xu M, Ji S, Pang S, Lu Y, Li S, Xu W (2025) Regulatory mechanisms of total soluble solids in tomato: From QTL mapping to gene editing. *Foods* 14 (21):3692

Ye J, Li W, Ai G, Li C, Liu G, Chen W, Wang B, Wang W, Lu Y, Zhang J (2019) Genome-wide association analysis identifies a natural variation in basic helix-loop-helix transcription factor regulating ascorbate biosynthesis via D-mannose/L-galactose pathway in tomato. *PLoS Genetics* 15 (5):e1008149

Ye J, Wang X, Hu T, Zhang F, Wang B, Li C, Yang T, Li H, Lu Y, Giovannoni JJ (2017) An InDel in the promoter of Al-ACTIVATED MALATE TRANSPORTER9 selected during tomato domestication determines fruit malate contents and aluminum tolerance. *The Plant Cell* 29 (9):2249–2268

Yeats TH, Martin LB, Viart HM, Isaacson T, He Y, Zhao L, Matas AJ, Buda GJ, Domozych DS, Clausen MH (2012) The identification of cutin synthase: Formation of the plant polyester cutin. *Nature Chemical Biology* 8 (7):609–611

Yuste-Lisbona FJ, Fernández-Lozano A, Pineda B, Bretones S, Ortíz-Atienza A, García-Sogo B, Müller NA, Angosto T, Capel J, Moreno V (2020) ENO regulates tomato fruit size through the floral meristem development network. *Proceedings of the National Academy of Sciences* 117 (14):8187–8195

Zhang B, Li Q, Keyhaninejad N, Taitano N, Sapkota M, Snouffer A, van der Knaap E (2023) A combinatorial TRM-OFP module bilaterally fine-tunes tomato fruit shape. *New Phytologist* 238 (6):2393–2409

Zhang J, Lyu H, Chen J, Cao X, Du R, Ma L, Wang N, Zhu Z, Rao J, Wang J (2024a) Releasing a sugar brake generates sweeter tomato without yield penalty. *Nature* 635 (8039):647–656

Zhang S, Wu S, Jia Z, Zhang J, Li Y, Ma X, Fan B, Wang P, Gao Y, Ye Z (2024b) Exploring the influence of a single-nucleotide mutation in EIN4 on tomato fruit firmness diversity through fruit pericarp microstructure. *Plant Biotechnology Journal* 22 (9):2379–2394

Zhang Y, Butelli E, Alseekh S, Tohge T, Rallapalli G, Luo J, Kawar PG, Hill L, Santino A, Fernie AR (2015) Multi-level engineering facilitates the production of phenylpropanoid compounds in tomato. *Nature Communications* 6 (1):8635

Zhang Y-Y, Li H-X, Shu W-B, Zhang C-J, Ye Z-B (2011) RNA interference of a mitochondrial APX gene improves vitamin C accumulation in tomato fruit. *Scientia Horticulturae* 129 (2):220–226

Zhu Q, Deng L, Chen J, Rodríguez GR, Sun C, Chang Z, Yang T, Zhai H, Jiang H, Topcu Y (2023) Redesigning the tomato fruit shape for mechanized production. *Nature Plants* 9 (10):1659–1674

9 Status of Breeding for Biotic Stress Resistance in Tomato

Dilip R. Panthee

9.1 INTRODUCTION

Tomato (*Solanum lycopersicum*) is among the most widely cultivated and economically important vegetable crops globally, with an annual production exceeding 180 million metric tons (FAOSTAT 2022). Valued for its rich content of vitamins, minerals, and antioxidants, tomato is a dietary staple in many regions. However, its productivity is significantly constrained by a multitude of biotic stresses, including bacterial, fungal, viral, and nematode pathogens, which can cause severe yield losses and quality deterioration (Cruz et al. 2010; Lebeau et al. 2011; Niederheitmann et al. 2015). It not only minimizes dependence on chemical pesticides but also addresses the growing global demand for resilient, high-yielding cultivars. Over the past three decades, host resistance breeding has evolved from traditional phenotypic selection to more sophisticated approaches incorporating molecular markers, quantitative trait locus (QTL) mapping, genetic transformation, and genome editing technologies (Foolad and Panthee 2012; Khojasteh et al. 2024; Nascimento et al. 2023).

Significant milestones underscore this progress. For example, resistance to late blight was successfully introgressed from *Solanum pimpinellifolium* into processing tomato lines, revealing the complex inheritance involving multiple genes and epistatic interactions (Kim and Mutschler 2005). Additionally, studies examining the interplay between biotic and abiotic stress responses have shown that salt stress can exacerbate susceptibility to diseases like powdery mildew, highlighting the need for integrated breeding strategies (Daware et al. 2016; Dodds and Rathjen 2010). Advances in genomics and functional biology—such as effectoromics, genome-wide association studies (GWAS), and CRISPR/Cas9-mediated gene editing—have significantly expanded our capacity to dissect plant–pathogen interactions and engineer durable resistance (Saini et al. 2025; Singh and Bokolia 2021). A case in point is bacterial spot, one of the most devastating foliar diseases of tomato, caused by several *Xanthomonas* species (*X. euvesicatoria*, *X. vesicatoria*, *X. perforans*, and *X. gardneri*). The disease leads to defoliation, fruit blemishing, and economic losses, particularly in warm and humid climates (Adhikari et al. 2020). Over the past three decades, resistance breeding for bacterial spot has evolved from conventional screening methods to marker-assisted selection (MAS), genetic transformation, and genome editing, leading to the development of more precise and efficient resistant cultivars (Hamza et al. 2010; Hutton et al. 2010; Scott et al. 2011). This chapter provides a comprehensive overview of advances in host resistance breeding for major tomato diseases, including bacterial speck, bacterial spot, bacterial wilt, early blight, Fusarium wilt, late blight, Septoria leaf spot, tomato mosaic virus, tomato spotted wilt virus, tomato yellow leaf curl virus, and root-knot nematodes. Emphasis is placed on pathogen diversity, available genetic resources, resistance mechanisms, major genes and QTLs, breeding strategies, and modern biotechnological tools that have revolutionized tomato disease resistance breeding in recent decades.

9.2 IMPACT OF BIOTIC STRESS ON TOMATO

Biotic stresses, including pathogens and pests, are major constraints in tomato cultivation. Diseases such as bacterial wilt, early blight, late blight, tomato yellow leaf curl virus, and root-knot

 DOI: 10.1201/9781003506539-9

nematodes can lead to substantial yield reductions. For instance, bacterial wilt caused by *Ralstonia solanacearum* can result in complete crop failure under favorable conditions (Kunwar et al. 2019; Lebeau et al. 2011). The economic losses due to these stresses are immense, affecting both smallholder and commercial farmers.

Tomato is one of the most widely cultivated vegetable crops globally, with annual production exceeding 190 million metric tons across more than 5 million hectares (FAOSTAT, 2022). However, biotic stresses—including bacterial, fungal, viral, and nematode pathogens—pose severe threats to both yield and quality, leading to multi-billion-dollar losses each year. The global economic impact of these diseases is substantial and continues to grow with climate change, evolving pathogen virulence, and increasing resistance to conventional control methods.

Bacterial wilt, caused by *Ralstonia solanacearum*, is one of the most devastating soil-borne diseases, particularly in tropical and subtropical regions of Asia, Africa, and South America. In areas with high disease pressure and conducive environmental conditions, yield losses can approach 100%, especially in susceptible cultivars and poorly managed soils. Globally, bacterial wilt alone is estimated to cause $1–2 billion in annual economic losses, considering yield reductions, replanting costs, and lost market opportunities. Its wide host range and soil persistence make management extremely difficult without resistant varieties.

Early blight (*Alternaria solani*) and late blight (*Phytophthora infestans*) are widespread fungal diseases that affect tomatoes in nearly all production zones. Late blight, in particular, has been responsible for historic crop failures and continues to cause losses ranging from 20% to 60%, depending on the level of control and cultivar susceptibility. Globally, late blight is estimated to result in losses of over $3 billion annually, including costs associated with fungicide applications and crop damage. Early blight, though often less catastrophic, contributes to steady annual losses of $500–800 million.

Tomato yellow leaf curl virus (TYLCV) is an emerging global threat, particularly in the Mediterranean basin, Asia, and the Americas. Transmitted by whiteflies, TYLCV can reduce yields by 30–100%, with early-season infections leading to near-total crop failure. Annual global losses from TYLCV exceed $1.5 billion, compounded by the cost of managing whitefly populations and crop losses in high-value production systems. The rapid spread of TYLCV underscores the need for integrated virus and vector management strategies, including host resistance.

Root-knot nematodes (*Meloidogyne* spp.) are prevalent in tropical and subtropical soils, affecting up to 50% of global tomato fields to varying degrees. Though less dramatic in visible symptoms, their cumulative impact on yield, plant vigor, and fruit size can be economically devastating. Yield reductions of 15–30% are common, leading to $800 million to $1.2 billion in losses annually. The problem is exacerbated in areas with limited access to resistant cultivars or sustainable nematode control options.

Altogether, global economic losses from these major biotic stresses are conservatively estimated at $8–10 billion annually, not including indirect costs such as increased pesticide use, labor, and post-harvest losses. These figures also likely underestimate total impact, as many smallholder losses go unreported.

These losses emphasize the urgent need for globally coordinated breeding efforts targeting durable, multi-disease resistance. Integration of traditional breeding, molecular tools, and regional screening platforms will be key to mitigating these persistent threats, ensuring productivity, profitability, and sustainability for tomato growers worldwide.

9.3 NEED FOR BREEDING FOR BIOTIC STRESS RESISTANCE

The development of resistant tomato varieties is a sustainable and environmentally friendly approach to managing biotic stresses. Breeding for resistance reduces the dependency on chemical control measures, lowers production costs, and minimizes environmental pollution. Moreover, resistant varieties contribute to food security by ensuring stable yields under biotic stress conditions (Bhattarai et al. 2017; Sim et al. 2015).

9.4 PRINCIPAL BIOTIC STRESSES AFFECTING TOMATO PRODUCTION

Tomato (*Solanum lycopersicum*) is susceptible to a wide array of biotic stresses that severely impact yield and fruit quality. These include bacterial, fungal, viral, and root-knot nematode pathogens, many of which exhibit high genetic diversity and adaptability, complicating management strategies. Resistance breeding has been a cornerstone of integrated disease management, leveraging both traditional and molecular approaches to combat these major threats (Devran et al. 2016; Hanson et al. 2016).

9.4.1 Bacterial Diseases

Bacterial speck, incited by *Pseudomonas syringae* pv. *tomato* (Pst), causes small, dark necrotic spots on leaves and fruits. It thrives in cool, moist climates and is classified into Race 0 and Race 1 based on interaction with the *Pto* resistance gene, which has been deployed in commercial cultivars with success (Pohronezny and Volin 1980; Schneider and Grogan 1977).

Bacterial spot, caused by multiple *Xanthomonas* species (*X. perforans*, *X. euvesicatoria*, *X. vesicatoria*, and *X. gardneri*), manifests as foliar lesions, defoliation, and fruit blemishes, especially under warm, humid conditions. Resistance is polygenic and quantitative in nature, often necessitating gene pyramiding to achieve durable control (Adhikari et al. 2020).

Bacterial wilt, caused by *Ralstonia solanacearum*, is a soil-borne vascular disease that leads to sudden wilting and plant death. The pathogen exhibits extensive strain diversity, requiring resistance sources from wild tomato relatives such as *S. habrochaites* (Khan et al. 2023; Kunwar et al. 2019).

9.4.2 Fungal and Oomycete Diseases

Early blight, caused by the fungus *Alternaria solani*, is characterized by concentric, necrotic lesions primarily on older foliage, leading to premature defoliation and fruit damage. The pathogen produces toxins and conidia, and recent variants exhibit differences in aggressiveness and fungicide sensitivity. Resistance is quantitatively inherited and involves the action of multiple genes (Adhikari et al. 2017; Singh et al. 2017).

Late blight, one of the most destructive diseases affecting tomatoes, is caused by the oomycete *Phytophthora infestans*. It can result in rapid plant collapse under cool and wet conditions. Epidemics are often driven by clonal lineages such as US-1, US-8, and EU-13-A2. Resistance genes from wild species, such as *S. pimpinellifolium*, are actively utilized in breeding programs (Sullenberger et al. 2018; Zhang et al. 2014).

Fusarium wilt, caused by *Fusarium oxysporum* f. sp. *lycopersici* (Fol), leads to yellowing and vascular wilting. It comprises three physiological races (1, 2, and 3), each requiring specific resistance genes—*I*, *I-2*, and *I-3*—that have been introgressed into commercial varieties (El Mohtar et al. 2007).

Septoria leaf spot, caused by *Septoria lycopersici*, results in small necrotic spots with dark margins on leaves. It often co-occurs with other foliar diseases and can lead to severe defoliation in conducive environments. Variability in virulence and environmental responses is noted among isolates (Pandey et al. 2024).

9.4.3 Viral Diseases

Tomato mosaic virus (ToMV), a tobamovirus, causes mottling, mosaic symptoms, and leaf distortion. The *Tm* series of resistance genes (e.g., *Tm-1*, *Tm-2*, *Tm-2*^{2}) has been effective against various ToMV strains, though resistance-breaking isolates have emerged (Panthee et al. 2013).

Tomato spotted wilt virus (TSWV), a tospovirus transmitted by thrips, is highly variable and possesses a wide host range. The emergence of resistance-breaking strains complicates control efforts, despite the deployment of resistance genes such as *Sw-5* (Gardner and Panthee 2012; Panthee and Gardner 2011; Robbins et al. 2010).

Tomato yellow leaf curl virus (TYLCV), a begomovirus transmitted by whiteflies, causes stunted growth, leaf curling, and yield loss. It exists as a complex of closely related viruses and recombinants, including TYLCV-IL and TYLCV-Mild. Resistance genes *Ty-1* through *Ty-5* have been identified and widely utilized in breeding programs (Sacco Botto et al. 2023; Verlaan et al. 2013).

9.4.4 Root-Knot Nematode

Root-knot nematodes (*Meloidogyne* spp.), particularly *M. incognita*, *M. javanica*, and *M. arenaria*, form galls on roots, disrupting nutrient uptake and stunting plant growth. The *Mi-1* gene has been effective in conferring resistance to several pathotypes, though virulent populations have been reported in some regions, necessitating continued exploration of durable sources (Barbary et al. 2014).

9.5 GENETIC RESOURCES FOR BIOTIC STRESS TOLERANCE IN TOMATO

The genetic improvement of tomato for biotic stress tolerance has relied extensively on the exploitation of wild relatives and landraces, which harbor a rich reservoir of resistance genes. Species such as *Solanum pimpinellifolium*, *S. peruvianum*, *S. habrochaites*, *S. chilense*, *S. pennellii*, and *S. lycopersicoides* have proven especially valuable in breeding programs targeting resistance to bacterial, fungal, viral, and nematode pathogens (El-Sappah et al. 2019; Engle et al. 2003). Resistance to major bacterial diseases—such as bacterial speck and bacterial spot—has been sourced primarily from *S. pimpinellifolium*, *S. chilense*, *S. pennellii*, and *S. lycopersicoides*. The *Pto* gene, conferring resistance to *Pseudomonas syringae* pv. *tomato* Race 0, was originally identified in the cultivar "Rio Grande-PtoR" and introgressed from *S. pimpinellifolium* (Martin et al. 1993a). Similarly, resistance to bacterial spot has been associated with key loci such as *Rx3*, *Rx4*, and *RxopJ4*, derived from multiple wild donors (Table 9.1). These loci offer varying degrees of resistance against different *Xanthomonas* races (Scott et al. 2003). However, their introgression into elite germplasm has been challenging due to issues such as linkage drag and poor agronomic traits in the donor species.

Fungal disease resistance, particularly to early blight (*Alternaria solani*) and Fusarium wilt (*Fusarium oxysporum* f. sp. *lycopersici*), has been obtained from several wild species. Sources such as *S. peruvianum* and *S. habrochaites* have contributed polygenic resistance traits, which offer more durable and broad-spectrum protection. Resistance to Fusarium wilt, in particular, has been effectively deployed through the use of race-specific genes (*I*, *I-2*, and *I-3*) in commercial varieties (Foolad and Ashrafi 2015b; Foolad and Panthee 2012). The development of near-isogenic lines (NILs), recombinant inbred lines (RILs), and introgression lines has facilitated the precise transfer and evaluation of these resistance traits. Virus resistance breeding has also benefited from wild tomato germplasm. The *Tm* gene series (*Tm-1*, *Tm-2*, *Tm-2*^{2}) has been instrumental in conferring resistance to Tomato mosaic virus (ToMV), while the *Sw-5* gene, derived from *S. peruvianum*, offers adequate protection against Tomato spotted wilt virus (TSWV). For Tomato yellow leaf curl virus (TYLCV), resistance genes *Ty-1* through *Ty-5* have been identified primarily in *S. chilense*, *S. habrochaites*, and *S. peruvianum*, and have been widely used in hybrid breeding programs (Ji et al. 2007a; Ji et al. 2007b; Lanfermeijer et al. 2005). Nematode resistance, particularly to root-knot nematodes (*Meloidogyne* spp.), has been derived from *S. peruvianum* and *S. arcanum*. The *Mi-1* gene, now widely deployed in commercial cultivars, confers resistance to multiple *Meloidogyne*

TABLE 9.1
Sources of Resistance for Major Tomato Diseases

Disease	Causal Organism	Source of Resistance (Accession No.)	Resistance Genes/Notes	References
Bacterial Spot	*Xanthomonas* spp.	PI 114490 (*S. pimpinellifolium*)	Quantitative resistance	Scott et al. (2003)
Bacterial Speck	*Pseudomonas syringae* pv. *tomato*	Hawaii 7981	*Pto* gene	Martin et al. (1993a)
Bacterial Wilt	*Ralstonia solanacearum*	LA1035 (*S. lycopersicoides*), LA1777 (*S. habrochaites*)	Polygenic resistance	Wang et al. (2000)
Early Blight	*Alternaria solani*	LA2157 (*S. arcanum*), LA1777 (*S. habrochaites*)	Quantitative resistance	Foolad et al. (2015)
Late Blight	*Phytophthora infestans*	LA1035 (S. *habrochaites*), LA2099 (*S. pimpinellifolium*)	*Ph-1, Ph-2, Ph-3*	Foolad and Ashrafi (2015a)
Fusarium Wilt (Race 1, 2, 3)	*Fusarium oxysporum f.sp. lycopersici*	PI 126915 (*S. lycopersicum*)	*I, I-2, I-3* genes	Foolad and Panthee (2012)
Verticillium Wilt	*Verticillium dahliae*	Cultivar "Motelle"	*Ve1* gene	Kawchuk et al. (1998), Kawchuk et al. (2001)
Tomato Yellow Leaf Curl Virus	Begomovirus	LA1969 (*S. chilense*), LA1932 (*S. peruvianum*)	*Ty-1* to *Ty-6*	Ji et al. (2007a, 2007b)
Tomato Mosaic Virus	Tobamovirus	PI 126445 (*S. lycopersicum*)	*Tm-1, Tm-2*, $Tm\text{-}2^2$	Lanfermeijer et al. (2003), Lanfermeijer et al. (2005)
Tomato Spotted Wilt Virus	Tospovirus	PI 126923 (*S. peruvianum*)	*Sw-5* gene	Gardner and Panthee (2012)
Root-Knot Nematode	*Meloidogyne* spp.	LA2822 (*S. peruvianum*)	*Mi-1.2* gene	Devran et al. (2016)

species, though virulence-breaking populations necessitate continued screening for additional sources (Devran et al. 2016).

The success of resistance breeding is closely tied to the availability and accessibility of diverse germplasm. Public repositories such as the Tomato Genetics Resource Center (TGRC) and GRIN have played a central role in conserving, characterizing, and distributing resistant accessions (Chetelat 2001). These resources have enabled extensive phenotypic and genotypic screening efforts, leading to the identification of both major resistance genes and quantitative trait loci (QTLs) associated with partial resistance. Advances in molecular tools, including genome-wide association studies (GWAS) and linkage mapping, have further accelerated the discovery and deployment of loci conferring resistance to biotic stress (Foolad and Panthee 2012; Robbins et al. 2010). In summary, the incorporation of resistance from wild relatives remains a cornerstone of tomato breeding for biotic stress tolerance. Continued efforts in pre-breeding, molecular mapping, and the development of structured populations are crucial for overcoming barriers such as linkage drag and enhancing the durability and effectiveness of resistance traits in commercial cultivars.

9.6 DISEASE RESISTANCE MECHANISM IN TOMATO PLANTS

Tomato plants defend themselves against a broad spectrum of pathogens using a multilayered immune system that includes both structural barriers and dynamic biochemical responses. At the core of these defenses are three major resistance strategies: effector-triggered immunity (ETI), pattern-triggered immunity (PTI), and quantitative or polygenic resistance (Hurley et al. 2014; Khojasteh et al. 2024; Sreekanta et al. 2015). Together, these mechanisms provide a framework for breeding tomato cultivars with durable resistance to various biotic stresses.

Effector-Triggered Immunity (ETI) represents a specialized defense mechanism in which plant resistance (R) genes recognize specific pathogen effectors, resulting in a robust and localized immune response often characterized by a hypersensitive response (HR). This rapid programmed cell death at the infection site limits pathogen proliferation. One of the best-characterized examples in tomato is the *Pto-Prf* signaling complex, which detects effectors *AvrPto* and *AvrPtoB* from *Pseudomonas syringae* pv. *tomato* via its type III secretion system. Activation of this pathway leads to HR, the production of reactive oxygen species (ROS), and the initiation of mitogen-activated protein kinase (MAPK) cascades (Abramovitch et al. 2003; Dong et al. 2009; Lin and Martin 2007). Similar ETI mechanisms are also involved in resistance to *Fusarium oxysporum* and *Xanthomonas* spp., where genes like *I*, *Rx3*, and *RxopJ4* recognize specific effectors to activate HR-based resistance (Hurley et al. 2014). In contrast, Pattern-Triggered Immunity (PTI) constitutes the plant's first line of active defense and is more broadly effective across pathogen species. PTI is initiated by the recognition of pathogen-associated molecular patterns (PAMPs), such as bacterial flagellin (*flg22*), by pattern recognition receptors (PRRs) on the plant cell surface. This interaction triggers downstream defense signaling, including the production of reactive oxygen species (ROS) bursts, callose deposition, and the activation of defense-related gene expression. PTI contributes to basal resistance and serves as a foundation upon which ETI is layered for more specific responses (Sreekanta et al. 2015).

Beyond ETI and PTI, quantitative resistance—governed by multiple genes or quantitative trait loci (QTLs)—plays a crucial role in conferring partial yet durable resistance, particularly against fungal pathogens such as *Alternaria solani* (early blight) and *Septoria lycopersici* (Septoria leaf spot). These mechanisms are less reliant on single R genes and involve traits such as increased cuticle thickness, heightened antioxidant enzyme activity, and enhanced expression of defense genes like those encoding pathogenesis-related (PR) proteins and antimicrobial phytoalexins (Khojasteh et al. 2024). Systemic Acquired Resistance (SAR) is another layer of defense, involving the induction of long-lasting resistance throughout the plant following localized infection. This systemic response often relies on signaling molecules such as salicylic acid and leads to the expression of PR proteins that contribute to broad-spectrum disease resistance (Vallad and Goodman 2004).

In the case of viral pathogens, resistance mechanisms include the inhibition of viral replication, movement, and systemic spread. Resistance genes, such as *Tm-1, Tm-2,* and *Tm-2*^{2}, act by blocking viral replication or interfering with the movement proteins. Similarly, the *Sw-5* gene confers resistance to Tomato spotted wilt virus (TSWV) by recognizing its movement protein and triggering HR. These responses are often oligogenic or polygenic, contributing to enhanced durability of resistance (Anikina et al. 2023). Resistance to nematodes, especially *Meloidogyne* spp., is conferred by genes like *Mi-1*, which activate HR at feeding sites, impairing nematode development. Additional resistance mechanisms include the disruption of giant cell formation and the activation of SAR (El-Sappah et al. 2019). Recent advances have also highlighted the role of susceptibility (S) genes, which pathogens exploit to facilitate infection. By knocking out or silencing these genes, such as SlJAZ2 in bacterial spot or SlDCL2 in viral diseases, plants can achieve resistance through a "loss-of-susceptibility" strategy. Genome editing tools, such as CRISPR/Cas9, are increasingly being employed to target these genes, offering novel avenues for durable resistance breeding (Jahan et al. 2025). In summary, tomato plants employ a complex and integrated defense system that involves ETI, PTI, SAR, quantitative resistance, and the manipulation of susceptibility genes. Understanding these mechanisms has not only enhanced our comprehension of plant immunity but has also enabled the strategic deployment of resistance genes in breeding programs, fostering the development of resilient tomato cultivars capable of withstanding evolving pathogen pressures.

9.7 MAJOR GENES AND QTLS ASSOCIATED WITH RESISTANCE

Tomato resistance to biotic stress is governed by a combination of primary resistance (R) genes and quantitative trait loci (QTLs) derived from cultivated lines and wild relatives. The deployment of both monogenic and polygenic resistance mechanisms enhances durability and broad-spectrum protection (Table 9.2).

Bacterial speck, caused by *Pseudomonas syringae* pv. *tomato*, is primarily controlled by the *Pto* gene, which encodes a serine/threonine kinase involved in effector-triggered immunity (ETI). *Pto* functions synergistically with the *Prf* gene to detect pathogen effectors AvrPto and AvrPtoB (Abramovitch et al. 2003; Dong et al. 2009; Lin and Martin 2007). Additional QTLs conferring partial resistance have been mapped to chromosomes 1, 5, and 11, suggesting a role for polygenic resistance in durable control.

Bacterial spot, caused by several *Xanthomonas* species, is a complex disease requiring multiple resistance sources. Major R genes identified include *Rx3*, *Rx4*, and *RxopJ4*, which confer race-specific resistance via ETI. Additionally, QTLs associated with resistance have been identified on chromosomes 6 and 11 (Adhikari et al. 2023). These loci have been introgressed from wild relatives, particularly *Solanum pennellii* and *S. chilense*, and are critical in resistance breeding programs due to the pathogen's genetic variability. Resistance to bacterial wilt caused by *Ralstonia solanacearum* is quantitatively inherited. Although no major R gene has been identified, several quantitative trait loci (QTLs) associated with resistance have been mapped to chromosomes 6 and 12 (Siddique et al. 2024; Sikirou et al. 2021). These regions contribute to partial resistance derived from *S. lycopersicoides* and *S. habrochaites*, and are useful in pyramiding efforts aimed at enhancing tolerance under tropical and subtropical conditions.

For early blight, caused by *Alternaria solani*, resistance is polygenic and lacks major R genes. Significant QTLs have been mapped on chromosomes 5, 6, and 11, primarily sourced from *S. habrochaites*. These loci are associated with traits such as reduced lesion development and delayed disease progression, and they form the foundation for developing moderately resistant cultivars (Adhikari et al. 2017). Fusarium wilt, caused by *Fusarium oxysporum* f. sp. *lycopersici*, exhibits race-specific resistance. The *I*, *I-2*, and *I-3* genes confer resistance to races 1, 2, and 3, respectively. *I* and *I-2* are located on chromosome 11, while *I-3* resides on chromosome 7 (El Mohtar et al.

TABLE 9.2
QTL Mapping Studies for Tomato Disease Resistance

Disease	QTL Name/ Region	Chromosome (Chr)	Resistance Source/Population	References
Bacterial Spot	*Rx-4*	Chr11	Hawaii 7981 × OH88119	Yang et al. (2005)
Bacterial Speck	*Pto* locus	Chr5	Hawaii 7981	Martin et al. (1993b)
Bacterial Wilt	*Bwr-6, Bwr-12*	Chr6, Chr12	*S. pimpinellifolium* (L285)	Wang et al. (2000)
Early Blight	*EbrI*	Chr5	*S. habrochaites* × *S. lycopersicum*	Foolad and Ashrafi (2015b)
Late Blight	*Ph-2, Ph-3*	Chr10	*S. pimpinellifolium*	Foolad et al. (2008)
Fusarium Wilt (Race 3)	*I-3*	Chr7	*S. pennellii* introgression	Lim et al. (2006)
Verticillium Wilt	*Ve1* locus	Chr9	Cultivar "Motelle"	Kawchuk et al. (2001)
TYLCV	*Ty-1, Ty-2, Ty-3*	Chr6, Chr11	*S. chilense, S. peruvianum*	Ji et al. (2007b)
ToMV	*Tm-2*^{2}	Chr9	*S. lycopersicum*	Lanfermeijer et al. (2003), Lanfermeijer et al. (2005)
TSWV	*Sw-5*	Chr9	*S. peruvianum*	Dockter et al. (2009), Stevens et al. (1992)
Root-Knot Nematode	*Mi-1.2*	Chr6	*S. peruvianum*	Devran et al. (2016)

2007; Lim et al. 2006). These genes are widely used in commercial hybrids, often in combination to achieve multi-race protection. Resistance to late blight, caused by *Phytophthora infestans*, is conferred by the *Ph-1*, *Ph-2*, and *Ph-3* genes, primarily introgressed from *S. pimpinellifolium* and *S. habrochaites*. These genes provide qualitative resistance, while additional QTLs mapped to chromosomes 5, 6, and 9 contribute to quantitative resistance, enhancing long-term durability (Panthee et al. 2017). Septoria leaf spot, caused by *Septoria lycopersici*, lacks major R genes and is controlled by polygenic resistance. QTLs conferring partial resistance have been identified on chromosomes 2, 4, and 6, mostly derived from wild germplasm (Sharon 2012; Siddique et al. 2023). These loci are associated with traits like reduced lesion size and increased cuticle thickness.

For tomato mosaic virus (ToMV), broad-spectrum resistance is mediated by *Tm-1*, *Tm-2*, and *Tm-2*^{2}, all located on chromosome 2 (Lanfermeijer et al. 2005). These genes inhibit viral replication and movement, providing durable resistance against multiple strains. Tomato spotted wilt virus (TSWV) resistance is governed by the *Sw-5* and *Sw-7 genes*, with *Sw-5* being the most widely used. Sourced from *S. peruvianum*, *Sw-5* is located on chromosome 9 and functions via ETI by recognizing the viral movement protein. *Sw-7*, a more recently identified gene, offers additional resistance coverage (Dockter et al. 2009). Resistance to tomato yellow leaf curl virus (TYLCV) is polygenic, with six key genes identified: *Ty-1* through *Ty-6*. *Ty-1* and *Ty-3*, located on chromosome 6, are among the most commonly deployed and offer broad-spectrum resistance. Other genes include *Ty-2* (chromosome 11), *Ty-4* (chromosome 3), *Ty-5* (chromosome 4), and *Ty-6* (chromosome 10), each contributing to varying levels of resistance against different virus strains (Ji et al. 2007b). Finally, root-knot nematode resistance is primarily mediated by the *Mi-1.2* gene, located on chromosome 6. This gene confers resistance to multiple *Meloidogyne* species and is widely utilized in both fresh market and processing tomato cultivars (Devran et al. 2016). Additional QTLs derived from wild relatives support the development of pyramided lines with enhanced resistance and durability.

9.8 BREEDING STRATEGIES FOR BIOTIC STRESS TOLERANCE

Breeding for biotic stress tolerance in tomato has evolved significantly from conventional methods to more advanced, genomics-assisted approaches. Traditional breeding strategies relied heavily on phenotypic selection through interspecific hybridization, recurrent backcrossing, and pedigree selection. These approaches were essential for the initial introgression of resistance genes such as *Pto*, *Mi-1.2*, and *I* genes from wild species into cultivated lines. However, they were often time-consuming and hindered by challenges like linkage drag, environmental variability, and genotype-by-environment interactions (Hanson et al. 2016). In the case of diseases such as bacterial spot and Fusarium wilt, early efforts used interspecific crosses followed by multiple generations of backcrossing and phenotypic screening. For example, resistance to bacterial spot involved introgression of *Rx* genes and the *Pto* gene, but durability was limited due to pathogen evolution and the emergence of new races. Similarly, monogenic resistance to Fusarium wilt, governed by *I*, *I-2*, and *I-3*, required careful management to avoid resistance breakdown and to ensure race coverage (Hanson et al. 2016). To address these limitations, modern breeding programs increasingly incorporate molecular tools to enhance selection efficiency and broaden the resistance spectrum. Marker-assisted selection (MAS) has become a standard approach, enabling precise introgression and pyramiding of major resistance genes such as *Ty-1* and *Ty-3* for TYLCV, *Ph* genes for late blight, and *Sw-5* for TSWV. Marker-assisted backcrossing (MABC) has also proven valuable in minimizing linkage drag while accelerating the development of near-isogenic lines (Devran et al. 2016).

Gene pyramiding is a core strategy in resistance breeding today, combining multiple R genes and QTLs to enhance durability and delay the breakdown of resistance. This has been applied across several pathosystems, including bacterial spot (e.g., *Rx3*, *Rx4*, *RxopJ4*), Fusarium wilt (*I*, *I-2*, *I-3*), and TYLCV (*Ty-1* through *Ty-6*). In cases where major R genes are lacking or insufficient, such as early blight and Septoria leaf spot, breeders focus on accumulating minor-effect QTLs through recurrent selection and genomic selection (GS), capitalizing on additive genetic variation for more durable quantitative resistance (Barone et al. 2005; Yamamoto et al. 2017). Pre-breeding using wild relatives remains vital for identifying novel resistance alleles. Species such as *Solanum habrochaites, S. pimpinellifolium, S. peruvianum*, and *S. chilense* have contributed to resistance against pathogens including *Alternaria*, *Phytophthora*, TYLCV, and root-knot nematodes (Haussmann et al. 2004). Backcrossing and MAS are then used to transfer these alleles into elite cultivars, often with the aid of advanced breeding populations such as recombinant inbred lines (RILs), near-isogenic lines (NILs), and advanced backcross-QTL (AB-QTL) mapping populations.

Genomics-assisted breeding (GAB) is playing an increasingly important role in managing complex traits, especially those governed by multiple loci. Genomic selection (GS) models utilize genome-wide marker data to predict breeding values, enabling the rapid selection of superior genotypes for traits such as early blight, bacterial wilt, and nematode resistance. Integration of genotyping-by-sequencing (GBS) and high-throughput phenotyping has further streamlined resistance breeding pipelines (Tiwari et al. 2022; Varshney and Dubey 2009). Rootstock breeding is another promising strategy, especially for managing soil-borne pathogens such as *Ralstonia solanacearum* and *Meloidogyne* spp. Resistant rootstocks derived from wild species or hybrids can protect susceptible scions while maintaining desirable horticultural traits (King et al. 2010; Thies and Panthee 2023). Combining rootstock resistance with scion-based R genes offers a multilayered approach to disease management. In summary, contemporary breeding strategies for biotic stress tolerance in tomato emphasize the integration of classical and molecular tools. Pyramiding of resistance genes, genomic selection for quantitative traits, use of wild relatives, and development of resistant rootstocks represent a comprehensive toolkit for developing cultivars with broad-spectrum and durable resistance.

9.9 MARKER-ASSISTED SELECTION FOR BIOTIC STRESS TOLERANCE IN TOMATO

Marker-assisted selection (MAS) has become a fundamental component of tomato breeding for biotic stress tolerance, enabling the precise and efficient incorporation of resistance genes into elite cultivars. By using molecular markers tightly linked to resistance loci, MAS allows breeders to select for target traits at early stages of plant development, significantly shortening breeding cycles and increasing accuracy (Foolad and Panthee 2012; Tiwari et al. 2022). Its application spans a wide range of pathogens, including bacteria, fungi, viruses, and nematodes (Table 9.3).

Early implementation of MAS in tomato focused on bacterial speck and spot resistance. Initial efforts used low-throughput markers such as RFLPs, RAPDs, and AFLPs, which provided preliminary linkage information but suffered from low reproducibility. These were eventually replaced by more robust and polymorphic markers, including SSRs, CAPS, and, more recently, high-resolution SNPs and gene-specific markers (Foolad and Panthee 2012; Tiwari et al. 2022). The successful cloning of *RxopJ4* from *S. pennellii* enabled the development of functional markers that are now routinely used in pyramiding bacterial spot resistance genes such as *Rx3*, *Rx4*, and *RxopJ4* into elite germplasm (Sharlach et al. 2013).

In the case of bacterial speck, MAS has been pivotal for deploying *Pto* and its signaling partner *Prf.* Early selection relied on CAPS and SCAR markers linked to *Pto*, while the advent of SNP-based assays has improved the precision and throughput of breeding efforts. High-throughput genotyping platforms now enable breeders to combine *Pto* with minor QTLs that confer broader, race-nonspecific resistance, thus improving field durability (Foolad and Panthee 2012).

MAS has also been successfully applied to manage fungal diseases such as Fusarium wilt and late blight. Resistance to different races of *Fusarium oxysporum* is governed by *I*, *I-2*, and *I-3*, each of which is tracked using tightly linked SCAR, CAPS, or SNP markers. The availability of these markers has streamlined the pyramiding of these race-specific resistance genes. Similarly, markers associated with *Ph-1*, *Ph-2*, and *Ph-3* are employed to pyramid late blight resistance in breeding

TABLE 9.3
Molecular Markers for Tomato Disease Resistance and Their Application in Marker-Assisted Selection

Disease	Marker Name/ Type	Forward/Reverse Primer (5′–3′)	Amplicon Size (bp)	Resistance vs. Susceptible Genotype Band	Application in MAS	References
Fusarium Wilt (I-2)	SSR—I2C-1	F: ATGGTGATGGGTTTATGGGA R: AATACGGCAGTGTGTGTTGG	550	Resistant: 550 bp, Susceptible: No band	Selection of I-2 gene for Fusarium wilt resistance	Ori et al. (1997)
Fusarium Wilt (I-3)	SCAR—I3/5	F: GAGGAGGATTTTGGAGATGC R: CCTCTGAGCCTTATCCTGGA	295	Resistant: 295 bp, Susceptible: No band	Introgression of I-3 gene into elite lines	Lim et al. (2006)
Verticillium Wilt	CAPS—Ve1	F: AAGGGCCTTCTTCATGTCTC R: TTTCATCTTTCTTGGGAGCG	438	Resistant: Digested band (300+138), Susceptible: 438 bp	Ve1 gene tracking	Kawchuk et al. (1998), Kawchuk et al. (2001)
TYLCV (Ty-1)	SCAR—Ty1–9	F: TTGGCGAGTGGTGATTTTTT R: GCGTTTTTCATCGGATGTTC	920	Resistant: 920 bp, Susceptible: No band	MAS for Ty-1 gene from *S. chilense*	Ji et al. (2007a)
TYLCV (Ty-3)	SCAR—Ty3-IN	F: GAGCTGTGGTTTTTGCATTT R: GGTTTATCGGATGTTCAGCA	550	Resistant: 550 bp, Susceptible: No band	Ty-3 gene introgression	Hanson et al. (2000)
Tomato Spotted Wilt Virus	SCAR—Sw5	F: GGGAACTTAGGGTTCCGAAA R: CTGTTCTCCATCAGGTTCGG	410	Resistant: 410 bp, Susceptible: No band	Sw5 gene selection	Stevens et al. (1995)
Root-Knot Nematode	SCAR—Mi23	F: GTGGTGATGATGGTGATGGA R: CTCAGCAGGATGTCATCTCA	380	Resistant: 380 bp, Susceptible: 450 bp	Marker for the *Mi-1* gene	Seah et al. (2007)
Powdery Mildew	SCAR—Ol-4	F: ATCTCCTCGATGATCTCGCC R: CTAGATCGGACCTCCAGTGC	750	Resistant: 750 bp, Susceptible: No band	*Ol-4* gene MAS	Bai and Lindhout (2005)

programs. For polygenic diseases like early blight and Septoria leaf spot, MAS is often combined with phenotypic screening, using QTL-linked markers to accumulate favorable alleles from wild relatives (Foolad and Panthee 2012; Hanson et al. 2016).

In virus resistance breeding, MAS has proven especially valuable. PCR-based markers are used to introgress and combine *Sw-5* (for TSWV), *Tm-1*, *Tm-2*, and *Tm-2*^{2} (for ToMV), as well as the *Ty* gene series (*Ty-1* to *Ty-6*) for TYLCV. Pyramiding of *Ty-1* and *Ty-3*, both of which are located on chromosome 6, is a common strategy to enhance the resistance spectrum and durability (Ji et al. 2007a; Tiwari et al. 2022). Functional markers linked to these genes enable precise selection and help avoid linkage drag and susceptibility alleles.

For nematode resistance, MAS targeting the *Mi-1.2* gene, located on chromosome 6, is a widely used approach. Closely linked DNA markers, such as SCAR and SNPs, facilitate the efficient screening and incorporation of this gene, which provides broad-spectrum resistance to multiple *Meloidogyne* species (Seah et al. 2007). Phenotypic validation under nematode pressure remains essential to confirm field performance, particularly when pyramiding *Mi-1.2* with other resistance quantitative trait loci (QTLs).

The integration of MAS with genomic tools and phenotypic data has further enhanced its utility. High-throughput SNP arrays and genotyping-by-sequencing (GBS) platforms enable genome-wide tracking of resistance loci, allowing breeders to manage background genome recovery in backcross programs and accelerate the development of near-isogenic lines (Tiwari et al. 2022). Coupling MAS with field data ensures that selected genotypes perform reliably across diverse environments and pathogen populations. In summary, MAS has revolutionized tomato breeding for biotic stress tolerance by enabling precise, efficient, and scalable deployment of resistance genes. Its integration with genomic selection, high-throughput genotyping, and phenotypic validation continues to drive progress in developing cultivars with durable and broad-spectrum disease resistance.

9.10 GENETIC TRANSFORMATION FOR BIOTIC STRESS TOLERANCE IN TOMATO

Genetic transformation has been explored as a strategy to enhance tomato resistance to a range of biotic stresses, particularly when the availability of resistance sources or lengthy breeding cycles limits conventional breeding (Chaudhary et al. 2019). Transgenic approaches enable the direct introduction of resistance genes into the tomato genome, bypassing reproductive barriers and linkage drag associated with traditional breeding. These strategies have primarily focused on enhancing resistance to bacterial, fungal, viral, and nematode pathogens through the expression of foreign genes, defense regulators, and RNA interference (RNAi) constructs (Singh et al. 2021).

For bacterial spot resistance, one notable example is the successful introduction of the *Bs2* gene from pepper into tomato. *Bs2* recognizes specific *Xanthomonas* effectors and triggers a strong hypersensitive response. Transgenic lines expressing *Bs2* have shown robust resistance in controlled environments (Horvath et al. 2015; Tai et al. 1999). Similarly, the overexpression of endogenous resistance genes, such as *Pto* and *Prf*, as well as defense-related transcription factors like *NPR1*, has conferred enhanced resistance to bacterial speck and other diseases. To broaden resistance, genes encoding antimicrobial peptides (AMPs), chitinases, glucanases, and other pathogenesis-related (PR) proteins have been introduced to reinforce basal defense mechanisms.

In viral disease management, transgenic tomatoes expressing viral coat protein genes or antisense RNA have shown resistance to viruses such as Tomato yellow leaf curl virus (TYLCV) and Tomato mosaic virus (ToMV). RNAi-based strategies targeting viral genomes or key replication elements have demonstrated high efficacy, with reduced viral accumulation and symptom severity (Chaudhary et al. 2019; Peng et al. 2014). These approaches offer a flexible and potentially broad-spectrum method of virus control, especially when combined with host resistance genes. For nematode resistance, host-delivered RNAi constructs targeting essential parasitism genes of root-knot nematodes have been developed. Additionally, transgenic lines expressing proteinase inhibitors and cystatins disrupt nematode digestion and development, resulting in reduced infection rates. These biotechnological solutions are particularly useful where natural resistance sources, such as *Mi-1.2*, are ineffective against emerging nematode populations or high-temperature conditions that suppress gene expression.

Despite promising results in laboratory and greenhouse conditions, the commercial deployment of transgenic tomatoes has faced significant hurdles. Regulatory constraints, high development costs, and ongoing public resistance to genetically modified organisms (GMOs) have limited field release and adoption. Furthermore, reliance on single transgenes raises concerns about durability, as pathogen populations can evolve to evade recognition or suppress transgene-mediated defenses (Peng et al. 2014). For instance, *Pto*-based resistance can be overcome by *P. syringae* strains lacking the corresponding *AvrPto* effector. To overcome these limitations, integrative approaches are being considered. Combining genetic transformation with marker-assisted selection, gene pyramiding, and genomic tools could enhance resistance durability and field relevance. Additionally, advances in genome editing technologies such as CRISPR/Cas offer new avenues to engineer resistance without introducing foreign DNA, potentially improving public acceptance and regulatory feasibility (Jahan et al. 2025). In summary, genetic transformation has expanded the toolbox for developing biotic stress-resistant tomato varieties. While current applications of these technologies remain largely experimental due to regulatory and social constraints, the continued refinement of transgenic and genome-editing technologies holds promise for future deployment in sustainable tomato production systems.

9.11 GENOME EDITING FOR BIOTIC STRESS TOLERANCE IN TOMATO

The emergence of genome editing technologies, particularly CRISPR/Cas9, has revolutionized plant breeding by enabling precise and efficient modifications of the tomato genome without introducing foreign DNA. This innovation has opened new avenues for developing biotic stress-tolerant tomato cultivars that are not only scientifically advanced but also more likely to meet regulatory standards and gain consumer acceptance compared to traditional transgenic approaches.

One of the most impactful applications of genome editing is the disruption of host susceptibility (S) genes—genes that pathogens exploit to facilitate their own infections. For example, knockout of *SlJAZ2*, a repressor in the jasmonate signaling pathway targeted by *Xanthomonas*, has conferred resistance to bacterial spot without negatively impacting plant growth or yield. Similar edits targeting *SlLOB1*, *SlMlo1*, and other negative regulators of immunity have also shown enhanced resistance to a range of pathogens, including *Pseudomonas syringae* and fungal agents (Saini et al. 2025). Genome editing has also been utilized to target key susceptibility genes involved in interactions between viruses and nematodes. Editing of *SlDCL2*, a gene associated with RNA silencing pathways, has improved resistance to RNA viruses such as TYLCV. Modifications of host factors like *eIF4E*, essential for viral replication, and root-associated genes involved in gall formation and nematode feeding have demonstrated potential to suppress virus and nematode infections (Dutta et al. 2023). These interventions can disrupt pathogen life cycles with high specificity and minimal trade-offs in plant performance.

Another promising application is the fine-tuning of defense-related genes by editing regulatory elements, such as promoters or enhancers, to modulate gene expression rather than completely knocking out gene function. This strategy allows for more nuanced control over plant defense responses, potentially avoiding the fitness penalties often associated with constitutive defense activation (Yamamoto et al. 2018). Such modifications have been employed to upregulate reactive oxygen species (ROS) signaling and transcriptional networks, thereby bolstering resistance without compromising productivity. Multiplex genome editing—targeting multiple loci simultaneously—has emerged as a powerful approach for stacking resistance traits without the complications of linkage drag associated with conventional breeding. This strategy enables the coordinated editing of several genes involved in immune signaling, receptor function, or defense modulation,

TABLE 9.4
Summary of Various Aspects in Tomato Breeding in Various Chronological Period

Aspect	1990s–2000s	Recent Advances
Genetic Resources	Wild relatives, major genes (e.g., *Rx3*)	Expanded wild germplasm, QTLs, cloned genes (*RxopJ4*)
Resistance Mechanisms	ETI via major R genes	ETI + PTI, S-gene targets, quantitative resistance
Breeding Strategies	Phenotypic selection, introgression	MAS, pyramiding, genomic selection
MAS	Low-resolution markers	SNPs, gene-specific markers for *Rx* genes
Genetic Transformation	AMP and PR gene expression	Limited adoption due to regulation
Genome Editing	Not available	CRISPR/Cas9 editing of S genes and defense regulators

thereby enhancing the breadth and durability of resistance (Blix et al. 2021; Gartland et al. 2017; Moradpour and Abdulah 2020). Efforts are also underway to engineer synthetic R genes or chimeric receptor constructs to expand recognition of pathogen effectors, offering a novel layer of resistance engineering.

In addition to targeting protein-coding genes, genome editing is being extended to non-coding regulatory regions that influence gene expression patterns during pathogen attack. This includes editing cis-elements that control the transcription of defense genes or miRNAs involved in immunity (Anupam et al. 2020; Kaur et al. 2021; Kumar et al. 2019). Such modifications can lead to heritable, stable enhancements in resistance pathways while maintaining the genetic integrity of the plant. Overall, genome editing represents a transformative tool for developing biotic stress-resistant tomato varieties. Its precision, flexibility, and potential to produce non-transgenic or cis-genic plants make it especially valuable in the context of evolving regulatory landscapes and public concerns over GMOs. As this technology matures, its integration into mainstream breeding programs will likely accelerate the release of tomato cultivars with broad-spectrum and durable resistance to major pathogens. A summary of technological advancement for biotic resistance breeding is presented in Table 9.4.

9.12 OMICS APPROACH FOR BIOTIC STRESS TOLERANCE IN TOMATO

Omics technologies have become indispensable tools for understanding the complex biological responses of tomato plants to biotic stresses (Van Weymers et al. 2016). These high-throughput approaches offer a systems-level understanding of the interactions between the host and various pathogens, facilitating the identification of resistance mechanisms and potential targets for crop improvement.

Genomics plays a foundational role by uncovering resistance (R) genes and associated quantitative trait loci (QTLs) across the tomato genome. Through comparative and structural genomics, researchers can identify genetic variations, alleles, and gene families involved in pathogen recognition and signal transduction (Rothan et al. 2016; Rothan et al. 2019; Van Weymers et al. 2016). Whole-genome sequencing and resequencing efforts, especially of wild relatives, have greatly expanded the repertoire of resistance genes available for breeding.

Transcriptomics provides insights into the dynamics of gene expression during pathogen attack. By profiling mRNA levels in infected versus healthy tissues, researchers can pinpoint differentially expressed genes involved in early defense responses, hormonal regulation, and secondary metabolism. RNA-seq studies have been pivotal in identifying transcription factors and signaling cascades that are activated or suppressed during specific pathogen interactions (Amoroso et al. 2023; Rosli and Martin 2015).

Proteomics complements transcriptomic data by revealing the actual protein products present during infection, including those that are post-translationally modified or differentially expressed. Since proteins are the functional molecules driving defense responses, this layer of information is crucial for identifying key enzymes, receptors, and pathogenesis-related (PR) proteins that mediate resistance (Hurley et al. 2014; Lautenbacher et al. 2022; Parker et al. 2013).

Metabolomics adds another dimension by capturing the biochemical shifts in plant metabolism under stress. Changes in primary and secondary metabolites, such as phenolics, alkaloids, and phytoalexins, reflect the plant's physiological adjustments and its readiness for defense. Metabolite profiling can also reveal biomarkers associated with specific resistance traits (Lopez-Gresa et al. 2010).

The integration of multi-omics data, linking genes, transcripts, proteins, and metabolites, enables a holistic understanding of defense networks and accelerates the identification of candidate genes for marker-assisted selection, genome editing, or transgenic interventions (Wang et al. 2022). This system biology approach can uncover novel resistance pathways and interactions that would be difficult to detect through single-layer studies.

9.13 CHALLENGES, LIMITATIONS, AND FUTURE PROSPECTS

Despite significant progress, several challenges persist in breeding tomatoes for biotic stress tolerance. One major obstacle is pathogen variability, as many pathogens evolve rapidly and can overcome single resistance genes, leading to breakdowns in previously effective defenses. Linkage drag is another concern, particularly when resistance genes are introgressed from wild species that also carry undesirable traits affecting yield or quality. Additionally, the use of advanced genetic tools often encounters regulatory hurdles, especially for genetically modified or gene-edited organisms. Public acceptance and complex approval processes can delay the deployment of improved cultivars, despite their potential benefits.

Looking forward, gene pyramiding, combining multiple resistance genes, offers a promising solution to enhance the durability and spectrum of resistance. Speed breeding, which utilizes controlled environments to reduce generation times, can expedite the development of resistant lines. Precision breeding techniques, particularly genome editing, provide a targeted approach to introduce or modify traits without affecting the overall genetic background. Importantly, these genetic strategies should be integrated into broader Integrated Pest Management (IPM) programs. Combining resistant cultivars with cultural practices, biological control agents, and informed chemical applications can create a more sustainable and resilient production system.

9.14 CONCLUSION

Breeding tomato cultivars with resistance to biotic stress remains a cornerstone of sustainable crop production. Over the past three decades, resistance breeding has transformed from classical phenotypic selection and single-gene introgression into a multi-disciplinary endeavor that harnesses advances in molecular biology, genomics, and biotechnology. This evolution reflects the urgent need to combat an array of pathogens, including bacteria, fungi, viruses, and nematodes, that threaten global tomato yields and quality. For bacterial diseases such as speck and spot, the integration of major resistance genes like *Pto* and *Bs2* with quantitative trait loci (QTLs), marker-assisted selection (MAS), and, more recently, CRISPR-based genome editing has enabled the development of cultivars with broader and more durable resistance. However, the continuous emergence of new pathogen races and the high genetic variability of pathogens underscore the need for ongoing surveillance and innovation in resistance strategies.

Similar advancements have been observed in breeding for resistance to fungal, viral, and nematode infections. The identification of key R genes, incorporation of wild germplasm, and application of gene pyramiding and precision editing techniques have significantly enhanced the breeding pipeline. Tools such as RNA interference (RNAi), multi-omics analyses, and advanced phenotyping are now integral to dissecting host-pathogen interactions and improving trait selection efficiency. Despite this progress, several challenges persist. Pathogen evolution, environmental variability, and linkage drag from wild relatives continue to complicate resistance deployment. Moreover, regulatory barriers and public skepticism regarding genetically modified organisms still hinder the field application of some biotechnological innovations. To overcome these obstacles, future breeding efforts must emphasize multi-layered resistance strategies, combining pattern-triggered immunity (PTI), effector-triggered immunity (ETI), and quantitative resistance, in conjunction with robust validation across diverse environments. Looking ahead, the integration of genomic, transcriptomic, proteomic, and metabolomic data will enhance our mechanistic understanding of plant immunity and guide the identification of novel resistance sources. The convergence of speed breeding, genome editing, and data-driven selection will accelerate the development of next-generation tomato cultivars tailored for resilience in a rapidly changing climate. Ultimately, sustained investment in research, germplasm characterization, and policy support, paired with responsible communication and public engagement, will be essential to ensure the long-term success and adoption of resistant tomato varieties. By uniting traditional knowledge with cutting-edge tools, the global tomato breeding community is well-positioned to deliver durable solutions to the persistent and evolving threat of biotic stress.

REFERENCES

Abramovitch RB, Kim YJ, Chen S, Dickman MB, Martin GB (2003) *Pseudomonas* type iii effector *AvrPtoB* induces plant disease susceptibility by inhibition of host programmed cell death. *EMBO J* 22:60–69. https://doi.org/10.1093/emboj/cdg006.

Adhikari P, Adhikari TB, Louws FJ, Panthee DR (2020) Advances and challenges in bacterial spot resistance breeding in tomato (*Solanum lycopersicum* L.). *Int J Mol Sci* 21. https://doi.org/10.3390/ijms21051734.

Adhikari P, Oh Y, Panthee DR (2017) Current status of early blight resistance in tomato: An update. *Int J Mol Sci* 18. https://doi.org/10.3390/ijms18102019.

Adhikari P, Siddique MI, Louws FJ, Panthee DR (2023) Identification of quantitative trait loci associated with bacterial spot race t4 resistance in intra-specific populations of tomato (*Solanum lycopersicum* L.). *PLoS ONE* 18:e0295551. https://doi.org/10.1371/journal.pone.0295551.

Amoroso CG, D'Esposito D, Aiese Cigliano R, Ercolano MR (2023) Comparison of tomato transcriptomic profiles reveals overlapping patterns in abiotic and biotic stress responses. *Int J Mol Sci* 24. https://doi.org/10.3390/ijms24044061.

Anikina I, Kamarova A, Issayeva K, Issakhanova S, Mustafayeva N, Insebayeva M, Mukhamedzhanova A, Khan SM, Ahmad Z, Lho LH, Han H, Raposo A (2023) Plant protection from virus: A review of different approaches. *Front Plant Sci* 14:1163270. https://doi.org/10.3389/fpls.2023.1163270.

Anupam, Dhillon NK, Kaur S, Jindal SK, Buttar HS (2020) Phenotypic and genotypic characterization of tomato genotypes for resistance to root-knot nematode, *Meloidogyne incognita*. *Phytoprotection* 100:28–33. https://doi.org/10.7202/1075563ar.

Bai Y, Lindhout P (2005) New challenges for durable resistance breeding in tomato. In *15th Meeting of the EUCARPIA-Tomato-Working-Group*, Bari, Italy, pp. 61–73.

Barbary A, Palloix A, Fazari A, Marteu N, Castagnone-Sereno P, Djian-Caporalino C (2014) The plant genetic background affects the efficiency of the pepper major nematode resistance genes me1 and me3. *Theor Appl Genet* 127:499–507. https://doi.org/10.1007/s00122-013-2235-1.

Barone A, Ercolano MR, Langella R, Monti L, Frusciante L (2005) Molecular marker-assisted selection for pyramiding resistance genes in tomato. *Adv Hortic Sci* 19:147–152.

Bhattarai K, Louws FJ, Williamson JD, Panthee DR (2017) Resistance to *Xanthomonas perforans* race T4 causing bacterial spot in tomato breeding lines. *Plant Pathol* 66:1103–1109. https://doi.org/10.1111/ppa.12656.

Blix TB, Dalmo RA, Wargelius A, Myhr AI (2021) Genome editing on finfish: Current status and implications for sustainability. *Rev Aquacult* 13:2344–2363. https://doi.org/10.1111/raq.12571.

Chaudhary J, Khatri P, Singla P, Kumawat S, Kumari A, R V, Vikram A, Jindal SK, Kardile H, Kumar R, Sonah H, Deshmukh R (2019) Advances in omics approaches for abiotic stress tolerance in tomato. *Biology (Basel)* 8. https://doi.org/10.3390/biology8040090.

Chetelat R (2001) Plant exploration in northern Chile to collect wild tomato species, with emphasis on *Solanum lycopersicoides* and *S. Sitiens*. Final report 6/28/2001, TGRC. http://tgrc.ucdavis.edu/chile.html.

Cruz L, Cruz J, Eloy M, Oliveira H, Vaz H, Tenreiro R (2010) First report of bacterial speck of tomato caused by *Pseudomonas syringae* pv. *Tomato* race 1 in Portugal. *Plant Dis* 94:1504. https://doi.org/10.1094/PDIS-06-10-0415.

Daware A, Das S, Srivastava R, Badoni S, Singh AK, Agarwal P, Parida SK, Tyagi AK (2016) An efficient strategy combining SSR markers- and advanced QTL-seq-driven QTL mapping unravels candidate genes regulating grain weight in rice. *Front Plant Sci* 7:1535. https://doi.org/10.3389/fpls.2016.01535.

Devran Z, Göknur A, Mesci L (2016) Development of molecular markers for the *Mi-1* gene in tomato using the KASP genotyping assay. *Hort Environ Biotechnol* 57:156–160. https://doi.org/10.1007/s13580-016-0028-6.

Dockter KG, O'Neil DS, Price DL, Scott J, Stevens MR (2009) Molecular mapping of the tomato spotted wilt virus resistance gene Sw-7 in tomato. *Hortscience* 44:1123–1123.

Dodds PN, Rathjen JP (2010) Plant immunity: Towards an integrated view of plant-pathogen interactions. *Nat Rev Genet* 11:539–548. https://doi.org/10.1038/nrg2812.

Dong J, Xiao F, Fan F, Gu L, Cang H, Martin GB, Chai J (2009) Crystal structure of the complex between *Pseudomonas* effector *AvrPtoB* and the tomato *Pto kinase* reveals both a shared and a unique interface compared with *AvrPto-Pto*. *Plant Cell* 21:1846–1859. https://doi.org/10.1105/tpc.109.066878.

Dutta TK, Ray S, Phani V (2023) The status of the *CRISPR-Cas9* research in plant-nematode interactions. *Planta* 258:103. https://doi.org/10.1007/s00425-023-04259-0.

El Mohtar CA, Atamian HS, Dagher RB, Abou-Jawdah Y, Salus MS, Maxwell DP (2007) Marker-assisted selection of tomato genotypes with the *I-2* gene for resistance to *Fusarium oxysporum* f. Sp. *Lycopersici* race 2. *Plant Dis* 91:758–762. https://doi.org/10.1094/PDIS-91-6-0758.

El-Sappah AH, M MI, H HE-A, Yan S, Qi S, Liu J, Cheng GT, Liang Y (2019) Tomato natural resistance genes in controlling the root-knot nematode. *Genes (Basel)* 10. https://doi.org/10.3390/genes10110925.

Engle LM, Shanmugasundaram S, Hanson P (2003) Evaluation and utilization of vegetable genetic resources for the tropics. In: PL Forsline, et al. (Eds.), *Plant Genetic Resources: The Fabric of Horticulture's Future*, pp. 263–274. University of Guelph, Canada.

FAOSTAT (2022) World production (tonnes) of tomatoes in 2020. Food and Agriculture Organization of the United Nations (FAO), Rome, Italy. http://www.fao.org/faostat/en/#data.

Foolad MR, Ashrafi H (2015a) Characterization of early blight resistance in a recombinant inbred line population of tomato: I. Heritability and trait correlations. *Adv Stud Biol* 7:131–148.

Foolad MR, Ashrafi H (2015b) Characterization of early blight resistance in a recombinant inbred line population of tomato: Ii. Identification of QTLS and their co-localization with candidate resistance genes. *Adv Stud Biol* 7:149–168.

Foolad MR, Merk HL, Ashrafi H (2008) Genetics, genomics and breeding of late blight and early blight resistance in tomato. *Crit Rev Plant Sci* 27:75–107. https://doi.org/10.1080/07352680802147353.

Foolad MR, Panthee DR (2012) Marker-assisted selection in tomato breeding. *Crit Rev Plant Sci* 31:93–123.

Foolad MR, Sullenberger MT, Ashrafi H (2015) Detached-leaflet evaluation of tomato germplasm for late blight resistance and its correspondence to field and greenhouse screenings. *Plant Dis* 99:718–722. https://doi.org/10.1094/PDIS-08-14-0794-RE.

Gardner RG, Panthee DR (2012) Tomato spotted wilt virus-resistant fresh-market tomato breeding lines: Nc 58s, nc 123s, nc 127s, and nc 132s. *HortScience* 47:531–532.

Gartland KMA, Dundar M, Beccari T, Magni MV, Gartland JS (2017) Advances in biotechnology: Genomics and genome editing. *EuroBiotech J* 1:2–9. https://doi.org/10.24190/issn2564-615x/2017/01.02.

Hamza AA, Robene-Soustrade I, Jouen E, Gagnevin L, Lefeuvre P, Chiroleu F, Pruvost O (2010) Genetic and pathological diversity among *Xanthomonas* strains responsible for bacterial spot on tomato and pepper in the Southwest Indian Ocean region. *Plant Dis* 94:993–999. https://doi.org/10.1094/PDIS-94-8-0993.

Hanson PM, Bernacchi D, Green S, Tanksley SD, Muniyappa V, Padmaja S, Chen HM, Kuo G, Fang D, Chen JT (2000) Mapping a wild tomato introgression associated with tomato yellow leaf curl virus resistance in a cultivated tomato line. *J Am Soc Hortic Sci* 125:15–20.

Hanson PM, Lu S-F, Wang J-F, Chen W, Kenyon L, Tan C-W, Tee KL, Wang Y-Y, Hsu Y-C, Schafleitner R, Ledesma D, Yang R-Y (2016) Conventional and molecular marker-assisted selection and pyramiding of genes for multiple disease resistance in tomato. *Sci Hortic* 201:346–354. https://doi.org/10.1016/j.scienta.2016.02.020.

Haussmann BIG, Parzies HK, Presterl T, Susic Z, Miedaner T (2004) Plant genetic resources in crop improvement. *Plant Genetic Resources Characterization and Utilization* 2:ISSN 1479–2621(print)| 1479–263x(electronic).

Horvath DM, Pauly MH, Hutton SF, Vallad GE, Scott JW, Jones JB, Stall RE, Dahlbeck D, Staskawicz BJ, Tricoli D, Deynze AV (2015) The pepper *bs2* gene confers effective field resistance to bacterial leaf spot and yield enhancement in Florida tomatoes. In: ML Paret, et al. (Eds.), *IV International Symposium on Tomato Diseases*, pp. 47–51. International Society for Horticultural Science.

Hurley B, Subramaniam R, Guttman DS, Desveaux D (2014) Proteomics of effector-triggered immunity (ETI) in plants. *Virulence* 5:752–760. https://doi.org/10.4161/viru.36329.

Hutton SF, Scott JW, Yang W, Sim SC, Francis DM, Jones JB (2010) Identification of QTL associated with resistance to bacterial spot race T4 in tomato. *Theor Appl Genet* 121:1275–1287. https://doi.org/10.1007/s00122-010-1387-5.

Jahan I, Angon PB, Mou SJ, Zannat M, Antu UB, Alam MM, Sweety Akivha a, Islam MS, Sakil MA (2025) The potential of CRISPR-Cas genome editing technologies to mitigate biotic stress in plants. *CABI Agric Biosci* 6. https://doi.org/10.1079/ab.2025.0005.

Ji Y, Salus MS, Van Betteray B, Smeets J, Jensen KS, Martin CT, Mejía L, Scott JW, Havey MJ, Maxwell DP (2007a) Co-dominant scar markers for detection of the *Ty-3* and *Ty-3a* loci from Solanum chilense at 25 cm of chromosome 6 of tomato. *Rep Tomato Genet Coop* 57:25–28.

Ji Y, Schuster DJ, Scott JW (2007b) *Ty-3*, a begomovirus resistance locus near the tomato yellow leaf curl virus resistance locus *Ty-1* on chromosome 6 of tomato. *Mol Breed* 20:271–284.

Kaur H, Pandey DK, Goutam U, Kumar V (2021) *CRISPR/Cas9*-mediated genome editing is revolutionizing the improvement of horticultural crops: Recent advances and future prospects. *Sci Hortic* 289. https://doi.org/10.1016/j.scienta.2021.110476.

Kawchuk LM, Hachey J, Lynch DR (1998) Development of sequence characterized DNA markers linked to a dominant verticillium wilt resistance gene in tomato. *Genome* 41:91–95.

Kawchuk LM, Hachey J, Lynch DR, Kulcsar F, Van Rooijen G, Waterer DR, Robertson A, Kokko E, Byers R, Howard RJ, Fischer R, Prufer D (2001) Tomato *Ve* disease resistance genes encode cell surface-like receptors. *Proc Natl Acad Sci U S A* 98:6511–6515. https://doi.org/10.1073/pnas.091114198.

Khan RaA, Alam SS, Najeeb S, Ali A, Ahmad A, Shakoor A, Tong L (2023) Mitigating cd and bacterial wilt stress in tomato plants through trico-synthesized silicon nanoparticles and trichoderma metabolites. *Environ Pollut* 333:122041. https://doi.org/10.1016/j.envpol.2023.122041.

Khojasteh M, Darzi Ramandi H, Taghavi SM, Taheri A, Rahmanzadeh A, Chen G, Foolad MR, Osdaghi E (2024) Unraveling the genetic basis of quantitative resistance to diseases in tomato: A meta-QTL analysis and mining of transcript profiles. *Plant Cell Rep* 43:184. https://doi.org/10.1007/s00299-024-03268-x.

Kim MJ, Mutschler MA (2005) Transfer to processing tomato and characterization of late blight resistance derived from *Solanum pimpinellifolium* L. L3708. *J Am Soc Horticult Sci* 130:877–884.

King SR, Davis AR, Zhang X, Crosby K (2010) Genetics, breeding and selection of rootstocks for Solanaceae and Cucurbitaceae. *Sci Hortic* 127:106–111. https://doi.org/10.1016/j.scienta.2010.08.001.

Kumar A, Jindal SK, Dhaliwal MS, Sharma A, Kaur S, Jain S (2019) Gene pyramiding for elite tomato genotypes against ToLCV (begomovirus spp.), late blight (*Phytophthora infestans*) and RKN (*Meloidogyne* spp.) for Northern India farmers. *Physiol Mol Biol Plants* 25:1197–1209. https://doi.org/10.1007/s12298-019-00700-5.

Kunwar S, Hsu YC, Lu SF, Wang JF, Jones JB, Hutton S, Paret M, Hanson P, Flachowsky H (2019) Characterization 3333 of tomato (*Solanum lycopersicum*) accessions for resistance to phylotype I and phylotype II strains of the *Ralstonia solanacearum* species complex under high temperatures. *Plant Breeding* 139:389–401. https://doi.org/10.1111/pbr.12767.

Lanfermeijer FC, Dijkhuis J, Sturre MJ, De Haan P, Hille J (2003) Cloning and characterization of the durable tomato mosaic virus resistance gene *Tm-2(2)* from *Lycopersicon esculentum*. *Plant Mol Biol* 52:1037–1049. https://doi.org/10.1023/a:1025434519282.

Lanfermeijer FC, Warmink J, Hille J (2005) The products of the broken *Tm-2* and the durable *Tm-2(2)* resistance genes from tomato differ in four amino acids. *J Exp Bot* 56:2925–2933. https://doi.org/10.1093/jxb/eri288.

Lautenbacher L, Samaras P, Muller J, Grafberger A, Shraideh M, Rank J, Fuchs ST, Schmidt TK, The M, Dallago C, Wittges H, Rost B, Krcmar H, Kuster B, Wilhelm M (2022) Proteomicsdb: Toward a fair open-source resource for life-science research. *Nucleic Acids Res* 50:D1541–D1552. https://doi.org/10.1093/nar/gkab1026.

Lebeau A, Daunay MC, Frary A, Palloix A, Wang JF, Dintinger J, Chiroleu F, Wicker E, Prior P (2011) Bacterial wilt resistance in tomato, pepper, and eggplant: Genetic resources respond to diverse strains in the *Ralstonia solanacearum* species complex. *Phytopathology* 101:154–165. https://doi.org/10.1094/PHYTO-02-10-0048.

Lim GTT, Wang GP, Hemming MN, Basuki S, Mcgrath DJ, Carroll BJ, Jones DA (2006) Mapping the *I-3* gene for resistance to Fusarium wilt in tomato: Application of an *I-3* marker in tomato improvement and progress towards the cloning of I-3. *Australas Plant Pathol* 35:671–680.

Lin NC, Martin GB (2007) *Pto-* and *Prf*-mediated recognition of *AvrPto* and *AvrPtoB* restricts the ability of diverse *Pseudomonas syringae* pathovars to infect tomato. *Mol Plant Microbe Interact* 20:806–815. https://doi.org/10.1094/MPMI-20-7-0806.

Lopez-Gresa MP, Maltese F, Belles JM, Conejero V, Kim HK, Choi YH, Verpoorte R (2010) Metabolic response of tomato leaves upon different plant-pathogen interactions. *Phytochem Anal* 21:89–94. https://doi.org/10.1002/pca.1179.

Martin GB, Brommonschenkel SH, Chunwongse J, Frary A, Ganal MW, Spivey R, Wu T, Earle ED, Tanksley SD (1993a) Map-based cloning of a protein kinase gene conferring disease resistance in tomato. *Science* 262:1432–1436. https://doi.org/10.1126/science.7902614.

Martin GB, De Vicente MC, Tanksley SD (1993b) High-resolution linkage analysis and physical characterization of the *Pto* bacterial resistance locus in tomato. *Mol Plant Microbe Interact* 6:26–26.

Moradpour M, Abdulah SNA (2020) *CRISPR/dCas9* platforms in plants: Strategies and applications beyond genome editing. *Plant Biotechnol J* 18:32–44. https://doi.org/10.1111/pbi.13232.

Nascimento FDS, Rocha AJ, Soares J, Mascarenhas MS, Ferreira MDS, Morais Lino LS, Ramos APS, Diniz LEC, Mendes TaO, Ferreira CF, Santos-Serejo JaD, Amorim EP (2023) Gene editing for plant resistance to abiotic factors: A systematic review. *Plants (Basel)* 12. https://doi.org/10.3390/plants12020305.

Niederheitmann M, Gonçalves J, Mat A, Bentley T, Green J (2015) Interaction between Fusarium wilt resistance genes in tomato and *Fusarium oxysporum* f. Sp. *lycopersici* races 1 and 2. *Acta Hortic*:77–80.

Ori N, Eshed Y, Paran I, Presting G, Aviv D, Tanksley S, Zamir D, Fluhr R (1997) The *I2C* family from the wilt disease resistance locus *I2* belongs to the nucleotide binding, leucine-rich repeat superfamily of plant resistance genes. *Plant Cell* 9:521–532. https://doi.org/10.1105/tpc.9.4.521.

Pandey A, Paudel R, Adhikari TB, Panthee DR, Louws FJ (2024) Septoria leaf spot of tomatoes: Historical insights, present challenges, and future prospects. *Horticulturae* 10. https://doi.org/10.3390/horticulturae10121299.

Panthee DR, Brown AF, Yousef GG, Ibrahem R, Anderson C, Havey M (2013) Novel molecular marker associated with *Tm2a* gene conferring resistance to tomato mosaic virus in tomato. *Plant Breeding* 132:413–416. https://doi.org/10.1111/pbr.12076.

Panthee DR, Gardner RG (2011) 'Mountain majesty': A tomato spotted wilt virus-resistant fresh-market hybrid tomato and its parents NC 714 and NC 1CS. *HortScience* 46:1321–1323.

Panthee DR, Piotrowski A, Ibrahem R (2017) Mapping quantitative trait loci (QTL) for resistance to late blight in tomato. *Int J Mol Sci* 18. https://doi.org/10.3390/ijms18071589.

Parker J, Koh J, Yoo MJ, Zhu N, Feole M, Yi S, Chen S (2013) Quantitative proteomics of tomato defense against Pseudomonas syringae infection. *Proteomics* 13:1934–1946. https://doi.org/10.1002/pmic.201200402.

Peng JC, Chen TC, Raja JA, Yang CF, Chien WC, Lin CH, Liu FL, Wu HW, Yeh SD (2014) Broad-spectrum transgenic resistance against distinct *Tospovirus* species at the genus level. *PLoS ONE* 9:e96073. https://doi.org/10.1371/journal.pone.0096073.

Pohronezny K, Volin RB (1980) *Bacterial Speck of Tomato Florida Cooperative Extension Service*. Institute of Food and Agricultural Sciences, University of Florida.

Robbins MD, Masud MaT, Panthee DR, Gardner RG, Francis DM, Stevens MR (2010) Marker-assisted selection for coupling phase resistance to *Tomato spotted wilt virus* and *Phytophthora infestans* (late blight) in tomato. *HortScience* 45:1424–1428.

Rosli HG, Martin GB (2015) Functional genomics of tomato for the study of plant immunity. *Brief Funct Genomics* 14:291–301. https://doi.org/10.1093/bfgp/elv007.

Rothan C, Bres C, Garcia V, Just D (2016) Tomato resources for functional genomics. In: M. Causse et al. (Eds.), *The Tomato Genome*, pp. 75–94. Compendium of Plant Genomes.

Rothan C, Diouf I, Causse M (2019) Trait discovery and editing in tomato. *Plant J* 97:73–90. https://doi.org/10.1111/tpj.14152.

Sacco Botto C, Matic S, Moine A, Chitarra W, Nerva L, D'Errico C, Pagliarani C, Noris E (2023) Tomato yellow leaf curl Sardinia virus increases drought tolerance of tomato. *Int J Mol Sci* 24. https://doi.org/10.3390/ijms24032893.

Saini H, Devrani A, Synrem G, Priyanka, Salem KFM (2025) Application of CRISPR technology in plant improvement: An update review. *Adv Agric* 2025. https://doi.org/10.1155/aia/4578877.

Schneider RW, Grogan RG (1977) Bacterial speck of tomato—sources of inoculum and establishment of a resident population. *Phytopathology* 67:388–394.

Scott JW, Francis DM, Miller SA, Somodi GC, Jones JB (2003) Tomato bacterial spot resistance derived from PI 114490; inheritance of resistance to race T2 and relationship across three pathogen races. *J Am Soc Hortic Sci* 128:698–703.

Scott JW, Vallad GE, Stall RE, Jeffrey BJ, Dahlbeck D, Staskawicz BJ (2011) Bacterial spot race T4 resistance and yield enhancement in tomatoes conferred by the pepper *BS2* gene in Florida field trials. *Acta Hortic*:437–440.

Seah S, Williamson VM, Garcia BE, Mejía L, Salus MS, Martin CT, Maxwell DP (2007) Evaluation of a co-dominant scar marker for detection of the MI-1 locus for resistance to root-knot nematode in tomato germplasm. *Report of the Tomato Genetics Cooperatives* 57:37–40.

Sharlach M, Dahlbeck D, Liu L, Chiu J, Jimenez-Gomez JM, Kimura S, Koenig D, Maloof JN, Sinha N, Minsavage GV, Jones JB, Stall RE, Staskawicz BJ (2013) Fine genetic mapping of *RXopJ4*, a bacterial spot disease resistance locus from *Solanum pennellii* LA716. *Theor Appl Genet* 126:601–609. https://doi.org/10.1007/s00122-012-2004-6.

Sharon SM (2012) *Septoria Leaf Spot of Tomato*. Grow: Plant Health Exchange, Scientific Societies.

Siddique MI, Blume J, Zhang JB, Panthee D (2023) Mapping QTL for septoria leaf spot resistance in tomato. *HortScience* 58:S242–S242.

Siddique MI, Silverman E, Louws F, Panthee DR (2024) Quantitative trait loci mapping for bacterial wilt resistance and plant height in tomatoes. *Plants (Basel)* 13. https://doi.org/10.3390/plants13060876.

Sikirou R, Dossoumou ME, Honfoga J, Afari-Sefa V, Srinivasan R, Paret M, Bihon W (2021) Screening of *Amaranthus* sp. Varieties for resistance to bacterial wilt caused by *Ralstonia solanacearum*. *Horticulturae* 7. https://doi.org/10.3390/horticulturae7110465.

Sim SC, Robbins MD, Wijeratne S, Wang H, Yang W, Francis DM (2015) Association analysis for bacterial spot resistance in a directionally selected complex breeding population of tomato. *Phytopathology* 105:1437–1445. https://doi.org/10.1094/PHYTO-02-15-0051-R.

Singh A, Bokolia M (2021) *CRISPR/Cas* for crop improvement a brief review. *Resonance J Sci Educ* 26:227–240. https://doi.org/10.1007/s12045-021-1121-4.

Singh AK, Rai N, Singh RK, Saha S, Rai RK, Singh RP (2017) Genetics of resistance to early blight disease in crosses of wild derivatives of tomato. *Sci Hortic* 219:70–78. https://doi.org/10.1016/j.scienta.2017.01.052.

Singh J, Aggarwal R, Bashyal BM, Darshan K, Parmar P, Saharan MS, Hussain Z, Solanke AU (2021) Transcriptome reprogramming of tomato orchestrate the hormone signaling network of systemic resistance induced by *Chaetomium globosum*. *Front Plant Sci* 12:721193. https://doi.org/10.3389/fpls.2021.721193.

Sreekanta S, Bethke G, Hatsugai N, Tsuda K, Thao A, Wang L, Katagiri F, Glazebrook J (2015) The receptor-like cytoplasmic kinase *PCRK1* contributes to pattern-triggered immunity against *Pseudomonas syringae* in *Arabidopsis thaliana*. *New Phytol* 207:78–90. https://doi.org/10.1111/nph.13345.

Stevens MR, Lamb EM, Rhoads DD (1995) Mapping the *Sw-5* locus for tomato spotted wilt virus resistance in tomatoes using RAPD and RFLP analysis. *Theor Appl Genet* 90:451–456.

Stevens MR, Scott SJ, Gergerich RC (1992) Inheritance of a gene for resistance to *Tomato spotted wilt virus* (TSWV) from *Lycopersicon peruvianum* Mill. *Euphytica* 59:9–17.

Sullenberger MT, Jia M, Gao S, Foolad MR, Wehling P (2018) Genetic analysis of late blight resistance in *Solanum pimpinellifolium* accession PI 270441: Heritability and response to selection. *Plant Breeding* 137:89–96. https://doi.org/10.1111/pbr.12561.

Tai TH, Dahlbeck D, Clark ET, Gajiwala P, Pasion R, Whalen MC, Stall RE, Staskawicz BJ (1999) Expression of the *Bs2* pepper gene confers resistance to bacterial spot disease in tomato. *Proc Natl Acad Sci U S A* 96:14153–14158. https://doi.org/10.1073/pnas.96.24.14153.

Thies JA, Panthee DR (2023) Editorial: Identification, development and use of rootstocks to improve pest and disease resistance of vegetable crops. *Front Plant Sci* 14:1320828. https://doi.org/10.3389/fpls.2023.1320828.

Tiwari JK, Yerasu SR, Rai N, Singh DP, Singh AK, Karkute SG, Singh PM, Behera TK (2022) Progress in marker-assisted selection to genomics-assisted breeding in tomato. *Crit Rev Plant Sci* 41:321–350. https://doi.org/10.1080/07352689.2022.2130361.

Vallad GE, Goodman RM (2004) Systemic acquired resistance and induced systemic resistance in conventional agriculture. *Crop Sci* 44:1920–1934. https://doi.org/10.2135/cropsci2004.1920.

Van Weymers PS, Baker K, Chen X, Harrower B, Cooke DE, Gilroy EM, Birch PR, Thilliez GJ, Lees AK, Lynott JS, Armstrong MR, Mckenzie G, Bryan GJ, Hein I (2016) Utilizing "omic" technologies to identify and prioritize novel sources of resistance to the oomycete pathogen *Phytophthora infestans* in potato germplasm collections. *Front Plant Sci* 7:672. https://doi.org/10.3389/fpls.2016.00672.

Varshney RK, Dubey A (2009) Novel genomic tools and modern genetic and breeding approaches for crop improvement. *J Plant Biochem Biotechnol* 18:127–138.

Verlaan MG, Hutton SF, Ibrahem RM, Kormelink R, Visser RG, Scott JW, Edwards JD, Bai Y (2013) The tomato yellow leaf curl virus resistance genes *Ty-1* and *Ty-3* are allelic and code for DFDGD-class RNA-dependent RNA polymerases. *PLoS Genet* 9:e1003399. https://doi.org/10.1371/journal.pgen.1003399.

Wang JF, Olivier J, Thoquet P, Mangin B, Sauviac L, Grimsley NH (2000) Resistance of tomato line Hawaii7996 to *Ralstonia solanacearum Pss4* in Taiwan is controlled mainly by a major strain-specific locus. *Mol Plant Microbe Interact* 13:6–13. https://doi.org/10.1094/MPMI.2000.13.1.6.

Wang JF, Xiao S, Zheng L, Pan Y, Zhao D, Zhang D, Li Q, Zhu J, Yang Z (2022) Multiomic approaches reveal novel lineage-specific effectors in the potato and tomato early blight pathogen *Alternaria solani*. *Phytopathol Res* 4. https://doi.org/10.1186/s42483-022-00135-z.

Yamamoto E, Matsunaga H, Onogi A, Ohyama A, Miyatake K, Yamaguchi H, Nunome T, Iwata H, Fukuoka H (2017) Efficiency of genomic selection for breeding population design and phenotype prediction in tomato. *Heredity (Edinb)* 118:202–209. https://doi.org/10.1038/hdy.2016.84.

Yamamoto T, Kashojiya S, Kamimura S, Kameyama T, Ariizumi T, Ezura H, Miura K (2018) Application and development of genome editing technologies to the *Solanaceae* plants. *Plant Physiol Biochem* 131:37–46. https://doi.org/10.1016/j.plaphy.2018.02.019.

Yang W, Miller SA, Francis DM, Sacks EJ, Lewis Ivey ML (2005) Resistance in *Lycopersicon esculentum* intraspecific crosses to race T1 strains of *Xanthomonas campestris* pv. *Vesicatoria* causing bacterial spot of tomato. *Phytopathology* 95:519–527.

Zhang C, Liu L, Wang X, Vossen J, Li G, Li T, Zheng Z, Gao J, Guo Y, Visser RG, Li J, Bai Y, Du Y (2014) The *Ph-3* gene from *Solanum pimpinellifolium* encodes CC-NBS-LRR protein conferring resistance to *Phytophthora infestans*. *Theor Appl Genet* 127:1353–1364. https://doi.org/10.1007/s00122-014-2303-1.

10 Status of Tomato Breeding for Abiotic Stress Tolerance

Md Jahid Hasan Jone and Dilip R. Panthee

10.1 INTRODUCTION

Tomato (*Solanum lycopersicum* L., Solanaceae, 2n = 2x = 24), a fleshy edible berry, is among the most economically and nutritionally significant vegetables consumed worldwide (J. Zhang et al. 2023). Tomatoes originated approximately 7,000 years ago in the Andean region (present-day Chile, Bolivia, Colombia, and Peru); however, the domestication history of tomatoes remains partially unknown, particularly regarding the role of human intervention (Razifard et al. 2020). From their native habitat, tomatoes were distributed to Central America, later to Europe, and eventually became a globally cultivated crop (Caspermeyer 2020).

Tomatoes are extensively consumed in fresh and processed forms, such as sauces, pastes, and juices, offering numerous health and nutritional benefits (Yong, Yong, and Ng 2023). They are an abundant source of beneficial bioactive compounds like lycopene, β-carotene, vitamin C, vitamin E, folic acid, flavonoids, and essential minerals such as potassium, calcium, and magnesium (Ali et al. 2020). These nutrients play a vital role in preventing chronic diseases, including cardiovascular conditions and various types of cancer, such as prostate, colorectal, stomach, and pancreatic cancer (Agarwal and Rao 2000; Jiménez Bolaño et al. 2024; Singh et al. 2023). Lycopene, in particular, along with other carotenoids like phytoene, phytofluene, α-carotene, γ-carotene, and neurosporene, contributes to the antioxidant capacity of tomatoes, which helps mitigate oxidative stress and reduces the risk of cancer development (Collins et al. 2022; Martí, Roselló, and Cebolla-Cornejo 2016). Additionally, organic acids like ascorbic acid, citric acid, and malic acid support metabolic health and immune function, further enhancing the nutritional profile of tomatoes (Yuan et al. 2024).

Economically, tomatoes contribute to 16% of global vegetable production, with an estimated market value of $58 billion annually (Hubab et al. 2025). In 2023, global tomato production reached 192.3 million tons, with China (70 million tons), India (20.4 million tons), Turkey (13.3 million tons), and the United States (12.37 million tons) leading as the major producers (FAO 2024). The high demand for tomatoes, with their multiple uses and nutritional benefits, makes them an essential crop for smallholder and commercial farmers. However, tomato production faces significant challenges, with studies indicating a 6% yield decline by 2050 due to global temperature increases and the detrimental consequences of climate change (Eyshi Rezaei and Webber 2022).

Abiotic stresses, including drought, salinity, and extreme temperatures, present substantial threats to tomato production, potentially reducing yields by up to 70% (Kumar 2020). These environmental stresses disrupt critical physiological and metabolic processes, including photosynthesis, growth, and nutrient accumulation, which ultimately degrade productivity and quality (Turan et al. 2023). Traditional tomato cultivars often lack resilience to such stresses, and the increasing frequency and intensity of stress due to climate change underscore the urgent need for developing stress-tolerant cultivars to ensure sustainable tomato production (Mustafa et al. 2025).

This chapter aims to provide an overview of the effects of abiotic stress on tomato growth, yield, and quality, as well as the current advancements and methodologies employed in tomato breeding programs aimed at enhancing abiotic stress tolerance. This chapter will synthesize current studies

DOI: 10.1201/9781003506539-10

and findings, emphasizing future directions and innovative technology that can enhance tomato resilience against changing environmental conditions.

10.2 MAJOR ABIOTIC STRESSES AND THEIR TOLERANCE MECHANISMS

10.2.1 Drought Stress

Drought stress adversely affects the root growth, nutrient uptake and transport, seed germination, seedling emergence, photosynthetic activity, carbon assimilation, dry matter accumulation, flowering, pollen formation, fertilization, and the yield of vegetable crops (Turan et al. 2023). In tomatoes, drought stress is expressed as wilted leaves, reduced stem diameter, blossom end rot, reduced fruit set, and reduced photosynthesis. Furthermore, it can affect the concentration of nutrients such as sugars, acids, and proteins in tomato fruit and lead to a lower yield and fruit quality (Weng et al. 2023). Drought stress affects various morphological, physiological, biochemical, and genetic traits of tomato plants. It increases the chance of pollen sterility affecting flower development and reduces seed production and fruit quality (Conti et al. 2023). At the physiological level, tomato plants recover from drought stress by modulating ABA and pigment levels, altering sugar metabolism, and adjusting photosynthesis (Conti et al. 2023). Interestingly, drought stress can enhance fruit quality by synthesizing different acids, such as ascorbic acid, citric acid, and malic acid (Nahar and Ullah 2017). Recent research has shown that biostimulants can ameliorate drought stress effects in tomato seedlings by improving morphological parameters, chlorophyll and carotenoid accumulation, and osmoprotectant levels (Selli et al. 2014; Turan et al. 2023). Tomato plants utilize various mechanisms to recover from drought stress, including:

- **Osmotic adjustment**: This involves the accumulation of compatible solutes, including proline, glucose, fructose, and sucrose. Studies have shown increases of glucose (100%), fructose (30%), sucrose (72%), and proline (345%) under severe water stress compared to control conditions. These osmolytes help maintain turgor pressure and protect cellular components under water-deficient conditions (Nahar and Ullah 2017).
- **Root system architecture modifications**: Tomato plants adapt their root systems to access deeper water sources and increase water uptake efficiency during drought periods (M'hamdi et al. 2023).
- **Stomatal regulation**: Plants control water loss by regulating stomatal aperture, which is mediated by ABA signaling pathways (Munemasa et al. 2015).
- **Antioxidant defenses**: Enhanced activity of antioxidants helps protect against drought-induced oxidative damage in tomato plants (Turan et al. 2023).

10.2.2 Salinity Stress

Salinity stress adversely affects tomato plants, reducing growth, yield, and fruit quality. High salinity can induce osmotic stress, nutritional imbalances, ion toxicity, and disrupted ion homeostasis, ultimately inhibiting numerous physiological processes (Roşca, Mihalache, and Stoleru 2023). Tomatoes are categorized as moderately salt-sensitive plants, and their salt resistance has been significantly diminished as a result of domestication (Guo et al. 2022). Salt stress in tomatoes is expressed as stunted growth, changes in leaf color (often bluish-green), and potentially premature leaf death. Additional symptoms include reduced root elongation, smaller leaves and stems, lower photosynthesis, chlorosis, necrosis, fruit cracking, decreased fruit size, and an overall decline in plant vigor and yield (Habibi et al. 2024). Under increased salinity stress, tomato plants accumulate higher amounts of Na^+ and Cl^-, while the concentrations of K^+ and Ca^{2+} decrease, compromising ionic homeostasis (Guo et al. 2022). Tomato plants exhibit a range of mechanisms to tolerate salinity stress:

- **Ion transporters and homeostasis**: The salt overly sensitive (SOS) pathway plays a vital role in salt tolerance. Decreased levels of *SOS1* lead to a salt-sensitive phenotype, while overexpression of tomato *SOS2* increases salt tolerance (Guo et al. 2022). The SOS pathway helps regulate Na^+ exclusion from the cytosol and compartmentalization into vacuoles.
- **NHX antiporters**: Transporters, particularly *AtNHX1*, help sequester Na^+ in the vacuole, reducing cytosolic Na^+ concentrations and improving tissue tolerance to salinity (Assaha et al. 2017; Wu 2018).
- **Osmotic adjustment**: Under salt stress, K^+, Na^+, Cl^-, and organic acids serve as the major osmolytes in tomato plants. Proline accumulation in salt-tolerant tomato varieties counteracts the osmotic stress caused by salinity (De la Torre-González et al. 2018; Wang et al. 2011).
- **Hormone signaling**: Abscisic acid, ethylene, and auxin are crucial in responding to osmotic stress. Endogenous ABA enhances short-term tolerance to osmotic stress through osmotic and hydraulic modifications (S. Li et al. 2022; Rowe et al. 2016).

10.2.3 Heat Stress

Heat stress negatively affects tomato plants, reducing vegetative growth and reproductive processes, ultimately leading to lower yields and reduced fruit quality. This stress can cause flower abscission, pollen abortion, and impaired fruit development, leading to fruit set failure. High temperatures can also accelerate fruit ripening, potentially leading to over-ripening and changes in fruit characteristics like size, color, and sunscald on fruits (Graci and Barone 2023). At the molecular level, high temperatures affect carotenoid accumulation via temperature-induced repression of genes, particularly through the regulation of light-sensing mechanisms such as the phytochrome-phytochrome interacting factor (PHY-PIF) tandem (Lymperopoulos, Msanne, and Rabara 2018). Tomato plants can exhibit heat stress tolerance through various mechanisms:

- **Synthesis of Heat Shock Proteins (HSPs)**: These proteins function as molecular chaperones, protecting cellular proteins against heat-induced denaturation and preserving membrane integrity (Hu et al. 2022).
- **Membrane thermostability**: Modifications in membrane lipid composition facilitate the retention of membrane fluidity and functionality at high temperatures.
- **Metabolite accumulation**: High temperature stimulates the accumulation of protective phenolic compounds and increases tocopherol content in tomatoes grown at temperatures around 32°C, which acts as an adaptive mechanism to heat stress (Alhaithloul, Galal, and Seufi 2021).
- **Altered gene expression**: Heat stress triggers specific transcriptional responses that activate protective mechanisms at the cellular level (Bakery et al. 2024).

10.2.4 Cold Stress (Chilling and Freezing)

Cold stress in tomatoes, particularly before blooming, can lead to misshapen or "cat-faced" fruits and affect plant growth and development. This stress can disrupt cell-to-cell communication, alter stem cell populations, and increase callose accumulation, contributing to malformed fruits (Lu et al. 2022). Cold stress in tomato plants is expressed as wilting, yellowing of leaves, leaf injury, reduced seedling vigor, stunted growth, and malformed fruits. These symptoms become more distinct when the temperature goes below 10°C and can negatively impact fruit development and overall yield if exposed for a long period (Adhikari et al. 2022). Cold stress reduces shoot and root water absorption, especially during germination and early growth stages. It adversely impacts seed germination and results in a notable reduction in seedling vigor, hence inhibiting plant growth and development (Hussain et al. 2018). Studies have documented negative consequences of cold stress on various crops, including necrotic lesions in leaves/stems, delayed leaf growth, wilting, and altered cell cycles leading to decreased cell division (Hussain et al. 2018). Tomato plants develop several mechanisms to tolerate cold stress:

- **Expression of CBF transcription factors**: The CBF (C-repeat binding factor) cold response pathway is initiated by rapid cold stimulation of CBF transcription factors, subsequently leading to the activation of regulon genes that induce cold tolerance. In Arabidopsis, three CBF genes are activated within 15 minutes after exposure to cold temperatures (Zarka et al. 2003).
- **Membrane lipid modification**: Alterations in membrane lipid composition facilitate the retention of membrane fluidity at low temperatures, preventing cellular damage (Thompson 1989).
- **Accumulation of cryoprotectants**: Cold acclimation involves the development of low molecular weight cryoprotective metabolites, including raffinose, sucrose, and proline, as well as the synthesis of cryoprotective polypeptides (Doherty et al. 2009).
- **Calcium signaling**: Alterations in cytosolic Ca^{2+} caused by low temperatures correlate with the expression of cold-responsive genes and the acquisition of freezing tolerance. In Arabidopsis, cold stress stimulus triggers Ca^{2+} release from extracellular and vacuolar compartments (Zarka et al. 2003).
- **Antioxidant enzyme activities**: Enhanced activities of antioxidant enzymes help mitigate cold-induced oxidative stress (Eom et al. 2022).

10.2.5 Oxidative Stress

Oxidative stress in tomatoes happens when there's an imbalance between ROS production and detoxification within a plant. This imbalance may be due to various factors, such as biotic and abiotic stresses, and can affect tomato growth, yield, and quality (Zhou et al. 2019). Symptoms may include cell membrane damage, lipid peroxidation, leaf damage, decreased chlorophyll and carotenoid content, and reduced germination and growth, particularly under stress conditions like salinity or water deficit (Sánchez-Rodríguez et al. 2010). ROS are continuously generated in chloroplasts, mitochondria, and peroxisomes through aerobic mechanisms, but their levels increase dramatically under various abiotic stresses (Sachdev et al. 2021). In tomato, salt stress increases oxidative stress in roots earlier and more sensitively than in leaves (Guo et al. 2022). The excessive generation of ROS results in oxidative damage, including lipid peroxidation of cellular membranes and potential cell death. Tomato plants possess several mechanisms to tolerate oxidative stress:

- **Enzymatic antioxidant systems**: These include superoxide dismutase (SOD), ascorbate peroxidase (APX), and catalase (CAT), collectively detoxify ROS. Increased anti-oxidase activity in salt-tolerant tomato genotypes protects against oxidative damage (Guo et al. 2022).
- **Non-enzymatic antioxidants**: These ROS scavengers include carotenoids, ascorbic acid, glutathione, and α-tocopherol.
- **Coordinated defense systems**: The antioxidant defense system and the glyoxalase system work together to detoxify salt-induced ROS and improve salt tolerance in tomatoes (Parvin et al. 2019).
- **Osmolyte accumulation**: Several osmolytes act as ROS scavengers and protect enzymes from oxidative damage, with their role in osmotic adjustment (Zarka et al. 2003).

10.2.6 Nutrient Stress

Nutrient stress in tomatoes occurs when the plant suffers from a deficiency of necessary macro/micronutrients (e.g., N, P, K, Fe). This can lead to various symptoms, including stunted growth, leaf curling, poor fruit development, lower biomass, and chlorosis (Kumar Shreevastav et al. 2022). Tomato plants utilize various mechanisms to tolerate nutrient stress:

- **Adjusting nitrogen remobilization**: Plants optimize the allocation of nitrogen resources to maintain essential functions under nutrient limitation.
- **Efficient nutrient uptake**: Enhanced expression of high-affinity transporters facilitates nutrient acquisition under deficient conditions (Singh et al. 2025).
- **Regulating osmosis**: Osmotic adjustment helps maintain water relations during nutrient stress.
- **Enhancing antioxidant defenses**: Increased activity of antioxidant systems mitigates oxidative damage associated with nutrient imbalances.
- **Altering ion homeostasis**: Plants regulate ion transport and compartmentalization, particularly in response to salinity, to maintain cellular functions under stress (Balasubramaniam et al. 2023).

10.2.7 Heavy Metal Stress

Heavy metals (Cd, Pb, Al) accumulation in soil can negatively impact tomato plant growth, leading to various stress responses and reduced yield (Naciri, Chtouki, and Oukarroum 2024). These metals can accumulate in the plant, interfering with essential processes like photosynthesis and enzyme activity, causing oxidative damage and growth inhibition. Symptoms include stunted growth, chlorosis, and potentially blackening of the root system. Other symptoms include reduced fruit yield, altered fruit quality (like stem-end yellowing in cadmium toxicity), and stunted plant growth (Krupa-Małkiewicz and Ochmian 2024). Tomato plants employ several tolerance mechanisms to cope with heavy metal stress:

- **Metal transporter proteins**: Metal homeostasis transporters (HMA, ABC transporters) help regulate metal uptake, translocation, and sequestration (Pasricha et al. 2021).
- **Avoidance strategies**: Plants can restrict metal absorption through root-level exclusion or binding metals to cell walls or root exudates (Dalvi and Bhalerao 2013).
- **Sequestration and compartmentalization**: Metals can be sequestered in vacuoles or bind to chelating agents such as phytochelatins and metallothioneins (Parida et al. 2024).
- **Antioxidative defense**: Improved antioxidant systems reduce oxidative stress caused by heavy metals. Tomato plants accumulate osmolytes such as proline and antioxidants, including polyphenols and flavonoids, in response to heavy metal stress (Ma et al. 2022).
- **Biochemical detoxification**: Plants can convert toxic metal forms to less toxic forms through various biochemical processes (Padmavathi and Jayalakshmi 2024).

10.2.8 Combined Stress Tolerance

Plants in natural environments rarely face a single stress factor alone but are often exposed to multiple stresses simultaneously. Understanding the mechanisms of combined stress tolerance is crucial for developing climate-resilient tomato varieties. Plants demonstrate stress-specific and shared responses to protect themselves from several environmental stresses (Pandey, Ramegowda, and Senthil-Kumar 2015). Gene expression induced by various stresses exhibits considerable overlap, as plants always experience osmotic and oxidative stress in response to salt, drought, and cold conditions (Ciarmiello et al. 2011). The prevention of osmotic stress is achieved through stress-specific and general tolerance mechanisms. Key aspects of combined stress tolerance include the following:

- **Crosstalk between signaling pathways**: Abiotic and biotic stress responses interact through hormone signaling pathways regulated by salicylic acid, ethylene, jasmonic acid, and abscisic acid (Yang et al. 2019). These hormones regulate several processes of plant growth, development, and stress responses.

- **Transcription factors**: Several transcription factors act as master regulators of multiple stress responses. For example, the CBF/DREB family of transcription factors participates in responses to both cold and drought stress (Wang et al. 2019).
- **Protein kinases**: Mitogen-activated protein kinases (MAPKs) participate in signaling related to development, hormones, abiotic stress, and biotic stress (Majeed et al. 2023). The activation of MAPK cascade components by several stress types indicates their role as intersections for various stress signaling pathways.
- **MicroRNAs**: Recent research has identified microRNAs as essential regulators of gene expression under multiple stresses, incorporating an additional layer of complexity into the understanding of stress tolerance mechanisms (Begum 2022; Q. Li et al. 2023).

Understanding these complicated relationships among various stress signaling pathways will be essential for developing tomato varieties with improved tolerance to multiple stresses, particularly in regard to climate change, which subjects plants to increasing combinations of abiotic stresses.

10.3 GENETIC RESOURCES FOR ABIOTIC STRESS TOLERANCE IN TOMATO

Abiotic stresses such as drought, high/low temperature, and high salinity significantly impact tomato production worldwide, affecting almost every stage of the tomato life cycle and causing up to 70% yield loss depending on stress duration and plant growth stage (Krishna et al. 2019). With climate change intensifying these challenges, exploiting diverse genetic resources for breeding stress-tolerant varieties has become essential for sustainable tomato production (Conti et al. 2023).

Wild relatives of cultivated tomato constitute an invaluable genetic pool for abiotic stress tolerance (Table 10.1). *Solanum pennellii* exhibits remarkable drought tolerance through superior water retention and reduced oxidative damage (Egea et al. 2018), while also having multiple salinity tolerance mechanisms (Yıldız, Doğanlar, and Frary 2025). Other wild species like *S. habrochaites*, *S. peruvianum*, and *S. pimpinellifolium* have demonstrated significant tolerance to various abiotic stresses (Ebert and Chou 2015). Introgression lines derived from these wild species have been successfully utilized in breeding programs to develop stress-tolerant cultivars (A. A. M. Ali et al. 2021).

TABLE 10.1
Wild Tomato Relatives and Their Tolerance to Abiotic Stresses

	Type of Tolerance			
Wild Relatives	**Drought**	**Salt**	**Heat**	**References**
S. habrochaites	Yes	Yes	Yes	Kapazoglou et al. (2023), Frary et al. (2010)
S. pennellii	Yes	Yes	Yes	Foolad and Chen (1998), Foolad and Jones (1993), Gonzalo et al. (2021), Rai and Rai (2006)
S. pimpinellifolium	Yes	Yes	Yes	Foolad and Chen (1998), Gonzalo et al. (2021), Krishna et al. (2022), O'Connell, Medina, and Sánchez Peña (2006)
S. cheesmanii	Yes	–	Yes	Faruq, Zakaria, and Arash (2012), Pailles et al. (2020)
S. chilense	Yes	–	–	Kapazoglou et al. (2023)
S. sitiens	Yes	–	–	Molitor et al. (2021)
S. parviflorum	–	Yes	–	Rao et al. (2015)
S. galapagense	–	Yes	–	Bonarota et al. (2025)
S. peruvianum	Yes	–	Yes	Lee et al. (2023)
S. chmielewskii	–	–	Yes	Nahar and Ullah (2011)

Traditional landraces and locally adapted varieties represent another crucial genetic resource, exhibiting resilience to environmental constraints through generations of natural and farmer selection. Mediterranean tomato landraces demonstrate superior drought adaptation compared to modern cultivars (Pompeiano et al. 2025), while Iranian landraces adapted to dry climates show enhanced tolerance to drought stress (Sousaraei et al. 2021). These traditional varieties often maintain yield stability and fruit quality under stress conditions, making them valuable breeding material (Meza et al. 2020).

Several international institutions maintain extensive tomato germplasm collections, facilitating global access to these genetic resources. The USDA Plant Genetic Resources Unit houses over 6,600 tomato accessions of 15 species, while the Tomato Genetics Resource Center at UC Davis specializes in wild relative collections. The World Vegetable Center (formerly AVRDC) maintains diverse tomato germplasm (8500) screened for abiotic stress tolerance, including salinity-tolerant *S. pimpinellifolium* accessions (Çelik et al. 2023; Ebert and Chou 2015). The efficient use of these genetic resources presents considerable promise for developing tomato varieties with improved tolerance to various abiotic stresses (Egea et al. 2022).

10.4 BREEDING STRATEGIES FOR ABIOTIC STRESS TOLERANCE

10.4.1 Conventional Breeding

10.4.1.1 Germplasm Evaluation and Selection

Germplasm evaluation and selection are an integral approach to crop improvement, particularly for abiotic stress tolerance. This process includes an organized exploration of the genetic diversity of different cultivars or landraces to identify individuals with a desirable trait, for example, drought, heat, or salinity tolerance (Mondal, Kumar, and Gnanesh 2023). When multiple germplasms are screened, superior genotypes that exceed others based on performance under specific environmental stresses can be identified. For example, drought-tolerance assessment of tomato genotypes can be achieved by field screening various morphological and physiological traits, including fruit set, leaf area, and photosynthetic rate under water-stressed conditions. Composite indices combining several traits have been successful in the selection of genotypes with good performance and stress tolerance (Chiwina et al. 2024; Grozeva et al. 2024). Moreover, controlled environment screening enables a more sensitive assessment of stress responses (e.g., stomatal conductance and leaf water potential), which leads to the understanding of tolerance mechanisms (Hameed et al. 2015). In tomato breeding, for example, the identification of landraces with improved drought resistance, with some accessions having shown reduced damage in terms of yield losses and superior physiological responses under water deficit conditions (Grozeva et al. 2024).

10.4.1.2 Hybridization

Hybridization is a basic plant breeding method, where two different varieties or species are crossed to produce hybrids exhibiting superior or new characteristics. Intraspecific hybridization is crossing different varieties of the same species (different tomato varieties) to bring desirable traits such as higher yield or disease resistance together. In interspecific crossbreeding, cultivated tomato species, including *Solanum lycopersicum*, are crossed with wild tomato relatives, such as *S. pimpinellifolium* or *S. peruvianum*, to incorporate beneficial traits, such as abiotic stress-tolerant genes (Kalloo 1991; Murray 2016; Vanlay et al. 2022). This method is significant as it diversifies the gene pool of the cultivated tomato, which is vital to improving tolerance to stresses. For instance, wild tomatoes can readily be crossed with cultivated types to impart resistance to diseases, such as Fusarium wilt, or pests, such as root knot nematodes (Khazaei and Madduri 2022; Klein-Lankhorst et al. 1991). Similar hybridization techniques enable breeders to develop tomato varieties with enhanced abiotic stress tolerance, higher yields, and better quality, thus improving productivity and environmental sustainability in tomato cultivation.

10.4.1.3 Backcross Breeding

Backcross breeding is a plant breeding technique aimed at introgressing a specific desirable trait from one variety (the donor parent) into another (the recurrent parent). This method is particularly effective for improving traits governed by a lower number of genes. The process involves crossing the hybrid offspring back with the recurrent parent for several generations, ensuring that the desired trait (from the donor parent) is retained while maintaining the favorable characteristics of the recurrent parent (Priyadarshan 2019). In tomato breeding, backcrossing has been widely utilized to improve various traits, including heat stress tolerance. By incorporating genes from wild tomato relatives like *S. pimpinellifolium*, which possess heat and drought tolerance, breeders can introduce heat-resistant traits into cultivated tomato varieties (Ayenan et al. 2021). For instance, a backcross introgression line (BIL) population derived from a cross between *S. pimpinellifolium* and *S. lycopersicum* has been evaluated for heat stress-related traits (Bashary et al. 2024). This method has led to the identification of lines with improved reproductive performance under heat stress conditions, offering valuable genetic resources for developing heat-tolerant tomato cultivars. The efficiency of backcross breeding allows for targeted improvements while preserving the elite traits of the recurrent parent, making it an essential tool in crop improvement programs.

10.4.2 Mutation Breeding

Conventional breeding techniques have typically been limited by the genetic variation of the existing cultivars and have thus contributed to a lower degree of trait enhancement. That is a problem with crops such as tomatoes, whose commercial varieties are genetically uniform (Desalew 2023). On the other hand, mutagenic agents (physical, chemical, or biological) have permitted breeders to induce mutations in the genome that can confer favorable phenotypic changes (Raina et al. 2016), introducing genetic variation at a faster pace, offering breeders a larger pool of genetic resources. This increased genetic pool facilitates the acquisition and incorporation of new traits that might not be available in the native (wild) gene pool, including improved tolerance to climate change. Consequently, mutation breeding is important to increase the genetic potential of crops, so that breeders can create varieties that can better withstand environmental stresses (Devi, Kumar, and Khah 2025).

Broadly, there are three types of mutagenesis. Physical mutagenesis (through high-energy radiation), chemical mutagenesis (using chemicals such as Ethyl Methanesulfonate and Sodium Azide), and insertional mutagenesis (in which the gene of interest is specifically inactivated using *Agrobacterium*-mediated T-DNA insertions) are the major techniques used for mutation breeding (Y. Çelik & Üstün, 2017). Following the induction of the mutations, the mutagenized plant population is grown and screened for plants having the desired trait(s), and it is monitored to confirm that such offspring stably appear across generations without losing the desired trait (Oladosu et al. 2016).

Tomato varieties have been developed with enhanced performance under stress conditions by mutagenesis at the molecular level. For example, tomato gamma-ray mutagenesis has been employed to develop drought-resistant mutants with increased yield and quality (Chaudhary et al. 2019), which also shows stability across fluctuating environmental conditions. By targeting individual genes related to stress tolerance, mutation breeding provides a solution to develop tomato varieties that yield high and adapt to diverse environments experiencing different climates.

10.4.3 Molecular Breeding

10.4.3.1 Marker-Assisted Selection and QTL Mapping

DNA markers play a vital role in modern plant breeding as they are helpful tools for genetic identification, diversity analysis, and the development of high-resolution genetic linkage maps (Kumar et al. 2024).

They are divided into PCR-based and non-PCR-based markers. PCR technology has revolutionized genetic studies by allowing effective and economical DNA amplification. Polymorphic PCR-based markers, including SNP, SCAR, and SSR markers, have remarkably promoted genomic studies. Molecular markers provide accurate locations of alleles and modes of inheritance, which are necessary for establishing a linkage map representative of genetic diversity and recombination in hybrid plants (Singh and Singh 2015).

QTL mapping and marker-assisted selection (MAS) are essential techniques in modern tomato breeding, especially for enhancing abiotic stress tolerance. QTL mapping identifies genomic regions linked to particular traits, like drought or salinity resistance, by genotyping segregating mapping populations, through the investigation of genetic linkage between molecular markers (such as EST and SNP) and phenotypic characteristics. This process enables the tracing of the precise location of QTLs on the chromosomes in the form of genetic and linkage maps. QTL mapping allows breeders to understand the genetic basis of complex and quantitative traits, thereby facilitating focused breeding programs. In tomatoes, QTL mapping has been widely used to identify loci linked to tolerance of abiotic stresses, including heat, drought, and salinity (Paliwal et al. 2021). For example, research has identified QTLs associated with traits like osmotic adjustment, root structure, and photosynthetic effectiveness in stressful environments (Table 10.2). These QTLs provide the genetic basis for understanding stress responses and act as a foundation for MAS.

MAS utilizes DNA markers that are strongly associated with favorable QTLs to expedite the breeding process. By choosing individuals with beneficial alleles, breeders can improve stress tolerance without depending exclusively on phenotypic assessments, which are frequently labor-intensive and affected by environmental fluctuations. In tomato, MAS has effectively been utilized to incorporate stress-tolerant features from wild relatives into cultivated types, enhancing resilience while preserving yield and quality (Foolad and Panthee 2012).

The integration of QTL mapping and MAS has revolutionized tomato breeding by enabling precision breeding for abiotic stress tolerance. These breeding techniques not only improve the effectiveness of breeding programs but also contribute to the development of climate-resilient tomato cultivars. With advancements in genomic technologies, the precision of QTL mapping and the effectiveness of MAS are anticipated to be enhanced, thus speeding up the development of stress-tolerant tomato varieties.

10.4.3.2 Genome-Wide Association Study

Advancements in high-throughput genotyping methods have greatly expanded the accessibility to huge genomic data from various germplasm resources in numerous crop varieties (Thomson 2014). Combining genomic and phenomic datasets creates a valuable framework for precise marker-trait association analysis and aids in identifying candidate genes related to crop development, adaptation, and productivity. This robust collection of genomic data, supplemented by strong and accurate phenotypic information, establishes the basis for successful genome-wide association studies (Rai et al. 2021).

GWAS has proven to be a precise method for analyzing the genetic basis of complex traits and is essential in genomics-based crop improvement initiatives. Utilizing genetic diversity among populations, GWAS facilitates the discovery of SNPs that are strongly linked to particular phenotypes (Sahito et al. 2024; Uffelmann et al. 2021). This approach has been effectively utilized in multiple crops, including tomato, to identify loci associated with important agronomic traits like plant height (Z. Liu, Jiang, and Li 2023), drought tolerance (Pessoa et al. 2023), salt resistance (Héreil et al. 2024), and others, demonstrating the effectiveness of GWAS in identifying genes related to significant traits in key crop species.

Additionally, numerous computational tools and software programs like GAPIT (Genome Association and Prediction Integrated Tool) and GEMMA (Genome-wide Efficient Mixed Model Association) are commonly used alongside traditional statistical methods to establish marker-trait

TABLE 10.2
Quantitative Trait Loci Linked to Abiotic Stress Tolerance in Tomato

Stress	Associated Traits	No of Loci	Markers	Chromosome/LG	Mapping population	R^2 %	References
Cold	Germination Index and Radicle Protrusion	5	RFLP	1, 4	BC_1S_1 (NC84173 × LA722)	11.9–43.8	Foolad, Chen, and Lin (1998b)
	Relative Germination Ratio, Chilling Index	9	SSR	1–4, 9, 12	RIL (XF98–7 × LA2184)	0.95–19.5	Liu et al. (2016)
	Chilling Index	4	–	1–3, 9	–	0.95–10.34	Liu et al. (2016)
	Relative Germination Ratio	5	–	1, 4, 9, 12	–	5.95–19.55	Liu et al. (2016)
	Seed germination	3	–	1, 4	BC_1	–	Foolad, Chen, and Lin (1998a)
		5	–	1–3, 12	RILs	–	Foolad (2007)
		6	RFLP	1, 4, 8	BC_1 (NC84173 × LA722)	5.76–30.9	Foolad, Subbiah, and Zhang (2007)
	Root ammonium uptake	10	–	1, 3, 5, 6, 7, 9, 11	BC_1	–	Truco et al. (2000)
Drought	Damage Index in the seedling stage	14	SSR	1–6, 8–10, 12	IL (LA2951, LA0716)	–	Lei, Yan, and JunMing (2011)
	Days to flowering, TSS, acid content	31	SNP	LG1, 2, 4–6, 9–12	141 Accessions	–	Albert et al. (2016)
	Seed germination	4	RFLP	1, 8, 9, 12	BC_1 (NC84173 × LA722)	–	Foolad, Zhang, and Subbiah (2003)
		8	–	1–4, 8, 9, 12	RILs	–	Foolad (2007)
	Survival rate at the seedling stage	15	SNP	1, 3, 4, 8, 9, 10, 12	RIL (Moneymaker × PI365967)	–	Li et al. (2024)
	Water use efficiency	3	–	–	BC_1, F_3	–	Martin et al. (1989)
Heat	Flower and fruit number per truss, stigma exertion, and pollen viability	12	SNP	2, 4, 6, 7, 11	RIL and ILs (Moneymaker × TO-937)	3.6–11.8	Gonzalo et al. (2020)
	Fruit set	6	AFLP	1, 3, 4, 6, 7, 9	F_2 (Jab-95 × Caribe)	32.8	Grilli, Braz, and Lemos (2007)
	Fruit yield, fruit set, TSS	19	SNP	1, 5, 6, 8, 9, 12	71 hybrid and pure-line	–	Elazazi et al. (2024)
	Heat tolerance during the reproductive stage	–	SNP	1, 7, 12	*S. lycopersicum* × *S. pimpinellifolium*	–	Gonzalo et al. (2020)
		–	SNP	1–3, 7, 8, 11	cv. Nagcarlang	–	Yunbi Xu et al. (2017)
	Heat tolerance during the vegetative stage	–	RNA-seq	1, 2	cv. Nagcarlang	–	Wen et al. (2019)
	Pollen viability	29	RAPD, SSR	LG3, LG7	RIL (CLN1621L × CA4)	–	Paliwal et al. (2021)
	Pollen viability and number, style and anther length, style protrusion, female fertility, inflorescence number	13	SNP	1–3, 7, 8, 11	F_2 (Nagcarlang × NCHS-1)	10.5–36.3	Jiemeng Xu, Wolters-Arts, et al. (2017), Yang Xu et al. (2017)
	Pollen viability, number of pollen and flowers, style protrusion, and length	13	SNP	1–3, 9, 11, 12	Advanced backcross populations	7.3–36.3	Ayenan et al. (2019)

(*Continued*)

TABLE 10.2 ***(Continued)***
Quantitative Trait Loci Linked to Abiotic Stress Tolerance in Tomato

Stress	Associated Traits	No of Loci	Markers	Chromosome/LG	Mapping population	R^2 %	References
	Relative EC, chlorophyll content, maximum photochemical quantum efficiency	5	SSR	1, 2	F_2 (LA1698 × LA2093)	1.0–16.48	Wen et al. (2019)
	Anther length	3	–	1, 2, 7	–	11.6–25.2	Jiemeng Xu, Driedonks, et al. (2017)
	Flowers per inflorescence	1	–	1	–	38.7	Jiemeng Xu, Driedonks, et al. (2017)
	Inflorescence number	2	–	1, 8	–	13.4–21.9	Jiemeng Xu, Driedonks, et al. (2017)
	Pollen number	1	–	7	–	18.6	Jiemeng Xu, Driedonks, et al. (2017)
	Pollen viability	1	–	11	–	36.3	Jiemeng Xu, Driedonks, et al. (2017)
	Style length	3	–	1–3	–	10.5–22.7	Jiemeng Xu, Driedonks, et al. (2017)
	Style protrusion	2	–	1, 3	–	19.5–28	Jiemeng Xu, Driedonks, et al. (2017)
Salinity	Fruit weight and number, stem diameter and length, number of leaves up to first flower, and chloride concentration	15	SSR	1, 2, 5, 6, 8, 10–12	F_7 (Cerasiforme × *S. pimpinellifolium*, Cerasiforme × *S. cheesmaniae*)	7–15	Villalta et al. (2007)
	Inositol phosphate pathway	–	SNP	–	*S. pimpinellifolium*	–	Razali et al. (2018)
	Ion accumulation	6	–	1, 2, 4–6, 12	F_2	–	Zamir and Tal (1987)
	Radicle Protrusion, Germination Index	7	RFLP	1, 2, 5, 7, 9, 12	BC_1S_1 (NC84173 × LA722)	6.5–15.6	Foolad et al. (1998b)
	Salinity tolerance during the reproductive stage	–	RFLP	–	*Lycopersicon pimpinellifolium*	–	Monforte, Asíns, and Carbonell (1996)
	Salinity tolerance during the vegetative stage	–	RFLP	1, 3, 5, 9	*S. lycopersicum* × *S. pimpinellifolium*	–	Foolad and Chen (1998)
		–	RFLP	1, 3, 5, 6, 11	*S. lycopersicum* × *S. pimpinellifolium*	–	Foolad, Zhang, and Lin (2001)
		–	SNP	1–3, 10, 11	*S. lycopersicum*, cv. MAGIC.	–	Diouf et al. (2018)
	Seed germination	–	RFLP	1–3, 7–9, 12	*S. pennellii*	–	Foolad et al. (1997)
		–	RFLP	1, 2, 5, 7, 9, 12	*S. pimpinellifolium*	–	Foolad et al. (1998b)
		7	–	1, 2, 5, 7, 9, 12	–	6.3–13.58	Foolad, Subbiah, and Zhang (2007)
	Seedling survival, leaf/plant damage	–	RAD	2, 7–9	*S. pennellii*	–	Bolger et al. (2014)
	Seedling-stage salt tolerance	10	SSR	4, 6, 7, 9, 11, 12	IL (LA716 × M82, LA2951 × VF36)	31.9–64.6	Jinquan Li et al. (2011), Junming Li et al. (2011)
		7	–	4, 6, 9, 12	–	61–64.8	Junming Li et al. (2011)
	Survival Rate at the seedling stage	5	SNP	1, 3, 5, 7, 12	RIL (Moneymaker × PI365967)	–	Li et al. (2024)
N Stress	Leaf N content, Leaf N Balance Index, Petiole NO^{3-} content	27	RNA-seq	1–4, 6–8, 11. 12	MAGIC	–	Desaint et al. (2024)

relationships in aid of GWAS experiments (Tan et al. 2020; Wang and Zhang 2021; Zhou and Stephens 2012).

10.4.3.3 Genomic Selection

Genomic selection (GS) is an emerging predictive breeding approach that uses genome-wide marker data (SNPs) to estimate the genetic potential or breeding value (a measure of genetic potential to transmit superior traits to its offspring) of an individual for complex traits without requiring phenotypic evaluation in each generation. Unlike marker-assisted selection, which targets specific loci, GS considers the cumulative effects of thousands of SNPs, making it particularly effective for quantitative traits governed by many genes with a small effect, such as different abiotic stress tolerance (Verma et al. 2024). The process involves developing a training population that is both genotyped and phenotyped, using this dataset to train statistical models (e.g., non-linear regression, GBLUP, Bayesian methods), and then applying these models to predict genomic estimated breeding values (GEBVs) in unphenotyped individuals (Budhlakoti et al. 2022; Crossa et al. 2025; Y. Zhang et al. 2023). GS enables selection at an earlier stage that can shorten breeding cycles and increase selection accuracy, leading to faster genetic improvement (Alemu et al. 2024).

In tomato breeding, GS has been shown to improve selection efficiency for key traits like fruit weight, firmness, and soluble solids content (Yeon et al. 2024). The application of GS in tomato not only accelerates the breeding cycle but also enhances selection accuracy, particularly under abiotic stress conditions such as drought and salinity (Cappetta et al. 2021). Integration with high-throughput genotyping and phenotyping platforms further strengthens its utility, allowing for environment-specific predictions and the early identification of superior genotypes. Thus, genomic selection offers a powerful strategy to expedite the development of stress-resilient tomato cultivars and improve overall genetic gain in breeding programs.

10.4.4 Genetic Engineering/Transgenic Approach

The limited genetic diversity of tomatoes presents significant challenges in conventional breeding programs that rely on phenotypic selection and require long breeding cycles (Tiwari, Singh, and Behera 2023). Traditional breeding approaches have contributed significantly to developing stress tolerance in crops; however, they may take several years, resulting in longer response times to evolving requirements of farming communities and industry (Kumar et al. 2023). Additionally, traditional methods like mutagenesis caused by chemical or physical mutagens are labor-intensive and time-consuming. Genome editing enables accurate modification of stress-responsive genes and their related molecular networks, facilitating sustainable management of abiotic stress conditions. These technologies enable targeted mutations, insertions, deletions, and specific sequence alterations with greater precision than conventional or marker-assisted breeding methods (Kaur et al. 2022; Kumar et al. 2023; Tiwari et al. 2023).

Three main genome editing technologies are Zinc-Finger Nucleases (ZFNs), Transcription Activator-Like Effector Nucleases (TALENs), and CRISPR/Cas (Clustered Regularly Interspaced Short Palindromic Repeats) systems (Gaj, Gersbach, and Barbas 2013). ZFNs were the first custom DNA endonucleases used to specifically recognize and cleave target DNA, with each zinc finger recognizing three to six nucleotide bases. A notable drawback of TALENs is their considerably greater size relative to ZFNs, with a standard cDNA encoding roughly 3 kb (Kumar et al. 2023; Zhou et al. 2022). However, TALENs offer greater flexibility in target site selection and potentially fewer off-target effects than ZFNs (Gupta and Musunuru 2014). When comparing these technologies, each has its own strengths and limitations, but CRISPR/Cas9 stands out for its simplicity and efficiency.

The simplicity, accessibility, adaptability, flexibility, and wide applicability of CRISPR/Cas9 have revolutionized tomato breeding (Kumar et al. 2023). Unlike ZFNs and TALENs, which require protein engineering for each target, CRISPR/Cas9 requires only the design of a guide RNA, making it significantly easier to implement. The system comprises CRISPR-associated protein 9 (Cas9) and a single guide RNA (sgRNA) that directs Cas9 to a specific DNA target site (Gupta and Musunuru 2014; Kumar et al. 2023). CRISPR/Cas9 compares favorably with ZFNs and TALENs regarding site selection, allowing researchers to target virtually any genomic location with a suitable PAM sequence.

Numerous traits have been successfully edited using CRISPR/Cas9 in tomato for abiotic stress tolerance (Table 10.3). For drought tolerance, editing of genes such as *SlMAPK3*, *SlLBD40*, *SlNPR1*, and *SlARF4* has shown promising results (M. Chen et al. 2021; Li et al. 2019; L. Liu et al. 2020; Wang et al. 2017). Knockout of *SlLBD40* by CRISPR/Cas9 enhanced drought tolerance compared to wild-type plants (L. Liu et al. 2020). For heat stress, genes like *SlAGL6*, *SlCPK28*, and *SlBZR1* have been targeted (Jia et al. 2021; Klap et al. 2017; X. Li et al. 2022). CRISPR/Cas9 editing of the *CPK28* gene improved thermotolerance in tomato (Hu et al. 2021). Beyond tomato, genome editing has been successfully applied in numerous crops, including rice, wheat, maize, and cotton, for abiotic stress management (Kumar et al. 2023).

Despite its potential, CRISPR/Cas9 technology faces limitations and controversies. Off-target effects, where unintended genome sites are modified, remain a concern (Chehelgerdi et al. 2024). Additionally, regulatory frameworks vary significantly across countries, with some, like Japan and Argentina, adopting genome-edited crops while others are still debating this subject (Hundleby and Harwood 2022; Kumar et al. 2023). The Court of Justice of the European Union categorizes genome-edited plants as transgenic, hindering their adoption (Tiwari et al. 2023).

To address these challenges, scientists have developed improved techniques like base editing and prime editing that offer higher precision without causing double-strand breaks (Kaya 2023). Transgene-free genome editing approaches using ribonucleoprotein complexes (RNPs) or transient expression systems can eliminate concerns about foreign DNA integration (Chehelgerdi et al. 2024). Public awareness campaigns explaining that genome-edited plants can be free of foreign genes, particularly through site-directed nucleases (*SDN1* and *SDN2*), could facilitate regulatory acceptance and consumer adoption (Tiwari et al. 2023).

10.4.5 Omics Approaches

10.4.5.1 Transcriptomics

Transcriptomics provides insights into gene expression patterns in response to different abiotic stresses, enabling the identification of key regulatory genes involved in tomato stress resilience. This approach involves analyzing whole RNA transcripts produced under specific conditions, allowing researchers to understand the changes in gene expression profiles during stress exposure (Pardo-Hernández, Arbona, et al. 2024; Pardo-Hernández, García-Pérez, et al. 2024). Transcriptomic studies in tomato have uncovered complex regulatory networks that respond to several abiotic stressors, including drought, salinity, heat, and cold (Table 10.4).

Recent transcriptomic studies have demonstrated that tomato plants exposed to drought stress exhibit significant changes in the expression patterns of genes related to osmotic regulation, antioxidant defense mechanisms, and phytohormone signaling pathways. For example, research on drought-responsive sORFs (small open reading frames) in MT1 tomato plants revealed 676,611 coding sORFs in intergenic regions, with 862 differentially expressed genes under drought conditions (Ng et al. 2025). Similarly, transcriptomic profiling of tomato under heat stress showed profound changes in chromatin organization affecting gene expression through promoter-enhancer contacts, with heat stress transcription factor *HSFA1a* playing a crucial role in this reorganization (Huang et al. 2023).

The specificity of transcriptional responses to different stresses is particularly noteworthy. A comparative analysis of gene expression in response to drought, heat, and salinity identified both common biomarkers and stress-specific signatures. In drought conditions, MYB, SAUR, and NAC

TABLE 10.3
Applications of Gene Editing in Enhancing Abiotic Stress Tolerance in Tomato

Method	Tolerance	Genotype	Gene	Gene ID	Transformation	References
CRISPR/Cas9	Cold Stress	Ailsa Craig	*CBF1*	AAS77820	*Agrobacterium*	Ran Li et al. (2018)
		–	*SlCBF1*	–	*Agrobacterium*	Rui Li et al. (2018)
		–	*SlNPR1*	–	*Agrobacterium*	Shu et al. (2023)
	Drought Stress	Ailsa Craig	*MAPK3*	AY261514	*Agrobacterium*	Wang et al. (2017)
			NPR1	KX198701	*Agrobacterium*	Li et al. (2019)
		Condine Red	*ALD1, FMO1*	Solyc11g044840, Solyc07g04243	*Agrobacterium*	Z. Wang et al. (2021)
		Micro-Tom	*LBD40*	Solyc02g085910	*Agrobacterium*	L. Liu et al. (2020)
		–	*SlARF4*	Solyc11g069190	*Agrobacterium*	S. Chen et al. (2021)
		–	*SlGID1*	–	–	Illouz-Eliaz et al. (2020)
		–	*SlLBD40*	Solyc02g085910	*Agrobacterium*	L. Liu et al. (2020)
		–	*SlMAPK3*	AY261514	*Agrobacterium*	WangWei et al. (2018)
		–	*SlNPR1*	KX198701	*Agrobacterium*	Li et al. (2019)
	Heat Stress	Condine Red	*BZR1*	Solyc04g079980	*Agrobacterium*	Yin et al. (2018)
			CPK28, APX2	Solyc02g083850, Solyc06g005150	*Agrobacterium*	Hu et al. (2019)
		MP-1	*AGL6*	Solyc01g093960	*Agrobacterium*	Klap et al. (2017)
		–	*SlAGL6*	Solyc01g093960	*Agrobacterium*	Klap et al. (2017)
		–	*SlBZR1*	Solyc04g079980	*Agrobacterium*	Yin et al. (2018)
		–	*SlCPK28*	Solyc02g083850	*Agrobacterium*	Hu et al. (2021)
		–	*SlMAPK3*	NM_001247431.2	*Agrobacterium*	Yu et al. (2019)
	Multiple Stresses	Rubion	*GRXS14/15/16/17*	Solyc02g082200, Solyc06g067960, Solyc09g005620, Solyc02g078360	*Agrobacterium*	Kakeshpour et al. (2021)
	Salinity Stress	Hongkwang	*HKT1;2, RAD51/54*	Solyc07g017540, Solyc04g056400	*Agrobacterium*	Vu et al. (2020)
		Hongkwang	*HyPRP1*	Solyc12g009650	*Agrobacterium*	Tran et al. (2021)
		S. lycopersicum, S. pimpinellifolium	*HAK20, SOS1*	Solyc04g008450, Solyc11g044540	*Agrobacterium*	P. Wang et al. (2021), Wang et al. (2020)
		–	*SlARF4*	–	–	Bouzroud et al. (2020)
		–	*StCoilin*	LOC102603469	PEG	Makhotenko et al. (2019)
RNAi	Chilling stress	–	*LeCOLD1*	–	–	L. Zhang et al. (2020)
		–	*LeERF2*	–	–	Zhang and Huang (2010)
	Multiple Stresses	–	*SlAMI*	–	–	Abuqamar et al. (2009)
	Oxidative stress	–	*SlHZ24*	–	–	Hu et al. (2016)
		–	*SlNL33*	–	–	Ye et al. (2019)
Transposon DNA tagging	Multiple Stresses	–	*SlARS1*	–	–	Campos et al. (2016)
VIGS	Drought	–	*SlSR1*	–	–	Liu et al. (2014)

TABLE 10.4
Key Transcription Factors Identified for Abiotic Stress Responses in Tomato

Transcription Factor	Stress Type	Function	References
HSFA1a	Heat	Modulates 3D chromatin architecture	Huang et al. (2023)
R2R3-MYB family	Combined salt/heat	Regulates flavonol biosynthetic pathway	Pardo-Hernández, Arbona, et al. (2024)
MYB	Drought	Enhances drought tolerance	Sadder et al. (2014)
NAC domain proteins	Drought	Regulates the dehydration response	Sadder et al. (2014)
AREB/ABF	Salt/Drought	Activates ABA-dependent pathway	Pardo-Hernández, García-Pérez, et al. (2024)
SlLBD40	Drought	Negative regulator of drought tolerance	Leporino et al. (2024)

transcription factor proteins were significantly increased, whereas proteinase inhibitors and heat shock proteins were prevalent during heat stress responses. Salinity stress specifically triggered the expression of proteins produced by phosphate deprivation (Li et al. 2020; Liu et al. 2024; Sadder, Alsadon, and Wahb-Allah 2014; Thirumalaikumar et al. 2017; Wu et al. 2012).

Importantly, transcriptomic analyses have revealed that combined stresses produce unique responses that differ from those caused by individual stresses. For instance, specific ABA-independent transcriptome reprogramming occurs when tomato plants face combinations of abiotic stressors. In ABA-deficient tomato mutants, 685 genes were particularly regulated by the combination of salt and heat, with 463 genes controlled by ABA-independent mechanisms, including six transcription factors from the R2R3-MYB family (Pardo-Hernández, Arbona, et al. 2024).

10.4.5.2 Genomics

Genomics-based approaches offer powerful tools for dissecting the genetic basis of abiotic stress tolerance in tomato and developing improved cultivars. The completion of the tomato genome sequencing has accelerated the identification and functional validation of genes associated with various stress responses (Razali et al. 2018). Through genomic analysis, researchers have successfully mapped QTLs for several abiotic stresses, particularly salinity, drought, and low temperatures, providing the foundation for marker-assisted selection in breeding programs (Chaudhary et al. 2019).

Recent genomic studies have identified numerous genes that confer enhanced tolerance to abiotic stresses when overexpressed or silenced. For instance, transcription factors from the GRAS family, particularly *SlGRAS10*, have been shown to play critical roles in abiotic stress responses (Sadder et al. 2025). RNA interference (RNAi)-mediated downregulation of *SlGRAS10* generated dwarf plants with shorter internodes that exhibited enhanced tolerance to salt, dehydration, and abscisic acid by improving osmotic potential, flavonoid biosynthesis, and ROS scavenging (Pardo-Hernández, García-Pérez, et al. 2024).

The CRISPR/Cas9 system has emerged as a revolutionary genomic tool for enhancing stress tolerance in tomato. For example, CRISPR/Cas9-mediated mutagenesis of *SlNPR1* reduced drought tolerance by increasing stomatal aperture and hydrogen peroxide levels (Chaudhary et al. 2019). Additionally, CRISPR/Cas9-mediated mutation of *SlLBD40*, a transcription factor associated with lateral organ borders, improved drought tolerance in tomato plants (Leporino et al. 2024).

Genomic approaches have also facilitated the characterization of gene families involved in stress responses (Table 10.5). Wang et al. (2022) identified 66 potential G2-like genes in tomatoes through genome-wide analysis, with expression patterns responsive to various abiotic stressors. Similarly, a comprehensive analysis of the PDI gene family revealed 19 PDI genes distributed across eight tomato chromosomes, with most PDI genes significantly activated by high temperature, salinity, and ABA treatments (Wai et al. 2021).

TABLE 10.5
Key Genes Enhancing Abiotic Stress Tolerance in Tomato

Tolerance	Gene	Function	Expression/ Regulation	References
Cold	*INVINH1*	Tolerance to cold stress	–	X. xia Xu et al. (2017)
	SlHY5	Tolerance to cold stress	Overexpression	Han et al. (2020)
	LeFAD3	Regulation of the unsaturation of fatty acids in membrane lipids	Overexpression	Yu et al. (2009)
	SlMPK7	Regulation of ROS homeostasis	Overexpression	Yu et al. (2016)
	LetAPX	Detoxify and stimulate the conversion of H_2O_2 into H_2O	Overexpression	Duan et al. (2012)
	SlHSP17.7	Control Cd Signaling and Phosphatidylglycerol Metabolism	Overexpression	N. Zhang et al. (2020)
	SlICE1	TF, transcriptional regulation	Overexpression	Miura et al. (2012)
Heat	*LeFAD7*	Regulation of fatty acid unsaturation of membrane lipids	Antisense expression	Liu et al. (2010)
	MT-sHSP	Involved in the accumulation of heat shock proteins	Expression	Nautiyal, Shono, and Egawa (2005)
	Solyc03g020030	Proteinase inhibitor-II	Gene silencing	Balyan et al. (2020)
	LeAN2	Upregulate anthocyanin biosynthesis and accumulation	Overexpression	Meng et al. (2015)
Drought	*SlLBD40*	Associated with JA signaling	CRISPR/Cas9 knockout	L. Liu et al. (2020)
	MdSWEET17	Drought stress response and the regulation of fructose.	Expression	LU et al. (2019)
	SLZF57	–	Gene silencing	Balyan et al. (2020)
	SlGRAS4	Drought stress and abscisic acid signaling	Overexpression	Liu et al. (2021)
	SlTIP2	Involved in plant water balance	Overexpression	Sade et al. (2009)
	SlWRKY6	Decrease MDA, ROS accumulation, and electrolyte leakage	Overexpression	Chen et al. (2024)
Salinity	*SlSOS1*	Maintain ion homeostasis, both Na^+ and K^+	Expression	Olías et al. (2009)
	SlABIG1	Salt stress negative regulator gene	Knockout	Ding et al. (2022)
	LeNHX2	Ion transport; compartmentalization of ions	Overexpression	Jia et al. (2021)
	PpSnRK1α	Accelerated ROS metabolism	Overexpression	Jia et al. (2021)
	SlBZR1D	Upregulate multiple stress-related gene expression	Overexpression	Zhu et al. (2018)
	SlERF3DRD	Ethylene-responsive TF; transcriptional regulation	Overexpression	Pan et al. (2010)
	SlMAPK3i	Remove ROS and increase ethylene signaling genes expression	Overexpression	Shu et al. (2022)

(Continued)

TABLE 10.5 ***(Continued)***
Key Genes Enhancing Abiotic Stress Tolerance in Tomato

Tolerance	Gene	Function	Expression/ Regulation	References
Heavy Metal	*SlSAM1*	Involved in alkali stress and ethylene biosynthesis	Overexpression	Gong et al. (2016)
	SlJMJ524	Control the expression of Cd transport proteins	Overexpression	Q. Li et al. (2022)
	SlHMA25	Heavy metal transportation	-	Nazmul Hasan et al. (2022)
	SlHMA32	Heavy metal transportation	-	Nazmul Hasan et al. (2022)
	SlHMA8	Heavy metal transportation	-	Nazmul Hasan et al. (2022)
Osmotic	*SlGRAS10*	Increase osmotic potential, flavonoid production, and ROS scavenging	Downregulation	Habib, Lwin, and Li (2021)
	TERF1	Ethylene-responsive transcription factor	Overexpression	Huang et al. (2004)
	SlSnRK2.2	Involved in ABA signaling and osmotic stress signal transduction	Overexpression	Yang et al. (2015)
	SlSnRK22.1		Overexpression	Yang et al. (2015)
Multiple	*SlDHN*	Cold and Drought tolerance	Overexpression	Guo et al. (2019)
	SlDEAD23	Abiotic (Cold, Salinity) stress responses	Overexpression	Pandey et al. (2019)
	SlDEAD35	Abiotic (Cold, Salinity) stress responses	Overexpression	Pandey et al. (2019)
	SlMAPK3	Resistance to chilling, Cd, and drought stresses	Overexpression	Muhammad et al. (2019)
	SlMPK7 i	Resistance to chilling, Cd, and drought stresses	Overexpression	Muhammad et al. (2019)
	SlbZIP1	Salt and drought stress tolerance	Expression	Chaudhary et al. (2018)
	SlMBP8	More tolerance to drought and salt stress	Gene silencing	Yin et al. (2017)
	SlAGO4A	Tolerance to salt, drought stress	Overexpression	Yin et al. (2017)
	SlHB2	Tolerance to salt, drought stress	Overexpression	Yin et al. (2017)
	SlMBP8	Tolerance to salt, drought stress	Overexpression	Yin et al. (2017)
	tas14	Accumulation of chaperone-like protein with detergent properties	Overexpression	Muñoz-Mayor et al. (2012)
	SlAREB1	TF, regulation of Drought and Salinity stress-related genes	Overexpression	Orellana et al. (2010)
	SDG34	Response to stress	Expression	Bvindi et al. (2022)
	SpWRKY1	Transcriptional regulation of stress-related genes	Overexpression	Li, Luan, and Liu (2015)
	SlAIM1	Salt and oxidative stress tolerance	Overexpression	Abuqamar et al. (2009), Wai et al. (2020)
	SlNL33	Ascorbate accumulation	Suppressed expression	Ye et al. (2019)
	BEL1-like genes	Involved in multiple biological processes	Tissue-specific expression	He et al. (2022)

The integration of genomics with other -omics approaches enables a more holistic understanding of stress responses. For example, Bvindi et al. (2022) investigated the activities of tomato histone H3 lysine methyltransferases (*SDG33* and *SDG34*), demonstrating their influence on H3K36 and H3K4 methylations and their impact on gene expression related to biotic and abiotic responses (Chaudhary et al. 2019).

10.4.5.3 Proteomics

Proteomics provides critical insights into the functional protein landscape that regulates plant responses to environmental stressors, revealing mechanisms that transcriptomics and genomics cannot fully explain alone (Table 10.6). This method involves a comprehensive analysis of the entire spectrum of proteins encoded by a genome, including post-translational modifications, protein-protein interactions, and patterns of cellular localization (Naik et al. 2023). For tomato under abiotic stress conditions, proteomics has revealed dynamic changes in protein abundance, localization, and function that directly impact stress adaptation (M. Liu et al. 2023).

Under abiotic stress, the tomato proteome undergoes substantial remodeling, with changes in proteins involved in ROS scavenging, stress signaling, detoxification, and chaperone functions (Naik et al. 2023). For example, thioredoxin, glyoxalase I, and *DnaK* and *HSP20* chaperones are upregulated during stress to prevent the aggregation of denatured proteins and promote refolding. Proteins involved in cell wall biosynthesis, particularly enzymes of the phenylpropanoid biosynthesis pathway, also play crucial roles in stress adaptation (Chaudhary et al. 2019).

Cellular localization of proteins provides important functional context during stress responses. Modern spatial proteomics techniques have enabled the identification of protein translocation between cellular compartments during stress conditions (Naik et al. 2023). Proteome profiling has revealed that stress-responsive proteins are distributed across four main cell fractions (nuclear, microsomal, mitochondrial, and plastidial), with each compartment exhibiting specific protein signatures during stress adaptation (Chaudhary et al. 2019).

Post-translational modifications (PTMs) represent another critical aspect of proteome regulation during stress. While hundreds of PTMs have been recognized, large-scale identification and quantification of PTMs during stress conditions have revealed how these modifications fine-tune protein function to enhance stress tolerance. For example, the phosphorylation status of key regulatory proteins dramatically changes during drought, salt, and temperature stress, altering their activity and interactions (Naik et al. 2023).

The integration of proteomics with other omics approaches has proven particularly valuable for identifying key regulatory nodes in stress response networks. Studies comparing proteome and transcriptome changes in drought-stressed tomato plants found good correlation between differential gene expression and protein abundance changes, with 3,765 differentially expressed genes corresponding to 294 differentially abundant proteins (M. Liu et al. 2023), highlighting the importance of multi-level regulation during stress adaptation.

10.4.5.4 Metabolomics

Metabolomics provides crucial insights into the biochemical adaptations that occur in tomato plants responding to environmental stresses, revealing the dynamic changes in primary and secondary metabolites that contribute to stress tolerance. This approach involves the thorough examination of metabolites, which are small molecules that serve as intermediates or final products of cellular processes, utilizing technologies such as liquid/gas chromatography combined with mass spectrometry, nuclear magnetic resonance, and Fourier transform infrared spectroscopy (Naik et al. 2023).

TABLE 10.6
Key Proteins Involved in Abiotic Stress Tolerance in Tomato

Protein	Function	Result	References
ABP19a	Multiple development processes and drought response	Drought tolerance	Oliver et al. (2011), Wang et al. (2016)
C2H2-Type Zinc Finger	Overall growth and development of plant tissues	Abiotic stress responses	Ming et al. (2020)
COR15 protein	Attaches to the chloroplast	Cold tolerance	Thalhammer and Hincha (2014), Zhang et al. (2021)
Dehydrins	-	Salinity and drought stress	Muñoz-Mayor et al. (2012)
E42 and LA3120	Improve plant growth and reduce stress	Better plant growth under water stress	Francesca et al. (2022)
Golden 2-Like	Plant development and reducing abiotic stress	Reduced drought stress tolerance	Wang et al. (2022)
Histidine kinase	Reduces abiotic stresses	Maintaining cellular Na+ homeostasis	Lohani, Singh, and Bhalla (2022)
LEA proteins	Prevents membrane leakage; membrane & protein stabilization	Water balance & ion sequestration	Olvera-Carrillo, Reyes, and Covarrubias (2011)
LeHSP 17.6	Involved in accumulation of heat shock proteins	Heat Stress Tolerance	Kadyrzhanova et al. (1998)
LeHSP 21.5	Improved tolerance to tunicamycin-ER stress inducer	Increased tolerance	Zhao et al. (2007)
MdVHA-B	-	Increased drought tolerance	Hu et al. (2012)
P5CS protein	Proline biosynthesis	Drought and salt tolerance	Karthikeyan, Pandian, and Ramesh (2011), Ogden et al. (2020)
RING zinc finger	Plant growth and reducing abiotic stresses	Reduced stress on the plant	Han et al. (2022)
SlbZIP1	ABA-mediated stress regulation	Reduced salt and drought stress	Zhu et al. (2018)
SlMAPK3	Improved chlorophyll content and root biomass accumulation	Tolerance to Cd and drought stress	Muhammad et al. (2019)
SlNAC1	Interactions with replication accessory proteins	Tolerance to lower temperatures	Ma et al. (2013), Selth et al. (2005)
SlNAC4	Fruit ripening and carotenoid accumulation	Oxidative stress response	Zhu et al. (2014)
SlNAM2	Flower-boundary morphogenesis	Inhibits extreme water loss	Hendelman et al. (2013)
Solyc04g014600	Protein profiling of phloem and its exudates, Universal stress protein	Drought tolerance	Irfan et al. (2021)
SpPKE1	Interact with an F-box protein	Drought tolerance	J. Li et al. (2018)
SpUSP	Prevent ROS generation	Drought tolerance	Loukehaich et al. (2012)
Systemin peptide	Na^+ exclusion, antioxidant activity, protease inhibitor	Salinity stress tolerance	Cirillo et al. (2022)

TABLE 10.7
Key Metabolites Involved in Abiotic Stress Responses in Tomato

Metabolites	Function	Result	References
2,3-Butanediol	–	Drought and chilling response	Park et al. (2022)
Flavonoids	Antioxidant defense	Heat, drought	Leporino et al. (2024)
Gamma-aminobutyric acid	Signaling molecule	Multiple stresses	Naik et al. (2023)
Glycine betaine	Stress protectant	Multiple stresses	Park, Jeknic, and Chen (2006)
Goblet	–	Reduce transpiration	Berger et al. (2009)
Inositol	Cell wall and membrane component	Drought	Asakura et al. (2021)
Lauric acid	Root exudate affecting rhizosphere	Water stress	Naik et al. (2023)
Melatonin	Mitigates damage to proteins and membranes	Abiotic stress tolerance	Martinez et al. (2018)
Osmotin	Osmotin accumulation (provides osmotolerance)	Osmotic Stress Tolerance	Patade et al. (2013)
Proline	Osmotic adjustment	Drought, salt	Naik et al. (2023)

Under abiotic stress conditions, tomato plants exhibit significant alterations in their metabolic profiles (Table 10.7). For instance, drought stress triggers the accumulation of various osmolytes and compatible solutes that help maintain cellular water balance and membrane stability (Shu et al. 2024). Salt stress similarly induces changes in sugar accumulation, as demonstrated by a study combining transcriptomic and metabolomic analyses, which revealed enhanced sugar accumulation in tomato fruits under salinity conditions. Cold stress notably affects flavor-related compounds, with chilling temperatures leading to depletion of important flavor volatiles due to transient cytosine methylations in fruit ripening-related genes (Naik et al. 2023).

Secondary metabolites play particularly important roles in stress adaptation. Phenolic compounds, including flavonoids, anthocyanins, and phenylpropanoids, increase under various stress conditions and contribute to antioxidant defense mechanisms (Naik et al. 2023). Metabolomic analysis of drought-stressed tomato plants revealed increased levels of metabolites associated with anthocyanin biosynthesis, which help scavenge ROS generated during stress (Chaudhary et al. 2019; Kumar et al. 2021).

Metabolomic studies have also revealed how phytohormones coordinate stress responses in tomato. Abscisic acid (ABA), jasmonic acid, and ethylene metabolites show distinctive patterns under different stress conditions, mediating specific adaptive responses (Khan 2025). A multi-omics study demonstrated that ABA-deficient tomato plants exhibit specific metabolic signatures under combined salinity and heat stresses, revealing ABA-dependent and ABA-independent metabolic pathways (Pardo-Hernández, García-Pérez, et al. 2024).

Recent advances in single-cell metabolomics allow for more precise spatial resolution of metabolic changes, enabling researchers to monitor stress responses at the cellular level using techniques such as picoliter pressure-probe electrospray-ionization mass spectrometry (Alam et al. 2025). These approaches are revealing how metabolic adaptations differ between cell types and tissues during stress exposure.

10.4.5.5 Ionomics

Ionomics provides crucial insights into how plants regulate mineral nutrient homeostasis under abiotic stress conditions (Luo et al. 2023). This high-throughput approach measures the complete set of mineral nutrients and trace elements in plant tissues using techniques such as inductively coupled plasma spectroscopy (ICP), revealing how elemental profiles change in response to abiotic stresses (S. Ali, Tyagi, and Bae 2021).

Abiotic stresses significantly alter tomato ionome, affecting both essential and non-essential elements. For example, drought stress modifies root architecture and function, directly impacting nutrient acquisition patterns. Salinity stress poses a particular challenge for tomato by disrupting ion homeostasis through Na^+ accumulation and K^+ depletion (Sánchez-Rodríguez et al. 2010). A comprehensive study of long-term salinity responses in advanced tomato breeding lines revealed that the salt-tolerant line L56 exhibited distinctive ionome profiles (Table 10.8), deploying more SOS paralogs and CBL-interacting kinase genes than salt-susceptible lines (Sadder et al. 2025).

The integration of ionomics with other -omics approaches has revealed complex regulatory networks controlling mineral homeostasis under stress. Transcriptomic analysis combined with ionomics has shown that genes encoding ion transporters, including those mediating Na^+/H^+ exchange and K^+ uptake, are differentially expressed during salt stress (Pardo-Hernández, García-Pérez, et al. 2024). Similarly, proteomics combined with ionomics has identified proteins involved in controlling ion fluxes during stress adaptation, such as the histidine kinase proteins that maintain cellular Na^+ homeostasis (Naik et al. 2023).

Genetic variations in ionome profiles under stress conditions provide valuable resources for crop improvement (Table 10.8). QTL mapping and GWAS have identified genomic regions related to favorable ionome profiles under stress conditions (S. Ali et al. 2021). These approaches have led to the identification of genes like *HAL1* from yeast (*Saccharomyces cerevisiae*), which enhances salt tolerance in transgenic tomatoes by improving intracellular K^+/Na^+ ratios through reduced long-term shoot Na^+ retention (Pardo-Hernández, García-Pérez, et al. 2024).

Recent advances in high-precision ionomic phenotyping combined with integrated omics analysis are particularly suitable for identifying genes conferring various traits for biotic and abiotic stress tolerance (S. Ali et al. 2021). These approaches have identified loci controlling element uptake and transport within tomato tissues, providing targets for breeding stress-resilient cultivars with improved nutritional value.

10.4.5.6 miRNA-Omics

Non-coding RNA, such as microRNA (miRNA) and long non-coding RNA (lncRNA), do not code any protein molecules but play a vital role in the regulation of gene expression, which may alter the response to various environmental stresses (Patra et al. 2023). miRNAs are a family of small (~20–24 nucleotides) noncoding RNAs responsible for mediating post-transcriptional gene silencing by targeting mRNAs for degradation or translational repression. Small RNAs play a vital role in regulating plant growth and development, particularly under stressful conditions (Mahto et al.

TABLE 10.8
Key Ionomic Responses and Associated Genes in Tomato under Abiotic Stress

Element	Stress Condition	Associated Gene/Protein	Function	References
K^+/Na^+ ratio	Salinity	*HAL1*	Improves K^+ retention under salt stress	Sadder et al. (2025)
Ca^{2+}, Zn^{2+}, Mn^{2+}	Multiple stresses	*ESB1*	Regulates endodermal barriers	S. Ali et al. (2021)
Na^+	Salinity	SOS pathway genes	Na^+ efflux and compartmentalization	Sadder et al. (2025)
K^+	Drought	*HAK5*	High-affinity potassium transporter	S. Ali et al. (2021)
Fe^{2+}	Low iron	Spermidine-responsive genes	Improves iron acquisition	Naik et al. (2023)
Multiple elements	Salt/Drought	*SlABIG1*	Salt stress negative regulator	Sadder et al. (2025)

2020). lncRNAs, in contrast, are a group of longer RNA molecules (over 200 nucleotides) that have been shown to play a role in regulating gene expression at different levels (transcriptional, post-transcriptional, and epigenetic) and through different mechanisms, including interactions with DNA, RNA, and proteins (Chen et al. 2018).

In tomato, miRNAs and lncRNAs participate in the response to abiotic stresses, including drought, salinity, heat/cold, and nutrient deficiency (Table 10.9). For example, several signal pathways that are responsive to stress, including ROS, stress hormones, and ion transporters, have been found to be regulated by miRNAs, and the response has been shown to increase plant stress tolerance to these stress factors. Similarly, lncRNAs also regulate stress tolerance pathways by acting

TABLE 10.9
miRNA and lncRNA Involved in Abiotic Stress Tolerance in Tomato

Type of Stress	miRNA or lncRNA	Function	Reference
Drought Stress	miR160, miR165, miR166, miR171, and miR9552	Regulates root development, stomatal movement, and dehydration response	Candar-Cakir, Arican, and Zhang (2016), Q. Li et al. (2023)
	miR156	Regulates stomatal movement through the ABA pathway regulation	Visentin et al. (2020)
	miR169c	Over-expression reduces stomatal opening and transpiration rate	Li et al. (2008)
	miR396	Improve WUE by regulating the GABA metabolic pathway	Fracasso et al. (2021)
	miR1916	Silencing maintains a balanced osmotic pressure and improves WUE	Chen, Meng, and Luan (2019)
Heat Stress	miR169	Overexpression inhibits the expression of the downstream *NF-YA9/10* gene	Rao et al. (2022)
	miR319	Overexpression can enhance the activity of antioxidant enzymes	Shi et al. (2019)
Cold Stress	miR166	Parthenocarpic fruit set by regulating the expression of *SlHB15*	Clepet et al. (2021)
	miR164a/b-5p	Targets the downstream *NAM3* gene; regulates the ethylene synthesis pathway	Dong et al. (2022)
	miR319d	Overexpression improves cold tolerance by regulating the ROS pathway	Shi et al. (2019)
Salt Stress	miR397	Target laccase enzymes, which catalyze phenolic compounds to reduce oxygen to water	Jeon et al. (2012)
	miR398	Targets *CSD* and *PR5*; involved in salt stress regulatory network	Luan, Wang, and Liu (2014)
Nutrient Stress	miR158, miR319, miR394, miR399, and miR862–3p	Regulates phosphate transporter genes, enhancing phosphate uptake	Gu et al. (2010)
	miR168	Modifies root architecture to improve K^+ uptake in low potassium conditions	X. Liu et al. (2020)
Combined Stress	miR167a	Functions as a reactive medium under diverse abiotic stressors	Jodder et al. (2018)
	miR1917	Inhibits the major regulator of the ethylene response *CONSTITUTIVE TRIPLE RESPONSE4*	Wang et al. (2018)
	lncRNA1459	Regulates ethylene synthesis, affecting fruit ripening in extreme temperatures	Ran Li et al. (2018)

ABA: Abscisic acid; GABA: Gamma Amino Butyric Acid; ROS: Reactive Oxygen Species; WUE: water use efficiency through miRNA sponging, protein activity, or perturbation of cell wall composition to cope with stresses like salinity and temperature extremes. Unravelling the regulatory network of these non-coding RNAs could be helpful to direct the construction of transgenic or breeding approaches for enhancing the stress tolerance of tomatoes (Q. Li et al. 2023).

10.5 PROGRESS AND SUCCESS STORIES IN TOMATO BREEDING FOR ABIOTIC STRESS TOLERANCE

Significant advancements have been made in developing cold-tolerant tomato varieties, particularly by identifying natural genetic variation in wild and cultivated germplasm (Table 10.10). Accessions such as TS-1, TS-48, and TS-183 from the *S. lycopersicum* group exhibit enhanced cold tolerance compared to wild relatives like *S. pimpinellifolium*. These lines demonstrated lower electrolyte leakage and higher survival rates under chilling stress (4°C), traits linked to a 27-bp promoter variation in the *SlBBX31* gene, which regulates cold-responsive pathways (Zhu et al. 2023). The domestication process favored these alleles, enabling improved cold adaptation in modern cultivars without compromising yield (Zhu et al. 2023).

Tomato breeding for heat tolerance has yielded numerous varieties capable of maintaining fruit set under sustained high temperatures. The Christy variety, evaluated in tropical conditions, showed reliable fruit production during warmer months (32°C daytime/26°C nighttime). Hybrids like F_1 2101, F_1 2201, and F_1 2501 achieved yields of 27.7–35.5 t/ha under subtropical heat, combining high productivity with robust fruit quality (Quamruzzaman et al. 2023). Pollen viability screening identified genotypes LA2854, LA1478, and LA0417 as thermo-tolerant, with viable pollen fractions exceeding 0.6 under 32°C stress (Paupière et al. 2017).

Specialized hybrids such as Heatmaster and Florida91 were bred for high-heat climates, retaining fruit set at temperatures above 32°C. Modern "heat-set" hybrids, including Summer Set, Solar Fire, and BHN 216, reliably produce fruit under prolonged heat (daytime >35°C, nighttime >21°C). Heritage varieties like Arkansas Traveler and Homestead 24 also exhibit inherent heat resilience, likely due to adaptive traits preserved in heirloom gene pools. Cherry tomatoes, such as Lollipop and Yellow Pear, further expand options for hot climates, yielding small, crack-resistant fruits under stress. In Nigeria, locally developed lines T1 205 and T1 539 have been adopted for extended growing seasons in regions with temperatures exceeding 40°C.

Wild tomato relatives remain pivotal in drought tolerance breeding. The LA 716 accession of *S. pennellii* outperforms cultivated varieties under water-limited conditions, maintaining leaf water potential and photosynthetic rates at 50% available soil water (Dariva et al. 2020; Pessoa et al. 2022). Introgression lines IL 1-4-18 and IL 10-1, derived from *S. pennellii*, showed superior germination rates and seedling vigor under osmotic stress (−0.3 MPa) (Pessoa et al. 2022).

Galapagos wild tomatoes (*S. cheesmaniae* and *S. galapagense*) exhibit exceptional salinity tolerance, with accessions LA0317 and LA1449 maintaining 55% higher fresh biomass than commercial cultivars under 150 mM NaCl (Pailles et al. 2020). These lines accumulate potassium (K^+) in roots and leaves, stabilizing ion homeostasis under stress (Pailles et al. 2020). The PS-10 cultivar demonstrated enhanced osmotic adjustment via proline accumulation (10.8-fold increase under 50 mM NaCl) and elevated antioxidant enzyme activity (CAT, SOD, APX), mitigating oxidative damage.

While phosphorus deficiency and heavy metal tolerance are less studied, preliminary efforts identified PI 121665 "Globonnie" as efficient in phosphorus uptake, while Sirin (AC7) and Karazi (AC5) showed reduced Cd accumulation in fruits under contaminated soils (Aazami et al. 2021). These examples underscore the potential for leveraging genetic diversity to address niche abiotic challenges.

TABLE 10.10
Tomato Varieties/Lines and Their Tolerance to Various Abiotic Stresses

Stress Type	Variety/Line	Notable Features	References
Cold	TS-1, TS-48, TS-183	Big accessions with enhanced cold tolerance	Zhu et al. (2023)
Heat	Christy	Suitable for cultivation during warmer months	
	F_1 2101, F_1 2201, F_1 2501	Hybrids with good yield performance	Quamruzzaman et al. (2023)
	LA2854, LA1478, LA0417	Genotypes identified through pollen viability screening	Paupière et al. (2017)
	Heatmaster	Bred specifically for high-heat climates	Paupière et al. (2017)
	Summer Set, Solar Fire, Florida 91	Hybrids with reliable fruit set under high temperatures	
	Arkansas Traveler, Homestead	Heritage varieties with proven heat tolerance	
	BHN 216, Florasette, Heatwave II	Modern heat-set hybrids	
	Sunchaser, Sun Leaper, Sunmaster, Sun Pride	Hybrids that continue fruit production during heat	
	Stupice	Cold-set variety that also performs well in heat	
	Black Cherry, Matt's Cherry, Lollipop, Yellow Pear	Cherry tomato varieties	
	T1 205, T1 539, T1 206, T1 204	Heat-tolerant varieties	
Drought	LA 716 (*Solanum pennellii*)	Wild accession closest to drought tolerance ideotype	Pessoa et al. (2022)
	IL 1-4-18, IL 2-3, IL 1-2, IL 9-2, IL 10-1	Most tolerant introgression lines at the germination stage	Pessoa et al. (2022)
Salinity	Galapagos (*S. cheesmaniae*, *S. galapagense*)	Greater tolerance than commercial lines	Pailles et al. (2020)
	LA0317, LA1449, LA1403	High tolerance based on growth performance under stress	Pailles et al. (2020)
	PS-10	Better osmotic adjustment and antioxidant enzyme activity	Aazami et al. (2021)
	Pakit	Most tolerant genotype to the tested salinity levels	Zhang et al. (2024)
P deficiency	PI 121665 "Globonnie" (crt/crt)	Efficient in phosphorus absorption/uptake	Marques et al. (2015)
	F1RC1 PI 121665 X TOM-584	Efficient in phosphorus absorption/uptake	Marques et al. (2015)
Heavy metal	Sirin (AC7), Karazi (AC5), Balami (AC31)	Highest tolerance to Cd stress; less Cd accumulation in fruits	Rasul and Tahir (2023)
Oxidative stress	PS-10	Higher activity of antioxidant enzymes under salinity stress	Aazami et al. (2021)

10.6 CHALLENGES, LIMITATIONS, AND FUTURE PROSPECTS

Tomato breeding for abiotic stress tolerance faces significant challenges rooted in biological complexity and practical constraints. A primary barrier is the polygenic nature of stress tolerance traits, which are often influenced by intricate gene-environment interactions. For instance, while wild tomato species like *Solanum pennellii* harbor alleles for drought tolerance, their introgression into cultivated lines can disrupt yield-related traits due to linkage drag, particularly in genomic regions with suppressed recombination (Diouf et al. 2018; N. Li et al. 2023). High-throughput phenotyping platforms, though promising for rapid trait screening, remain underutilized in large-scale breeding programs due to infrastructure costs and technical expertise requirements (Marko et al. 2018; Smith, Potgieter, and Chapman 2021). Even when tolerant genotypes are identified, yield penalties may arise if stress-adaptive traits conflict with agronomic performance. For example, heat escape mechanisms that accelerate flowering can reduce fruit set if not synchronized with optimal growing windows (Osei-Bonsu et al. 2022).

Public skepticism toward genetically modified (GM) crops further complicates adoption. Despite scientific consensus on GM safety, cultural values and misinformation, particularly in Europe and Africa, hinder acceptance of transgenic solutions like drought-resistant lines engineered with *DREB* or *Osmotin* genes (Hui et al. 2024). Additionally, the underutilization of wild genetic diversity persists, as only 10% of tomato wild relatives have been systematically integrated into breeding pipelines, leaving valuable stress-adaptive alleles untapped. (Arrones et al. 2024; N. Li et al. 2023).

Advancements in genomic selection and speed breeding offer transformative potential. By combining genomic prediction models with accelerated generation cycles, achieved through controlled photoperiod and temperature, breeders can rapidly fix beneficial alleles while maintaining genetic diversity (N. Li et al. 2023). The development of transgene-free CRISPR/Cas9-edited lines, such as *SlHyPRP1* variants with enhanced salinity and heat tolerance, addresses both regulatory and consumer concerns while enabling precise trait modulation (Tiwari et al. 2023; Tran et al. 2023).

Prioritizing multi-stress tolerance is critical as climate change exacerbates concurrent abiotic pressures. Pioneering efforts, including the eight-way MAGIC population derived from *S. pimpinellifolium* and *S. lycopersicum* var. *cerasiforme*, demonstrate the feasibility of pyramiding traits like drought resilience and nutrient efficiency (Egea et al. 2022). Complementing genetic gains, rhizosphere microbiome engineering presents a novel avenue to enhance stress adaptation. Beneficial microbes, such as *Pseudomonas* spp., improve nutrient uptake and osmotic adjustment under drought, reducing reliance on genetic modifications alone (Sierra et al. 2024).

Global collaboration is essential to overcome these challenges. Initiatives like the tomato super-pangenome project, which integrates genomic data from 11 wild and cultivated accessions, provide open-access resources for identifying stress-responsive structural variants (N. Li et al. 2023). By fostering international partnerships and democratizing data, the tomato breeding community can accelerate the development of resilient cultivars tailored to diverse agroecological zones.

10.7 CONCLUSION

Enhancing abiotic stress resistance is essential for sustaining tomato production against escalating global temperatures and erratic climatic circumstances. This chapter has thoroughly reviewed the major abiotic factors impacting tomato plants, such as drought, salinity, heat, cold, and oxidative stress, while explaining the diverse tolerance mechanisms that tomato plants utilize to address these challenges. Understanding these mechanisms is crucial for developing stress-resilient tomato cultivars, especially via breeding, genetic engineering, and advanced omics methodologies.

Abiotic stress adversely affects tomato growth, yield, and fruit quality by disrupting important physiological processes such as photosynthesis, nutrient uptake, and reproduction. Traditional breeding techniques progressed significantly with the identification of stress-tolerant genotypes through germplasm evaluation, hybridization, and backcross breeding. However, these approaches often face limitations, particularly in the face of rapid climate change. As such, molecular breeding methods such as MAS, QTL mapping, and GWAS are becoming increasingly vital in precision breeding efforts, allowing for faster development of stress-tolerant cultivars with better resilience and yield stability.

Furthermore, genetic engineering methods, especially the utilization of CRISPR/Cas9, have emerged as effective tools for the precise alteration of stress-responsive genes in tomato plants. This method facilitates accurate and rapid genetic modifications that may greatly enhance stress resilience. The integration of multiple omics technologies, including transcriptomics, proteomics, metabolomics, ionomics, and miRNA/lncRNA studies, has considerably enhanced our understanding of the complex processes associated with stress tolerance. These methodologies provide significant insights into the molecular and metabolic mechanisms that may be utilized to develop tomatoes resilient to various abiotic stressors.

Despite these advancements, challenges remain, including the complex nature of gene-environment interactions and the need for sustainable and cost-effective breeding practices. Future research should focus on exploring the combined effects of multiple abiotic stresses and further refining breeding strategies to develop climate-resilient tomato varieties. Ultimately, the integration of genetic resources, biotechnological tools, and innovative breeding strategies will be crucial for ensuring the continued production of tomatoes.

REFERENCES

Aazami, Mohammad Ali, Farzad Rasouli, and Asghar Ebrahimzadeh. 2021. "Oxidative Damage, Antioxidant Mechanism and Gene Expression in Tomato Responding to Salinity Stress under in Vitro Conditions and Application of Iron and Zinc Oxide Nanoparticles on Callus Induction and Plant Regeneration." *BMC Plant Biology* 21(1). https://doi.org/10.1186/S12870-021-03379-7.

Abuqamar, Synan, Hongli Luo, Kristin Laluk, Michael V. Mickelbart, and Tesfaye Mengiste. 2009. "Crosstalk between Biotic and Abiotic Stress Responses in Tomato Is Mediated by the AIM1 Transcription Factor." *Plant Journal* 58(2):347–360. https://doi.org/10.1111/J.1365-313X.2008.03783.X;SUBPAGE:STRING:FULL.

Adhikari, Laxman, Rudra Baral, Dev Paudel, Doohong Min, Shiva O. Makaju, Hari P. Poudel, Janam P. Acharya, and Ali M. Missaoui. 2022. "Cold Stress in Plants: Strategies to Improve Cold Tolerance in Forage Species." *Plant Stress* 4:100081. https://doi.org/10.1016/J.STRESS.2022.100081.

Agarwal, S., and A. V. Rao. 2000. "Tomato Lycopene and Its Role in Human Health and Chronic Diseases." *CMAJ: Canadian Medical Association Journal* 163(6):739.

Alam, Rafiqul, Jung Hyun Lee, Doyun Shin, Il Kyu Choi, Sunghwan Kim, Heejin Lim, and Min Sik Kim. 2025. "Recent Advances in Single-Cell Metabolomics Using Mass Spectrometry: Emerging Challenges and Future Perspectives." *Applied Spectroscopy Reviews*. https://doi.org/10.1080/05704928.2025.2483996;SUBPAGE:STRING:ACCESS.

Albert, Elise, Vincent Segura, Justine Gricourt, Julien Bonnefoi, Laurent Derivot, and Mathilde Causse. 2016. "Association Mapping Reveals the Genetic Architecture of Tomato Response to Water Deficit: Focus on Major Fruit Quality Traits." *Journal of Experimental Botany* 67(22):6413–6430. https://doi.org/10.1093/JXB/ERW411.

Alemu, Admas, Johanna Åstrand, Osval A. Montesinos-López, Julio Isidro y Sánchez, Javier Fernández-Gónzalez, Wuletaw Tadesse, Ramesh R. Vetukuri, Anders S. Carlsson, Alf Ceplitis, José Crossa, Rodomiro Ortiz, and Aakash Chawade. 2024. "Genomic Selection in Plant Breeding: Key Factors Shaping Two Decades of Progress." *Molecular Plant* 17(4):552–578. https://doi.org/10.1016/J.MOLP.2024.03.007.

Alhaithloul, Haifa A. S., Fatma H. Galal, and Alaa Eddeen M. Seufi. 2021. "Effect of Extreme Temperature Changes on Phenolic, Flavonoid Contents and Antioxidant Activity of Tomato Seedlings (*Solanum lycopersicum* L.)." *PeerJ* 9:e11193. https://doi.org/10.7717/PEERJ.11193/SUPP-9.

Ali, Ahmed Abdelrahim Mohamed, Walid Ben Romdhane, Mohamed Tarroum, Mohammed Al-Dakhil, Abdullah Al-Doss, Abdullah A. Alsadon, and Afif Hassairi. 2021. "Analysis of Salinity Tolerance in Tomato Introgression Lines Based on Morpho-Physiological and Molecular Traits." *Plants* 10(12). https://doi.org/10.3390/PLANTS10122594.

Ali, Md Yousuf, Abu Ali Ibn Sina, Shahad Saif Khandker, Lutfun Neesa, E. M. Tanvir, Alamgir Kabir, Md Ibrahim Khalil, and Siew Hua Gan. 2020. "Nutritional Composition and Bioactive Compounds in Tomatoes and Their Impact on Human Health and Disease: A Review." *Foods* 10(1):45. https://doi.org/10.3390/FOODS10010045.

Ali, Sajad, Anshika Tyagi, and Hanhong Bae. 2021. "Ionomic Approaches for Discovery of Novel Stress-resilient Genes in Plants." *International Journal of Molecular Sciences* 22(13). https://doi.org/10.3390/IJMS22137182.

Arrones, Andrea, Oussama Antar, Leandro Pereira-Dias, Andrea Solana, Paola Ferrante, Giuseppe Aprea, Mariola Plazas, Jaime Prohens, María José Díez, Giovanni Giuliano, Pietro Gramazio, and Santiago

Vilanova. 2024. "A Novel Tomato Interspecific (*Solanum Lycopersicum* Var. *Cerasiforme* and *Solanum Pimpinellifolium*) MAGIC Population Facilitates Trait Association and Candidate Gene Discovery in Untapped Exotic Germplasm." *Horticulture Research* 11(7). https://doi.org/10.1093/HR/UHAE154.

Asakura, Hiroko, Takashi Yamakawa, Tomoko Tamura, Reiko Ueda, Shu Taira, Yoshikazu Saito, Keiko Abe, and Tomiko Asakura. 2021. "Transcriptomic and Metabolomic Analyses Provide Insights into the Upregulation of Fatty Acid and Phospholipid Metabolism in Tomato Fruit under Drought Stress." *Journal of Agricultural and Food Chemistry* 69(9):2894–2905. https://doi.org/10.1021/ACS.JAFC.0C06168/ASSET/IMAGES/LARGE/JF0C06168_0009.JPEG.

Assaha, Dekoum V. M., Akihiro Ueda, Hirofumi Saneoka, Rashid Al-Yahyai, and Mahmoud W. Yaish. 2017. "The Role of Na^+ and K^+ Transporters in Salt Stress Adaptation in Glycophytes." *Frontiers in Physiology* 8(July):509. https://doi.org/10.3389/FPHYS.2017.00509.

Ayenan, Mathieu Anatole Tele, Agyemang Danquah, Peter Hanson, Charles Ampomah-Dwamena, Frejus Ariel Kpedetin Sodedji, Isaac K. Asante, and Eric Yirenkyi Danquah. 2019. "Accelerating Breeding for Heat Tolerance in Tomato (*Solanum lycopersicum* L.): An Integrated Approach." *Agronomy* 9(11):720. https://doi.org/10.3390/AGRONOMY9110720.

Ayenan, Mathieu Anatole Tele, Agyemang Danquah, Peter Hanson, Isaac K. Asante, and Eric Y. Danquah. 2021. "Identification of New Sources of Heat Tolerance in Cultivated and Wild Tomatoes." *Euphytica* 217(3). https://doi.org/10.1007/S10681-021-02772-5.

Bakery, Ayat, Stavros Vraggalas, Boushra Shalha, Harsh Chauhan, Moussa Benhamed, and Sotirios Fragkostefanakis. 2024. "Heat Stress Transcription Factors as the Central Molecular Rheostat to Optimize Plant Survival and Recovery from Heat Stress." *New Phytologist* 244(1):51–64. https://doi.org/10.1111/NPH.20017.

Balasubramaniam, Thuvaraki, Guoxin Shen, Nardana Esmaeili, and Hong Zhang. 2023. "Plants' Response Mechanisms to Salinity Stress." *Plants* 12(12):2253. https://doi.org/10.3390/PLANTS12122253.

Balyan, Sonia, Sombir Rao, Sarita Jha, Chandni Bansal, Jaishri Rubina Das, and Saloni Mathur. 2020. "Characterization of Novel Regulators for Heat Stress Tolerance in Tomato from Indian Sub-Continent." *Plant Biotechnology Journal* 18(10):2118–2132. https://doi.org/10.1111/PBI.13371.

Bashary, Neta, Golan Miller, Tzion Doitsch-Movshovits, Avital Beery, Bo Ouyang, and Michal Lieberman-Lazarovich. 2024. "New Population of *Solanum Pimpinellifolium* Backcross Inbred Lines as a Resource for Heat Stress Tolerance in Tomato." *Frontiers in Plant Science* 15:1386824. https://doi.org/10.3389/fpls.2024.1386824

Begum, Yasmin. 2022. "Regulatory Role of MicroRNAs (MiRNAs) in the Recent Development of Abiotic Stress Tolerance of Plants." *Gene* 821:146283. https://doi.org/10.1016/J.GENE.2022.146283.

Berger, Yael, Smadar Harpaz-Saad, Arnon Brand, Hadas Melnik, Neti Sirding, John Paul Alvarez, Michael Zinder, Alon Samach, Yuval Eshed, and Naomi Ori. 2009. "The NAC-Domain Transcription Factor GOBLET Specifies Leaflet Boundaries in Compound Tomato Leaves." *Development* 136(5):823–832. https://doi.org/10.1242/DEV.031625.

Bolger, Anthony, Federico Scossa, Marie E. Bolger, Christa Lanz, Florian Maumus, Takayuki Tohge, Hadi Quesneville, Saleh Alseekh, Iben Sørensen, Gabriel Lichtenstein, Eric A. Fich, Mariana Conte, Heike Keller, Korbinian Schneeberger, Rainer Schwacke, Itai Ofner, Julia Vrebalov, Yimin Xu, Sonia Osorio, Saulo Alves Aflitos, Elio Schijlen, José M. Jiménez-Goméz, Malgorzata Ryngajllo, Seisuke Kimura, Ravi Kumar, Daniel Koenig, Lauren R. Headland, Julin N. Maloof, Neelima Sinha, Roeland C. H. J. Van Ham, René Klein Lankhorst, Linyong Mao, Alexander Vogel, Borjana Arsova, Ralph Panstruga, Zhangjun Fei, Jocelyn K. C. Rose, Dani Zamir, Fernando Carrari, James J. Giovannoni, Detlef Weigel, Björn Usadel, and Alisdair R. Fernie. 2014. "The Genome of the Stress-Tolerant Wild Tomato Species *Solanum Pennellii*." *Nature Genetics* 46(9):1034–1038. https://doi.org/10.1038/ng.3046.

Bonarota, Maria-Sole, Sean Fenstemaker, Hans Vasquez-Gross, Juli Petereit, Patricia Santos, David M. Francis, and Felipe H. Barrios-Masias. 2025. "Physiology of Salt Tolerance Introgressions from *Solanum Galapagense* in the Domesticated Tomato." *Frontiers in Plant Science* 16:1568851. https://doi.org/10.3389/FPLS.2025.1568851.

Bouzroud, Sarah, Karla Gasparini, Guojian Hu, Maria Antonia Machado Barbosa, Bruno Luan Rosa, Mouna Fahr, Najib Bendaou, Mondher Bouzayen, Agustin Zsögön, Abdelaziz Smouni, and Mohamed Zouine. 2020. "Down Regulation and Loss of Auxin Response Factor 4 Function Using CRISPR/Cas9 Alters Plant Growth, Stomatal Function and Improves Tomato Tolerance to Salinity and Osmotic Stress." *Genes* 11(3):272. https://doi.org/10.3390/GENES11030272.

Budhlakoti, Neeraj, Amar Kant Kushwaha, Anil Rai, K. K. Chaturvedi, Anuj Kumar, Anjan Kumar Pradhan, Uttam Kumar, Rajeev Ranjan Kumar, Philomin Juliana, D. C. Mishra, and Sundeep Kumar. 2022. "Genomic Selection: A Tool for Accelerating the Efficiency of Molecular Breeding for Development of Climate-Resilient Crops." *Frontiers in Genetics* 13:832153. https://doi.org/10.3389/FGENE.2022.832153/XML/NLM.

Bvindi, Carol, Liang Tang, Sanghun Lee, Ryan M. Patrick, Zheng Rong Yee, Tesfaye Mengiste, and Ying Li. 2022. "Histone Methyltransferases SDG33 and SDG34 Regulate Organ-Specific Nitrogen Responses in Tomato." *Frontiers in Plant Science* 13:1005077. https://doi.org/10.3389/FPLS.2022.1005077

Campos, Juan F., Beatriz Cara, Fernando Pérez-Martín, Benito Pineda, Isabel Egea, Francisco B. Flores, Nieves Fernandez-Garcia, Juan Capel, Vicente Moreno, Trinidad Angosto, Rafael Lozano, and Maria C. Bolarin. 2016. "The Tomato Mutant Ars1 (Altered Response to Salt Stress 1) Identifies an R1-Type MYB Transcription Factor Involved in Stomatal Closure under Salt Acclimation." *Plant Biotechnology Journal* 14(6):1345–1356. https://doi.org/10.1111/PBI.12498;WGROUP:STRING:PUBLICATION.

Candar-Cakir, Bilgin, Ercan Arican, and Baohong Zhang. 2016. "Small RNA and Degradome Deep Sequencing Reveals Drought-and Tissue-Specific Micrornas and Their Important Roles in Drought-Sensitive and Drought-Tolerant Tomato Genotypes." *Plant Biotechnology Journal* 14(8):1727–1746. https://doi.org/10.1111/PBI.12533.

Cappetta, Elisa, Giuseppe Andolfo, Anna Guadagno, Antonio Di Matteo, Amalia Barone, Luigi Frusciante, and Maria Raffaella Ercolano. 2021. "Tomato Genomic Prediction for Good Performance under High-Temperature and Identification of Loci Involved in Thermotolerance Response." *Horticulture Research* 8(1):1–16. https://doi.org/10.1038/S41438-021-00647-3;SUBJMETA=208,2661,2665,449,631,711;KWRD=ABIOTIC,PLANT+BREEDING.

Çelik, Ibrahim, Nergiz Gürbüz Çolak, Sami Doğanlar, and Anne Frary. 2023. "Tomato: Genetics, Genomics, and Breeding of Health-Related Traits." *Compendium of Crop Genome Designing for Nutraceuticals*:1217–1267. https://doi.org/10.1007/978-981-19-4169-6_49.

Çelik, Yusuf, and Ahmet Üstün. 2017. "Üniversite Son Sinif Öğretmen Adaylarinin Tükenmişlik Düzeylerinin Çeşitli Değişkenler Açisindan Incelenmesi." *Hitit Üniversitesi Sosyal Bilimler Enstitüsü Dergisi* 10(2):1825–1846. https://doi.org/10.17218/HITITSOSBIL.320178.

Chaudhary, Juhi, Alisha Alisha, Vacha Bhatt, Sonali Chandanshive, Nirbhay Kumar, Humira Sonah, Rupesh Deshmukh, Zahoor Mir, Ashwini Kumar, Satish Yadav, and S. M. S. Shivaraj. 2019. "Mutation Breeding in Tomato: Advances, Applicability and Challenges." *Plants* 8(5):128. https://doi.org/10.3390/PLANTS8050128.

Chaudhary, Poonam, Ashita Sharma, Balwinder Singh, and Avinash Kaur Nagpal. 2018. "Bioactivities of Phytochemicals Present in Tomato." *Journal of Food Science and Technology* 55(8):2833–2849. https://doi.org/10.1007/S13197-018-3221-Z/FIGURES/2.

Chehelgerdi, Mohammad, Matin Chehelgerdi, Milad Khorramian-Ghahfarokhi, Marjan Shafieizadeh, Esmaeil Mahmoudi, Fatemeh Eskandari, Mohsen Rashidi, Asghar Arshi, and Abbas Mokhtari-Farsani. 2024. "Comprehensive Review of CRISPR-Based Gene Editing: Mechanisms, Challenges, and Applications in Cancer Therapy." *Molecular Cancer* 23(1):1–45. https://doi.org/10.1186/S12943-023-01925-5.

Chen, Haoting, Yu Shi, Lu An, Xiaohui Yang, Jie Liu, Zemin Dai, Yi Zhang, Tianlai Li, and Golam Jalal Ahammed. 2024. "Overexpression of SlWRKY6 Enhances Drought Tolerance by Strengthening Antioxidant Defense and Stomatal Closure via ABA Signaling in *Solanum Lycopersicum* L." *Plant Physiology and Biochemistry* 213. https://doi.org/10.1016/j.plaphy.2024.108855.

Chen, Lei, Jun Meng, and Yushi Luan. 2019. "MiR1916 Plays a Role as a Negative Regulator in Drought Stress Resistance in Tomato and Tobacco." *Biochemical and Biophysical Research Communications* 508(2):597–602. https://doi.org/10.1016/J.BBRC.2018.11.165.

Chen, Mengyi, Xiaoyang Zhu, Xiaojuan Liu, Caiyu Wu, Canye Yu, Guojian Hu, Lin Chen, Riyuan Chen, Mondher Bouzayen, Mohammed Zouine, and Yanwei Hao. 2021. "Knockout of Auxin Response Factor SlARF4 Improves Tomato Resistance to Water Deficit." *International Journal of Molecular Sciences* 22(7):3347. https://doi.org/10.3390/IJMS22073347/S1.

Chen, Siyu, Na Zhang, Ganghua Zhou, Saddam Hussain, Sajjad Ahmed, Hainan Tian, and Shucai Wang. 2021. "Knockout of the Entire Family of AITR Genes in Arabidopsis Leads to Enhanced Drought and Salinity Tolerance without Fitness Costs." *BMC Plant Biology* 21(1):1–15. https://doi.org/10.1186/S12870-021-02907-9/FIGURES/9.

Chen, Xiaochang, Yunmei Sun, Rui Cai, Guoqiang Wang, Xiaoyan Shu, and Weijun Pang. 2018. "Long Noncoding RNA: Multiple Players in Gene Expression." *BMB Reports* 51(6):280–289. https://doi.org/10.5483/BMBREP.2018.51.6.025.

Chiwina, Kenani E., Gehendra Bhattarai, Haizheng Xiong, Neelendra K. Joshi, Ryan W. Dickson, Theresa M. Phiri, Ibtisam Alatawi, Yilin Chen, Zachary Stansell, Kai Shu Ling, and Ainong Shi. 2024. "Evaluation of Drought Tolerance in USDA Tomato Germplasm at Seedling Stage." *Agronomy* 14(2):380. https://doi.org/10.3390/AGRONOMY14020380/S1.

Ciarmiello, Loredana F., Pasqualina Woodrow, Amodio Fuggi, Giovanni Pontecorvo, Petronia Carillo, Loredana F. Ciarmiello, Pasqualina Woodrow, Amodio Fuggi, Giovanni Pontecorvo, and Petronia Carillo. 2011. "Plant Genes for Abiotic Stress." *Abiotic Stress in Plants—Mechanisms and Adaptations*. https://doi.org/10.5772/22465.

Cirillo, Valerio, Donata Molisso, Anna Maria Aprile, Albino Maggio, and Rosa Rao. 2022. "Systemin Peptide Application Improves Tomato Salt Stress Tolerance and Reveals Common Adaptation Mechanisms to Biotic and Abiotic Stress in Plants." *Environmental and Experimental Botany* 199:104865. https://doi.org/10.1016/J.ENVEXPBOT.2022.104865.

Clepet, Christian, Ravi Sureshbhai Devani, Rachid Boumlik, Yanwei Hao, Halima Morin, Fabien Marcel, Marion Verdenaud, Brahim Mania, Gwilherm Brisou, Sylvie Citerne, Gregory Mouille, Jean Christophe Lepeltier, Shai Koussevitzky, Adnane Boualem, and Abdelhafid Bendahmane. 2021. "The MiR166–SlHB15A Regulatory Module Controls Ovule Development and Parthenocarpic Fruit Set under Adverse Temperatures in Tomato." *Molecular Plant* 14(7):1185–1198. https://doi.org/10.1016/J.MOLP.2021.05.005/ATTACHMENT/D1C753C3-8CAF-4079-8033-60395F98EA52/MMC2.PDF.

Collins, Edward J., Cressida Bowyer, Audrey Tsouza, and Mridula Chopra. 2022. "Tomatoes: An Extensive Review of the Associated Health Impacts of Tomatoes and Factors That Can Affect Their Cultivation." *Biology* 11(2):239. https://doi.org/10.3390/BIOLOGY11020239.

Conti, Veronica, Luigi Parrotta, Marco Romi, Stefano Del Duca, and Giampiero Cai. 2023. "Tomato Biodiversity and Drought Tolerance: A Multilevel Review." *International Journal of Molecular Sciences* 24(12). https://doi.org/10.3390/IJMS241210044.

Crossa, José, Johannes W. R. Martini, Paolo Vitale, Paulino Pérez-Rodríguez, Germano Costa-Neto, Roberto Fritsche-Neto, Daniel Runcie, Jaime Cuevas, Fernando Toledo, H. Li, Pasquale De Vita, Guillermo Gerard, Susanne Dreisigacker, Leonardo Crespo-Herrera, Carolina Saint Pierre, Alison Bentley, Morten Lillemo, Rodomiro Ortiz, Osval A. Montesinos-López, and Abelardo Montesinos-López. 2025. "Expanding Genomic Prediction in Plant Breeding: Harnessing Big Data, Machine Learning, and Advanced Software." *Trends in Plant Science*. https://doi.org/10.1016/J.TPLANTS.2024.12.009.

Dalvi, Amita A., and Satish A. Bhalerao. 2013. "Response of Plants towards Heavy Metal Toxicity: An Overview of Avoidance, Tolerance and Uptake Mechanism." *Annals of Plant Sciences* 2(9):362–368.

Dariva, Françoise Dalprá, Mariane Gonçalves Ferreira Copati, Herika Paula Pessoa, Flávia Maria Alves, Felipe de Oliveira Dias, Edgard Augusto de Toledo Picoli, Fernando França da Cunha, and Carlos Nick. 2020. "Evaluation of Anatomical and Physiological Traits of *Solanum Pennellii* Cor. Associated with Plant Yield in Tomato Plants under Water-Limited Conditions." *Scientific Reports* 10(1):1–13. https://doi.org/10.1038/S41598-020-73004-4;SUBJMETA=449,631,711;KWRD=PLANT+BREEDING,PLANT+SCIENCES.

De la Torre-González, A., D. Montesinos-pereira, B. Blasco, and J. M. Ruiz. 2018. "Influence of the Proline Metabolism and Glycine Betaine on Tolerance to Salt Stress in Tomato (*Solanum lycopersicum* L.) Commercial Genotypes." *Journal of Plant Physiology* 231:329–336. https://doi.org/10.1016/J.JPLPH.2018.10.013.

Desaint, Henri, Alexandre Héreil, Javier Belinchon-Moreno, Yolande Carretero, Esther Pelpoir, Michel Pascal, Marie Brault, Doriane Dumont, François Lecompte, Patricia Laugier, Renaud Duboscq, Frederique Bitton, Mara Grumic, Christophe Giraud, Paola Ferrante, Giovanni Giuliano, Francesco Sunseri, and Mathilde Causse. 2024. "Integration of QTL and Transcriptome Approaches for the Identification of Genes Involved in Tomato Response to Nitrogen Deficiency." *Journal of Experimental Botany* 75(18):5880–5896. https://doi.org/10.1093/JXB/ERAE265.

Desalew, Fikir. 2023. "Mutation Breeding Application and Future Prospects." *International Journal of Research Studies in Agricultural Sciences* 9(8). https://doi.org/10.20431/2454-6224.0908002.

Devi, Ashni, Kewal Kumar, and Mushtaq Ahmad Khah. 2025. "The Potential of Mutation Breeding for Ensuring Sustainable Food Security." *International Journal of Advanced Research in Science and Technology* 14(3):1533–1541. https://doi.org/10.62226/ijarst20252525.

Ding, Fei, Xin Qiang, Zhiqi Jia, Lili Li, Jianbin Hu, Mengru Yin, Shaoying Xia, Bin Chen, Jieyu Qi, Qiong Li, Yanna Gao, Shiwen Zhang, Shouru Sun, Changsheng Ma, Luming Yang, and Ying Li. 2022. "Knockout of a Novel Salt Responsive Gene SlABIG1 Enhance Salinity Tolerance in Tomato." *Environmental and Experimental Botany* 200:104903. https://doi.org/10.1016/J.ENVEXPBOT.2022.104903.

Diouf, Isidore A., Laurent Derivot, Frédérique Bitton, Laura Pascual, and Mathilde Causse. 2018. "Water Deficit and Salinity Stress Reveal Many Specific QTL for Plant Growth and Fruit Quality Traits in Tomato." *Frontiers in Plant Science* 9:304902. https://doi.org/10.3389/FPLS.2018.00279/BIBTEX.

Doherty, Colleen J., Heather A. Van Buskirk, Susan J. Myers, and Michael F. Thomashow. 2009. "Roles for Arabidopsis CAMTA Transcription Factors in Cold-Regulated Gene Expression and Freezing Tolerance." *The Plant Cell* 21(3):972–984. https://doi.org/10.1105/TPC.108.063958.

Dong, Yufei, Mingjia Tang, Zelan Huang, Jianing Song, Jin Xu, Golam Jalal Ahammed, Jingquan Yu, and Yanhong Zhou. 2022. "The MiR164a-NAM3 Module Confers Cold Tolerance by Inducing Ethylene Production in Tomato." *The Plant Journal* 111(2):440–456. https://doi.org/10.1111/TPJ.15807.

Duan, Ming, Hai Long Feng, Li Yan Wang, Dong Li, and Qing Wei Meng. 2012. "Overexpression of Thylakoidal Ascorbate Peroxidase Shows Enhanced Resistance to Chilling Stress in Tomato." *Journal of Plant Physiology* 169(9):867–877. https://doi.org/10.1016/J.JPLPH.2012.02.012.

Ebert, A. W., and Y. Y. Chou. 2015. "The Tomato Collection Maintained by AVRDC—The World Vegetable Center: Composition, Germplasm Dissemination and Use in Breeding." *Acta Horticulturae* (1101):169–176. https://doi.org/10.17660/ACTAHORTIC.2015.1101.26
.

Egea, Isabel, Irene Albaladejo, Victoriano Meco, Belén Morales, Angel Sevilla, Maria C. Bolarin, and Francisco B. Flores. 2018. "The Drought-Tolerant *Solanum Pennellii* Regulates Leaf Water Loss and Induces Genes Involved in Amino Acid and Ethylene/Jasmonate Metabolism under Dehydration." *Scientific Reports* 8(1):1–14. https://doi.org/10.1038/S41598-018-21187-2;TECHMETA=38,61,90;SUBJMETA=2146,2661,449,631;KWRD=DROUGHT,PLANT+SCIENCES.

Egea, Isabel, Yanira Estrada, Francisco B. Flores, and Maria C. Bolarín. 2022. "Improving Production and Fruit Quality of Tomato under Abiotic Stress: Genes for the Future of Tomato Breeding for a Sustainable Agriculture." *Environmental and Experimental Botany* 204:105086. https://doi.org/10.1016/J.ENVEXPBOT.2022.105086.

Elazazi, Elsayed, Laura Ziems, Tariq Mahmood, Naeema Eltanger, Maryam Al-Qahtani, Nafeesath Shahsil, Aisha Al-Kuwari, Mohammed M. Metwally, Richard Trethowan, and Chongmei Dong. 2024. "Genotypic Selection Using Quantitative Trait Loci for Better Productivity under High Temperature Stress in Tomato (*Solanum lycopersicum* L.)." *Horticulturae* 10(8):874. https://doi.org/10.3390/HORTICULTURAE10080874/S1.

Eom, Seung Hee, Min A. Ahn, Eunhui Kim, Hee Ju Lee, Jin Hyoung Lee, Seung Hwan Wi, Sung Kyeom Kim, Heung Bin Lim, and Tae Kyung Hyun. 2022. "Plant Response to Cold Stress: Cold Stress Changes Antioxidant Metabolism in Heading Type Kimchi Cabbage (*Brassica Rapa* L. Ssp. *Pekinensis*)." *Antioxidants* 11(4). https://doi.org/10.3390/ANTIOX11040700.

Eyshi Rezaei, Ehsan, and Heidi Webber. 2022. "Processing Tomatoes under Climate Change." *Nature Food* 3(6):404–405. https://doi.org/10.1038/S43016-022-00520-Z;SUBJMETA=106,449,631,704;KWRD=CLIMATE+SCIENCES,PLANT+SCIENCES.

FAO. 2024. *Agricultural Production Statistics 2010–2023*. FAO.

Faruq, G., H. P. Zakaria, and N. Arash. 2012. "Heat Tolerance in Tomato." *Life Science Journal* 9(4):1097–1135.

Foolad, Majid R. 2007. "Genome Mapping and Molecular Breeding of Tomato." *International Journal of Plant Genomics* 2007. https://doi.org/10.1155/2007/64358.

Foolad, M. R., and F. Q. Chen. 1998. "RAPD Markers Associated with Salt Tolerance in an Interspecific Cross of Tomato (*Lycopersicon Esculentum* x L. *Pennellii*)." *Plant Cell Reports* 17(4):306–312. https://doi.org/10.1007/S002990050398/METRICS.

Foolad, M. R., F. Q. Chen, and G. Y. Lin. 1998a. "RFLP Mapping of QTLs Conferring Cold Tolerance during Seed Germination in an Interspecific Cross of Tomato." *Molecular Breeding* 4(6):519–529. https://doi.org/10.1023/A:1009648112491/METRICS.

Foolad, M. R., F. Q. Chen, and G. Y. Lin. 1998b. "RFLP Mapping of QTLs Conferring Salt Tolerance during Germination in an Interspecific Cross of Tomato." *Theoretical and Applied Genetics* 97(7):1133–1144. https://doi.org/10.1007/S001220051002/METRICS.

Foolad, M. R., and R. A. Jones. 1993. "Mapping Salt-Tolerance Genes in Tomato (*Lycopersicon Esculentum*) Using Trait-Based Marker Analysis." *Theoretical and Applied Genetics: International Journal of Plant Breeding Research* 87(1):184–192. https://doi.org/10.1007/BF00223763/METRICS.

Foolad, Majid R., and Dilip R. Panthee. 2012. "Marker-Assisted Selection in Tomato Breeding." *Critical Reviews in Plant Sciences* 31(2):93–123. https://doi.org/10.1080/07352689.2011.616057.

Foolad, M. R., T. Stoltz, C. Dervinis, R. L. Rodriguez, and R. A. Jones. 1997. "Mapping QTLs Conferring Salt Tolerance during Germination in Tomato by Selective Genotyping." *Molecular Breeding* 3(4):269–277. https://doi.org/10.1023/A:1009668325331/METRICS.

Foolad, Majid R., Prakash Subbiah, and Liping Zhang. 2007. "Common QTL Affect the Rate of Tomato Seed Germination under Different Stress and Nonstress Conditions." *International Journal of Plant Genomics* 2007(1):097386. https://doi.org/10.1155/2007/97386.

Foolad, M. R., L. P. Zhang, and G. Y. Lin. 2001. "Identification and Validation of QTLs for Salt Tolerance during Vegetative Growth in Tomato by Selective Genotyping." *Genome* 44(3):444–454. https://doi.org/10.1139/G01-030/ASSET/IMAGES/APPROX.GIF.

Foolad, M. R., L. P. Zhang, and P. Subbiah. 2003. "Genetics of Drought Tolerance during Seed Germination in Tomato: Inheritance and QTL Mapping." *Genome* 46(4):536–545. https://doi.org/10.1139/G03-035.

Fracasso, Alessandra, Marta Vallino, Alessia Staropoli, Francesco Vinale, Stefano Amaducci, and Andrea Carra. 2021. "Increased Water Use Efficiency in MiR396-Downregulated Tomato Plants." *Plant Science* 303:110729. https://doi.org/10.1016/J.PLANTSCI.2020.110729.

Francesca, Silvana, Sabri Najai, Rong Zhou, Guillaume Decros, Cedric Cassan, Frederic Delmas, Carl Otto Ottosen, Amalia Barone, and Maria Manuela Rigano. 2022. "Phenotyping to Dissect the Biostimulant Action of a Protein Hydrolysate in Tomato Plants under Combined Abiotic Stress." *Plant Physiology and Biochemistry* 179:32–43. https://doi.org/10.1016/J.PLAPHY.2022.03.012.

Frary, Anne, Deniz Göl, Davut Keleş, Bilal Ökmen, Hasan Pinar, Hasan T. Şiğva, Ahmet Yemenicioğlu, and Sami Doğanlar. 2010. "Salt Tolerance in *Solanum Pennellii*: Antioxidant Response and Related QTL." *BMC Plant Biology* 10(1):1–16. https://doi.org/10.1186/1471-2229-10-58/FIGURES/5.

Gaj, Thomas, Charles A. Gersbach, and Carlos F. Barbas. 2013. "ZFN, TALEN and CRISPR/Cas-Based Methods for Genome Engineering." *Trends in Biotechnology* 31(7):397. https://doi.org/10.1016/J.TIBTECH.2013.04.004.

Gong, Biao, Xiufeng Wang, Min Wei, Fengjuan Yang, Yan Li, and Qinghua Shi. 2016. "Overexpression of S-Adenosylmethionine Synthetase 1 Enhances Tomato Callus Tolerance to Alkali Stress through Polyamine and Hydrogen Peroxide Cross-Linked Networks." *Plant Cell, Tissue and Organ Culture* 124(2):377–391. https://doi.org/10.1007/S11240-015-0901-5/FIGURES/8.

Gonzalo, Maria José, Yi Cheng Li, Kai Yi Chen, David Gil, Teresa Montoro, Inmaculada Nájera, Carlos Baixauli, Antonio Granell, and Antonio José Monforte. 2020. "Genetic Control of Reproductive Traits in Tomatoes Under High Temperature." *Frontiers in Plant Science* 11:505845. https://doi.org/10.3389/FPLS.2020.00326/BIBTEX.

Gonzalo, María José, Inmaculada Nájera, Carlos Baixauli, David Gil, Teresa Montoro, Vicky Soriano, Fabrizio Olivieri, Maria Manuela Rigano, Daniela Ganeva, Stanislava Grozeva-Tileva, Galina Pevicharova, Amalia Barone, Antonio Granell, and Antonio José Monforte. 2021. "Identification of Tomato Accessions as Source of New Genes for Improving Heat Tolerance: From Controlled Experiments to Field." *BMC Plant Biology* 21(1):1–28. https://doi.org/10.1186/S12870-021-03104-4/TABLES/11.

Graci, Salvatore, and Amalia Barone. 2023. "Tomato Plant Response to Heat Stress: A Focus on Candidate Genes for Yield-Related Traits." *Frontiers in Plant Science* 14:1245661. https://doi.org/10.3389/FPLS.2023.1245661/XML/NLM.

Grilli, G. V. G., L. T. Braz, and E. G. M. Lemos. 2007. "QTL Identification for Tolerance to Fruit Set in Tomato by FAFLP Markers." *Crop Breeding and Applied Biotechnology* 7(3):234–241.

Grozeva, Stanislava, Elena Topalova, Daniela Ganeva, and Ivanka Tringovska. 2024. "Evaluation of Tomato Landraces for Tolerance to Drought Stress Using Morphological and Physiological Traits." *International Journal of Plant Biology* 15(4):1391–1404. https://doi.org/10.3390/IJPB15040096.

Gu, Mian, Ke Xu, Aiqun Chen, Yiyong Zhu, Guiliang Tang, and Guohua Xu. 2010. "Expression Analysis Suggests Potential Roles of MicroRNAs for Phosphate and Arbuscular Mycorrhizal Signaling in *Solanum Lycopersicum*." *Physiologia Plantarum* 138(2):226–237. https://doi.org/10.1111/J.1399-3054.2009.01320.X.

Guo, Meng, Xin Sheng Wang, Hui Dan Guo, Sheng Yi Bai, Abid Khan, Xiao Min Wang, Yan Ming Gao, and Jian She Li. 2022. "Tomato Salt Tolerance Mechanisms and Their Potential Applications for Fighting Salinity: A Review." *Frontiers in Plant Science* 13:949541. https://doi.org/10.3389/FPLS.2022.949541.

Guo, Xinyong, Li Zhang, Xiaozhen Wang, Minhuan Zhang, Yuxin Xi, Aiying Wang, and Jianbo Zhu. 2019. "Overexpression of *Saussurea Involucrata* Dehydrin Gene SiDHN Promotes Cold and Drought Tolerance in Transgenic Tomato Plants." *PLoS ONE* 14(11):e0225090. https://doi.org/10.1371/JOURNAL.PONE.0225090.

Gupta, Rajat M., and Kiran Musunuru. 2014. "Expanding the Genetic Editing Tool Kit: ZFNs, TALENs, and CRISPR-Cas9." *The Journal of Clinical Investigation* 124(10):4154. https://doi.org/10.1172/JCI72992.

Habib, Sidra, Yee Yee Lwin, and Ning Li. 2021. "Article Down-Regulation of Slgras10 in Tomato Confers Abiotic Stress Tolerance." *Genes* 12(5):623. https://doi.org/10.3390/GENES12050623/S1.

Habibi, Nasratullah, Naoki Terada, Atsushi Sanada, and Kaihei Koshio. 2024. "Alleviating Salt Stress in Tomatoes through Seed Priming with Polyethylene Glycol and Sodium Chloride Combination." *Stresses* 4(2):210–224. https://doi.org/10.3390/STRESSES4020012.

Hameed, Muhammed, Claudia Keitel, Nabil Ahmad, Tariq Mahmood, and Richard Trethowan. 2015. "Screening of Tomatoes Germplasm for Heat Stress Tolerance Under Controlled Conditions." *Procedia Environmental Sciences* 29:173–174. https://doi.org/10.1016/J.PROENV.2015.07.245.

Han, Guoliang, Ziqi Qiao, Yuxia Li, Zongran Yang, Chengfeng Wang, Yuanyuan Zhang, Lili Liu, and Baoshan Wang. 2022. "RING Zinc Finger Proteins in Plant Abiotic Stress Tolerance." *Frontiers in Plant Science* 13:877011. https://doi.org/10.3389/FPLS.2022.877011/XML/NLM.

Han, Nani, Shuya Fan, Tiantian Zhang, Hao Sun, Yongxing Zhu, Haijun Gong, and Jia Guo. 2020. "SlHY5 Is a Necessary Regulator of the Cold Acclimation Response in Tomato." *Plant Growth Regulation* 91(1):1–12. https://doi.org/10.1007/S10725-020-00583-7/FIGURES/6.

He, Yu, Tongwen Yang, Siwei Yan, Shaobo Niu, and Yan Zhang. 2022. "Identification and Characterization of the BEL1-like Genes Reveal Their Potential Roles in Plant Growth and Abiotic Stress Response in Tomato." *International Journal of Biological Macromolecules* 200:193–205. https://doi.org/10.1016/J.IJBIOMAC.2021.12.175.

Hendelman, Anat, Ran Stav, Hanita Zemach, and Tzahi Arazi. 2013. "The Tomato NAC Transcription Factor SlNAM2 Is Involved in Flower-Boundary Morphogenesis." *Journal of Experimental Botany* 64(18):5497–5507. https://doi.org/10.1093/JXB/ERT324.

Héreil, A., M. Guillaume, R. Duboscq, Y. Carretero, E. Pelpoir, F. Bitton, C. Giraud, R. Karlova, C. Testerink, R. Stevens, and M. Causse. 2024. "Characterization of a Major QTL for Sodium Accumulation in Tomato Shoot." *BioRxiv*. https://doi.org/10.1101/2024.05.21.595121.

Hu, Chen, Jing Yang, Ziping Qi, Hong Wu, Beilei Wang, Fengming Zou, Husheng Mei, Jing Liu, Wenchao Wang, and Qingsong Liu. 2022. "Heat Shock Proteins: Biological Functions, Pathological Roles, and Therapeutic Opportunities." *MedComm* 3(3):e161. https://doi.org/10.1002/MCO2.161.

Hu, Da Gang, Shu Hui Wang, Hua Luo, Qi Jun Ma, Yu Xin Yao, Chun Xiang You, and Yu Jin Hao. 2012. "Overexpression of MdVHA-B, a V-ATPase Gene from Apple, Confers Tolerance to Drought in Transgenic Tomato." *Scientia Horticulturae* 145:94–101. https://doi.org/10.1016/J.SCIENTA.2012.08.010.

Hu, Nan, Zhiqiang Xian, Ning Li, Yudong Liu, Wei Huang, Fang Yan, Deding Su, Jingxuan Chen, and Zhengguo Li. 2019. "Rapid and User-Friendly Open-Source CRISPR/Cas9 System for Single- or Multi-Site Editing of Tomato Genome." *Horticulture Research* 6(1):7. https://doi.org/10.1038/S41438-018-0082-6/41973646/41438_2018_ARTICLE_82.PDF.

Hu, Tixu, Jie Ye, Peiwen Tao, Hanxia Li, Junhong Zhang, Yuyang Zhang, and Zhibiao Ye. 2016. "The Tomato HD-Zip i Transcription Factor SlHZ24 Modulates Ascorbate Accumulation through Positive Regulation of the d-Mannose/l-Galactose Pathway." *Plant Journal* 85(1):16–29. https://doi.org/10.1111/TPJ.13085; JOURNAL:JOURNAL:1365313X;PAGE:STRING:ARTICLE/CHAPTER.

Hu, Zhangjian, Jianxin Li, Shuting Ding, Fei Cheng, Xin Li, Yuping Jiang, Jingquan Yu, Christine H. Foyer, and Kai Shi. 2021. "The Protein Kinase CPK28 Phosphorylates Ascorbate Peroxidase and Enhances Thermotolerance in Tomato." *Plant Physiology* 186(2):1302–1317. https://doi.org/10.1093/PLPHYS/KIAB120.

Huang, Ying, Jing An, Sanchari Sircar, Clara Bergis, Chloé Dias Lopes, Xiaoning He, Barbara Da Costa, Feng Quan Tan, Jeremie Bazin, Javier Antunez-Sanchez, Maria Florencia Mammarella, Ravi sureshbhai Devani, Rim Brik-Chaouche, Abdelhafid Bendahmane, Florian Frugier, Chongjing Xia, Christophe Rothan, Aline V. Probst, Zouine Mohamed, Catherine Bergounioux, Marianne Delarue, Yijing Zhang, Shaojian Zheng, Martin Crespi, Sotirios Fragkostefanakis, Magdy M. Mahfouz, Federico Ariel, Jose Gutierrez-Marcos, Cécile Raynaud, David Latrasse, and Moussa Benhamed. 2023. "HSFA1a Modulates Plant Heat Stress Responses and Alters the 3D Chromatin Organization of Enhancer-Promoter Interactions." *Nature Communications* 14(1):1–15. https://doi.org/10.1038/s41467-023-36227-3.

Huang, Zejun, Zhijin Zhang, Xiuling Zhang, Hongbo Zhang, Dafang Huang, and Rongfeng Huang. 2004. "Tomato TERF1 Modulates Ethylene Response and Enhances Osmotic Stress Tolerance by Activating Expression of Downstream Genes." *FEBS Letters* 573(1–3):110–116. https://doi.org/10.1016/J.FEBSLET.2004.07.064;JOURNAL:JOURNAL:18733468.

Hubab, Muhammad, Niloufar Lorestani, Rogayah Akram Mheisin Al-Awabdeh, and Farzin Shabani. 2025. "Climate Change-Driven Shifts in the Global Distribution of Tomato and Potato Crops and Their Associated Bacterial Pathogens." *Frontiers in Microbiology* 16:1520104. https://doi.org/10.3389/FMICB.2025.1520104/BIBTEX.

Hui, Xu, Randy Kwaku Amponsah, Samuel Antwi, Patrick Kweku Gbolonyo, Moses Agyemang Ameyaw, Geoffrey Bentum-Micah, and Edward Oppong Adjei. 2024. "Understanding the Societal Dilemma of Genetically Modified Food Consumption: A Stimulus-Organism-Response Investigation." *Frontiers in Sustainable Food Systems* 8:1364052. https://doi.org/10.3389/FSUFS.2024.1364052/BIBTEX.

Hundleby, Penny, and Wendy Harwood. 2022. "Regulatory Constraints and Differences of Genome-Edited Crops Around the Globe." In *Genome Editing: Current Technology Advances and Applications for Crop Improvement*, 319–341. https://doi.org/10.1007/978-3-031-08072-2_17/FIGURES/1.

Hussain, Hafiz A., Saddam Hussain, Abdul Khaliq, Umair Ashraf, Shakeel A. Anjum, Shengnan Men, and Longchang Wang. 2018. "Chilling and Drought Stresses in Crop Plants: Implications, Cross Talk, and Potential Management Opportunities." *Frontiers in Plant Science* 9:393. https://doi.org/10.3389/FPLS.2018.00393.

Illouz-Eliaz, Natanella, Idan Nissan, Ido Nir, Uria Ramon, Hagai Shohat, and David Weiss. 2020. "Mutations in the Tomato Gibberellin Receptors Suppress Xylem Proliferation and Reduce Water Loss under Water-Deficit Conditions." *Journal of Experimental Botany* 71(12):3603–3612. https://doi.org/10.1093/JXB/ERAA137.

Irfan, Mohammad, Pankaj Kumar, Irshad Ahmad, and Asis Datta. 2021. "Unraveling the Role of Tomato Bcl-2-Associated Athanogene (BAG) Proteins during Abiotic Stress Response and Fruit Ripening." *Scientific Reports* 11(1):1–15. https://doi.org/10.1038/S41598-021-01185-7;SUBJMETA=1659,2661,447,449,631;KWRD=PLANT+BIOTECHNOLOGY,PLANT+MOLECULAR+BIOLOGY,PLANT+STRESS+RESPONSES.

Jeon, Jong Rok, Petr Baldrian, Kumarasamy Murugesan, and Yoon Seok Chang. 2012. "Laccase-Catalysed Oxidations of Naturally Occurring Phenols: From in Vivo Biosynthetic Pathways to Green Synthetic Applications." *Microbial Biotechnology* 5(3):318–332. https://doi.org/10.1111/J.1751-7915.2011.00273.X.

Jia, Chengguo, Shengke Zhao, Tingting Bao, Peiqing Zhao, Kuan Peng, Qingxun Guo, Xiang Gao, and Jianchun Qin. 2021. "Tomato BZR/BES Transcription Factor SlBZR1 Positively Regulates BR Signaling and Salt Stress Tolerance in Tomato and Arabidopsis." *Plant Science* 302:110719. https://doi.org/10.1016/J.PLANTSCI.2020.110719.

Jiménez Bolaño, Diana Carolina, Daniel Insuasty, Juan David Rodríguez Macías, and Carlos David Grande-Tovar. 2024. "Potential Use of Tomato Peel, a Rich Source of Lycopene, for Cancer Treatment." *Molecules* 29(13):3079. https://doi.org/10.3390/MOLECULES29133079.

Jodder, Jayanti, Rohit Das, Deepti Sarkar, Payel Bhattacharjee, and Pallob Kundu. 2018. "Distinct Transcriptional and Processing Regulations Control MiR167a Level in Tomato during Stress." *RNA Biology* 15(1):130–143. https://doi.org/10.1080/15476286.2017.1391438.

Joseph Caspermeyer. 2020. "Evolution on the Vine: A History of Tomato Domestication in Latin America." *Molecular Biology and Evolution* 37(4):1241–1242. https://doi.org/10.1093/MOLBEV/MSAA021.

Kadyrzhanova, Dina K., Konstantinos E. Vlachonasios, Philippos Ververidis, and David R. Dilley. 1998. "Molecular Cloning of a Novel Heat Induced/Chilling Tolerance Related cDNA in Tomato Fruit by Use of mRNA Differential Display." *Plant Molecular Biology* 36(6):885–895. https://doi.org/10.1023/A:1005954909011.

Kakeshpour, Tayebeh, Tej Man Tamang, Gergely Motolai, Zachary Wayne Fleming, Jung Eun Park, Qingyu Wu, and Sunghun Park. 2021. "CGFS-Type Glutaredoxin Mutations Reduce Tolerance to Multiple Abiotic Stresses in Tomato." *Physiologia Plantarum* 173(3):1263–1279. https://doi.org/10.1111/PPL.13522.

Kalloo, G. 1991. Interspecific and Intergeneric Hybridization in Tomato. In: Kalloo, G. (eds) *Genetic Improvement of Tomato. Monographs on Theoretical and Applied Genetics*, vol 14. Springer, Berlin, Heidelberg. https://doi.org/10.1007/978-3-642-84275-7_7

Kapazoglou, Aliki, Maria Gerakari, Efstathia Lazaridi, Konstantina Kleftogianni, Efi Sarri, Eleni Tani, and Penelope J. Bebeli. 2023. "Crop Wild Relatives: A Valuable Source of Tolerance to Various Abiotic Stresses." *Plants* 12(2):328. https://doi.org/10.3390/PLANTS12020328.

Karthikeyan, Alagarsamy, Shumugiah Karutha Pandian, and Manikandan Ramesh. 2011. "Transgenic Indica Rice Cv. ADT 43 Expressing a Δ^1-Pyrroline-5-Carboxylate Synthetase (P5CS) Gene from *Vigna Aconitifolia* Demonstrates Salt Tolerance." *Plant Cell, Tissue and Organ Culture* 107(3):383–395. https://doi.org/10.1007/S11240-011-9989-4/FIGURES/5.

Kaur, Navdeep, Shubham Sharma, Mirza Hasanuzzaman, and Pratap Kumar Pati. 2022. "Genome Editing: A Promising Approach for Achieving Abiotic Stress Tolerance in Plants." *International Journal of Genomics* 2022(1):5547231. https://doi.org/10.1155/2022/5547231.

Kaya, Hilal Betul. 2023. "Base Editing and Prime Editing." In *A Roadmap for Plant Genome Editing*, 17–39. https://doi.org/10.1007/978-3-031-46150-7_2/TABLES/2.

Khan, Naeem. 2025. "Decoding Phytohormone Signaling in Plant Stress Physiology: Insights, Challenges, and Future Directions." *Environmental and Experimental Botany* 231:106099. https://doi.org/10.1016/J.ENVEXPBOT.2025.106099.

Khazaei, Hamid, and Adithya Madduri. 2022. "The Role of Tomato Wild Relatives in Breeding Disease-Free Varieties." *Genetic Resources* 3(6):64–73. https://doi.org/10.46265/GENRESJ.PSES6766.

Klap, Chen, Ester Yeshayahou, Anthony M. Bolger, Tzahi Arazi, Suresh K. Gupta, Sara Shabtai, Björn Usadel, Yehiam Salts, and Rivka Barg. 2017. "Tomato Facultative Parthenocarpy Results from SlAGAMOUS-LIKE 6 Loss of Function." *Plant Biotechnology Journal* 15(5):634–647. https://doi.org/10.1111/PBI.12662.

Klein-Lankhorst, R., P. Rietveld, B. Machiels, R. Verkerk, R. Weide, C. Gebhardt, M. Koornneef, and P. Zabel. 1991. "RFLP Markers Linked to the Root Knot Nematode Resistance Gene Mi in Tomato." *Theoretical and Applied Genetics* 81(5):661–667. https://doi.org/10.1007/BF00226734.

Krishna, Ram, Waquar Akhter Ansari, P. S. Soumia, Akhilesh Yadav, Durgesh Kumar Jaiswal, Sudhir Kumar, Achuit Kumar Singh, Major Singh, and Jay Prakash Verma. 2022. "Biotechnological Interventions in Tomato (*Solanum lycopersicum*) for Drought Stress Tolerance: Achievements and Future Prospects." *BioTech* 11(4):48. https://doi.org/10.3390/BIOTECH11040048.

Krishna, Ram, Suhas G. Karkute, Waquar A. Ansari, Durgesh Kumar Jaiswal, Jay Prakash Verma, and Major Singh. 2019. "Transgenic Tomatoes for Abiotic Stress Tolerance: Status and Way Ahead." *3 Biotech* 9(4). https://doi.org/10.1007/S13205-019-1665-0.

Krupa-Małkiewicz, Marcelina, and Ireneusz Ochmian. 2024. "Sustainable Approaches to Alleviate Heavy Metal Stress in Tomatoes: Exploring the Role of Chitosan and Nanosilver." *Agronomy* 14(11):2477. https://doi.org/10.3390/AGRONOMY14112477.

Kumar, Manoj, Manish Kumar Patel, Navin Kumar, Atal Bihari Bajpai, and Kadambot H. M. Siddique. 2021. "Metabolomics and Molecular Approaches Reveal Drought Stress Tolerance in Plants." *International Journal of Molecular Sciences* 22(17):9108. https://doi.org/10.3390/IJMS22179108.

Kumar, Manoj, Manas Ranjan Prusty, Manish K. Pandey, Prashant Kumar Singh, Abhishek Bohra, Baozhu Guo, and Rajeev K. Varshney. 2023. "Application of CRISPR/Cas9-Mediated Gene Editing for Abiotic Stress Management in Crop Plants." *Frontiers in Plant Science* 14:1157678. https://doi.org/10.3389/FPLS.2023.1157678/XML/NLM.

Kumar, Rahul, Sankar Prasad Das, Burhan Uddin Choudhury, Amit Kumar, Nitish Ranjan Prakash, Ramlakhan Verma, Mridul Chakraborti, Ayam Gangarani Devi, Bijoya Bhattacharjee, Rekha Das, Bapi Das, H. Lembisana Devi, Biswajit Das, Santoshi Rawat, and Vinay Kumar Mishra. 2024. "Advances in Genomic Tools for Plant Breeding: Harnessing DNA Molecular Markers, Genomic Selection, and Genome Editing." *Biological Research* 57(1):1–23. https://doi.org/10.1186/S40659-024-00562-6.

Kumar, Shreevastav, Chandan, Swastika Subedi, Sudha Gajurel, and Pratibha Basnet. 2022. "A Review on Nutrient Deficiency Symptoms and Effects on Tomato Plant." *Food and Agri Economics Review* 2(1):34–36. https://doi.org/10.26480/faer.01.2022.34.36.

Kumar, Suresh. 2020. "Abiotic Stresses and Their Effects on Plant Growth, Yield and Nutritional Quality of Agricultural Produce." *International Journal of Food Science and Agriculture* 4(4):367–378. https://doi.org/10.26855/IJFSA.2020.12.002.

Lee, Chungkeun, Joshua T. Harvey, Asmita Nagila, Kuan Qin, and Daniel I. Leskovar. 2023. "Thermotolerance of Tomato Plants Grafted onto Wild Relative Rootstocks." *Frontiers in Plant Science* 14:1252456. https://doi.org/10.3389/FPLS.2023.1252456/BIBTEX.

Lei, L., L. L. Liu, S. Y. Song, L. Yan, and J. M. Li. 2011. "Mapping of QTLs for Drought Tolerance during Seedling Stage Using Introgression Line Populations in Tomato." https://doi.org/10.5555/20123032340.

Leporino, Marzia, Youssef Rouphael, Paolo Bonini, Giuseppe Colla, and Mariateresa Cardarelli. 2024. "Protein Hydrolysates Enhance Recovery from Drought Stress in Tomato Plants: Phenomic and Metabolomic Insights." *Frontiers in Plant Science* 15:1357316. https://doi.org/10.3389/FPLS.2024.1357316/BIBTEX.

Li, Cheng, Xinyi Liu, Hui Ruan, Jingyao Zhang, Fengbin Xie, Junyi Gai, and Shouping Yang. 2020. "GmWRKY45 Enhances Tolerance to Phosphate Starvation and Salt Stress, and Changes Fertility in Transgenic Arabidopsis." *Frontiers in Plant Science* 10:492174. https://doi.org/10.3389/FPLS.2019.01714/BIBTEX.

Li, Jing Bin, Yu Shi Luan, and Zhen Liu. 2015. "SpWRKY1 Mediates Resistance to Phytophthora Infestans and Tolerance to Salt and Drought Stress by Modulating Reactive Oxygen Species Homeostasis and Expression of Defense-Related Genes in Tomato." *Plant Cell, Tissue and Organ Culture* 123(1):67–81. https://doi.org/10.1007/S11240-015-0815-2/FIGURES/7.

Li, Jinhua, Yaling Wang, Juanjuan Wei, Yu Pan, Chenggang Su, and Xingguo Zhang. 2018. "A Tomato Proline-, Lysine-, and Glutamic-Rich Type Gene SpPKE1 Positively Regulates Drought Stress Tolerance." *Biochemical and Biophysical Research Communications* 499(4):777–782. https://doi.org/10.1016/J.BBRC.2018.03.222.

Li, Jinquan, Ann Katrin Lühmann, Knuth Weißleder, and Benjamin Stich. 2011. "Genome-Wide Distribution of Genetic Diversity and Linkage Disequilibrium in Elite Sugar Beet Germplasm." *BMC Genomics* 12(1):1–10. https://doi.org/10.1186/1471-2164-12-484/TABLES/2.

Li, Junming, Lei Liu, Yuling Bai, Pujuan Zhang, Richard Finkers, Yongchen Du, Richard G. F. Visser, and Adriaan W. van Heusden. 2011. "Seedling Salt Tolerance in Tomato." *Euphytica* 178(3):403–414. https://doi.org/10.1007/S10681-010-0321-X/FIGURES/4.

Li, Ning, Qiang He, Juan Wang, Baike Wang, Jiantao Zhao, Shaoyong Huang, Tao Yang, Yaping Tang, Shengbao Yang, Patiguli Aisimutuola, Ruiqiang Xu, Jiahui Hu, Chunping Jia, Kai Ma, Zhiqiang Li, Fangling Jiang, Jie Gao, Haiyan Lan, Yongfeng Zhou, Xinyan Zhang, Sanwen Huang, Zhangjun Fei, Huan Wang, Hongbo Li, and Qinghui Yu. 2023. "Super-Pangenome Analyses Highlight Genomic Diversity and Structural Variation across Wild and Cultivated Tomato Species." *Nature Genetics* 55(5):852–860. https://doi.org/10.1038/s41588-023-01340-y.

Li, Qian, Heng Shen, Shoujuan Yuan, Xigang Dai, and Changxian Yang. 2023. "miRNAs and lncRNAs in Tomato: Roles in Biotic and Abiotic Stress Responses." *Frontiers in Plant Science* 13:1094459. https://doi.org/10.3389/FPLS.2022.1094459/XML/NLM.

Li, Qian, Weiyue Sun, Chao Chen, Dingxiao Dong, Yaoliang Cao, Yanlong Dong, Lijie Yu, Zhonghui Yue, and Xiaoxia Jin. 2022. "Overexpression of Histone Demethylase Gene SlJMJ524 from Tomato Confers Cd Tolerance by Regulating Metal Transport-Related Protein Genes and Flavonoid Content in Arabidopsis." *Plant Science* 318. https://doi.org/10.1016/j.plantsci.2022.111205.

Li, Ran, Daqi Fu, Benzhong Zhu, Yunbo Luo, and Hongliang Zhu. 2018. "CRISPR/Cas9-Mediated Mutagenesis of lncRNA1459 Alters Tomato Fruit Ripening." *Plant Journal* 94(3):513–524. https://doi.org/10.1111/TPJ.13872;WGROUP:STRING:PUBLICATION.

Li, Rui, Chunxue Liu, Ruirui Zhao, Liu Wang, Lin Chen, Wenqing Yu, Shujuan Zhang, Jiping Sheng, and Lin Shen. 2019. "CRISPR/Cas9-Mediated SlNPR1 Mutagenesis Reduces Tomato Plant Drought Tolerance." *BMC Plant Biology* 19(1):1–13. https://doi.org/10.1186/S12870-018-1627-4/FIGURES/7.

Li, Rui, Lixing Zhang, Liu Wang, Lin Chen, Ruirui Zhao, Jiping Sheng, and Lin Shen. 2018. "Reduction of Tomato-Plant Chilling Tolerance by CRISPR-Cas9-Mediated SlCBF1 Mutagenesis." *Journal of Agricultural and Food Chemistry* 66(34):9042–9051. https://doi.org/10.1021/ACS.JAFC.8B02177/SUPPL_FILE/JF8B02177_SI_001.PDF.

Li, Shenglan, Xizi Wang, Xiaojuan Liu, Andrew J. Thompson, and Fulai Liu. 2022. "Elevated CO_2 and High Endogenous ABA Level Alleviate PEG-Induced Short-Term Osmotic Stress in Tomato Plants." *Environmental and Experimental Botany* 194:104763. https://doi.org/10.1016/J.ENVEXPBOT.2021.104763.

Li, Wen Xue, Youko Oono, Jianhua Zhu, Xin Jian He, Jian Min Wu, Kei Iida, Xiao Yan Lu, Xinping Cui, Hailing Jin, and Jian Kang Zhu. 2008. "The Arabidopsis NFYA5 Transcription Factor Is Regulated Transcriptionally and Posttranscriptionally to Promote Drought Resistance." *The Plant Cell* 20(8):2238–2251. https://doi.org/10.1105/TPC.108.059444.

Li, Xiaohan, Siyan Xu, Martina Bianca Fuhrmann-Aoyagi, Shaoze Yuan, Takeru Iwama, Misaki Kobayashi, and Kenji Miura. 2022. "CRISPR/Cas9 Technique for Temperature, Drought, and Salinity Stress Responses." *Current Issues in Molecular Biology* 44(6):2664. https://doi.org/10.3390/CIMB44060182.

Li, Xin, Xiyan Liu, Feng Pan, Junling Hu, Yunhao Han, Ripu Bi, Chen Zhang, Yan Liu, Yong Wang, Zengwen Liang, Can Zhu, Yanmei Guo, Zejun Huang, Xiaoxuan Wang, Yongchen Du, Lei Liu, and Junming Li. 2024. "Dissection of Major QTLs and Candidate Genes for Seedling Stage Salt/Drought Tolerance in Tomato." *BMC Genomics* 25(1):1–14. https://doi.org/10.1186/S12864-024-11101-8/FIGURES/6.

Liu, Bo, Zhigang Ouyang, Yafen Zhang, Xiaohui Li, Yongbo Hong, Lei Huang, Shixia Liu, Huijuan Zhang, Dayong Li, and Fengming Song. 2014. "Tomato NAC Transcription Factor SlSRN1 Positively Regulates Defense Response against Biotic Stress but Negatively Regulates Abiotic Stress Response." *PLoS ONE* 9(7):e102067. https://doi.org/10.1371/JOURNAL.PONE.0102067.

Liu, Lun, Jialong Zhang, Jiayi Xu, Yafei Li, Luqin Guo, Zhirong Wang, Xichun Zhang, Bing Zhao, Yang Dong Guo, and Na Zhang. 2020. "CRISPR/Cas9 Targeted Mutagenesis of SlLBD40, a Lateral Organ Boundaries Domain Transcription Factor, Enhances Drought Tolerance in Tomato." *Plant Science* 301:110683. https://doi.org/10.1016/J.PLANTSCI.2020.110683.

Liu, Minmin, Gangjun Zhao, Xin Huang, Ting Pan, Wenjie Chen, Mei Qu, Bo Ouyang, Min Yu, and Sergey Shabala. 2023. "Candidate Regulators of Drought Stress in Tomato Revealed by Comparative Transcriptomic and Proteomic Analyses." *Frontiers in Plant Science* 14:1282718. https://doi.org/10.3389/FPLS.2023.1282718/BIBTEX.

Liu, Xin, Chunchang Tan, Xin Cheng, Xiaoming Zhao, Tianlai Li, and Jing Jiang. 2020. "MiR168 Targets Argonaute1A Mediated miRNAs Regulation Pathways in Response to Potassium Deficiency Stress in Tomato." *BMC Plant Biology* 20(1):1–17. https://doi.org/10.1186/S12870-020-02660-5/FIGURES/11.

Liu, X., J. H. Yang, B. Li, X. M. Yang, and Q. W. Meng. 2010. "Antisense Expression of Tomato Chloroplast Omega-3 Fatty Acid Desaturase Gene (LeFAD7) Enhances the Tomato High-Temperature Tolerance through Reductions of Trienoic Fatty Acids and Alterations of Physiological Parameters." *Photosynthetica* 48(1):59–66. https://doi.org/10.1007/S11099-010-0009-4/METRICS.

Liu, Yang, Tengxia Zhou, Haiyan Ge, Wen Pang, Lijie Gao, Li Ren, and Huoying Chen. 2016. "SSR Mapping of QTLs Conferring Cold Tolerance in an Interspecific Cross of Tomato." *International Journal of Genomics* 2016. https://doi.org/10.1155/2016/3219276.

Liu, Yudong, Yuan Shi, Deding Su, Wang Lu, and Zhengguo Li. 2021. "SlGRAS4 Accelerates Fruit Ripening by Regulating Ethylene Biosynthesis Genes and SlMADS1 in Tomato." *Horticulture Research* 8(1):1–11. https://doi.org/10.1038/s41438-020-00431-9.

Liu, Zengbing, Jingbin Jiang, and Jingfu Li. 2023. "GWAS Characterization of the Genetic Regions Associated with Nine Important Agronomic Traits in Tomato (*Solanum lycopersicum* L.)." *Agronomy* 13(5):1191. https://doi.org/10.3390/AGRONOMY13051191/S1.

Liu, Zhouyuan, Jianan Li, Shuang Li, Yan Li, Min Miao, Tingting Fan, and Xiaofeng Tang. 2024. "The 1r-Myb Transcription Factor Slmyb11 Modulates Drought Tolerance Via an Aba-Dependent Pathway in Tomato." https://doi.org/10.2139/SSRN.5047071.

Lohani, Neeta, Mohan B. Singh, and Prem L. Bhalla. 2022. "Biological Parts for Engineering Abiotic Stress Tolerance in Plants." *BioDesign Research* 2022:9819314. https://doi.org/10.34133/2022/9819314.

Loukehaich, Rachid, Taotao Wang, Bo Ouyang, Khurram Ziaf, Hanxia Li, Junhong Zhang, Yongen Lu, and Zhibiao Ye. 2012. "SpUSP, an Annexin-Interacting Universal Stress Protein, Enhances Drought Tolerance in Tomato." *Journal of Experimental Botany* 63(15):5593–5606. https://doi.org/10.1093/JXB/ERS220.

Lu, Jing, Mei Hong Sun, Qi Jun Ma, Hui Kang, Ya Jing Liu, Yu Jin Hao, and Chun Xiang You. 2019. "MdSWEET17, a Sugar Transporter in Apple, Enhances Drought Tolerance in Tomato." *Journal of Integrative Agriculture* 18(9):2041–2051. https://doi.org/10.1016/S2095-3119(19)62695-X.

Lu, Tao, Yangfan Song, Hongjun Yu, Qiang Li, Jingcheng Xu, Yong Qin, Guanhua Zhang, Yuhong Liu, and Weijie Jiang. 2022. "Cold Stress Resistance of Tomato (*Solanum lycopersicum*) Seedlings Is Enhanced by Light Supplementation From Underneath the Canopy." *Frontiers in Plant Science* 13:831314. https://doi.org/10.3389/FPLS.2022.831314/BIBTEX.

Luan, Yushi, Weichen Wang, and Ping Liu. 2014. "Identification and Functional Analysis of Novel and Conserved MicroRNAs in Tomato." *Molecular Biology Reports* 41(8):5385–5394. https://doi.org/10.1007/S11033-014-3410-4/FIGURES/4.

Luo, Qiuhong, Bin Bai, Yunhe Xie, Dongping Yao, Zhe Chen, Dongmeng Zhang, Yu Liu, Yinghui Xiao, Yinghong Yu, and Jun Wu. 2023. "Spatial Ionomics Provides New Insights into the Accumulation and Transport of Mineral Ions in Rice (*Oryza Sativa* L.) under Cadmium Stress." *Environmental and Experimental Botany* 208:105267. https://doi.org/10.1016/J.ENVEXPBOT.2023.105267.

Lymperopoulos, Panagiotis, Joseph Msanne, and Roel Rabara. 2018. "Phytochrome and Phytohormones: Working in Tandem for Plant Growth and Development." *Frontiers in Plant Science* 9:359649. https://doi.org/10.3389/FPLS.2018.01037/XML/NLM.

Ma, Jing, Muhammad Hamzah Saleem, Baber Ali, Rizwan Rasheed, Muhammad Arslan Ashraf, Humera Aziz, Sezai Ercisli, Sana Riaz, Mohsen Mohamed Elsharkawy, Iqbal Hussain, Sadeq K. Alhag, Ahmed Ezzat Ahmed, Dan C. Vodnar, Sahar Mumtaz, and Romina Alina Marc. 2022. "Impact of Foliar Application of Syringic Acid on Tomato (*Solanum lycopersicum* L.) under Heavy Metal Stress-Insights into Nutrient Uptake, Redox Homeostasis, Oxidative Stress, and Antioxidant Defense." *Frontiers in Plant Science* 13:950120. https://doi.org/10.3389/FPLS.2022.950120/BIBTEX.

Ma, Na Na, Yan Qiu Zuo, Xiao Qing Liang, Bo Yin, Guo Dong Wang, and Qing Wei Meng. 2013. "The Multiple Stress-Responsive Transcription Factor SlNAC1 Improves the Chilling Tolerance of Tomato." *Physiologia Plantarum* 149(4):474–486. https://doi.org/10.1111/PPL.12049.

Mahto, Binod Kumar, Amit Katiyar, Sangram K. Lenka, and Kailash C. Bansal. 2020. "Small RNA Technology for Plant Abiotic Stress Tolerance." In *Plant Small RNA: Biogenesis, Regulation and Application*, 521–541. https://doi.org/10.1016/B978-0-12-817112-7.00023-7.

Majeed, Yasir, Xi Zhu, Ning Zhang, Noor ul-Ain, Ali Raza, Fasih Ullah Haider, and Huaijun Si. 2023. "Harnessing the Role of Mitogen-Activated Protein Kinases against Abiotic Stresses in Plants." *Frontiers in Plant Science* 14:932923. https://doi.org/10.3389/FPLS.2023.932923/XML/NLM.

Makhotenko, A. V., A. V. Khromov, E. A. Snigir, S. S. Makarova, V. V. Makarov, T. P. Suprunova, N. O. Kalinina, and M. E. Taliansky. 2019. "Functional Analysis of Coilin in Virus Resistance and Stress Tolerance of Potato *Solanum Tuberosum* Using CRISPR-Cas9 Editing." *Doklady Biochemistry and Biophysics* 484(1):88–91. https://doi.org/10.1134/S1607672919010241/FIGURES/2.

Marko, D., N. Briglia, S. Summerer, A. Petrozza, F. Cellini, and R. Iannacone. 2018. "High-Throughput Phenotyping in Plant Stress Response: Methods and Potential Applications to Polyamine Field." *Methods in Molecular Biology* 1694:373–388. https://doi.org/10.1007/978-1-4939-7398-9_31.

Marques, Douglas José, Mozart Martins Ferreira, Mariana Ferreira Dias, Marlon José Figueiredo Pereira Junior, Elen Favaron, and Fabio Augusto Ishimoto. 2015. "Characterization of Tomato Genotypes under Deficiency of Phosphorus (1)." *Open Access Journal of Agricultural Research* 3(2):000153.

Martí, Raúl, Salvador Roselló, and Jaime Cebolla-Cornejo. 2016. "Tomato as a Source of Carotenoids and Polyphenols Targeted to Cancer Prevention." *Cancers* 8(6):58. https://doi.org/10.3390/CANCERS8060058.

Martin, Bjorn, James Nienhuis, Gretchen King, and Alison Schaefer. 1989. "Restriction Fragment Length Polymorphisms Associated with Water Use Efficiency in Tomato." *Science* 243(4899):1725–1728. https://doi.org/10.1126/SCIENCE.243.4899.1725.

Martinez, Vicente, Manuel Nieves-Cordones, Maria Lopez-Delacalle, Reyes Rodenas, Teresa C. Mestre, Francisco Garcia-Sanchez, Francisco Rubio, Pedro A. Nortes, Ron Mittler, and Rosa M. Rivero. 2018. "Tolerance to Stress Combination in Tomato Plants: New Insights in the Protective Role of Melatonin." *Molecules* 23(3):535. https://doi.org/10.3390/MOLECULES23030535.

Meng, Xia, Jie Ru Wang, Guo Dong Wang, Xiao Qing Liang, Xiao Dong Li, and Qing Wei Meng. 2015. "An R2R3-MYB Gene, LeAN2, Positively Regulated the Thermo-Tolerance in Transgenic Tomato." *Journal of Plant Physiology* 175:1–8. https://doi.org/10.1016/J.JPLPH.2014.09.018.

Meza, Silvia L. R., Isabel Egea, Isabel L. Massaretto, Belén Morales, Eduardo Purgatto, José M. Egea-Fernández, María C. Bolarin, and Francisco B. Flores. 2020. "Traditional Tomato Varieties Improve Fruit Quality Without Affecting Fruit Yield Under Moderate Salt Stress." *Frontiers in Plant Science* 11:587754. https://doi.org/10.3389/FPLS.2020.587754/BIBTEX.

M'hamdi, Oussama, Márton Égei, Zoltán Pék, Riadh Ilahy, Eszter Nemeskéri, Lajos Helyes, and Sándor Takács. 2023. "Root Development Monitoring under Different Water Supply Levels in Processing Tomato Plants." *Plants* 12(20):3517. https://doi.org/10.3390/PLANTS12203517.

Ming, Nan, Nana Ma, Baozhen Jiao, Wei Lv, and Qingwei Meng. 2020. "Genome Wide Identification of C2H2-Type Zinc Finger Proteins of Tomato and Expression Analysis Under Different Abiotic Stresses." *Plant Molecular Biology Reporter* 38(1):75–94. https://doi.org/10.1007/S11105-019-01182-1/FIGURES/5.

Miura, Kenji, Hayato Shiba, Masaru Ohta, Seung Won Kang, Ayaka Sato, Takashi Yuasa, Mari Iwaya-Inoue, Hiroshi Kamada, and Hiroshi Ezura. 2012. "SlICE1 Encoding a MYC-Type Transcription Factor Controls

Cold Tolerance in Tomato, *Solanum Lycopersicum*." *Plant Biotechnology* 29(3):253–260. https://doi.org/10.5511/PLANTBIOTECHNOLOGY.12.0303A.

Molitor, Corentin, Tomasz J. Kurowski, Pedro M. Fidalgo De Almeida, Pramod Eerolla, Daniel J. Spindlow, Sarvesh P. Kashyap, Bijendra Singh, H. C. Prasanna, Andrew J. Thompson, and Fady R. Mohareb. 2021. "De Novo Genome Assembly of *Solanum Sitiens* Reveals Structural Variation Associated with Drought and Salinity Tolerance." *Bioinformatics* 37(14):1941–1945. https://doi.org/10.1093/BIOINFORMATICS/BTAB048.

Mondal, Raju, Amit Kumar, and Belaghihalli N. Gnanesh. 2023. "Crop Germplasm: Current Challenges, Physiological-Molecular Perspective, and Advance Strategies towards Development of Climate-Resilient Crops." *Heliyon* 9(1):e12973. https://doi.org/10.1016/J.HELIYON.2023.E12973.

Monforte, A. J., M. J. Asíns, and E. A. Carbonell. 1996. "Salt Tolerance in Lycopersicon Species. IV. Efficiency of Marker-Assisted Selection for Salt Tolerance Improvement." *Theoretical and Applied Genetics* 93(5–6):765–772. https://doi.org/10.1007/BF00224074/METRICS.

Muhammad, Tayeb, Jie Zhang, Yalin Ma, Yushun Li, Fei Zhang, Yan Zhang, and Yan Liang. 2019. "Overexpression of a Mitogen-Activated Protein Kinase SlMAPK3 Positively Regulates Tomato Tolerance to Cadmium and Drought Stress." *Molecules* 24(3):556. https://doi.org/10.3390/MOLECULES24030556.

Munemasa, Shintaro, Felix Hauser, Jiyoung Park, Rainer Waadt, Benjamin Brandt, and Julian I. Schroeder. 2015. "Mechanisms of Abscisic Acid-Mediated Control of Stomatal Aperture." *Current Opinion in Plant Biology* 28:154. https://doi.org/10.1016/J.PBI.2015.10.010.

Muñoz-Mayor, Alicia, Benito Pineda, Jose O. Garcia-Abellán, Teresa Antón, Begoña Garcia-Sogo, Paloma Sanchez-Bel, Francisco B. Flores, Alejandro Atarés, Trinidad Angosto, Jose A. Pintor-Toro, Vicente Moreno, and Maria C. Bolarin. 2012. "Overexpression of Dehydrin Tas14 Gene Improves the Osmotic Stress Imposed by Drought and Salinity in Tomato." *Journal of Plant Physiology* 169(5):459–468. https://doi.org/10.1016/J.JPLPH.2011.11.018.

Murray, B. G. 2016. "Hybridization and Plant Breeding." *Encyclopedia of Applied Plant Sciences* 2:168–173. https://doi.org/10.1016/B978-0-12-394807-6.00191-X.

Mustafa, Mohammed, Ruth W. Mwangi, Zita Szalai, Noémi Kappel, and László Csambalik. 2025. "Sustainable Responses to Open Field Tomato (*Solanum lycopersicum* L.) Stress Impacts." *Journal of Agriculture and Food Research* 21:101825. https://doi.org/10.1016/J.JAFR.2025.101825.

Naciri, Rachida, Mohamed Chtouki, and Abdallah Oukarroum. 2024. "Mechanisms of Cadmium Mitigation in Tomato Plants under Orthophosphate and Polyphosphate Fertilization Regimes." *Ecotoxicology and Environmental Safety* 274:116219. https://doi.org/10.1016/J.ECOENV.2024.116219.

Nahar, K., and S. M. Ullah. 2011. "Effect of Water Stress on Moisture Content Distribution in Soil and Morphological Characters of Two Tomato (*Lycopersicon Esculentum Mill*) Cultivars." *Journal of Scientific Research* 3(3):677–682. https://doi.org/10.3329/JSR.V3I3.7000.

Nahar, Kamrun, and S. M. Ullah. 2017. "Fruit Quality and Osmotic Adjustment of Four Tomato Cultivars under Drought Stress." *Asian Journal of Soil Science and Plant Nutrition* 2(1):1–8. https://doi.org/10.9734/AJSSPN/2017/36861.

Naik, Bindu, Vijay Kumar, Sheikh Rizwanuddin, Mansi Chauhan, Megha Choudhary, Arun Kumar Gupta, Pankaj Kumar, Vivek Kumar, Per Erik Joakim Saris, Muzamil Ahmad Rather, Shuvam Bhuyan, Panchi Rani Neog, Sadhna Mishra, and Sarvesh Rustagi. 2023. "Genomics, Proteomics, and Metabolomics Approaches to Improve Abiotic Stress Tolerance in Tomato Plant." *International Journal of Molecular Sciences* 24(3). https://doi.org/10.3390/IJMS24033025.

Nautiyal, Prakash C., Mariko Shono, and Yoshinobu Egawa. 2005. "Enhanced Thermotolerance of the Vegetative Part of MT-SHSP Transgenic Tomato Line." *Scientia Horticulturae* 105(3):393–409. https://doi.org/10.1016/J.SCIENTA.2005.02.001.

Nazmul Hasan, Md, Shiful Islam, Fahmid H. Bhuiyan, Shahrear Arefin, Hammadul Hoque, Nurnabi Azad Jewel, Ajit Ghosh, and Shamsul H. Prodhan. 2022. "Genome Wide Analysis of the Heavy-Metal-Associated (HMA) Gene Family in Tomato and Expression Profiles under Different Stresses." *Gene* 835. https://doi.org/10.1016/j.gene.2022.146664.

Ng, Wei Quan, Nur Ardiyana Rejab, Sima Taheri, and Chee How Teo. 2025. "Elucidating the Roles of Small Open Reading Frames towards Drought Stress in *Solanum lycopersicum*." *Crop Breeding and Applied Biotechnology* 25(1):e510625112. https://doi.org/10.1590/1984-70332025V25N1A12.

O'Connell, Mary A., Andrea L. Medina, and Pedro Sánchez Peña. 2006. "Molecular Genetics of Drought Resistance Response in Tomato and Related Species." In *Genetic Improvement of Solanaceous*

Crops Volume 2: Tomato, 261–283. https://doi.org/10.1201/B10744-9/APPLICATIONS-GENETIC-ENGINEERING-TOMATO-RAJAM-MADHULATHA-PANDEY-HAZARIKA-RAZDAN.

Ogden, Aaron J., Jishnu J. Bhatt, Heather M. Brewer, Jack Kintigh, Samwel M. Kariuki, Sairam Rudrabhatla, Joshua N. Adkins, and Wayne R. Curtis. 2020. "Phloem Exudate Protein Profiles during Drought and Recovery Reveal Abiotic Stress Responses in Tomato Vasculature." *International Journal of Molecular Sciences* 21(12):4461. https://doi.org/10.3390/IJMS21124461.

Oladosu, Yusuff, Mohd Y. Rafii, Norhani Abdullah, Ghazali Hussin, Asfaliza Ramli, Harun A. Rahim, Gous Miah, and Magaji Usman. 2016. "Principle and Application of Plant Mutagenesis in Crop Improvement: A Review." *Biotechnology and Biotechnological Equipment* 30(1):1–16. https://doi.org/10.1080/13102818.2015.1087333;REQUESTEDJOURNAL:JOURNAL:TBEQ20;WGROUP:STRING:PUBLICATION.

Olías, Raquel, Zakia Eljakaoui, Jun Li, Paz Alvarez De Morales, Mari Carmen MarÍn-Manzano, Jose M. Pardo, and Andrés Belver. 2009. "The Plasma Membrane Na^+/H^+ Antiporter SOS1 Is Essential for Salt Tolerance in Tomato and Affects the Partitioning of Na^+ between Plant Organs." *Plant, Cell and Environment* 32(7):904–916. https://doi.org/10.1111/J.1365-3040.2009.01971.x.

Oliver, Melvin J., Renuka Jain, Tiago S. Balbuena, Ganesh Agrawal, Franscisco Gasulla, and Jay J. Thelen. 2011. "Proteome Analysis of Leaves of the Desiccation-Tolerant Grass, *Sporobolus Stapfianus*, in Response to Dehydration." *Phytochemistry* 72(10):1273–1284. https://doi.org/10.1016/J.PHYTOCHEM.2010.10.020.

Olvera-Carrillo, Yadira, José Luis Reyes, and Alejandra A. Covarrubias. 2011. "Late Embryogenesis Abundant Proteins." *Plant Signaling & Behavior* 6(4):586–589. https://doi.org/10.4161/PSB.6.4.15042.

Orellana, Sandra, Mónica Yañez, Analía Espinoza, Isabel Verdugo, Enrique González, Simón Ruiz-Lara, and José A. Casaretto. 2010. "The Transcription Factor SlAREB1 Confers Drought, Salt Stress Tolerance and Regulates Biotic and Abiotic Stress-Related Genes in Tomato." *Plant, Cell and Environment* 33(12):2191–2208. https://doi.org/10.1111/J.1365-3040.2010.02220.x.

Osei-Bonsu, I., M. K. Osei, R. Y. Agyare, J. Adjebeng-Danquah, K. Asare Bediako, J. Gyau, J. Adomako, P. Ofori, R. N. A. Prempeh, and Myeong Cheoul Cho. 2022. "Assessing the Heat Stress Tolerance Potential of Tomato Lines under Poly-House and Open Field Conditions." *Cogent Food and Agriculture* 8(1). https://doi.org/10.1080/23311932.2022.2115665;WGROUP:STRING:PUBLICATION.

Padmavathi, S., and G. Jayalakshmi. 2024. "Application of Bioremediation in Mitigation of Heavy Metal Stress on Plants." *International Journal of Advanced Research* 12(04):509–517. https://doi.org/10.21474/IJAR01/18576.

Pailles, Yveline, Mariam Awlia, Magdalena Julkowska, Luca Passone, Khadija Zemmouri, Sónia Negrão, Sandra M. Schmöckel, and Mark Tester. 2020. "Diverse Traits Contribute to Salinity Tolerance of Wild Tomato Seedlings from the Galapagos Islands." *Plant Physiology* 182(1):534–546. https://doi.org/10.1104/PP.19.00700.

Paliwal, Rajneesh, Gagandeep Singh, Reyazul Rouf Mir, and Badara Gueye. 2021. "Genomic-Assisted Breeding for Abiotic Stress Tolerance in Horticultural Crops." In *Stress Tolerance in Horticultural Crops: Challenges and Mitigation Strategies*, 91–118. https://doi.org/10.1016/B978-0-12-822849-4.00007-3.

Pan, I. Chun, Chia Wen Li, Ruey Chih Su, Chiu Ping Cheng, Choun Sea Lin, and Ming Tsair Chan. 2010. "Ectopic Expression of an EAR Motif Deletion Mutant of SlERF3 Enhances Tolerance to Salt Stress and *Ralstonia Solanacearum* in Tomato." *Planta* 232(5):1075–1086. https://doi.org/10.1007/S00425-010-1235-5/TABLES/1.

Pandey, Prachi, Venkategowda Ramegowda, and Muthappa Senthil-Kumar. 2015. "Shared and Unique Responses of Plants to Multiple Individual Stresses and Stress Combinations: Physiological and Molecular Mechanisms." *Frontiers in Plant Science* 6(September):153997. https://doi.org/10.3389/FPLS.2015.00723/EPUB.

Pandey, Saurabh, Mehanathan Muthamilarasan, Namisha Sharma, Vaishali Chaudhry, Priya Dulani, Shweta Shweta, Sarita Jha, Saloni Mathur, and Manoj Prasad. 2019. "Characterization of DEAD-Box Family of RNA Helicases in Tomato Provides Insights into Their Roles in Biotic and Abiotic Stresses." *Environmental and Experimental Botany* 158:107–116. https://doi.org/10.1016/J.ENVEXPBOT.2018.11.018.

Pardo-Hernández, Miriam, Vicent Arbona, Inmaculada Simón, and Rosa M. Rivero. 2024. "Specific ABA-Independent Tomato Transcriptome Reprogramming under Abiotic Stress Combination." *Plant Journal* 117(6):1746–1763. https://doi.org/10.1111/TPJ.16642;CTYPE:STRING:JOURNAL.

Pardo-Hernández, Miriam, Pascual García-Pérez, Luigi Lucini, and Rosa M. Rivero. 2024. "Multi-Omics Exploration of the Involvement of ABA in Identifying Unique Molecular Markers for Single and

Combined Stresses in Tomato Plants" edited by Y. Gibon. *Journal of Experimental Botany*. https://doi.org/10.1093/JXB/ERAE372.

Parida, Tamanna, Shaik Riyazuddin, Suresh Kumar Kolli, Anindita Chakraborty, Namuduri Srinivas, Pritha Kundu, Sayan Bhattacharya, Chandra Shekhar Seth, and Jayanta Kumar Biswas. 2024. "Phytoremediation of Heavy Metal(Loid)s with Integral Involvement of the Endogenous Metal-Chelators: Present State-of-the-Art and Future Prospect." *Discover Environment* 2(1):1–15. https://doi.org/10.1007/S44274-024-00170-X.

Park, Ae Ran, Jongmun Kim, Bora Kim, Areum Ha, Ji Yeon Son, Chan Woo Song, Hyohak Song, and Jin Cheol Kim. 2022. "Exogenous Bio-Based 2,3-Butanediols Enhanced Abiotic Stress Tolerance of Tomato and Turfgrass under Drought or Chilling Stress." *Journal of Microbiology and Biotechnology*. 32(5):582–593. https://doi.org/10.4014/JMB.2201.01025.

Park, Eung Jun, Zoran Jeknic, and Tony H. H. Chen. 2006. "Exogenous Application of Glycinebetaine Increases Chilling Tolerance in Tomato Plants." *Plant and Cell Physiology* 47(6):706–714. https://doi.org/10.1093/PCP/PCJ041.

Parvin, Khursheda, Mirza Hasanuzzaman, M. H. M. Borhannuddin Bhuyan, Kamrun Nahar, Sayed Mohammad Mohsin, and Masayuki Fujita. 2019. "Comparative Physiological and Biochemical Changes in Tomato (*Solanum lycopersicum* L.) under Salt Stress and Recovery: Role of Antioxidant Defense and Glyoxalase Systems." *Antioxidants* 8(9):350. https://doi.org/10.3390/ANTIOX8090350.

Pasricha, Sharda, Vartika Mathur, Arushi Garg, Satyajit Lenka, Kavita Verma, and Surbhi Agarwal. 2021. "Molecular Mechanisms Underlying Heavy Metal Uptake, Translocation and Tolerance in Hyperaccumulators-an Analysis: Heavy Metal Tolerance in Hyperaccumulators." *Environmental Challenges* 4:100197. https://doi.org/10.1016/J.ENVC.2021.100197.

Patade, Vikas Yadav, Deepti Khatri, Maya Kumari, Atul Grover, Sanjay Mohan Gupta, and Zakwan Ahmed. 2013. "Cold Tolerance in Osmotin Transgenic Tomato (*Solanum lycopersicum* L.) Is Associated with Modulation in Transcript Abundance of Stress Responsive Genes." *SpringerPlus* 2(1):1–7. https://doi.org/10.1186/2193-1801-2-117/TABLES/2.

Patra, Gyanendra K., Divya Gupta, Gyana Ranjan Rout, and Sanjib Kumar Panda. 2023. "Role of Long Non Coding RNA in Plants under Abiotic and Biotic Stresses." *Plant Physiology and Biochemistry* 194:96–110. https://doi.org/10.1016/J.PLAPHY.2022.10.030.

Paupière, Marine J., Pauline van Haperen, Ivo Rieu, Richard G. F. Visser, Yury M. Tikunov, and Arnaud G. Bovy. 2017. "Screening for Pollen Tolerance to High Temperatures in Tomato." *Euphytica* 213(6):130. https://doi.org/10.1007/S10681-017-1927-Z.

Pessoa, Herika Paula, João Romero Do Amaral Santos de Carvalho Rocha, Flávia Maria Alves, Mariane Gonçalves Ferreira Copati, Françoise Dalprá Dariva, Laércio Junio da Silva, Pedro Crescêncio Carneiro, and Carlos Nick Gomes. 2022. "Multi-Trait Selection of Tomato Introgression Lines under Drought-Induced Conditions at Germination and Seedling Stages." *Acta Scientiarum. Agronomy* 44:e55876. https://doi.org/10.4025/ACTASCIAGRON.V44I1.55876.

Pessoa, Herika Paula, Françoise Dalprá Dariva, Mariane Gonçalves Ferreira Copati, Ramon Gonçalves de Paula, Felipe de Oliveira Dias, and Carlos Nick Gomes. 2023. "Uncovering Tomato Candidate Genes Associated with Drought Tolerance Using *Solanum Pennellii* Introgression Lines." *PLoS ONE* 18(6 June). https://doi.org/10.1371/JOURNAL.PONE.0287178.

Pompeiano, A., T. M. Moles, L. Mariotti, A. Santaniello, D. Di Baccio, A. Scartazza, T. Huarancca Reyes, and L. Guglielminetti. 2025. "Tomato Biodiversity Reveals Landrace Enhanced Drought-Adaptive Strategy." *Plant Physiology and Biochemistry* 220:109495. https://doi.org/10.1016/J.PLAPHY.2025.109495.

Priyadarshan, P. M. 2019. "Backcross Breeding." In *Plant Breeding: Classical to Modern*, 203–221. https://doi.org/10.1007/978-981-13-7095-3_10.

Quamruzzaman, Akm, Limu Akter, Ferdouse Islam. 2023. "Yield and Pest Performance of High-Temperature Tolerant Tomato (*Lycopersicon Esculentum*) Lines for Year-Round Tomato Production in Bangladesh." *Agricultural Sciences* 14(7):855–863. https://doi.org/10.4236/AS.2023.147057.

Rai, Krishna Kumar, Ajay Kumar, Ashutosh Rai, Ved Prakash Rai, and Avinash Chandra Rai. 2021. "Conventional Breeding Approaches for Abiotic Stress Management in Horticultural Crops." In *Stress Tolerance in Horticultural Crops: Challenges and Mitigation Strategies*, 21–32. https://doi.org/10.1016/B978-0-12-822849-4.00017-6.

Rai, N., and M. Rai. 2006. *Heterosis Breeding in Vegetable Crops*. New India Publishing Agency.

Raina, Aamir, Rafiul Laskar, Shahnawaz Khursheed, Ruhul Amin, Younas Tantray, Kouser Parveen, and Samiullah Khan. 2016. "Role of Mutation Breeding in Crop Improvement-Past, Present and Future." *Asian Research Journal of Agriculture* 2(2):1–13. https://doi.org/10.9734/ARJA/2016/29334.

Rao, Eguru Sreenivasa, Palchamy Kadirvel, Rachael C. Symonds, Subramaniam Geethanjali, Ramadihalli N. Thontadarya, and Andreas W. Ebert. 2015. "Variations in DREB1A and VP1.1 Genes Show Association with Salt Tolerance Traits in Wild Tomato (*Solanum pimpinellifolium*)." *PLoS ONE* 10(7):e0132535. https://doi.org/10.1371/JOURNAL.PONE.0132535.

Rao, Sombir, Apoorva Gupta, Chandni Bansal, Celine Sorin, Martin Crespi, and Saloni Mathur. 2022. "A Conserved HSF:MiR169:NF-YA Loop Involved in Tomato and Arabidopsis Heat Stress Tolerance." *The Plant Journal* 112(1):7–26. https://doi.org/10.1111/TPJ.15963.

Rasul, Kamaran Salh, and Nawroz Abdul-Razzak Tahir. 2023. "Response of Different Tomato Accessions to Biotic and Abiotic Stresses." *SSRN Electronic Journal*. https://doi.org/10.2139/SSRN.4517869.

Razali, Rozaimi, Salim Bougouffa, Mitchell J. L. Morton, Damien J. Lightfoot, Intikhab Alam, Magbubah Essack, Stefan T. Arold, Allan A. Kamau, Sandra M. Schmöckel, Yveline Pailles, Mohammed Shahid, Craig T. Michell, Salim Al-Babili, Yung Shwen Ho, Mark Tester, Vladimir B. Bajic, and Sónia Negrão. 2018. "The Genome Sequence of the Wild Tomato *Solanum Pimpinellifolium* Provides Insights into Salinity Tolerance." *Frontiers in Plant Science* 9:390082. https://doi.org/10.3389/FPLS.2018.01402/XML/NLM.

Razifard, Hamid, Alexis Ramos, Audrey L. Della Valle, Cooper Bodary, Erika Goetz, Elizabeth J. Manser, Xiang Li, Lei Zhang, Sofia Visa, Denise Tieman, Esther Van Der Knaap, and Ana L. Caicedo. 2020. "Genomic Evidence for Complex Domestication History of the Cultivated Tomato in Latin America." *Molecular Biology and Evolution* 37(4):1118–1132. https://doi.org/10.1093/MOLBEV/MSZ297.

Roşca, Mihaela, Gabriela Mihalache, and Vasile Stoleru. 2023. "Tomato Responses to Salinity Stress: From Morphological Traits to Genetic Changes." *Frontiers in Plant Science* 14:1118383. https://doi.org/10.3389/FPLS.2023.1118383.

Rowe, James H., Jennifer F. Topping, Junli Liu, and Keith Lindsey. 2016. "Abscisic Acid Regulates Root Growth under Osmotic Stress Conditions via an Interacting Hormonal Network with Cytokinin, Ethylene and Auxin." *New Phytologist* 211(1):225–239. https://doi.org/10.1111/NPH.13882.

Sachdev, Swati, Shamim Akhtar Ansari, Mohammad Israil Ansari, Masayuki Fujita, and Mirza Hasanuzzaman. 2021. "Abiotic Stress and Reactive Oxygen Species: Generation, Signaling, and Defense Mechanisms." *Antioxidants* 10(2):277. https://doi.org/10.3390/ANTIOX10020277.

Sadder, Monther, Abdullah Alsadon, and Mahmoud Wahb-Allah. 2014. "Transcriptomic Analysis of Tomato Lines Reveals Putative Stress-Specific Biomarkers." *Turkish Journal of Agriculture and Forestry* 38(5):700–715. https://doi.org/10.3906/tar-1312-17.

Sadder, Monther T., Ahmad Abdelrahim Mohamed Ali, Abdullah A. Alsadon, and Mahmoud A. Wahb-Allah. 2025. "Long-Term Salinity-Responsive Transcriptome in Advanced Breeding Lines of Tomato." *Plants* 14(1). https://doi.org/10.3390/PLANTS14010100.

Sade, Nir, Basia J. Vinocur, Alex Diber, Arava Shatil, Gil Ronen, Hagit Nissan, Rony Wallach, Hagai Karchi, and Menachem Moshelion. 2009. "Improving Plant Stress Tolerance and Yield Production: Is the Tonoplast Aquaporin SlTIP2;2 a Key to Isohydric to Anisohydric Conversion?" *New Phytologist* 181(3):651–661. https://doi.org/10.1111/J.1469-8137.2008.02689.X.

Sahito, Javed Hussain, Hao Zhang, Zeeshan Ghulam Nabi Gishkori, Chenhui Ma, Zhihao Wang, Dong Ding, Xuehai Zhang, and Jihua Tang. 2024. "Advancements and Prospects of Genome-Wide Association Studies (GWAS) in Maize." *International Journal of Molecular Sciences* 25(3):1918. https://doi.org/10.3390/IJMS25031918.

Sánchez-Rodríguez, Eva, MªMar Rubio-Wilhelmi, Luis M. Cervilla, Begoña Blasco, Juan J. Rios, Miguel A. Rosales, Luis Romero, and Juan M. Ruiz. 2010. "Genotypic Differences in Some Physiological Parameters Symptomatic for Oxidative Stress under Moderate Drought in Tomato Plants." *Plant Science* 178(1):30–40. https://doi.org/10.1016/J.PLANTSCI.2009.10.001.

Selli, Serkan, Hasim Kelebek, Mehmet Turan Ayseli, and Habip Tokbas. 2014. "Characterization of the Most Aroma-Active Compounds in Cherry Tomato by Application of the Aroma Extract Dilution Analysis." *Food Chemistry* 165:540–546. https://doi.org/10.1016/J.FOODCHEM.2014.05.147.

Selth, Luke A., Satish C. Dogra, M. Saif Rasheed, Helen Healy, John W. Randles, and M. Ali Rezaian. 2005. "A NAC Domain Protein Interacts with Tomato Leaf Curl Virus Replication Accessory Protein and Enhances Viral Replication." *The Plant Cell* 17(1):311–325. https://doi.org/10.1105/TPC.104.027235.

Shi, Xiaopu, Fangling Jiang, Junqin Wen, and Zhen Wu. 2019. "Overexpression of *Solanum Habrochaites* microRNA319d (Sha-miR319d) Confers Chilling and Heat Stress Tolerance in Tomato (*Solanum lycopersicum*)." *BMC Plant Biology* 19(1):1–17. https://doi.org/10.1186/S12870-019-1823-X/FIGURES/8.

Shu, Jinshuai, Lili Zhang, Guiming Liu, Xiaoxuan Wang, Fuzhong Liu, Ying Zhang, and Yuhui Chen. 2024. "Transcriptome Analysis and Metabolic Profiling Reveal the Key Regulatory Pathways in Drought Stress Responses and Recovery in Tomatoes." *International Journal of Molecular Sciences* 25(4):2187. https://doi.org/10.3390/IJMS25042187/S1.

Shu, Pan, Yujing Li, Ziye Li, Jiping Sheng, and Lin Shen. 2022. "SlMAPK3 Enhances Tolerance to Salt Stress in Tomato Plants by Scavenging ROS Accumulation and Up-Regulating the Expression of Ethylene Signaling Related Genes." *Environmental and Experimental Botany* 193:104698. https://doi.org/10.1016/J.ENVEXPBOT.2021.104698.

Shu, Pan, Yujing Li, Lanting Xiang, Jiping Sheng, and Lin Shen. 2023. "SlNPR1 Modulates Chilling Stress Resistance in Tomato Plant by Alleviating Oxidative Damage and Affecting the Synthesis of Ferulic Acid." *Scientia Horticulturae* 307:111486. https://doi.org/10.1016/J.SCIENTA.2022.111486.

Sierra, Arévalo, Bartholomew Saanu Adeleke, Parul Chaudhary, Modupe Stella Ayilara, Funmilola Mabel Ojo, Sakiru Morenikeji Erinoso, Viabhav Kumar Upadhayay, Adeyemi Isaiah Adeyemo, and Saheed Adekunle Akinola. 2024. "Rhizosphere Microbiomes Mediating Abiotic Stress Mitigation for Improved Plant Nutrition." *Ecologies* 5(3):375–401. https://doi.org/10.3390/ECOLOGIES5030024.

Singh, Ajey, Shubhra Khare, Niharika, and Praveen Gupta. 2025. "Sulfur and Phosphorus Transporters in Plants: Integrating Mechanisms for Optimized Nutrient Supply." *Plant Physiology and Biochemistry* 224:109918. https://doi.org/10.1016/J.PLAPHY.2025.109918.

Singh, B. D., and A. K. Singh. 2015. "Linkage Mapping of Molecular Markers and Oligogenes." In *Marker-Assisted Plant Breeding: Principles and Practices*, 151–183. https://doi.org/10.1007/978-81-322-2316-0_6.

Singh, Sachin Kumar, Kiran Thakur, Vishal Sharma, Mamta Saini, Diksha Sharma, Sukriti Vishwas, Violina Kakoty, Rashmi Saxena Pal, M. V. N. L. Chaitanya, M. Ravindra Babu, Saurabh Gupta, Zia ur Rehman, Smriti, Madhav Singla, Gaurav Gupta, Vikash Jakhmola, Terezinha de Jesus Andreoli Pinto, Popat Kumbhar, John Disouza, Vandana Patravale, Kamal Dua, and Manoj M. Gadewar. 2023. "Exploring the Multifaceted Potential of Chlorogenic Acid: Journey from Nutraceutical to Nanomedicine." *South African Journal of Botany* 159:658–677. https://doi.org/10.1016/J.SAJB.2023.06.038.

Smith, Daniel T., Andries B. Potgieter, and Scott C. Chapman. 2021. "Scaling up High-Throughput Phenotyping for Abiotic Stress Selection in the Field." *Theoretical and Applied Genetics* 134(6):1845–1866. https://doi.org/10.1007/S00122-021-03864-5.

Sousaraei, Naeimeh, Kambiz Mashayekhi, Seyyed Javad Mousavizadeh, Vahid Akbarpour, Joaquín Medina, and Sasan Aliniaeifard. 2021. "Screening of Tomato Landraces for Drought Tolerance Based on Growth and Chlorophyll Fluorescence Analyses." *Horticulture Environment and Biotechnology* 62(4):521–535. https://doi.org/10.1007/S13580-020-00328-5.

Tan, Shun Ling, Ying Jie Yang, Tao Liu, Shi Bao Zhang, and Wei Huang. 2020. "Responses of Photosystem I Compared with Photosystem II to Combination of Heat Stress and Fluctuating Light in Tobacco Leaves." *Plant Science* 292:110371. https://doi.org/10.1016/J.PLANTSCI.2019.110371.

Thalhammer, Anja, and Dirk K. Hincha. 2014. "A Mechanistic Model of COR15 Protein Function in Plant Freezing Tolerance: Integration of Structural and Functional Characteristics." *Plant Signaling and Behavior* 9(12):e977722–e977723. https://doi.org/10.4161/15592324.2014.977722;WGROUP:STRING:PUBLICATION.

Thirumalaikumar, Venkatesh P., Vikas Devkar, Nikolay Mehterov, Shawkat Ali, Rengin Ozgur, Ismail Turkan, Bernd Mueller-Roeber, and Salma Balazadeh. 2017. "NAC Transcription Factor JUNGBRUNNEN1 Enhances Drought Tolerance in Tomato." *Plant Biotechnology Journal* 16(2):354. https://doi.org/10.1111/PBI.12776.

Thompson, G. A. 1989. "Molecular Changes in Membrane Lipids During Cold Stress." In *Environmental Stress in Plants*, 249–257. https://doi.org/10.1007/978-3-642-73163-1_23.

Thomson, Michael J. 2014. "High-Throughput SNP Genotyping to Accelerate Crop Improvement." *Plant Breeding and Biotechnology* 2(3):195–212. https://doi.org/10.9787/PBB.2014.2.3.195.

Tiwari, Jagesh Kumar, Anand Kumar Singh, and Tusar Kanti Behera. 2023. "CRISPR/Cas Genome Editing in Tomato Improvement: Advances and Applications." *Frontiers in Plant Science* 14:1121209. https://doi.org/10.3389/FPLS.2023.1121209/XML/NLM.

Tran, Mil Thi, Duong Thi Hai Doan, Jihae Kim, Young Jong Song, Yeon Woo Sung, Swati Das, Eun Jung Kim, Geon Hui Son, Sang Hee Kim, Tien Van Vu, and Jae Yean Kim. 2021. "CRISPR/Cas9-Based Precise

Excision of SlHyPRP1 Domain(s) to Obtain Salt Stress-Tolerant Tomato." *Plant Cell Reports* 40(6):999–1011. https://doi.org/10.1007/S00299-020-02622-Z/FIGURES/7.

Tran, Mil Thi, Geon Hui Son, Young Jong Song, Ngan Thi Nguyen, Seonyeong Park, Thanh Vu Thach, Jihae Kim, Yeon Woo Sung, Swati Das, Dibyajyoti Pramanik, Jinsu Lee, Ki Ho Son, Sang Hee Kim, Tien Van Vu, and Jae Yean Kim. 2023. "CRISPR-Cas9-Based Precise Engineering of SlHyPRP1 Protein Towards Multi-Stress Tolerance in Tomato." *Frontiers in Plant Science* 14:1186932. https://doi.org/10.3389/FPLS.2023.1186932/FULL.

Truco, M. J., L. B. Randall, A. J. Bloom, and D. A. St. Clair. 2000. "Detection of QTLs Associated with Shoot Wilting and Root Ammonium Uptake under Chilling Temperatures in an Interspecific Backcross Population from *Lycopersicon Esculentum* x *L. Hirsutum*." *Theoretical and Applied Genetics* 101(7):1082–1092. https://doi.org/10.1007/S001220051583/METRICS.

Turan, Metin, Melek Ekinci, Sanem Argin, Mihail Brinza, and Ertan Yildirim. 2023. "Drought Stress Amelioration in Tomato (*Solanum lycopersicum* L.) Seedlings by Biostimulant as Regenerative Agent." *Frontiers in Plant Science* 14:1211210. https://doi.org/10.3389/FPLS.2023.1211210/BIBTEX.

Uffelmann, Emil, Qin Qin Huang, Nchangwi Syntia Munung, Jantina de Vries, Yukinori Okada, Alicia R. Martin, Hilary C. Martin, Tuuli Lappalainen, and Danielle Posthuma. 2021. "Genome-Wide Association Studies." *Nature Reviews Methods Primers* 1(1):1–21. https://doi.org/10.1038/S43586-021-00056-9;SUBJMETA=1647,2000,208,2138,2217,631,727;KWRD=GENETIC+PREDISPOSITION+TO+DISEASE,GENOME-WIDE+ASSOCIATION+STUDIES.

Vanlay, Mean, Song Samnang, Hee Jong Jung, Phillip Choe, Kwon Kyoo Kang, and Ill Sup Nou. 2022. "Interspecific and Intraspecific Hybrid Rootstocks to Improve Horticultural Traits and Soil-Borne Disease Resistance in Tomato." *Genes* 13(8):1468. https://doi.org/10.3390/GENES13081468.

Verma, Shulbhi, A. R. S. S. H. Gupta, Santhoshi Yalla, Shreya, Parth J. Patel, Ritu Sharma, A. V. Anjana, and Akhil Donga. 2024. "Integrating Marker-Assisted (MAS) and Genomic Selection (GS) for Plant Functional Trait Improvement." In *Plant Functional Traits for Improving Productivity*, 203–215. https://doi.org/10.1007/978-981-97-1510-7_11.

Villalta, I., G. P. Bernet, E. A. Carbonell, and M. J. Asins. 2007. "Comparative QTL Analysis of Salinity Tolerance in Terms of Fruit Yield Using Two *Solanum* Populations of F7 Lines." *Theoretical and Applied Genetics* 114(6):1001–1017. https://doi.org/10.1007/S00122-006-0494-9/FIGURES/2.

Visentin, Ivan, Chiara Pagliarani, Eleonora Deva, Alessio Caracci, Veronika Turečková, Ondrej Novák, Claudio Lovisolo, Andrea Schubert, and Francesca Cardinale. 2020. "A Novel Strigolactone-miR156 Module Controls Stomatal Behaviour during Drought Recovery." *Plant, Cell & Environment* 43(7):1613–1624. https://doi.org/10.1111/PCE.13758.

Vu, Tien Van, Velu Sivankalyani, Eun Jung Kim, Duong Thi Hai Doan, Mil Thi Tran, Jihae Kim, Yeon Woo Sung, Minwoo Park, Yang Jae Kang, and Jae Yean Kim. 2020. "Highly Efficient Homology-Directed Repair Using CRISPR/Cpf1-Geminiviral Replicon in Tomato." *Plant Biotechnology Journal* 18(10):2133–2143. https://doi.org/10.1111/PBI.13373.

Wai, Antt Htet, Aung Htay Naing, Do Jin Lee, Chang Kil Kim, and Mi Young Chung. 2020. "Molecular Genetic Approaches for Enhancing Stress Tolerance and Fruit Quality of Tomato." *Plant Biotechnology Reports* 14(5):515–537. https://doi.org/10.1007/S11816-020-00638-1.

Wai, Antt Htet, Muhammad Waseem, A. B. M. Mahbub Morshed Khan, Ujjal Kumar Nath, Do Jin Lee, Sang Tae Kim, Chang Kil Kim, and Mi Young Chung. 2021. "Genome-Wide Identification and Expression Profiling of the Pdi Gene Family Reveals Their Probable Involvement in Abiotic Stress Tolerance in Tomato (*Solanum lycopersicum* l.)." *Genes* 12(1):1–24. https://doi.org/10.3390/GENES12010023.

Wang, Guodong, Xinping Xu, Hao Wang, Qi Liu, Xiaotong Yang, Lixiang Liao, and Guohua Cai. 2019. "A Tomato Transcription Factor, SlDREB3 Enhances the Tolerance to Chilling in Transgenic Tomato." *Plant Physiology and Biochemistry* 142:254–262. https://doi.org/10.1016/J.PLAPHY.2019.07.017.

Wang, Jiabo, and Zhiwu Zhang. 2021. "GAPIT Version 3: Boosting Power and Accuracy for Genomic Association and Prediction." *Genomics, Proteomics & Bioinformatics* 19(4):629. https://doi.org/10.1016/J.GPB.2021.08.005.

Wang, Liu, Lin Chen, Rui Li, Ruirui Zhao, Meijing Yang, Jiping Sheng, and Lin Shen. 2017. "Reduced Drought Tolerance by CRISPR/Cas9-Mediated SlMAPK3 Mutagenesis in Tomato Plants." *Journal of Agricultural and Food Chemistry* 65(39):8674–8682. https://doi.org/10.1021/ACS.JAFC.7B02745.

Wang, Ping, Qian Luo, Weicheng Yang, Golam Jalal Ahammed, Shuting Ding, Xingyu Chen, Jiao Wang, Xiaojian Xia, and Kai Shi. 2021. "A Novel Role of Pipecolic Acid Biosynthetic Pathway in Drought

Tolerance through the Antioxidant System in Tomato." *Antioxidants* 10(12):1923. https://doi.org/10.3390/ANTIOX10121923/S1.

Wang, Wei, Qianli Pan, Fei He, Alina Akhunova, Shiaoman Chao, Harold Trick, and Eduard Akhunov. 2018. "Transgenerational CRISPR-Cas9 Activity Facilitates Multiplex Gene Editing in Allopolyploid Wheat." *CRISPR Journal* 1(1):65–74. https://doi.org/10.1089/CRISPR.2017.0010.

Wang, Xiaoli, Xiaofeng Cai, Chenxi Xu, Quanhua Wang, and Shaojun Dai. 2016. "Drought-Responsive Mechanisms in Plant Leaves Revealed by Proteomics." *International Journal of Molecular Sciences* 17(10):1706. https://doi.org/10.3390/IJMS17101706.

Wang, Xiaoping, Shujuan Geng, Yong Jun Ri, Donghui Cao, Jie Liu, Decheng Shi, and Chunwu Yang. 2011. "Physiological Responses and Adaptive Strategies of Tomato Plants to Salt and Alkali Stresses." *Scientia Horticulturae* 130(1):248–255. https://doi.org/10.1016/J.SCIENTA.2011.07.006.

Wang, Yanling, Wenxiong Zou, Yan Xiao, Lina Cheng, Yudong Liu, Song Gao, Zihang Shi, Yun Jiang, Mingfang Qi, Tao Xu, and Tianlai Li. 2018. "microRNA1917 Targets CTR4 Splice Variants to Regulate Ethylene Responses in Tomato." *Journal of Experimental Botany* 69(5):1011–1025. https://doi.org/10.1093/JXB/ERX469.

Wang, Zhen, Yechun Hong, Yumei Li, Huazhong Shi, Juanjuan Yao, Xue Liu, Fuxing Wang, Sanwen Huang, Guangtao Zhu, and Jian Kang Zhu. 2021. "Natural Variations in SlSOS1 Contribute to the Loss of Salt Tolerance during Tomato Domestication." *Plant Biotechnology Journal* 19(1):20–22. https://doi.org/10.1111/PBI.13443;JOURNAL:JOURNAL:14677652;WGROUP:STRING:PUBLICATION.

Wang, Zhen, Yechun Hong, Guangtao Zhu, Yumei Li, Qingfeng Niu, Juanjuan Yao, Kai Hua, Jinjuan Bai, Yingfang Zhu, Huazhong Shi, Sanwen Huang, and Jian-Kang Zhu. 2020. "Loss of Salt Tolerance during Tomato Domestication Conferred by Variation in a Na^+/K^+ Transporter." *The EMBO Journal* 39(10). https://doi.org/10.15252/EMBJ.2019103256/SUPPL_FILE/EMBJ2019103256-SUP-0014-DATASETEV5.ZIP.

Wang, Zi Yu, Shuang Zhao, Jun Fang Liu, Hai Yan Zhao, Xu Ying Sun, Tai Ru Wu, Tong Pei, Yue Wang, Qi Feng Liu, Huan Huan Yang, He Zhang, Jing Bin Jiang, Jing Fu Li, Ting Ting Zhao, and Xiang Yang Xu. 2022. "Genome-Wide Identification of Tomato Golden 2-Like Transcription Factors and Abiotic Stress Related Members Screening." *BMC Plant Biology* 22(1):1–17. https://doi.org/10.1186/S12870-022-03460-9/FIGURES/9.

Wen, Junqin, Fangling Jiang, Yiqun Weng, Mintao Sun, Xiaopu Shi, Yanzhao Zhou, Lu Yu, and Zhen Wu. 2019. "Identification of Heat-Tolerance QTLs and High-Temperature Stress-Responsive Genes through Conventional QTL Mapping, QTL-Seq and RNA-Seq in Tomato." *BMC Plant Biology* 19(1):1–17. https://doi.org/10.1186/S12870-019-2008-3/FIGURES/6.

Weng, Shizhuang, Junjie Ma, Wentao Tao, Yujian Tan, Meijing Pan, Zixi Zhang, Linsheng Huang, Ling Zheng, and Jinling Zhao. 2023. "Drought Stress Identification of Tomato Plant Using Multi-Features of Hyperspectral Imaging and Subsample Fusion." *Frontiers in Plant Science* 14:1073530. https://doi.org/10.3389/FPLS.2023.1073530/BIBTEX.

Wu, Honghong. 2018. "Plant Salt Tolerance and Na^+ Sensing and Transport." *The Crop Journal* 6(3):215–225. https://doi.org/10.1016/J.CJ.2018.01.003.

Wu, Jian, Songyu Liu, Yanjun He, Xiaoyan Guan, Xiangfei Zhu, Lin Cheng, Jie Wang, and Gang Lu. 2012. "Genome-Wide Analysis of SAUR Gene Family in Solanaceae Species." *Gene* 509(1):38–50. https://doi.org/10.1016/J.GENE.2012.08.002.

Xu, Jiemeng, Nicky Driedonks, Marc J. M. Rutten, Wim H. Vriezen, Gert Jan de Boer, and Ivo Rieu. 2017. "Mapping Quantitative Trait Loci for Heat Tolerance of Reproductive Traits in Tomato (*Solanum lycopersicum*)." *Molecular Breeding* 37(5):1–9. https://doi.org/10.1007/S11032-017-0664-2/

Xu, Jiemeng, Mieke Wolters-Arts, Celestina Mariani, Heidrun Huber, and Ivo Rieu. 2017. "Heat Stress Affects Vegetative and Reproductive Performance and Trait Correlations in Tomato (*Solanum lycopersicum*)." *Euphytica* 213(7):1–12. https://doi.org/10.1007/S10681-017-1949-6/TABLES/2.

Xu, Xiao xia, Qin Hu, Wan nian Yang, and Ye Jin. 2017. "The Roles of Call Wall Invertase Inhibitor in Regulating Chilling Tolerance in Tomato." *BMC Plant Biology* 17(1):1–13. https://doi.org/10.1186/S12870-017-1145-9/FIGURES/9.

Xu, Yang, Shi rong Guo, Sheng Shu, Yan Ren, and Jin Sun. 2017. "Construction of a Genetic Linkage Map of Rootstock-Used Pumpkin Using SSR Markers and QTL Analysis for Cold Tolerance." *Scientia Horticulturae* 220:107–113. https://doi.org/10.1016/J.SCIENTA.2017.03.051.

Xu, Yunbi, Ping Li, Cheng Zou, Yanli Lu, Chuanxiao Xie, Xuecai Zhang, Boddupalli M. Prasanna, and Michael S. Olsen. 2017. "Enhancing Genetic Gain in the Era of Molecular Breeding." *Journal of Experimental Botany* 68(11):2641–2666. https://doi.org/10.1093/JXB/ERX135.

Yang, Jing, Guihua Duan, Chunqin Li, Lin Liu, Guangyu Han, Yaling Zhang, and Changmi Wang. 2019. "The Crosstalks Between Jasmonic Acid and Other Plant Hormone Signaling Highlight the Involvement of Jasmonic Acid as a Core Component in Plant Response to Biotic and Abiotic Stresses." *Frontiers in Plant Science* 10:458580. https://doi.org/10.3389/FPLS.2019.01349/XML/NLM.

Yang, Yang, Ning Tang, Zhiqiang Xian, and Zhengguo Li. 2015. "Two SnRK2 Protein Kinases Genes Play a Negative Regulatory Role in the Osmotic Stress Response in Tomato." *Plant Cell, Tissue and Organ Culture* 122(2):421–434. https://doi.org/10.1007/S11240-015-0779-2/FIGURES/9.

Ye, Jie, Genzhong Liu, Weifang Chen, Fengxia Zhang, Hanxia Li, Zhibiao Ye, and Yuyang Zhang. 2019. "Knockdown of SlNL33 Accumulates Ascorbate, Enhances Disease and Oxidative Stress Tolerance in Tomato (*Solanum lycopersicum*)." *Plant Growth Regulation* 89(1):49–58. https://doi.org/10.1007/S10725-019-00512-3/FIGURES/5.

Yeon, Jeyun, Thuy Tien Phan Nguyen, Minkyung Kim, and Sung Chur Sim. 2024. "Prediction Accuracy of Genomic Estimated Breeding Values for Fruit Traits in Cultivated Tomato (*Solanum lycopersicum* L.)." *BMC Plant Biology* 24(1):1–12. https://doi.org/10.1186/S12870-024-04934-8/FIGURES/5.

Yin, Wencheng, Zongli Hu, Baolu Cui, Xuhu Guo, Jingtao Hu, Zhiguo Zhu, and Guoping Chen. 2017. "Suppression of the MADS-Box Gene SlMBP8 Accelerates Fruit Ripening of Tomato (*Solanum lycopersicum*)." *Plant Physiology and Biochemistry* 118:235–244. https://doi.org/10.1016/J.PLAPHY.2017.06.019.

Yin, Yanling, Kezhen Qin, Xuewei Song, Qihao Zhang, Yanhong Zhou, Xiaojian Xia, and Jingquan Yu. 2018. "BZR1 Transcription Factor Regulates Heat Stress Tolerance Through FERONIA Receptor-Like Kinase-Mediated Reactive Oxygen Species Signaling in Tomato." *Plant and Cell Physiology* 59(11):2239–2254. https://doi.org/10.1093/PCP/PCY146.

Yıldız, Hatice Şelale, Sami Doğanlar, and Anne Frary. 2025. "Salt Tolerance Potential of Selected *S. Pennellii* Introgression Lines: Unique Shoot and Root Responses." https://doi.org/10.2139/SSRN.5197793.

Yong, Ke Tian, Phaik Har Yong, and Zhi Xiang Ng. 2023. "Tomato and Human Health: A Perspective from Post-Harvest Processing, Nutrient Bio-Accessibility, and Pharmacological Interaction." *Food Frontiers* 4(4):1702–1719. https://doi.org/10.1002/FFT2.299;REQUESTEDJOURNAL:JOURNAL:26438429;WGROUP:STRING:PUBLICATION.

Yu, Chao, Hua Sen Wang, Sha Yang, Xian Feng Tang, Ming Duan, and Qing Wei Meng. 2009. "Overexpression of Endoplasmic Reticulum Omega-3 Fatty Acid Desaturase Gene Improves Chilling Tolerance in Tomato." *Plant Physiology and Biochemistry* 47(11–12):1102–1112. https://doi.org/10.1016/J.PLAPHY.2009.07.008.

Yu, Li, Jun Yan, Yanjuan Yang, Lizhong He, and Weimin Zhu. 2016. "Enhanced Tolerance to Chilling Stress in Tomato by Overexpression of a Mitogen-Activated Protein Kinase, SlMPK7." *Plant Molecular Biology Reporter* 34(1):76–88. https://doi.org/10.1007/S11105-015-0897-3/TABLES/2.

Yu, Wenqing, Liu Wang, Ruirui Zhao, Jiping Sheng, Shujuan Zhang, Rui Li, and Lin Shen. 2019. "Knockout of SlMAPK3 Enhances Tolerance to Heat Stress Involving ROS Homeostasis in Tomato Plants." *BMC Plant Biology* 19(1):1–13. https://doi.org/10.1186/S12870-019-1939-Z/FIGURES/9.

Yuan, Liangdan, Wenxian Gai, Xiaoxiao Xuan, John Kojo Ahiakpa, Fangman Li, Pingfei Ge, Xingyu Zhang, Jinbao Tao, Yang Yang, and Yuyang Zhang. 2024. "Advances in Improving Tomato Fruit Quality by Gene Editing." *Horticultural Plant Journal*. https://doi.org/10.1016/J.HPJ.2024.04.008.

Zamir, D., and M. Tal. 1987. "Genetic Analysis of Sodium, Potassium and Chloride Ion Content in Lycopersicon." *Euphytica* 36(1):187–191. https://doi.org/10.1007/BF00730663/METRICS.

Zarka, Daniel G., Jonathan T. Vogel, Daniel Cook, and Michael F. Thomashow. 2003. "Cold Induction of Arabidopsis CBF Genes Involves Multiple ICE (Inducer of CBF Expression) Promoter Elements and a Cold-Regulatory Circuit That Is Desensitized by Low Temperature." *Plant Physiology* 133(2):910. https://doi.org/10.1104/PP.103.027169.

Zhang, Jing, Sitian Liu, Xiumei Zhu, Youlin Chang, Cheng Wang, Ning Ma, Junwen Wang, Xiaodan Zhang, Jian Lyu, and Jianming Xie. 2023. "A Comprehensive Evaluation of Tomato Fruit Quality and Identification of Volatile Compounds." *Plants* 12(16):2947. https://doi.org/10.3390/PLANTS12162947/S1.

Zhang, Li, Xinyong Guo, Yujie Qin, Bin Feng, Yating Wu, Yaling He, Aiying Wang, and Jianbo Zhu. 2020. "The Chilling Tolerance Divergence 1 Protein Confers Cold Stress Tolerance in Processing Tomato." *Plant Physiology and Biochemistry* 151:34–46. https://doi.org/10.1016/J.PLAPHY.2020.03.007.

Zhang, Li, Xinyong Guo, Zexing Zhang, Aiying Wang, and Jianbo Zhu. 2021. "Cold-Regulated Gene LeCOR413PM2 Confers Cold Stress Tolerance in Tomato Plants." *Gene* 764:145097. https://doi.org/10.1016/J.GENE.2020.145097.

Zhang, Ning, Huaiyin Zhao, Jiewei Shi, Yuanyuan Wu, and Jing Jiang. 2020. "Functional Characterization of Class I SlHSP17.7 Gene Responsible for Tomato Cold-Stress Tolerance." *Plant Science* 298:110568. https://doi.org/10.1016/J.PLANTSCI.2020.110568.

Zhang, Yingyu, Ansar Hussain, Muhammad Arif, Jawaher Alkahtani, Bandar M. AlMunqedhi, and Cheng Song. 2024. "Genetic Variability for Salinity Tolerance of Tomato (*Solanum lycopersicon* MILL.) Genotypes Determined by Stress Tolerance Indices." *Journal of King Saud University—Science* 36(9):103386. https://doi.org/10.1016/J.JKSUS.2024.103386.

Zhang, Yuanyuan, Mengchen Zhang, Junhua Ye, Qun Xu, Yue Feng, Siliang Xu, Dongxiu Hu, Xinghua Wei, Peisong Hu, and Yaolong Yang. 2023. "Integrating Genome-Wide Association Study into Genomic Selection for the Prediction of Agronomic Traits in Rice (*Oryza Sativa* L.)." *Molecular Breeding* 43(11):1–17. https://doi.org/10.1007/S11032-023-01423-Y/METRICS.

Zhang, Zhijin, and Rongfeng Huang. 2010. "Enhanced Tolerance to Freezing in Tobacco and Tomato Overexpressing Transcription Factor TERF2/LeERF2 Is Modulated by Ethylene Biosynthesis." *Plant Molecular Biology* 73(3):241–249. https://doi.org/10.1007/S11103-010-9609-4/FIGURES/6.

Zhao, Chunmei, Mariko Shono, Aiqing Sun, Shuying Yi, Minghui Li, and Jian Liu. 2007. "Constitutive Expression of an Endoplasmic Reticulum Small Heat Shock Protein Alleviates Endoplasmic Reticulum Stress in Transgenic Tomato." *Journal of Plant Physiology* 164(7):835–841. https://doi.org/10.1016/J.JPLPH.2006.06.004.

Zhou, Rong, Lingpeng Kong, Xiaqing Yu, Carl Otto Ottosen, Tongmin Zhao, Fangling Jiang, and Zhen Wu. 2019. "Oxidative Damage and Antioxidant Mechanism in Tomatoes Responding to Drought and Heat Stress." *Acta Physiologiae Plantarum* 41(2):1–11. https://doi.org/10.1007/S11738-019-2805-1/FIGURES/7.

Zhou, Weilin, Jinrong Yang, Yalan Zhang, Xiaoyi Hu, and Wei Wang. 2022. "Current Landscape of Gene-Editing Technology in Biomedicine: Applications, Advantages, Challenges, and Perspectives." *MedComm* 3(3):e155. https://doi.org/10.1002/MCO2.155.

Zhou, Xiang, and Matthew Stephens. 2012. "Genome-Wide Efficient Mixed-Model Analysis for Association Studies." *Nature Genetics* 44(7):821–824. https://doi.org/10.1038/NG.2310;SUBJMETA=114,205,208,2138,2397,631;KWRD=COMPUTATIONAL+MODELS,GENOME-WIDE+ASSOCIATION+STUDIES.

Zhu, Mingku, Guoping Chen, Jianling Zhang, Yanjie Zhang, Qiaoli Xie, Zhiping Zhao, Yu Pan, and Zongli Hu. 2014. "The Abiotic Stress-Responsive NAC-Type Transcription Factor SlNAC4 Regulates Salt and Drought Tolerance and Stress-Related Genes in Tomato (*Solanum lycopersicum*)." *Plant Cell Reports* 33(11):1851–1863. https://doi.org/10.1007/S00299-014-1662-Z/FIGURES/5.

Zhu, Mingku, Xiaoqing Meng, Jing Cai, Ge Li, Tingting Dong, and Zongyun Li. 2018. "Basic Leucine Zipper Transcription Factor SlbZIP1 Mediates Salt and Drought Stress Tolerance in Tomato." *BMC Plant Biology* 18(1):1–14. https://doi.org/10.1186/S12870-018-1299-0/FIGURES/7.

Zhu, Yingfang, Guangtao Zhu, Rui Xu, Zhixin Jiao, Junwei Yang, Tao Lin, Zhen Wang, Sanwen Huang, Leelyn Chong, and Jian Kang Zhu. 2023. "A Natural Promoter Variation of SlBBX31 Confers Enhanced Cold Tolerance during Tomato Domestication." *Plant Biotechnology Journal* 21(5):1033. https://doi.org/10.1111/PBI.14016.

11 Tomato Breeding for Vertical Farming

Krishna Bhattarai and Sadikshya Sharma

11.1 INTRODUCTION

Rapidly changing global trends due to natural and anthropogenic factors are increasing food-security and nutritional challenges to open-field agriculture and food production systems (Vos and Bellù 2019). Rapid rate of urbanization, unpredictable climate patterns, biodiversity loss, and existing and emerging biotic stresses are challenging the food supply chain (van Delden et al. 2021). Feeding an estimated global population of nearly 10 billion by 2050 will not be a trivial status quo (Lee 2011). Controlled environment agriculture (CEA), along with open field crop production, can ensure food security for the rising population (Specht et al. 2014). CEA is a climate-resilient production system that capitalizes on the environmental manipulation to achieve extended season or year-round crop production. CEA production systems include facilities that offer partial or modified growing conditions in the form of climate, growing substrate, or production methods. Greenhouses are the most common form of CEA facilities, offering a wide range of climate control based on the technological applications to control temperature, day length, humidity, irrigation, and other growing conditions. Indoor vertical farms are the most advanced form of CEA facilities, where most of the crop production parameters are controlled to create the closest to the ideal growing conditions required for the crop. Large variations of greenhouses and vertical farms, depending on the technological applications installed, have already emerged in the industry.

Furthermore, rising consumer awareness and preference for locally grown, fresh, and quality food drive the need for increased crop production, including in areas where the crops were not grown before. The implementation of CEA is sought to realize local food production, particularly in urban regions and areas where climatic conditions do not allow for open field production of the desired crops. CEA production facilities require plant materials with specific traits to thrive and produce food. For optimization of the higher investment and to achieve greater returns, it is imperative to grow plants consisting of premium traits. Conventional breeding targeted at open field crop production, and cultivars developed are not tailored to be grown in CEA. Further crop improvement through breeding is needed to develop new cultivars that are optimized for CEA. Although some premium traits have been bred into new cultivars of horticultural crops for field production, further exploration and trait integration are required to optimize them in CEA and make the industry profitable.

A wide range of horticultural crops has been successfully grown in single-tiered greenhouses. However, as an evolving form of food production, vertical farms are still limited to leafy greens like lettuce that are only economically profitable yet vulnerable, whereas crops like strawberries at ordinary scales are not yet economically viable except in a few countries (Zhuang et al. 2022). For the sustainability of vertical farms, economic sustainability is critical, which can be achieved by increasing crop diversity (Zhuang et al. 2024). Incorporation of fruiting crops into the vertical farms is the logical next step to increase the assortment of food crops to offer (Zelkind et al. 2022). However, improvements in both technological and biological systems need to be made to optimize the production of diverse crops in vertical farms. Due to this limitation, not all crops can be grown at present in the vertical farms. Among fruiting crops, a few rise in the list with elevated potential,

DOI: 10.1201/9781003506539-11

with slight improvements, to be successful in vertical farms, such as tomato and strawberries. Tomato (*Solanum lycopersicum*) has been profitably grown in greenhouses for a while and has the potential to be grown in vertical farms with a few traits optimized (D. Zhou et al. 2021). Based on a study, it is necessary to achieve a 75 $kg/m^2/year$ fruit yield to be economically viable for a small-sized tomato at $15/kg (Zhuang et al. 2024). In a high-tech greenhouse, a dwarf tomato cultivar, Micro-Tom, yields as high as 167 $kg/m^2/year$ (Terada et al. 2023). At present, the maximum yield reported from vertical farms are reported up to 60 $kg/m^2/year$, which is lower than the economically sustainable level (Zhuang et al. 2024). Recently, a gene-edited Ailsa Craig developed for production in vertical farms produces 150 to 300 kg/m^2 (Marston 2024). The photosynthetic conversion efficiency of this plant was reported to be high, with a 300 g plant producing 1 kg of fruit (Marston 2024). Tomato production in high-tech greenhouses in Washington, US, increased by 6.4 times when compared to open-field production (Maureira et al. 2022). In hydroponics in vertical farm research, increased tomato production has been observed as compared to hydroponic production in greenhouses (Lim et al. 2024). Despite higher production, the cost associated with tomato production in CEA is higher due to increased energy usage (Maureira et al. 2022). Extreme environmental temperatures drive up the production costs in CEA due to increased energy use. During the extreme high temperatures in 2022–2023 growing season in Türki'ye, growers' survey reported a 29.32% increased water use for irrigation and 19% increase in electricity consumption (Kürklü et al. 2025). Productivity is crucial in vertical farming, where leafy greens thrive due to their short life cycle, compact growth, and fully edible aerial parts, all of which maximize yield per unit area. Although tomatoes are not as perfect fit as leafy greens for vertical farming, a dwarf canopy and shorter life cycle offer increased productivity for a high-value crop. There is an opportunity to increase tomato fruit yield per plant, especially in dwarf tomatoes. Further increase in yield can be achieved by increasing the yield-determining parameters: yield per plant, plant density, and number of crops per year (Zhuang et al. 2024). Improving cultivation practices can contribute to an increment in yield; however, achieving double or triple-fold yield requires the application of genetics and breeding (D. Zhou et al. 2021). Breeding improvements to enable shorter plant, compact growth, rapid flowering, and early ripening can increase the yield-determining factors and further contribute toward making tomato production in vertical farming economically sustainable (Zhuang et al. 2024).

11.2 TOMATO IN CEA

Tomatoes have long been considered fit for production in CEA because of their availability in a wide range of growth habits. The indeterminate growth habit has been harnessed for high-tech greenhouse production in many regions of the world, especially in Europe. In the United States, both determinate and indeterminate tomatoes have been grown in high-tech greenhouses. Indeterminate plants are grown mainly for smaller fruit sizes, including cherry, grape, and cocktail types, whereas determinate plants are grown for larger sizes like Roma and beefsteak types. In vertical farms, smaller plant canopies help in maximizing space use, facilitating increased plant density because of their compact architecture and ability to thrive in limited space. Therefore, extremely determinate and dwarf plant types are most suitable for multi-tier systems, with the dwarf type being the preferred one. The advantage of dwarf plants is the small fruit sizes that the plants can hold without external support. Larger fruit sizes in dwarf plants may require external support like string tying, stakes, or clamps. However, a dwarf plant with larger fruit size is yet to be available commercially for CEA farms.

The fitness of tomatoes in vertical farms is also associated with fruit traits. The large fruit size originated from *S. lycopersicum* var. *lycopersicum*, and the smaller fruit size originated from *S. lycopersicum* var. *cerasiforme* and *S. pimpinellifolium* (Gao et al. 2019; Blanca et al. 2012). The high-tech greenhouses grow all tomato types. However, until now, most of the dwarf tomatoes grown in vertical farms have produced small fruit sizes. Cherry and grape-type tomatoes have high soluble solids, and aroma and flavors as compared to Roma and beefsteak types. These tomatoes can also be consumed fresh as salads. Higher soluble solids, flavor, and aroma make them an excellent ingredient in salads and in culinary forms that help in increasing the market premiums of these tomato fruits as compared to leafy greens. Further improvement in tomato fruit quality opens up the fruit value in niche markets. Some of these qualities include developing specific aromatic or flavor compounds, sweetness, and acid balance and increasing lycopene and other beneficial bioactive compounds, post-harvest shelf life, and firmness.

Tomato has been validated to grow well in CEA systems like high-tech greenhouses and high tunnels. Dwarf plants of Micro-tom produced twice the total fruit weight as compared to greenhouse-grown plants in nutrient film technique and ebb and flow hydroponic systems (Lim et al. 2024). Although tomatoes from vertical farms had a lower 10-fruit weight, they produced more fruits than greenhouse-grown tomatoes (Lim et al. 2024). However, there is a large variation in tomato yield depending on the type of CEA farms. Various forms of CEA farms have been designed based on light sources. While high-tech greenhouse systems rely on sunlight supplemented with LED lighting, indoor vertical farms use sole-source lighting using LED lights. Numerous investigations have been done to optimize light in CEA for tomato production, focusing on transplant production, plant growth and development, fruit quality, and post-harvest alterations (Ngcobo et al. 2021; Gómez and Mitchell 2015; Ali et al. 2024; Appolloni et al. 2023; Zheng et al. 2021).

The technological aspects of CEA, especially sensor-based monitoring systems, production platforms, overall facility structures, and environmental control, have been researched, and new features are emerging. On the other hand, the progress on the plant aspect has been slow and needs attention. Interest in the seed industry, especially in northern countries, has led to CEA-optimized leafy green crops such as lettuce. Similarly, in conjunction with grafting, numerous tomato varieties have been released for high-tech greenhouses. However, tomato crops that fit in a vertical setup are yet to be released. Some of the traits that help optimization of tomato in vertical farms are discussed.

11.3 CRITICAL TRAITS OF IMPORTANCE

Beyond technological innovations for energy and cost reduction, the success of vertical farming lies in the development of crop cultivars optimized to the growing conditions (Yu et al. 2025b). Development of tomato cultivars that can thrive in vertical farms needs improvement in several aspects that help to maximize fruit production, enhance resource management, and cater to consumer preferences (Teo and Yu 2024). The traits that are associated with these aspects are discussed.

11.3.1 Yield

Increasing fruit yield is critical for vertical farming because the majority of the current cultivars with the potential to be grown in these systems do not produce enough to make it economically profitable. Identification and utilization of genes associated with higher fruit yield can help enhance overall productivity. Yield-related genes in tomato have been previously reviewed by Ariizumi et al. (Ariizumi et al. 2013). A flowering gene, *SINGLE FLOWER TRUSS (SFT)*, in tomato increases fruit yield by 60% through heterosis and overdominance (Krieger et al. 2010). While higher yields can be beneficial, it is important to consider the threshold of the plant in bearing fruits in CEA. If the growing system is not equipped to support the heavily fruit-loaded plants, increasing yields to extremely high levels is not going to be applicable. Instead, the breeding may need to focus on early flowering and uniform ripening, so the production cycle can be shortened while early produce receives higher premiums.

Early flowering and uniform ripening traits in tomato can save resources and shorten the crop cycle while minimizing the risk factors, such as diseases and pests that occur in the later stages of crop cycles (Bhattarai et al. 2025). Genes *IAA9*, *ARF7/8*, *GA20ox1*, and *FW2.2* are known to modulate hormone signaling and cell division and help the tomato plant set and develop fruits early (Hu et al. 2023). Similarly, early ripening trait will enable early harvesting of the fruits, limiting disease and pest damage risks in the late season. Genes *ACS2/4*, *ACO1*, *PSY1*, and *PG2a* control ethylene biosynthesis, color, and fruit softening (J. Li et al. 2025). Overly soft ripe fruits are prone to bruising, physical damage, and pathogen infection, especially in the production system that will likely see mechanical harvesting as the technology gets refined. To balance the ripening and fruit softness, genes *RIN*, *NOR*, and *CNR* can be used (R. Wang et al. 2020).

11.3.2 Compact Growth

A wide pattern of growth habits is found in tomatoes, ranging from indeterminate to dwarf. Most of the single-tier CEA systems, like high-tech greenhouses, have capitalized on the indeterminate growth and grow plants for an extended time to achieve higher yields. In vertical farms and their existing concepts, the stacking of plants limits the use of indeterminate plants; hence, the need for a smaller canopy is high. Dwarf growth habits, shorter internodes, and dense foliage enable higher plant counts within a given area. Concepts of using indeterminate tomatoes have also been proposed, where the fruit clusters are trained to stay within the space assigned to each plant within the vertical stack. With rapidly evolving technologies, the production of indeterminate plants may be feasible in these systems with structural adjustments or improvements in plants. Until vertical farms gain the flexibility to accommodate indeterminate tomato, short-statured plants will mainly be grown in these platforms. The limitation with the dwarf and determinate plants is the need for support when loaded with fruits to prevent the fruits from touching the substrate or ground. Numerous dwarf and compact tomatoes are commercially available (https://heritageseedmarket.com/index.php/product-category/dwarf-tomatoes/, https://victoryseeds.com/collections/tomato/44-124/). However, the fitness of these cultivars in vertical farms has not been completely explored. The smaller plant architecture is associated with early flowering and harvest, allowing increased frequency of crop cycles in CEA farms. In tomatoes, *SELF PRUNING* (*SP*), *SELF PRUNING 5G* (*SP5G*), and *ERECTA* (*SlER*) genes control flower timing and plant architecture, increasing their fitness in CEA (Kwon et al. 2020). Genes including *DWARF* (*D*), miniature (*mnt*), and *Dumpy* (*dpy*) contribute to reducing tomato plant height (Bishop et al. 1996; Pnueli et al. 1998; Koka et al. 2000; Marti 2006).

11.3.3 Adaptation to Supplemental and Sole-Source Lighting

CEA farms use multiple light sources depending on the design of the farm and the technology installed. Some vertical farms utilize transparent plastic or glass on the walls and roofs to use sunlight as the main light source. Supplemental lighting is required when sunlight is used, especially in the rows near the bottom of the stacked platforms, as the plant canopy on the upper rows blocks the sunlight, limiting the required amount of light reaching the lower plants. Supplemental lighting is provided through light-emitting diodes (LEDs). In indoor vertical farms, sole-source lighting from LEDs is used as sunlight is not an option. The use of sole-source or supplemental lighting can be used to increase photosynthetic efficiency, synchronous maturity and ripening, higher production of bioactive compounds, and control of pests and diseases. Genes *HY5* and *CRY1* enhance photosynthetic capacity and antioxidant content when they are grown under blue and red LED lights as compared to white light (Kreslavski et al. 2021). Different genes regulate plant responses to respond to different wavelengths of light spectrum. Genes *PHYB1* and *PHYB2* regulate stem elongation, flowering, and leaf development under red light, whereas *CRYs* and *PHOTs* help tomato adapt to blue light regulation of photomorphogenesis and stomatal opening (Fantini and Facella 2020). There are phytochrome-interacting factors (PIF 1/3/4) that modulate shade avoidance. Harnessing/overexpressing these PIFs can help in the adaptation of plants to low light. Light spectrum influences the bioactive compound production in tomato. Production of compounds like lycopene, flavonoids, and vitamin C is upregulated under red and blue wavelengths. Genes Phytoene synthase (*PSY1*) and lycopene β-cyclase (*LCY-B*) expressions are upregulated under red and blue lights for carotenoid biosynthesis (Falcone 2010).

11.3.4 Fruit Quality

Fruit quality is the most anticipated trait that makes the overall CEA and vertical farming hopeful in leading to economic sustainability of the production system. Production of high-value fruits that deliver prized traits, including better taste, aroma and flavors, health/nutritional compounds. Development of a wide array of fruit colors could also improve the market, as bright colors and color combinations tend to draw attention and desire to taste the fruits. Additional qualities that come along with CEA production are freshness, clean production, and extended post-harvest shelf life. Although tomatoes are a predominantly self-pollinating crop, regular movement or shaking of flowers can result in improved fruit set and yield in CEA production. Varieties that show a high self-pollination rate and fruit set in CEA are useful, and selecting this germplasm helps contribute to the traits.

Tomato above the Brix level of 8% is perceived as premium by consumers and can enter the internet luxury market (Zhuang et al. 2024). Tomato with 10% Brix can command a 50% higher market price than that containing 6% Brix (Hemming et al. 2020). The economic benefits from an increase in Brix from 6% to 10% could be as large as that of a 50% yield increase (Zhuang et al. 2024). Selection for higher Brix is not trivial, as there are often trade-offs associated with organic acid content, vitamin C, and yield (Hemming et al. 2020; Roohanitaziani et al. 2020). Furthermore, Brix is also affected by growing conditions (Roohanitaziani et al. 2020). Oishii released Rubī Tomato, claimed to be a premium product, sweet and containing umami, and non-GMO (Oishii 2025). Traditional tomato breeding has used *S. pimpinellifolium* extensively to improve fruit quality such as lycopene content, vitamins, soluble solids, flavors, and resistance to biotic and abiotic stress (X. Wang et al. 2020). Zhou et al. (2022) identified *TREHALOSE-PHOSPHATE PHOSPHATASE*, which is critical for sugar accumulation. A deletion in promoter of the sugar transporter *STP1* plays a role in changing fruit soluble sugar content (H. Wang et al. 2023). A high fructose to glucose level in tomato has been mapped to the sugar transporter SWEET gene in *S. habrochaites* (Shammai et al. 2018). Similarly, genetic control of acid metabolism in tomatoes is also known. InDel in *TOMATO FRUIT MALATE 6* and in the promoter region of a *bHLH* control malic acid and ascorbic acid content (Ye et al. 2017, 2019). Metabolic studies revealed high levels of citrate and malate as well as threonine and tryptophan in *S. habrochaites* (Schauer et al. 2005).

Enhancement of sugar content and total soluble solids increases fruit quality in tomato fruits. The gene *LIN5* produces cell wall invertase that converts sucrose to glucose and fructose, leading to increased soluble sugar content and sweeter fruit. Similarly, sucrose transporter 1 (*SUT1*) regulates sugar transport from leaves (source) to developing fruits (sink). Organic acids are the second-most important component governing the taste of tomato after sugar. Brix-acid ratio provides the ultimate balanced taste. Genes aluminum-activated malate transporter 9 (*ALMT9*) and phosphoenolpyruvate carboxylase (*PEPC*) are involved in the accumulation of malic acid and biosynthesis of organic acids. Both soluble sugars and organic acid content can be useful to leverage the taste for niche markets and pursue premium prices. For instance, the American market prefers sweeter fruits, whereas the oriental market prefers a balance of sweet and sour taste. Tomato aroma is especially valuable when consumed fresh as salad or in sandwich. The presence of volatile esters, aldehydes, and alcohols affects the aroma of the fruits. Genes *LeE8*, *LeAAT1*, and *TomLoxC* are involved in modulating fruit aroma.

The bright red color of tomato is due to carotenoid pigments, mainly lycopene and β-carotene. Anthocyanins are involved in red, blue, and purple colors. However, there are small-sized tomato fruits available in many colors, such as red, orange, yellow, pink, brown, green, purple, and even white, because of varying levels of pigments. Tomatoes with pink and brown colors are the result of the mutation of *MYB12* and Stay Green 1 (*SGR1*) genes. Although not many, mixed colored fruits are also available in the market, mostly from heirloom cultivars. Some fruits demonstrate color differences in skin, flesh, and gel within a tomato fruit, giving rise to novel forms. These fruits could be valuable in the regions where red colored fruits have high demand, while anthocyanin-loaded gels could be developed for increased nutritional value.

Fruit texture and firmness are desirable traits for consumption and extended post-harvest shelf life. Cell wall degradation plays an important role in fruit softening through pectin degradation, breaking down of cellulose and hemicellulose, and loosening of the cell wall. Gene polygalacturonase (*PG2a*) is responsible for pectic degradation, Expansin 1 (*EXP1*) plays role in cell wall loosening, whereas cellulase and β-galactosidase are involved in breaking down cellulose and hemicellulose. On the other hand, there are genes that inhibit the ripening process and extend the post-harvest shelf life. Non-ripening (*NOR*) is a transcription factor that activates the ripening pathway, and its mutation works as a silencing of the ripening pathway. The ripening inhibitor (*RIN*) gene has been widely used in tomato breeding to prevent fruit softening and prolong shelf life. Genes involved in the ethylene pathway, *ACS2* and *ACO1,* play an important role in regulating the ripening speed.

11.3.5 Disease and Insect Resistance

Enclosed structures in CEA production systems prevent most of the pests and pathogens from freely moving from outside to inside these farms. However, there are a few pests and pathogens that have been difficult to avoid because of the favorable environmental conditions for these organisms. Pests that enter and adapt in these farms explode in populations due to their polyphagous feeding habit, survival in a wide range of temperatures and humidity, lack of predators, and rapid proliferation cycles. Most problematic pests in greenhouses and vertical farms are whiteflies, thrips, aphids, spider mites, fungus gnats, leaf miners, and mealybugs. These pests, once established, are difficult to get rid of due to the installed growing platforms, rapid rate of resistance development against insecticides (spider mites), cryptic lifestyle (thrips), and difficulty in detection (leaf miners). Some of these insects also cause further damage as they carry pathogens or viruses. Whiteflies transmit tomato yellow leaf curl virus. A multispecies screening of tomato germplasm identified multiple accessions in *S. chilense*, *S. arcanum*, *S. corneliomulleri*, *S. habrochaites*, *S. huaylasense*, *S. neorickii*, *S. peruvianum*, and one accession in *S. pimpinellifolium*, *S. chmielewskii*, *S. pennellii* were symptomless (Yan et al. 2018). *S. chilense* and *S. peruvianum* are the most common sources of TYLCV resistance in cultivated tomatoes. *S. chilense* accessions LA1938 and LA1971 are allelic to *Ty-1* gene; LA1932 and LA1938 carry *Ty-4* and *Ty-6*, respectively. LA0130, LA1960, and LA2737 show TYLCV resistance due to *Ty-1*/*Ty-3* alleles. Further, *S. habrochaites* accessions LA1777 and LA1718 and *S. peruvianum* accessions CGN15530, CGN15532, and PI 266375 are the sources for *Ty-2* and *Ty-4* genes (Yan et al. 2018).

Aphids carry numerous viruses, among which Potato virus Y (PVY), Cucumber mosaic virus (CMV), and Alfalfa mosaic virus (AMV) are problematic in tomato. Thrips transmit Tomato spotted wilt virus (TSWV). Fungus gnats are different than other insects in a way that they carry fungus, such as *Pythium,* one of the most problematic fungal pathogens in CEA systems.

Tobacco mosaic virus (TMV), previously prevalent Tobamovirus in major tomato growing regions, has been managed through introgression of *Tm-1*, *Tm-2,* and *Tm-2*^{2} from *S. habrochaites* and *S. peruvianum* (Alexander 1971). Recently discovered Tobamovirus, tomato brown rugose fruit virus (ToBRFV) that overcame resistance conferred by *Tm-2*^{2}, is a threat to tomato production in both open fields and CEA, as it is transmitted mechanically, by pollinators, and infested soils (Rochsar et al. 2025). Several ToBRFV resistance sources have been reported in *S. peruvianum* and *S. habrochaites* (Jewehan et al. 2022). A QTL in chromosome 2 next to *tomato mosaic-1* (*Tm-1*) has been discovered in *S. pennellii* that can be used in breeding for ToBRFV resistance (Rochsar et al. 2025).

Breeding host resistance to insect pests is not trivial in both open-field and CEA growing conditions. Rapid overcoming of monogenic resistance by insects like whiteflies and aphids remains the most hindering factor in host resistance. Improvement of agronomic traits could have eliminated genetic sources of insect pest resistance in tomato. Alternatively, incorporation of host resistance could lead to linkage drag, resulting in loss or inhibition of agronomic traits or losses in adaptational traits.

Crop improvement for CEA is fairly recent as compared to open field breeding and has mainly focused on adaptational traits, among which pest resistance has not been considered. Therefore, it is important to prioritize pest resistance, especially against those problematic in tomato. The need for pest resistance is amplified in CEA because there are no predators naturally available. Application of chemical insecticides is cumbersome due to infrastructural complexity, difficult to spot pests due to LED lighting, and difficult to eliminate due to the polyphagous nature of the pests.

Similarly, the pathogens that survive and affect the plants in CEA systems also possess the ability to survive in a wide range of temperature and humidity conditions. Some of the reported pathogens in CEA are root rot (*Pythium* spp.), powdery mildew (*Oidium neolycopersici*), bacterial spot (*Xanthomonas* spp.), bacterial speck (*Pseudomonas syringae*), TYLCV, Tomato Mosaic Virus, and Tomato spotted wilt virus. Recently, Tomato brown rugose virus has been reported to become problematic in CEA in most parts of the world.

Powdery mildew is one of the most common diseases in CEA, although it is not among the leading challenges in open-field production. Therefore, integration of the resistance genes in tomato varieties targeted for CEA is critical. Five dominant resistance genes (*Ol*-genes) have been identified from wild species. Genes *Ol-1,* identified from *S. habrochaites* G1.1560; *Ol-3,* identified from *S. habrochaites* G1.1290; and *Ol-5,* identified from *S. habrochaites* PI247087 offer partial resistance through a delayed hypersensitive response. Gene *Ol-4* was identified from *S. peruvianum* LA2172, and *ol-2*, a recessive gene, was identified in *S. lycopersicum* var. *cerasiforme* LA1230. However, the origin of *Ol-6* is not known. Genes *Ol-4* and *Ol-6* provide rapid HR response and offer race-specific resistance (Bai et al. 2005). Gene *ol-2*, a homologue of PM susceptibility gene in barley, *MILDEW LOCUS O* (*MLO*), offers a broad-spectrum resistance to multiple PM species (Bai et al. 2008). These genes have already been introgressed into advanced breeding lines. Therefore, traditional breeding can be implemented to introgress these genes into germplasm developed for CEA and ultimately into new varieties.

Gray mold, caused by *Botrytis cinerea*, is a major disease in all tomato production systems, including in the postharvest stages. With the wider host range and necrotrophic habit, *B cinerea* is difficult to manage in protected culture, highlighting the need for the development of host resistance. Fungal diseases are primarily managed by applying chemical fungicides. However, development of fungicide resistance has been reported in *B. cinerea* in a greenhouse production system (Rodríguez et al. 2014). Previously, quantitative genetic resistance to *B. cinerea* conferred by 10 QTLs has been reported in F_2 mapping populations and inbred lines from *S. habrochaites* LYC4 (N. Li et al. 2023). A regulatory mechanism involving Ethylene-MPK8-ERF.C1-PR gene complex was shown to confer *B. cinerea* resistance in tomato fruits through gene knockouts and overexpression studies (Deng et al. 2024). Although no varieties completely resistant to *B. cinerea* are available, varieties that show high tolerance are available for open field production. The major QTLs and gene complex conferring resistance to *B. cinerea* can be introgressed into germplasm optimized for CEA to develop resistant varieties for CEA. However, strong resistance and markers for gray mold breeding still await discovery.

Root rot caused by *Pythium* spp. is a challenging disease in CEA and hydroponic systems because multiple species cause root rot or seedling damping-off. Although no *Pythium*-resistant variety is available, partial tolerance has been identified in *S. habrochaites*, *S. peruvianum*, and *S. pennellii* (Lukyanenko 1991). Two anti-fungal genes against oomycetes encoding pathogenesis-related (PR) proteins, *PR1b1 and PR1a2*, could be investigated to evaluate their role in combating the disease in tomato (Tornero et al. 1997; Ma et al. 2023).

Fusarium wilt caused by *Fusarium oxysporum* f. sp. *Lycopersici* (*FOL*) is another leading problem in greenhouse tomato production. The fungus colonizes xylem tissues, interrupting the water movement and resulting in water stress and wilt. In greenhouse tomato production and fields infested with the disease, grafting *Fusarium*-resistant rootstocks with scions containing good agronomic traits has been commonly used. Four *R*-genes, *I-1*, *I-2*, *I-3*, and *I-7*, that confer race-specific resistance to *Fusarium* have been discovered and used in breeding programs. Genes *I-1* and *I-2* conferring resistance to races 1 and 2 were identified in *S. pimpinellifolium*, and genes *I-3* and *I-7* were identified in *S. pennellii,* which confer resistance to race 3 (Gonzalez-Cendales et al. 2016; Bohn and Tucker 1939; Alexander and Hoover 1955; McGrath et al. 1987). Another species, *Fusarium oxysporum* f. sp. *radicis-lycopersici* (*FORL*), causes crown and root rot in tomato. A single dominant locus, *Frl*, has been identified from *S. peruvianum* (Vakalounakis 1988). Molecular markers for these resistance genes have been developed and can be used to introgress them into CEA-optimized germplasm and ultimately new varieties. Recently, a new species, *Fusarium falciforme*, able to evade the resistance conferred by previously discovered genes, has been reported in Mexico, the United States, and Brazil (Severo et al. 2024; Vega-Gutiérrez et al. 2019). No sources of resistance have been publicly known, and exploratory research in search of host resistance is needed.

Root and crown diseases in tomato are mainly managed in greenhouse production systems by pyramiding resistance genes in one plant, which is eventually used as rootstock grafted to a scion that contains good yield and fruit quality. Grafted tomatoes may not be the top choice where efforts are to maximize vertical space utilization, a short growing season is preferred, and smaller plant canopies are considered. However, grafted tomatoes could be a preferred choice if the vertical farming system adopts a similar growing practice to that currently adopted by greenhouses.

Leaf mold caused by the obligate biotroph *Fulvia fulva* is problematic in CEA, particularly under high temperatures and humid growing conditions. Numerous physiological races of *F. fulva* have been discovered globally (S. Li et al. 2015; Yoshida et al. 2021; Scholtens-Toma et al. 1989). Numerous leaf mold resistance genes have been reported originating from *S. pimpinellifolium*, *S. esculentum* var. *cerasiforme*, *S. habrochaites,* and *S. peruvianum* (Kanwar et al. 1981). Five *Cf* genes, *Cf-2*, *Cf-4*, *Cf-4E*, *Cf-5*, and *Cf-9*, have been cloned, with *Cf-9* being widely used globally in modern varieties (de la Rosa et al. 2024; Dixon et al. 1996; Thomas et al. 1997; Dixon et al. 1998; Jones et al. 1993). Races of *F. fulvia* capable of evading resistance conferred by the known host resistance genes are emerging, stressing the need for new host resistance genes to be introgressed into tomato varieties for commercial production (de la Rosa et al. 2024).

Bacterial diseases such as bacterial canker (*Clavibacter michiganensis* subsp. *michiganensis*), bacterial speck (*Pseudomonas syringae* pv. *tomato*), and bacterial spot (*Xanthomonas* spp.) are seedborne and can enter the CEA system through infected seeds or transplants, whereas bacterial wilt (*Ralstonia solanacearum*) and pith necrosis (*Pseudomonas* spp.) are transmitted through contaminated media and water, mechanically or diseased plant parts. Therefore, incorporation of bacterial resistance is critical as these diseases are difficult to eliminate from the production system when they get established.

Bacterial canker caused by *Clavibacter michiganensis* subsp. *michiganensis* (*Cmm*) resistance is known to be polygenic and inherited quantitatively. The sources of resistance have been identified in *S. habrochaites*, *S. arcanum*, and *S. pimpinellifolium* (Sandbrink et al. 1995; Coaker and Francis 2004; Berry et al., 1989). Multiple QTLs from these wild relatives have been identified and bred into advanced germplasm, and new resistant varieties have been developed (Abebe et al. 2022).

Different races of *Xanthomonas*, *X. euvesicatoria* (race 1), *X. vesicatoria* (race 2), *X. perforans* (race T3 and T4), and *X. gardneri* are known to cause bacterial spot, while *X. perforans* is prevalent at this time in southeast United States (Sharma and Bhattarai 2019; Hutton et al. 2010). Hypersensitive and field resistance to these races have been identified in *S. lycopersicum*, *S. pennellii*, *S. habrochaites*, *S. pimpinellifolium*, and *S. lycopersicum* var. *cerasiforme* (Liabeuf et al. 2015; Bhattarai et al. 2017; Berrueta et al. 2016; Yang and Francis 2005). Breeding for host resistance to bacterial spot has been detailed in Sharma et al. (Sharma and Bhattarai 2019). These sources of resistance will be important in mitigating bacterial spot in CEA tomato production.

Bacterial speck, caused by *P. syringae* pv. *tomato*, is another seedborne disease that can be a potential threat to CEA tomato production. A *Pto* gene originating from *S. pimpinellifoliun* confers an HR restricting the growth and spread of the pathogen (Martin et al. 1993). Integration of *Pto* in tomato germplasm optimized for CEA can help in combating bacterial speck and in the development of resistant varieties.

11.4 GENOMICS

Tomato domestication has gone through severe bottlenecks, with cultivated tomato accounting for only 4.5% of the total genetic variation among tomato species (Ashrafi et al. 2009). There are 12 wild tomato species with diverse germplasm resources that can be exploited for traits important for vertical farms, including enhancing volatile compounds, disease resistance, and abiotic stress tolerance (Knapp and Peralta 2016). The first tomato reference genome was sequenced from *S. lycopersicum* cultivar Heinz 1706 (Sato et al. 2012). Heinz 1706 was selected for good flavor, high vitamins, and disease-resistant traits. The recent reference genome is SL4.0, which was assembled using PacBio long reads using Hi-C contact maps (Hosmani et al. 2019). More than 70 genomes of tomato, including 25 cultivated, 14 *S. lycopersicum* var. *cerasiforme,* and 29 wild species, are available (N. Li et al. 2023). Genome sequencing of numerous wild species has also been conducted, including *S. pimpinellifolium*, *S. cerasiforme*, *S. habrochaites*, *and S. galapagense* (X. Wang et al. 2020; Takei et al. 2021; N. Li et al. 2023; X. Yu et al. 2022; Han et al. 2024; Gupta et al. 2020; Benoit et al. 2025; L. Zhang et al. 2025; Y. Wang et al. 2024). These genomes have aided in understanding the evolutionary process, identifying genes for important traits including fruit flavors, discover unique in/dels and structural variants, salinity tolerance, stress tolerance, and terpene biosynthesis. A reference genome from one accession is not able to represent the genetic variation of a crop. Pan-genomes are able to capture the genetic variation at the comprehensive level from species or higher taxonomic levels. Several pan-genomes and super-pan-genomes at the genus levels are now available for tomato and *Solanum* species (Gao et al. 2019; N. Li et al. 2023; J. Yu et al. forthcoming; Y. Zhou et al. 2022). Eighty-one tomato breeding and germplasm accessions related to fresh-market consumption, including Micro-Tom, were previously sequenced using Illumina (Kim and Lee 2024). These genome sequencing approaches have identified 59,066 orthologous groups with 13,932 core orthologous groups (Yu et al. 2025a). The recent super pan-genome revealed genes *HAI1*, *CYP711A1/MAX1*, *WRKY9*, *CNGC15*, and *Hsp90* that are involved in plant growth, development, stress responses, and signal transduction are undergoing rapid evolution and positive selection (Yu et al. 2025a). Genes in *FAS* family have undergone sequence divergence, contraction, and expansion in the regulation of fruit size and locule numbers (Yu et al., 2025a). This diversity can be used in creating fruit of different sizes in CEA production that fit the local market demand. Pangenomes have revealed that genes that help in fruit storage (*CER1–1*), shelf life (*ARF*), disease resistance (*Hsp90* and *PR*), and stress responses (*WIP*, *ALDH,* and *WES*) have not been selected for and were only identified in wild relatives (Yu et al., 2025a). The application of artificial intelligence in genome annotation has facilitated automated gene structure annotation or transposable elements classification. QTL-seq, a combination of bulk segregant analysis and whole-genome resequencing, has been applied to discover a QTL, *Rcm6*, that confers resistance to bacterial canker and associated InDel markers that can be used for marker-assisted selection (Abebe et al. 2022).

These whole genome sequences can be leveraged to gain a deep understanding of genetic control of traits pertaining to CEA production and leveraging them in cultivar development. These resources have made conducting genetic analyses at the individual plant level, identification and discovery of genetic components for specific traits, and development of associated markers that can be used in tomato breeding possible. The availability of genomic resources within tomato and CRISPR-mediated gene editing has enabled the possibility of validating gene functionality and using them for tailoring tomato plants to specific needs and breeding targets. The concept of realizing the re-domestication using known beneficial traits and other traits that were lost due to negative selection has emerged to bring increased flavors and aroma. Furthermore, application of AI-based approaches will help in efficiently generating precise phenotypic data, improve trait prediction models, optimize germplasm selection strategies, and speed up breeding activities, especially for complex traits. For genomic resources-rich crops like tomato, implementation of AI is facilitating in predicting gene expression is taking shape, and increased genomic data will further improve AI-based tools for prediction accuracy (Rose et al. 2025).

Genotyping has been made possible in tomato through SNP arrays. Initially, SolCAP SNP array through Illumina Infinium identified 7,720 scorable SNPs were developed using 62,576 SNPs from the transcriptomic database (Sim et al. 2012). A 51K Axiom Tomato SNP Array was developed by Affymetrix designed from 625 tomato accessions (Lin et al. 2014). A reduced version of 20K Axiom Tomato Genotyping array by Affymetrix was created to genotype elite breeding lines for breeding programs (Lin et al. 2014). A 1K SNP chip has been used to genotype 2,726 tomato samples by AgriPlex Genomics, and GenoBaits Tomato 10K panel has been recently developed for genetic analysis (Adedze et al. 2025). These platforms and genomic tools can be used to genotype tomato germplasm, conduct GWAS, bi- and multi-parental, and advanced inbred line mapping studies, and use in introgressing targeted genes and parental line and hybrid development for CEA.

11.5 GENOME EDITING

Availability of pangenomes and super-pangenomes has enabled the identification of important variants and genes that were not previously identified using single genomes. This technology has enabled revisiting the wild accessions to harness the genes and variants and introgress them precisely into advanced tomato using genome editing techniques like CRISPR/Cas9. Genome editing has been performed on domesticated traits, *SP*, *OVATE*, *FAS*, *FW2.2*, *MULT*, *CycB*, *CLV3*, to reconstruct the domestication of tomato, which resulted in three-fold increase in fruit size and 10-fold increase in fruit number as compared to *S. pimpinellifolium* (Zsögön et al. 2018). Application of genome editing can expedite the domestication and create crops while retaining all favorable alleles that could help in regaining traits such as flavors and disease resistance that were lost during domestication and selective breeding.

Knocking out of *SlGA20ox1* resulted in short stems and compact canopy, which saved 75% space, allowing higher planting density and achieving 38–69% yield increase in indoor vertical farms and hydroponics (Yu et al. 2025b). Further, stacking of *slga20ox1* with *SP* and *SP5G* genes produced compact plant architecture with accelerated flowering and synchronized fruit ripening (Yu et al. 2025b). The triple mutant line *sp sp5g slga20ox1* reduced 85% space occupation, shortened life cycle by 16%, and increased effective fruit yield up to 180% by allowing higher density planting. Gene editing has been used to create a mini Ailsa Craig that is one-sixth the size of a conventional tomato for vertical farms (Marston 2024). Genome editing has been used to broaden color diversity. Recently, a multiplexed CRISPR/Cas9-mediated gene editing of *PSY1*, *MYB12*, and *SGR1* resulted in seven colors of fruits, including yellow, brown, pink, light-yellow, pink-brown, yellow-green, and light green (T. Yang et al. 2023). With the refinement of multiplexed genome editing techniques, numerous traits that are of interest in vertical farms can be precisely targeted, and a desirable phenotype can be obtained.

To enhance PM resistance in tomato, *MILDEW LOCUS O* (*MLO*) orthologous genes, Solyc04g049090 and *pSolyc04g049090* have been targeted in *S. peruvianum* (Nekrasov et al. 2017). Knockout of the *MLO1*-like genes conferred resistance to *O. neolycopersici* without the occurrence of undesirable phenotype (Nekrasov 2019). Similarly, knockout of two orthologous genes of *DOWNY MILDEW RESISTANT 6* (*DMR6*), *DMR6–1* (*Solyc03g080190*), and *DMR6–2 (Solyc06g073080)* showed enhanced disease resistance to bacterial, fungal, and oomycete pathogens in tomato (Thomazella et al. 2021).

Regulatory changes for genome-edited crops are required to streamline the production of gene-edited tomatoes in vertical farms. While genome-edited foods are already available in the United States, progress elsewhere is being made. Recently, CRISPR-edited crops without foreign DNA are generally not regulated in the United States and Canada, although approval is granted on a case-by-case basis (Ahmad et al. 2021). Most countries that have proposed rules for genome-edited crops have put forth regulations when the final product consists of foreign DNA. Post-Brexit, the UK has allowed GE-crops to be trialed and exempts CRISPR-edited crops without containing foreign DNA from GMO regulation; however, the restriction on commercialization exists. The European Union currently considers genome-edited crops as GMOs, and efforts to reform regulatory process are under consideration for approval in agriculture by 2026–2030 (Ahmad et al. 2021). In Asia, South Korea, plant materials without foreign genes are considered low-risk LMOs and are exempt from risk review and approval on import, production, and contained use (Sprink et al. 2022). Japan exempts CRISPR-edited crops without foreign DNA from regulation, which, however, requires developer notification (Sprink et al. 2022). India exempts products that do not carry exogenous DNA and are comparable to naturally occurring events from GMO legislation, with a requirement of evaluating phenotypic equivalence on a case-by-case basis (Sprink et al. 2022). While China has created four groups for genome-edited crops for regulatory purposes, classification of products in each category is dependent on the information, including molecular characterization, methods used, targeted gene, vector sequence residue, and off-target analysis (Sprink et al. 2022). In South American nations, including Argentina and Brazil, each case is handled uniquely, and prior consultation is required for deregulation decisions. In Africa, while Nigeria does not regulate genome-edited crops unless novel genetic combinations are present and are handled on a case-by-case basis, South Africa groups all genome-edited crops into a regulated group (Sprink et al. 2022). The use of genome editing in plant breeding will keep changing in different regions of the world as nations or regions decide the regulatory fate of the technology and products developed.

11.6 BREEDING STRATEGY

CEA tomato production is the application of structural construction, traditional agriculture, technological advancements, and economic sustainability. These factors need to be considered when developing breeding strategies for CEA. While these factors are also an integral component of open-field agriculture, their scope of application is broad in CEA due to its developmental stage. Increased infrastructural and system-dependent operational costs drive the bar high for breeding to deliver tomato varieties that have higher consumer and economic value. The opportunity prevails to leverage breeding knowledge and resources gained from open-field breeding, which eliminates the need for reinventing the wheel and promotes to use of the existing and evolving information and technology rationally to meet breeding objectives. The multidisciplinary integration of existing and novel technologies that can be useful in breeding tomato for CEA is shown in Figure 11.1.

As one of the highly produced and studied crops, tomato diversity has been collected and retained in multiple locations for conservation, breeding, and research. Global tomato collections of wild, domesticated, and advanced materials are available at various locations, including USDA-ARS Plant Genetic Resources Unit, Geneva, New York; Tomato Genetics Resource Center, Davis, California, in the United States; World Vegetable Center, Tainan, Taiwan; Center for Genetic Resources, Wageningen, the Netherlands; Leibniz Institute of Plant Genetics Gatersleben Genebank, Gatersleben, Germany; Centro de Conservación y Mejora de la Agrodiversidad Valenciana, Valencia, Spain; and Embrapa Hortaliças, Brasília, Brazil. These materials can be investigated and used in bringing the traits of interest to develop optimized varieties for CEA. Previous breeding efforts have made tremendous progress in transferring important genes from wild tomato into advanced materials. When accessible, these advanced materials can be utilized to

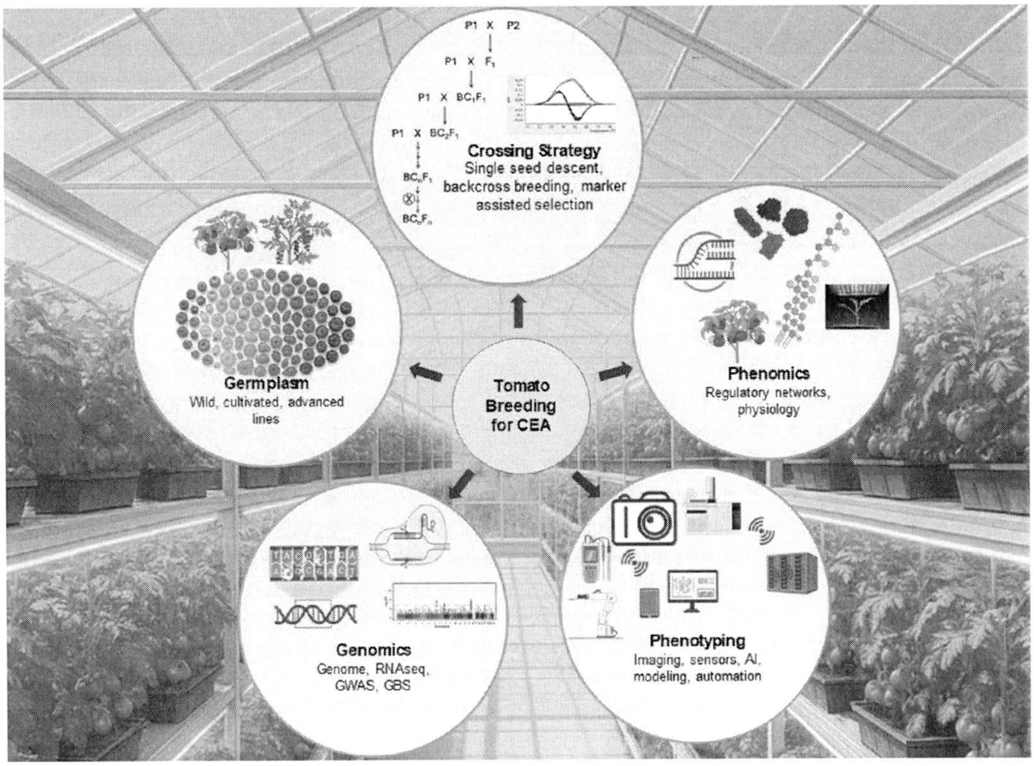

FIGURE 11.1 Tomato breeding for controlled environment agriculture integrating multidisciplinary approach.

speed up the breeding process as compared to when using wild lines and associated linkage drag. Traits including fruit size and canopy have been greatly improved in modern germplasm and can be readily introgressed to CEA germplasm from accessible advanced materials from repositories. In other traits including resistance to emerging diseases like ToBRFV and RB-TSWV, enhanced flavor and aromatic compounds, investigating wild accessions may be needed in search of novel genes and pathways.

Whole-genome and reduced-genome sequencing are actively taking place, and new genetic information is emerging at a fast rate in tomato. The genomic sequences, nucleotide polymorphisms, and genetic diversity can be used to develop germplasm comprised of necessary trait diversity and maintain the collection size. Available pangenomes and super-pangenomes can be useful in making decisions on retaining accessions harboring genes of interest and genetic diversity.

Many public and private breeding programs are now exploring their existing germplasm that was shelved because of the traits that were not important for open-field production; however, they have potential in CEA. Dwarf canopy offering high-density planting is among those traits that are not readily useful for fresh-market production due to their staking and stringing-based production method in the southeastern United States. For CEA, dwarf canopy fits very well, especially in a vertical setup. Similarly, higher total soluble solids (TSS) with moderate yield when transferred to a dwarf variety can lead to a desirable ideotype for CEA, where good fruit quality is needed. Therefore, germplasm for breeding for CEA can benefit from the existing germplasm that breeders have spent years developing and improving.

Due to specific traits of interest, the germplasm diversity for CEA may be specialized early on in a breeding program. Therefore, creating and maintaining germplasm diversity remains very critical. As the industry grows, expanding the diversity will be critical to deliver higher quality and improved products. Phenotypic evaluation and genotypic characterization of the germplasm reveal the germplasm structure and diversity status. For instance, tomato genetics that help plants thrive, yield, and produce high-quality fruits under supplemental or sole-source LED lighting are important in CEA. Therefore, germplasm needs to be evaluated under different light conditions and included accordingly in the program. Introgression breeding from primary, secondary, and tertiary gene pools can be used to increase diversity. Techniques like ploidy manipulation, embryo rescue, and tissue culture can assist in distant crosses. Genome editing and transgenic approaches can be used for gene insertion, knockouts, and transfer. Mutation breeding through chemical reagents such as ethyl methane sulfonate, physical agents such as gamma irradiation, and fast neutron bombardment alters nucleotides, creating novel genetic variation.

Development of parental lines is key to creating optimized hybrids for CEA. Selection and generation advancement can be done using single-seed descent, backcrossing coupled with marker-assisted selection for qualitative traits like *R*-gene-mediated disease resistance. For qualitative traits such as flavors and yield, QTLs associated with MAS and genomic selection in early generation can expedite the selection of favorable combinations. Given that CEA offers near-uniform growing conditions for an extended period or year-round, three to four generations can be achieved in a year. Hence, breeding for CEA often resembles speed breeding. Several SNP arrays and genotyping platforms are available, including the SolCAP Tomato SNP Array (Sim et al. 2012), 20K Tomato Infinium SNP array, 8K SolCAP Expanded Array, and recently GenoBaits Tomato10K panel (Adedze et al. 2025; Sim et al. 2012; Víquez-Zamora et al. 2013). Genotyping-by-sequencing, targeted amplicon sequencing, and whole-genome resequencing are common methods used to identify high-density SNPs, allele mining, and genotyping SNPs around targeted genes or loci.

Phenomics-integrated breeding in conjunction with phenotyping based on imaging, sensors, and AI-based computational tools can provide comprehensive assessment of breeding materials at a wide spectrum from cellular to whole plant, populations, and community level (Kaya 2025). Phenomics aids in understanding the regulatory networks and physiological processes involved in a given phenotypic trait and can be a key technique to use while selecting for traits, including taste, flavor, and abiotic stresses. High-throughput genotyping has been established as a key technique in MAS, and efforts like image-based phenotyping using unmanned aerial vehicles (UAVs) are being increasingly investigated and applied in field crops with a goal to use the data in making breeding decisions. While larger UAVs may not be feasible in smaller greenhouses or indoor vertical farms, mobile imaging system sensors and monitoring devices will provide real-time and accurate data that are cumbersome and labor- and cost-intensive when done manually.

Breeding methods that increase accuracy and expedite the selection process are crucial in creating optimized crops for CEA. Genomic selection utilizes markers distributed across whole genome to estimate breeding values and facilitate the selection of multiple alleles in a short time. MAS reduces the need for phenotypically evaluating populations for the trait of interest and can be implemented at the seed or early stages. Genome editing using techniques such as CRISPR/Cas9 can be used to edit the genome by altering, inserting, and deleting the nucleotides of interest while keeping the rest of the genome intact. Genome editing has been predominantly used in mutating tomato genes such as *SP5G*, *SlMLO1*, and GABA-TP to enhance PM resistance and nutritional quality (Soyk et al. 2017; R. Li et al. 2017; Takayama and Ezura 2015). CRISPR/Cas9 has been improved to multiplex genome editing that targets multiple genes using multiple guide RNAs (R. Li et al. 2017; Xie et al. 2015). Gene-edited plants are evaluated for phenotypic traits expression, backcrossed to remove the foreign nucleotide segments, and move forward with regulatory considerations or release. A CRISPR-edited Sicilian Rouge tomato that has enriched GABA expression and supports lower blood pressure and promotes relaxation has been released in the Japanese market (Waltz 2021). Gene-edited *SlCDPK27* and *SlCDPK26* knockouts showed 30% increased sweetness,

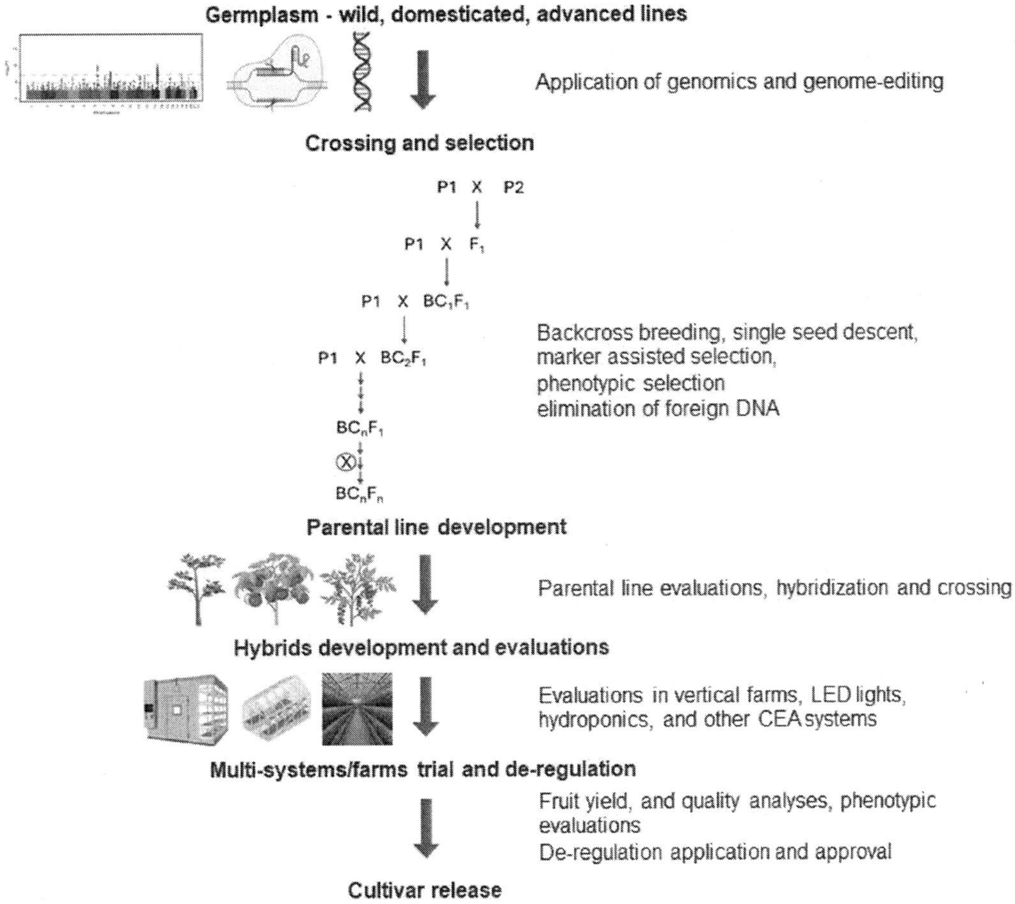

FIGURE 11.2 A simplified breeding strategy for tomato breeding targeted to controlled environment agriculture.

and the technology is now being used to commercially release sweeter tomatoes in China (Zhang et al. 2024a, 2024b). To sum up, a simplified version of a tomato breeding strategy is presented in Figure 11.2.

11.7 CONCLUSION

Diversification of crop portfolio from leafy greens to deliver a complete nutritional package to consumers is needed in the CEA industry to make it a complete and sustainable system and maximize profits. The interdisciplinary advancement in tomato breeding, genomics, genome editing, and molecular markers has facilitated the breeding process. Progress in technologies such as multiplexed genome editing, phenotyping, phenomics, and sensory studies has opened new avenues to make genetic gains with higher precision. Application of these technologies and breeding tools in other crops has led to significant progress in other crops and can make a significant impact in developing CEA-optimized tomato varieties to make the industry economically sustainable. While the industry is rapidly evolving, it is yet to see how it develops as a production system aiding food security and delivering high-quality produce to consumers all over the world. The breeding strategies and knowledge discussed will help in precisely harnessing host genetics to improve germplasm for CEA and develop CEA-optimized varieties.

REFERENCES

Abebe, Alebel Mekuriaw, Chang-Sik Oh, Hyoung Tae Kim, et al. 2022. "QTL-Seq Analysis for Identification of Resistance Loci to Bacterial Canker in Tomato." *Frontiers in Plant Science* 12 (January). https://doi.org/10.3389/fpls.2021.809959.

Adedze, Yawo Mawunyo Nevame, Yanfen Xu, Song Liu, et al. 2025. "An Advanced 10K SNP Panel for Genotyping Tomato (*Solanum Lycopersicum* L.) via Targeted Genome Sequencing." *Frontiers in Plant Science* 16 (May). https://doi.org/10.3389/fpls.2025.1582241.

Ahmad, Aftab, Nayla Munawar, Zulqurnain Khan, et al. 2021. "An Outlook on Global Regulatory Landscape for Genome-Edited Crops." *International Journal of Molecular Sciences* 22 (21): 11753. https://doi.org/10.3390/ijms222111753.

Alexander, Leonard Jay. 1971. "Host-Pathogen Dynamics of Tobacco Mosaic Virus on Tomato." *Phytopathology* 61 (61): 1–617.

Alexander, Leonard Jay, and Max M. Hoover. 1955. *Disease Resistance in Wild Species of Tomato: Report of the National Screening Committee.* https://kb.osu.edu/bitstreams/5def9be1-f366-5a8c-bf1d-af43eab65046/download.

Ali, Awais, Viviana Cavallaro, Piero Santoro, Jacopo Mori, Antonio Ferrante, and Giacomo Cocetta. 2024. "Quality and Physiological Evaluation of Tomato Subjected to Different Supplemental Lighting Systems." *Scientia Horticulturae* 323 (January): 112469. https://doi.org/10.1016/j.scienta.2023.112469.

Appolloni, E., G. Pennisi, I. Paucek, et al. 2023. "Potential Application of Pre-Harvest LED Interlighting to Improve Tomato Quality and Storability." *Postharvest Biology and Technology* 195 (January): 112113. https://doi.org/10.1016/j.postharvbio.2022.112113.

Ariizumi, Tohru, Yoshihito Shinozaki, and Hiroshi Ezura. 2013. "Genes That Influence Yield in Tomato." *Breeding Science* 63 (1): 3–13. https://doi.org/10.1270/jsbbs.63.3.

Ashrafi, Hamid, Matthew Kinkade, and Majid R. Foolad. 2009. "A New Genetic Linkage Map of Tomato Based on a Solanum Lycopersicum × S. Pimpinellifolium RIL Population Displaying Locations of Candidate Pathogen Response Genes." *Genome* 52 (11): 935–956. https://doi.org/10.1139/G09-065.

Bai, Yuling, Ron van der Hulst, Guusje Bonnema, et al. 2005. "Tomato Defense to Oldium Neolycopersici: Dominant OI Genes Confer Isolate-Dependent Resistance Via a Different Mechanism Than Recessive oI-2." *Molecular Plant-Microbe Interactions®* 18 (4): 354–362. https://doi.org/10.1094/MPMI-18-0354.

Bai, Yuling, Stefano Pavan, Zheng Zheng, et al. 2008. "Naturally Occurring Broad-Spectrum Powdery Mildew Resistance in a Central American Tomato Accession Is Caused by Loss of Mlo Function." *Molecular Plant-Microbe Interactions®* 21 (1): 30–39. https://doi.org/10.1094/MPMI-21-1-0030.

Benoit, Matthias, Katharine M. Jenike, James W. Satterlee, et al. 2025. "Solanum Pan-Genetics Reveals Paralogues as Contingencies in Crop Engineering." *Nature* 640 (8057): 135–145. https://doi.org/10.1038/s41586-025-08619-6.

Berrueta, María C., Gustavo Giménez, Guillermo A. Galván, and Alejandra Borges. 2016. "New Sources of Partial Resistance to Bacterial Spot Race T2 in Processing Tomatoes." *Horticultura Brasileira* 34 (September): 326–332. https://doi.org/10.1590/S0102-05362016003004.

Berry, S. Z., G. G. Madumadu, M. R. Uddin, and D. L. Coplin. 1989. "Virulence Studies and Resistance to Clavibacter Michiganensis Ssp. Michiganensis in Tomato Germplasm." HortScience. *HortScience* 24 (2): 362–365. https://doi.org/10.21273/HORTSCI.24.2.362.

Bhattarai, K., F. J. Louws, J. D. Williamson, and D. R. Panthee. 2017. "Resistance to Xanthomonas Perforans Race T4 Causing Bacterial Spot in Tomato Breeding Lines." *Plant Pathology* 66 (7): 1103–1109. https://doi.org/10.1111/ppa.12656.

Bhattarai, Krishna, Andrew B. Ogden, Sudeep Pandey, et al. 2025. "Improvement of Crop Production in Controlled Environment Agriculture through Breeding." *Frontiers in Plant Science* 15 (January): 1524601. https://doi.org/10.3389/fpls.2024.1524601.

Bishop, G. J., K. Harrison, and J. D. Jones. 1996. "The Tomato Dwarf Gene Isolated by Heterologous Transposon Tagging Encodes the First Member of a New Cytochrome P450 Family." *The Plant Cell* 8 (6): 959–969. https://doi.org/10.1105/tpc.8.6.959.

Blanca, Jose, Joaquín Cañizares, Laura Cordero, Laura Pascual, María José Diez, and Fernando Nuez. 2012. "Variation Revealed by SNP Genotyping and Morphology Provides Insight into the Origin of the Tomato." *PLoS ONE* 7 (10): e48198. https://doi.org/10.1371/journal.pone.0048198.

Bohn, G. W., and C. M. Tucker. 1939. "Immunity to Fusarium Wilt in the Tomato." *Science* 89 (2322): 603–604. https://doi.org/10.1126/science.89.2322.603.

Coaker, G. L., and D. M. Francis. 2004. "Mapping, Genetic Effects, and Epistatic Interaction of Two Bacterial Canker Resistance QTLs from Lycopersicon Hirsutum." *Theoretical and Applied Genetics* 108 (6): 1047–1055. https://doi.org/10.1007/s00122-003-1531-6.

Deng, Heng, Yangang Pei, Xin Xu, et al. 2024. "Ethylene-MPK8-ERF.C1-PR Module Confers Resistance against Botrytis Cinerea in Tomato Fruit without Compromising Ripening." *New Phytologist* 242 (2): 592–609. https://doi.org/10.1111/nph.19632.

Dixon, Mark S., Kostas Hatzixanthis, David A. Jones, Kate Harrison, and Jonathan D. G. Jones. 1998. "The Tomato Cf-5 Disease Resistance Gene and Six Homologs Show Pronounced Allelic Variation in Leucine-Rich Repeat Copy Number." *The Plant Cell* 10 (11): 1915–1925. https://doi.org/10.1105/tpc.10.11.1915.

Dixon, Mark S., David A. Jones, James S. Keddie, Colwyn M. Thomas, Kate Harrison, and Jonathan D. G. Jones. 1996. "The Tomato Cf-2 Disease Resistance Locus Comprises Two Functional Genes Encoding Leucine-Rich Repeat Proteins." *Cell* 84 (3): 451–459. https://doi.org/10.1016/S0092-8674(00)81290-8.

Falcone, Giulia. 2010. *Genetic Evolution of the Pathway Controlling Fruit Carotenoid Content in Tomato (S. Lycopersicum) and Its Wild Relatives*. https://tesidottorato.depositolegale.it/handle/20.500.14242/112202.

Fantini, Elio, and Paolo Facella. 2020. "Cryptochromes in the Field: How Blue Light Influences Crop Development." *Physiologia Plantarum* 169 (3): 336–346. https://doi.org/10.1111/ppl.13088.

Gao, Lei, Itay Gonda, Honghe Sun, et al. 2019. "The Tomato Pan-Genome Uncovers New Genes and a Rare Allele Regulating Fruit Flavor." *Nature Genetics* 51 (6): 1044–1051. https://doi.org/10.1038/s41588-019-0410-2.

Gómez, Celina, and Cary A. Mitchell. 2015. "Growth Responses of Tomato Seedlings to Different Spectra of Supplemental Lighting." *HortScience* 50 (1): 112–118. https://doi.org/10.21273/HORTSCI.50.1.112.

Gonzalez-Cendales, Yvonne, Ann-Maree Catanzariti, Barbara Baker, Des J. Mcgrath, and David A. Jones. 2016. "Identification of I-7 Expands the Repertoire of Genes for Resistance to Fusarium Wilt in Tomato to Three Resistance Gene Classes." *Molecular Plant Pathology* 17 (3): 448–463. https://doi.org/10.1111/mpp.12294.

Gupta, Prateek, Pankaj Singh Dholaniya, Sameera Devulapalli, Nilesh Ramesh Tawari, Yellamaraju Sreelakshmi, and Rameshwar Sharma. 2020. "Reanalysis of Genome Sequences of Tomato Accessions and Its Wild Relatives: Development of Tomato Genomic Variation (TGV) Database Integrating SNPs and INDELs Polymorphisms." *Bioinformatics* 36 (20): 4984–4990. https://doi.org/10.1093/bioinformatics/btaa617.

Han, Hongyu, Xiuhong Li, Tianze Li, et al. 2024. "Chromosome-Level Genome Assembly of Solanum Pimpinellifolium." *Scientific Data* 11 (1): 577. https://doi.org/10.1038/s41597-024-03442-6.

Hemming, Silke, Feije de Zwart, Anne Elings, Anna Petropoulou, and Isabella Righini. 2020. "Cherry Tomato Production in Intelligent Greenhouses—Sensors and AI for Control of Climate, Irrigation, Crop Yield, and Quality." *Sensors* 20 (22): 22. https://doi.org/10.3390/s20226430.

Hosmani, Prashant S., Mirella Flores-Gonzalez, Henri van de Geest, et al. 2019. "An Improved de Novo Assembly and Annotation of the Tomato Reference Genome Using Single-Molecule Sequencing, Hi-C Proximity Ligation and Optical Maps." Preprint, *bioRxiv*, September 14. https://doi.org/10.1101/767764.

Hu, Jianhong, Xiao Li, and Tai-ping Sun. 2023. "Four Class A-Auxin Response Factors Promote Tomato Fruit Growth despite Suppressing Its Initiation." *Nature Plants* 9 (5): 706–719. https://doi.org/10.1038/s41477-023-01396-y.

Hutton, Samuel F., Jay W. Scott, Wencai Yang, Sung-Chur Sim, David M. Francis, and Jeffrey B. Jones. 2010. "Identification of QTL Associated with Resistance to Bacterial Spot Race T4 in Tomato." *Theoretical and Applied Genetics* 121 (7): 1275–1287.

Jewehan, Ahmad, Nida Salem, Zoltán Tóth, Pál Salamon, and Zoltán Szabó. 2022. "Evaluation of Responses to Tomato Brown Rugose Fruit Virus (ToBRFV) and Selection of Resistant Lines in Solanum Habrochaites and Solanum Peruvianum Germplasm." *Journal of General Plant Pathology* 88 (3): 187–196. https://doi.org/10.1007/s10327-022-01055-8.

Jones, D. A., M. J. Dickinson, P. J. Balint-Kurti, M. S. Dixon, and J. D. G. Jones. 1993. "Two Complex Resistance Loci Revealed in Tomato by Classical and RFLP Mapping of the Cf-2, Cf-4, Cf-5, and Cf-9 Genes for Resistance to Cladosporium Fulvum." *Molecular Plant Microbe Interactions* 6 (1993): 348–348.

Kanwar, J. S., P. M. Harney, and E. A. Kerr. 1981. *Allelic Relationships of Genes for Resistance to Tomato Leaf Mold, Cladosporium Fulvum Cke*. https://www.cabidigitallibrary.org/doi/full/10.5555/19811600044.

Kaya, Cengiz. 2025. "Optimizing Crop Production with Plant Phenomics Through High-Throughput Phenotyping and AI in Controlled Environments." *Food and Energy Security* 14 (1): e70050. https://doi.org/10.1002/fes3.70050.

Kim, Juhee, and Tong Geon Lee. 2024. "Genome Sequence Data of the Contemporary Fresh-Market Tomatoes." *BMC Genomic Data* 25 (1): 65. https://doi.org/10.1186/s12863-024-01249-3.

Knapp, Sandra, and Iris Edith Peralta. 2016. "The Tomato (Solanum Lycopersicum L., Solanaceae) and Its Botanical Relatives." In *The Tomato Genome*, edited by Mathilde Causse, Jim Giovannoni, Mondher Bouzayen, and Mohamed Zouine. Springer. https://doi.org/10.1007/978-3-662-53389-5_2.

Koka, Chala V., R. Eric Cerny, Randy G. Gardner, et al. 2000. "A Putative Role for the Tomato Genes DUMPY and CURL-3 in Brassinosteroid Biosynthesis and Response1." *Plant Physiology* 122 (1): 85–98. https://doi.org/10.1104/pp.122.1.85.

Kreslavski, Vladimir D., Aleksandra Yu Khudyakova, Valeria V. Strokina, et al. 2021. "Impact of High Irradiance and UV-B on the Photosynthetic Activity, pro-/Antioxidant Balance and Expression of Light-Activated Genes in *Arabidopsis Thaliana Hy4* Mutants Grown under Blue Light." *Plant Physiology and Biochemistry* 167 (October): 153–162. https://doi.org/10.1016/j.plaphy.2021.07.030.

Krieger, Uri, Zachary B. Lippman, and Dani Zamir. 2010. "The Flowering Gene Single Flower Truss Drives Heterosis for Yield in Tomato." *Nature Genetics* 42 (5): 459–463. https://doi.org/10.1038/ng.550.

Kürklü, Ahmet, Simon Pearson, and Turgut Felek. 2025. "Climate Change Impacts on Tomato Production in High-Tech Soilless Greenhouses in Türkiye." *BMC Plant Biology* 25 (1): 339. https://doi.org/10.1186/s12870-025-06307-1.

Kwon, Choon-Tak, Jung Heo, Zachary H. Lemmon, et al. 2020. "Rapid Customization of Solanaceae Fruit Crops for Urban Agriculture." *Nature Biotechnology* 38 (2): 182–188. https://doi.org/10.1038/s41587-019-0361-2.

Lee, Ronald. 2011. "The Outlook for Population Growth." *Science* 333 (6042): 569–573. https://doi.org/10.1126/science.1208859.

Li, Jinyan, Yao Lu, Ke Cheng, et al. 2025. "ACS4 Exerts a Pivotal Role in Ethylene Biosynthesis during the Ripening of Tomato Fruits in Comparison to ACS2." *The Plant Journal* 121 (5): e70043. https://doi.org/10.1111/tpj.70043.

Li, Ning, Qiang He, Juan Wang, et al. 2023. "Super-Pangenome Analyses Highlight Genomic Diversity and Structural Variation across Wild and Cultivated Tomato Species." *Nature Genetics* 55 (5): 852–860. https://doi.org/10.1038/s41588-023-01340-y.

Li, Rui, Ran Li, Xindi Li, et al. 2017. "Multiplexed CRISPR/Cas9-mediated Metabolic Engineering of Γ-aminobutyric Acid Levels in Solanum Lycopersicum." *Plant Biotechnology Journal* 16 (2): 415–427. https://doi.org/10.1111/pbi.12781.

Li, Shuai, Tingting Zhao, Huijia Li, Xiangyang Xu, and Jingfu Li. 2015. "First Report of Races 2.5 and 2.4.5 of Cladosporium Fulvum (Syn. Passalora Fulva), Causal Fungus of Tomato Leaf Mold Disease in China." *Journal of General Plant Pathology* 81 (2): 162–165. https://doi.org/10.1007/s10327-015-0577-z.

Liabeuf, D., D. M. Francis, and S. Sim, S. 2015. "Screening Cultivated and Wild Tomato Germplasm for Resistance to Xanthomonas Gardneri." *Acta Horticulture* 1069: 65–70.

Lim, Yoonseo, Myeong-Gyun Seo, Sehee Lee, et al. 2024. "Comparative Yield Evaluation of Mini-Tomato Cultivar in Two Hydroponic Systems." *Horticulture, Environment, and Biotechnology* 65 (2): 239–250. https://doi.org/10.1007/s13580-023-00564-5.

Lin, Tao, Guangtao Zhu, Junhong Zhang, et al. 2014. "Genomic Analyses Provide Insights into the History of Tomato Breeding." *Nature Genetics* 46 (11): 1220–1226. https://doi.org/10.1038/ng.3117.

Lukyanenko, A. N. 1991. "Breeding Tomato for Mechanized Harvesting." In *Genetic Improvement of Tomato*, edited by G. Kalloo. Springer. https://doi.org/10.1007/978-3-642-84275-7_17.

Ma, Minxiao, Paul W. J. Taylor, Deli Chen, Niloofar Vaghefi, and Ji-Zheng He. 2023. "Major Soilborne Pathogens of Field Processing Tomatoes and Management Strategies." *Microorganisms* 11 (2): 263. https://doi.org/10.3390/microorganisms11020263.

Marston, Jennifer. 2024. "Phytoform Says Its Gene-Edited Tomato Could Produce 'up to 400% More Fruit' in a Vertical Farm." *AgFunderNews*, November 4. https://agfundernews.com/phytoform-says-its-gene-edited-tomato-could-produce-up-to-400-more-fruit-in-a-vertical-farm.

Marti, E. 2006. "Genetic and Physiological Characterization of Tomato Cv. Micro-Tom." *Journal of Experimental Botany* 57 (9): 2037–2047. https://doi.org/10.1093/jxb/erj154.

Martin, Gregory B., Sergio H. Brommonschenkel, Julapark Chunwongse, et al. 1993. "Map-Based Cloning of a Protein Kinase Gene Conferring Disease Resistance in Tomato." *Science* 262 (5138): 1432–1436. https://doi.org/10.1126/science.7902614.

Maureira, Fidel, Kirti Rajagopalan, and Claudio O. Stöckle. 2022. "Evaluating Tomato Production in Open-Field and High-Tech Greenhouse Systems." *Journal of Cleaner Production* 337 (February): 130459. https://doi.org/10.1016/j.jclepro.2022.130459.

McGrath, D. J., D. Gillespie, and L. Vawdrey. 1987. "Inheritance of Resistance to Fusarium Oxysporum f. Sp. Lycopersici Races 2 and 3 in Lycopersicon Pennellii." *Australian Journal of Agricultural Research* 38 (4): 729–733. https://doi.org/10.1071/ar9870729.

Nekrasov, Vladimir. 2019. "Sequence-Specific Nucleases as Tools for Enhancing Disease Resistance in Crops." *Transgenic Research* 28 (2): 75–80. https://doi.org/10.1007/s11248-019-00137-2.

Nekrasov, Vladimir, Congmao Wang, Joe Win, Christa Lanz, Detlef Weigel, and Sophien Kamoun. 2017. "Rapid Generation of a Transgene-Free Powdery Mildew Resistant Tomato by Genome Deletion." *Scientific Reports* 7 (1): 482. https://doi.org/10.1038/s41598-017-00578-x.

Ngcobo, B. L., Bertling, I., and A. D. and Clulow. 2021. "Post-Harvest Alterations in Quality and Health-Related Parameters of Cherry Tomatoes at Different Maturity Stages Following Irradiation with Red and Blue LED Lights." *The Journal of Horticultural Science and Biotechnology* 96 (3): 383–391. https://doi.org/10.1080/14620316.2020.1847696.

Oishii. 2025. "The Rubi Tomato." Accessed July 16, 2025. https://oishii.com/pages/the-rubi-tomato.

Pnueli, Lilac, Lea Carmel-Goren, Dana Hareven, et al. 1998. "The *SELF-PRUNING* Gene of Tomato Regulates Vegetative to Reproductive Switching of Sympodial Meristems and Is the Ortholog of *CEN* and *TFL1*." *Development* 125 (11): 1979–1989. https://doi.org/10.1242/dev.125.11.1979.

Rochsar, Edan, Shai Torgeman, Kfir Bandel, et al. 2025. "Tissue-Specific Resistance and Susceptibility to the Tomato Brown Rugose Fruit Virus (ToBRFV) Conferred by Solanum Pennellii Loci." *BMC Plant Biology* 25 (1): 51. https://doi.org/10.1186/s12870-024-05989-3.

Rodríguez, A., A. Acosta, and C. Rodríguez. 2014. "Fungicide Resistance of Botrytis Cinerea in Tomato Greenhouses in the Canary Islands and Effectiveness of Non-Chemical Treatments against Gray Mold." *World Journal of Microbiology and Biotechnology* 30 (9): 2397–2406. https://doi.org/10.1007/s11274-014-1665-5.

Roohanitaziani, Raana, Ruud A. de Maagd, Michiel Lammers, et al. 2020. "Exploration of a Resequenced Tomato Core Collection for Phenotypic and Genotypic Variation in Plant Growth and Fruit Quality Traits." *Genes* 11 (11): 11. https://doi.org/10.3390/genes11111278.

Rosa, Silvia de la, Christiaan R. Schol, Ángeles Ramos Peregrina, et al. 2024. "Sequential Breakdown of the Cf-9 Leaf Mould Resistance Locus in Tomato by Fulvia Fulva." *New Phytologist* 243 (4): 1522–1538. https://doi.org/10.1111/nph.19925.

Rose, Laura Ellen, Zahra Zangishei, Alisdair R Fernie, and Björn Usadel. 2025. "Tomato Genomics and Its Relatives Are Breaking the Genomics Barriers." *Journal of Experimental Botany*, July 3, eraf298. https://doi.org/10.1093/jxb/eraf298.

Sandbrink, J. M., J. W. Van Ooijen, C. C. Purimahua, et al. 1995. "Localization of Genes for Bacterial Canker Resistance in Lycopersicon Peruvianum Using RFLPs." *Theoretical and Applied Genetics* 90 (3–4): 444–450. https://doi.org/10.1007/bf00221988.

Sato, Shusei, Satoshi Tabata, Hideki Hirakawa, et al. 2012. "The Tomato Genome Sequence Provides Insights into Fleshy Fruit Evolution." *Nature* 485 (7400): 635–641. https://doi.org/10.1038/nature11119.

Schauer, Nicolas, Dani Zamir, and Alisdair R. Fernie. 2005. "Metabolic Profiling of Leaves and Fruit of Wild Species Tomato: A Survey of the Solanum Lycopersicum Complex." *Journal of Experimental Botany* 56 (410): 297–307. https://doi.org/10.1093/jxb/eri057.

Scholtens-Toma, I. M. J., G. J. M. Wit, and P. J. G. M. Wit. 1989. "Characterization of Elicitor Activities of Apoplastic Fluids Isolated from Tomato Lines Infected with New Races of Cladosporium Fulvum." *Netherlands Journal of Plant Pathology* 95 (S1): 161–168. https://doi.org/10.1007/BF01974295.

Severo, Robinson, Leandro Jun Soki Shibutani, Gilvan Ferreira da Silva, et al. 2024. "Fusarium Species Causing Root Rot and Wilt in Tomato in Brazil." *Journal of Phytopathology* 172 (1): e13261. https://doi.org/10.1111/jph.13261.

Shammai, Arik, Marina Petreikov, Yelena Yeselson, et al. 2018. "Natural Genetic Variation for Expression of a SWEET Transporter among Wild Species of Solanum Lycopersicum (Tomato) Determines the Hexose Composition of Ripening Tomato Fruit." *The Plant Journal* 96 (2): 343–357. https://doi.org/10.1111/tpj.14035.

Sharma, Sadikshya, and Krishna Bhattarai. 2019. "Progress in Developing Bacterial Spot Resistance in Tomato." *Agronomy* 9 (1): 1. https://doi.org/10.3390/agronomy9010026.

Sim, Sung-Chur, Gregor Durstewitz, Jörg Plieske, et al. 2012. "Development of a Large SNP Genotyping Array and Generation of High-Density Genetic Maps in Tomato." *PLoS ONE* 7 (7): e40563. https://doi.org/10.1371/journal.pone.0040563.

Soyk, Sebastian, Niels A. Müller, Soon Ju Park, et al. 2017. "Variation in the Flowering Gene SELF PRUNING 5G Promotes Day-Neutrality and Early Yield in Tomato." *Nature Genetics* 49 (1): 162–168. https://doi.org/10.1038/ng.3733.

Specht, Kathrin, Rosemarie Siebert, Ina Hartmann, et al. 2014. "Urban Agriculture of the Future: An Overview of Sustainability Aspects of Food Production in and on Buildings." *Agriculture and Human Values* 31: 33–51.

Sprink, Thorben, Ralf Wilhelm, and Frank Hartung. 2022. "Genome Editing around the Globe: An Update on Policies and Perceptions." *Plant Physiology* 190 (3): 1579–1587. https://doi.org/10.1093/plphys/kiac359.

Takayama, Mariko, and Hiroshi Ezura. 2015. "How and Why Does Tomato Accumulate a Large Amount of GABA in the Fruit?" *Frontiers in Plant Science* 6 (August). https://doi.org/10.3389/fpls.2015.00612.

Takei, Hitomi, Kenta Shirasawa, Kosuke Kuwabara, et al. 2021. "De Novo Genome Assembly of Two Tomato Ancestors, Solanum Pimpinellifolium and Solanum Lycopersicum Var. Cerasiforme, by Long-Read Sequencing." *DNA Research* 28 (1): dsaa029. https://doi.org/10.1093/dnares/dsaa029.

Teo, Zhi Wei Norman, and Hao Yu. 2024. "Genetic Breeding for Indoor Vertical Farming." *NPJ Sustainable Agriculture* 2 (1): 1–9. https://doi.org/10.1038/s44264-024-00021-5.

Terada, Naoki, Kalara Dissanayake, Chiharu Okada, Atsushi Sanada, and Kaihei Koshio. 2023. "Micro-Tom Tomato Response to Fertilization Rates and the Effect of Cultivation Systems on Fruit Yield and Quality." *Horticulturae* 9 (3): 3. https://doi.org/10.3390/horticulturae9030367.

Thomas, C. M., D. A. Jones, M. Parniske, et al. 1997. "Characterization of the Tomato Cf-4 Gene for Resistance to Cladosporium Fulvum Identifies Sequences That Determine Recognitional Specificity in Cf-4 and Cf-9." *The Plant Cell* 9 (12): 2209–2224. https://doi.org/10.1105/tpc.9.12.2209.

Thomazella, Daniela Paula de Toledo, Kyungyong Seong, Rebecca Mackelprang, et al. 2021. "Loss of Function of a DMR6 Ortholog in Tomato Confers Broad-Spectrum Disease Resistance." *Proceedings of the National Academy of Sciences* 118 (27): e2026152118. https://doi.org/10.1073/pnas.2026152118.

Tornero, Pablo, José Gadea, Vicente Conejero, and Pablo Vera. 1997. "Two PR-1 Genes from Tomato Are Differentially Regulated and Reveal a Novel Mode of Expression for a Pathogenesis-Related Gene During the Hypersensitive Response and Development." *Molecular Plant-Microbe Interactions®* 10 (5): 624–634. https://doi.org/10.1094/MPMI.1997.10.5.624.

Vakalounakis, D. J. 1988. "The Genetic Analysis of Resistance to Fusarium Crown and Root Rot of Tomato." *Plant Pathology* 37 (1): 71–73. https://doi.org/10.1111/j.1365-3059.1988.tb02197.x.

Van Delden, S. H., M. Sharath Kumar, M. Butturini, L. J. A. Graamans, E. Heuvelink, M. Kacira, E. Kaiser, R. S. Klamer, L. Klerkx, G. Kootstra, and A. Loeber. 2021. "Current Status and Future Challenges in Implementing and Upscaling Vertical Farming Systems." *Nature Food* 2 (12): 944–956.

Vega-Gutiérrez, T. A., C. A. López-Orona, G. A. López-Urquídez, et al. 2019. "Foot Rot and Wilt in Tomato Caused by Fusarium Falciforme (FSSC 3 + 4) in Mexico." *Plant Disease* 103 (1): 157. https://doi.org/10.1094/PDIS-06-18-1001-PDN.

Víquez-Zamora, Marcela, Ben Vosman, Henri van de Geest, et al. 2013. "Tomato Breeding in the Genomics Era: Insights from a SNP Array." *BMC Genomics* 14 (1): 354. https://doi.org/10.1186/1471-2164-14-354.

Vos, Rob, and Lorenzo Giovanni Bellù. 2019. "Chapter 2—Global Trends and Challenges to Food and Agriculture into the 21st Century." In *Sustainable Food and Agriculture*, edited by Clayton Campanhola and Shivaji Pandey. Academic Press. https://doi.org/10.1016/B978-0-12-812134-4.00002-9.

Waltz, Emily. 2021. "GABA-Enriched Tomato Is First CRISPR-Edited Food to Enter Market." *Nature Biotechnology* 40 (1): 9–11. https://doi.org/10.1038/d41587-021-00026-2.

Wang, Heng, Hongjun Yu, Lin Chai, et al. 2023. "Exogenous Sucrose Confers Low Light Tolerance in Tomato Plants by Increasing Carbon Partitioning from Stems to Leaves." *Journal of Agricultural and Food Chemistry* 71 (51): 20625–20642. https://doi.org/10.1021/acs.jafc.3c05985.

Wang, Rufang, Michiel Lammers, Yury Tikunov, Arnaud G. Bovy, Gerco C. Angenent, and Ruud A. de Maagd. 2020. "The *Rin*, *nor* and *Cnr* Spontaneous Mutations Inhibit Tomato Fruit Ripening in Additive and Epistatic Manners." *Plant Science* 294 (May): 110436. https://doi.org/10.1016/j.plantsci.2020.110436.

Wang, Xin, Lei Gao, Chen Jiao, et al. 2020. "Genome of Solanum Pimpinellifolium Provides Insights into Structural Variants during Tomato Breeding." *Nature Communications* 11 (1): 5817. https://doi.org/10.1038/s41467-020-19682-0.

Wang, Yan, Chuanlong Sun, Zhibiao Ye, Chuanyou Li, Sanwen Huang, and Tao Lin. 2024. "The Genomic Route to Tomato Breeding: Past, Present, and Future." *Plant Physiology* 195 (4): 2500–2514. https://doi.org/10.1093/plphys/kiae248.

Xie, Kabin, Bastian Minkenberg, and Yinong Yang. 2015. "Boosting CRISPR/Cas9 Multiplex Editing Capability with the Endogenous tRNA-Processing System." *Proceedings of the National Academy of Sciences* 112 (11): 3570–3575. https://doi.org/10.1073/pnas.1420294112.

Yan, Zhe, Ana Pérez-de-Castro, Maria J. Díez, et al. 2018. "Resistance to Tomato Yellow Leaf Curl Virus in Tomato Germplasm." *Frontiers in Plant Science* 9 (August). https://doi.org/10.3389/fpls.2018.01198.

Yang, Tianxia, Muhammad Ali, Lihao Lin, et al. 2023. "Recoloring Tomato Fruit by CRISPR/Cas9-Mediated Multiplex Gene Editing." *Horticulture Research* 10 (1): uhac214. https://doi.org/10.1093/hr/uhac214.

Yang, Wencai, and David M. Francis. 2005. "Marker-Assisted Selection for Combining Resistance to Bacterial Spot and Bacterial Speck in Tomato." *Journal of the American Society for Horticultural Science* 130 (5): 716–721. https://doi.org/10.21273/jashs.130.5.716.

Ye, Jie, Wangfang Li, Guo Ai, et al. 2019. "Genome-Wide Association Analysis Identifies a Natural Variation in Basic Helix-Loop-Helix Transcription Factor Regulating Ascorbate Biosynthesis via D-Mannose/L-Galactose Pathway in Tomato." *PLOS Genetics* 15 (5): e1008149. https://doi.org/10.1371/journal.pgen.1008149.

Ye, Jie, Xin Wang, Tixu Hu, et al. 2017. "An InDel in the Promoter of Al-ACTIVATED MALATE TRANSPORTER9 Selected during Tomato Domestication Determines Fruit Malate Contents and Aluminum Tolerance." *The Plant Cell* 29 (9): 2249–2268. https://doi.org/10.1105/tpc.17.00211.

Yoshida, Kandai, Shunsuke Asano, Hirotoshi Sushida, and Yuichiro Iida. 2021. "Occurrence of Tomato Leaf Mold Caused by Novel Race 2.4.9 of Cladosporium Fulvum in Japan." *Journal of General Plant Pathology* 87 (1): 35–38. https://doi.org/10.1007/s10327-020-00963-x.

Yu, Jingyin, Qionglin Chen, Lu Yuan, Shouli Feng, Miaomiao Huang, Peng Zheng, Guang Chen, et al. 2025a. "Phylogenomic and Super-Pangenome Analyses Unveil the Genetic Landscape of Tomato Evolution and Domestication." *Plant Biotechnology Journal*. https://doi.org/10.1111/pbi.70199.

Yu, Xiaofen, Minghao Qu, Yanna Shi, et al. 2022. "Chromosome-Scale Genome Assemblies of Wild Tomato Relatives Solanum Habrochaites and Solanum Galapagense Reveal Structural Variants Associated with Stress Tolerance and Terpene Biosynthesis." *Horticulture Research* 9 (January): uhac139. https://doi.org/10.1093/hr/uhac139.

Yu, Xuchen, Zuoyao Li, Yongfang Yang, Shujia Li, Yezi Lu, Yang Li, Xinyu Zhang, Fan Chen, and Cao Xu. 2025b "Harnessing Green Revolution Genes to Optimize Tomato Production Efficiency for Vertical Farming." *Journal of Integrative Plant Biology*. https://doi.org/10.1111/jipb.13927.

Zelkind, Mike, Tisha Livingston, and Victor Verlage. 2022. "Chapter 16—Indoor Production of Tomatoes." In *Plant Factory Basics, Applications and Advances*, edited by Toyoki Kozai, Genhua Niu, and Joseph Masabni. Academic Press. https://doi.org/10.1016/B978-0-323-85152-7.00008-2.

Zhang, Jinzhe, Hongjun Lyu, Jie Chen, et al. 2024. "Releasing a Sugar Brake Generates Sweeter Tomato without Yield Penalty." *Nature* 635 (8039): 647–656. https://doi.org/10.1038/s41586-024-08186-2.

Zhang, Lingkui, Yuanhang Liu, Yile Huang, et al. 2025. "Solanaceae Pan-Genomes Reveal Extensive Fractionation and Functional Innovation of Duplicated Genes." *Plant Communications* 6 (3). https://doi.org/10.1016/j.xplc.2024.101231.

Zhang, Zhonghui, Weizhen Ye, Chun Li, et al. 2024. "Volatilomics-Based Discovery of Key Volatiles Affecting Flavor Quality in Tomato." *Foods* 13 (6): 6. https://doi.org/10.3390/foods13060879.

Zheng, Jianfeng, Peidian Gan, Fang Ji, Dongxian He, and Po Yang. 2021. "Growth and Energy Use Efficiency of Grafted Tomato Transplants as Affected by LED Light Quality and Photon Flux Density." *Agriculture* 11 (9): 9. https://doi.org/10.3390/agriculture11090816.

Zhou, Dianfan, Holger Meinke, Matthew Wilson, Leo F. M. Marcelis, and Ep Heuvelink. 2021. "Towards Delivering on the Sustainable Development Goals in Greenhouse Production Systems." *Resources, Conservation and Recycling* 169 (June): 105379. https://doi.org/10.1016/j.resconrec.2020.105379.

Zhou, Yao, Zhiyang Zhang, Zhigui Bao, et al. 2022. "Graph Pangenome Captures Missing Heritability and Empowers Tomato Breeding." *Nature* 606 (7914): 527–534. https://doi.org/10.1038/s41586-022-04808-9.

Zhuang, Yunfei, Na Lu, Masao Kikuchi, Michiko Takagaki, and Tetsu Tamashiro. 2024. "Productivity Potential and Economic Feasibility of Small-Sized Tomato Production in Plant Factories with Artificial Lighting: A Comparative Study with High-Tech Greenhouse Production." *Journal of Cleaner Production* 470 (September): 143171. https://doi.org/10.1016/j.jclepro.2024.143171.

Zhuang, Yunfei, Na Lu, Shigeharu Shimamura, Atsushi Maruyama, Masao Kikuchi, and Michiko Takagaki. 2022. "Economies of Scale in Constructing Plant Factories with Artificial Lighting and the Economic Viability of Crop Production." *Frontiers in Plant Science* 13 (September). https://doi.org/10.3389/fpls.2022.992194.

Zsögön, Agustin, Tomáš Čermák, Emmanuel Rezende Naves, et al. 2018. "De Novo Domestication of Wild Tomato Using Genome Editing." *Nature Biotechnology* 36 (12): 1211–1216. https://doi.org/10.1038/nbt.4272.

Index

Note: Page numbers in *italics* indicate a figure and page numbers in **bold** indicate a table on the corresponding page.

N

O

P

Q

R

S

For Product Safety Concerns and Information please contact our
EU representative GPSR@taylorandfrancis.com Taylor & Francis
Verlag GmbH, Kaufingerstraße 24, 80331 München, Germany